Thermal Recovery

Michael Prats
Senior Research Associate
Shell Development Co.

SPE Monograph Series, Volume 7

Henry L. Doherty Memorial Fund of AIME
Society of Petroleum Engineers
Richardson, TX USA

Dedication

To my family and friends, who waited patiently for my return to a normal life.

Disclaimer

This book was prepared by members of the Society of Petroleum Engineers and their well-qualified colleagues from material published in the recognized technical literature and from their own individual experience and expertise. While the material presented is believed to be based on sound technical knowledge, neither the Society of Petroleum Engineers nor any of the authors or editors herein provide a warranty either expressed or implied in its application. Correspondingly, the discussion of materials, methods, or techniques that may be covered by letters patents implies no freedom to use such materials, methods, or techniques without permission through appropriate licensing. Nothing described within this book should be construed to lessen the need to apply sound engineering judgment nor to carefully apply accepted engineering practices in the design, implementation, or application of the techniques described herein.

Printed in the United States of America.

ISBN 978-1-61399-548-8

Society of Petroleum Engineers
222 Palisades Creek Drive
Richardson, TX 75080-2040 USA

http://www.spe.org/store
books@spe.org
1.972.952.9393

Contents

SPE Monograph Series

The Monograph Series of the Society of Petroleum Engineers of AIME was established in 1965 by action of the SPE Board of Directors. The Series is intended to provide an authoritative, up-to-date treatment of the fundamental principles and state of the art in selected fields of technology. The work is directed by the Society's Monograph Committee, one of more than 50 Society-wide committees, through a committee member designated as Monograph Coordinator. Technical evaluation is provided by the Monograph Review Committee. Below is a listing of those who have been most closely involved with the preparation of this monograph.

Monograph Coordinators

William Vasilauskas, Chevron Overseas Petroleum Inc.
Aziz Odeh, Mobil Research & Development Corp.

Monograph Review Committee

L.A. Wilson, chairman, Gulf Research & Development Co., Pittsburgh
John Havenstrite, Exxon Co., U.S.A., New Orleans
Ted R. Blevins, Chevron Oil Field Research Co., La Habra, CA
W.L. Martin, Continental Oil Co., Ponca City, OK

Acknowledgments

I thank Shell Development Co. for allowing me to use support facilities and company time in the preparation of this monograph. Of the support staff, three groups did yeoman work: the library staff, who hunted and delivered even more references than were used; the drafting department, who had their hands full drawing and modifying figures; and the typists, who cheerfully processed drafts over and over again.

I also appreciate the help of various engineers (SPE members as well as nonmembers and the Monograph Committee) in providing helpful suggestions on the organization of this monograph and in checking details.

Lastly, I also appreciate the great effort provided by the staff of SPE Headquarters and their consultants in coordinating the editing and handling of the manuscript.

Foreword to Second Printing

This second printing of the *Thermal Recovery* monograph has provided an opportunity not only to correct typographical errors and to clarify the text (as advised by a half dozen engineers to whom I am indebted), but also to summarize here the major changes and trends in the field since the text was written. Except for these corrections and clarifications, the text itself has not been altered, not even to update statistical records. Revising the text to reflect today's state of the art would have required substantial modifications, for thermal recovery is the most active of the EOR processes. Instead, this Foreword identifies the major changes and trends in thermal recovery and provides a few additional references.

Many steam-injection projects started as steam cyclic operations to take advantage of the early production response obtained by this process. As repeated cycles yielded poorer returns, these have been converted (where warranted by other factors) to steamdrives. Steam breakthrough now has occurred in many field projects, while in others the process is sufficiently advanced to consider how best to wind down these mature steamdrives. A current topic of general interest is the operation of steamdrives after steam breakthrough, especially those of mature projects. Items under active consideration include development of guidelines for (1) reducing the steam injection rate,[1,2] temperature, and quality after steam breakthrough and (2) determining whether steam, water, some other fluid, or a combination of fluids should be injected during the final phase of a project.

In addition to the consideration of injection of fluids—such as air, solvents, and CO_2—during the wind-down phase of steamdrives, efforts continue on the use of additives (such as foams, emulsifying and spreading agents, solvents, and CO_2) during the normal course of a steamdrive to improve project economics through increases in either the displacing or sweep efficiencies. In combustion, the recent trend is in the injection of oxygen-rich streams.

Reservoir modeling has advanced in two separate directions. The advent of vector array processors and improved solution algorithms has led to more common usage of numerical simulation as a tool to analyze reservoir performance. Color-display facilities, which help visualize and analyze results, have eased significantly the task of interpreting the large volumes of information arising from typical numerical simulation studies. In a second and complementary direction, an increased effort has been made in developing simple models for both steam and combustion processes that recognize the difference in importance of producing mechanisms before and after breakthrough of the injected fluid.[1-3]

A wealth of literature continues to be published on the various thermal recovery processes. Recent books in English include those of White[4] and of Burger *et al.*,[5] with at least two others scheduled for publication this year. Surveys on the state of the art of steamdrives and in-situ combustion have appeared recently,[6,7] as has a survey of EOR activity (mostly thermal) for the Athabasca oil sands.[8] Also, the *Oil and Gas J.* traditionally publishes a survey of worldwide EOR activity in one of the late-winter issues of even-numbered years.

In operations, special attention has been given to reduce steam generation cost. To this end, a limited number of cogeneration plants[9] and steam generators able to burn multiple grades of fuels,[10] including solids, have been designed and put into operation. For conventional wet-steam generators, recommended practices now have been issued by the API.[11] Meeting minimum pollution requirements continues to be a problem in many areas of the world. Steam generators continue to be improved to meet more exacting requirements. Many different types of scrubbing and particulate-removal units are widely available. Controls placed on the amounts and type of effluents that can be vented from wellheads and field equipment, as well as the interaction between vent closing and production response, have led to the design and implementation of a number of casing vent recovery systems. But there are still no easy, reliable methods used in the oil patch to determine the steam quality at any point in the surface lines where the steam stream is split at least once.

Unusual operations that appear to receive a fair amount of attention include the use of surface wells that end on a horizontal trajectory, primarily in the Cold Lake area.[8] Problems related to these types of well completions and how they should be used in thermal operations remain to be resolved. Drainage of heated oil into these wells is a subject of active investigation.[12] Horizontal wells or conduits drilled from vertical shafts[8] fall into the same category.

A recent study on the effect of temperature on the oil/water relative permeabilities[13] casts doubt on the conclusions of the references on this subject cited in the monograph (where refined oils were used). The recent study, which appears to have been done extremely carefully, implies that there may not be any temperature effect caused by fundamental flow properties, although interactions in real systems between components in the crude and mineral surfaces could result in a temperature dependence.

There appears to be a growing awareness that thermal systems are not inert. Several studies have addressed the thermal alteration of crudes (including those of asphaltenes) through visbreaking reactions, reactions involving water (aquathermolysis), and those leading to the formation of acid gases.

Properties of steam and hot water are included in Appendix B. Ref. 14 provides other useful graphs of steam and water properties, of the gas deviation factor of steam, and of the FVF of pure liquid water as a function of both temperature and pressure, which are not included in Appendix B.

References

1. Vogel, J.V.: "Simplified Heat Calculations for Steamfloods," *J. Pet. Tech.* (July 1984) 1127–1136.
2. Farouq-Ali, S.M.: "Steam Injection Theories—A Unified Approach," paper SPE 10746 presented at 1982 SPE California Regional Meeting, San Francisco, March 24–26.
3. Neuman, C.H.: "A Gravity Override Model of Steamdrive," *J. Pet. Tech.* (Jan. 1985) 163–169.
4. White, P.D.: *Thermal Recovery Methods*, PennWell Publishing Co., Tulsa (1983).
5. Burger, J., Sourieau, P., and Combarnous, M.: *Thermal Methods of Oil Recovery*, Gulf Publishing, Houston (1985).
6. White, P.D.: "In-Situ Combustion Appraisal and Status," *J. Pet. Tech.* (Nov. 1985) 1943–1949.
7. Chu, C.: "State-of-the-Art Review of the Steamflood Field Projects," *J. Pet. Tech.* (Oct. 1985) 1887–1902.
8. Redford, D.A.: "In-Situ Recovery from the Athabasca Oil Sands—Past Experience and Future Potential," *J. Cdn. Pet. Tech.* (May–June 1985) 52–62.
9. Stambler, I.: "Plans for 300 MW and 800 MW Oil Field Cogeneration Plants," *Gas Turbine World & Cogeneration* (July–Aug. 1983) 38–41.
10. Plass, L. and Daradimos, G.: "Combustion of Heavy Oil and Upgrading of Residues in the Circulating Fluid Bed for Steam Production," *Proc.*, Second Intl. Conference on the Future of Heavy Crude and Tar Sands, McGraw-Hill Book Co. Inc., New York City (1984) 638–652.
11. "API Recommended Practice for Installation and Operation of Wet Steam Generators," RP 11T, API, Washington, DC (March 1983).
12. Butler, R.M.: "A New Approach to the Modelling of Steam-Assisted Gravity Drainage," CIM 84-35-21, 35th Annual Technical Meeting of the Petroleum Soc. of the CIM, Calgary, June 10–13, 1984.
13. Miller, M.A. and Ramey, H.J. Jr.: "Effect of Temperature on Oil/Water Relative Permeabilities of Unconsolidated and Consolidated Sands," *Soc. Pet. Eng. J.* (Dec. 1985) 945–953.
14. Ramey, H.J. Jr.: "Thermodynamic and Hydrodynamic Properties of Hydrothermal Systems," Stanford Geothermal Program Report SGP-TR-6, Stanford U., Stanford, CA (April 1974).

Chapter 1

Introduction

1.1 Scope of the Monograph

The purpose of this monograph is to present the state of the art in thermal recovery processes. Publications are used as a basis, supplemented by discussion and interpretation to provide additional insight wherever appropriate. Thermal recovery is defined as a process in which heat is introduced intentionally into a subsurface accumulation of organic compounds for the purpose of recovering fuels through wells. This definition covers, to the best of my knowledge, all practiced and proposed methods for recovering oil and combustible gases from the subsurface by thermal means.

Thousands of papers and articles have been published since 1865 on the introduction of heat into subsurface reservoirs to improve or accelerate oil recovery. This literature reflects the great variety in which thermal energy has been and is being used or considered to solve or improve many different types of problems associated with the production of oil. Thermal recovery is used in preference to other recovery methods for a number of reasons. In the case of viscous oils, which is the case of most current interest, heat is used to improve the displacement and recovery efficiencies. The reduction in crude oil viscosity that accompanies a temperature increase not only allows the oil to flow more freely but also results in a more favorable mobility ratio. This monograph emphasizes the reservoir aspects of conventional thermal recovery processes – combustion, steam, hot water, and hot gases – and considers in detail and at length the application of these processes to oil reservoirs. One chapter, however, is devoted to the use of unusual thermal recovery processes and to innovative applications seriously being considered for coal and oil-shale deposits. Other chapters, or parts of chapters, briefly discuss selected topics on operational problems, well completions, surface facilities, pollution, health and safety, and economics.

1.2 Objectives of the Monograph

The sole objective of this monograph is a simple, valid presentation of the state of knowledge, prediction, and practice of thermal recovery processes. Although some aspects of thermal recovery are difficult to explain in simple terms, every effort has been made to do so. Mathematical expressions, in particular, are often difficult to understand by those who are unfamiliar with their physical meaning; yet they are essential to the subject. Our intent is to reach those engineers, particularly reservoir engineers, who are responsible for the design and operation of different aspects of a field project. It is expected that parts of the monograph also will be useful to those persons who, day after day, are responsible for overseeing and troubleshooting the project.

It is premised that an understanding of the various mechanisms occurring during a thermal displacement process, as well as of the properties of the reservoir and its fluids, is important in the selection, design, operation, surveillance, and interpretation of that process. Accordingly, key concepts are discussed at length, sometimes (where appropriate) with the aid of mathematics, but always in terms of the physics and chemistry of the system. Field experience is summarized, analyzed, and contrasted with performance predictions. Example calculations are provided. Procedures are included for determining the effect of temperature on fluid and formation properties.

The intent of the monograph, then, is to provide whatever information is necessary for an engineer to make a preliminary design of a thermal recovery project. The reader should not expect to find specific solutions to specific problems. He should expect, however, some discussion, references to publications that treat specific problems in more detail, or procedures that are likely to be useful in planning, implementing, and interpreting field operations.

The monograph is not intended for specialized researchers or other bona fide experts in thermal recovery, although they, like anyone else, might use it as a general source of information on thermal recovery and might refer to it for material outside a narrow area of specialization.

1.3 Early History of Thermal Recovery

The oldest known means of introducing heat into reservoirs is downhole heaters. One of the earliest references to them is in a patent issued to Perry and Warner[1] in 1865. The primary purpose of downhole heaters is to reduce the viscosity and, thus, increase the production rate of viscous crudes, but occasionally they also are used to maintain the crude at

TABLE 1.1 – EFFLUENT GAS COMPOSITION (VOLUME PERCENT) FROM AIR INJECTION PROJECTS*

Gas Component	Name of Oil Sand			
	First Cow Run	Mitchell	Macksburg 500	Peeker
Carbon dioxide	1.07	1.20	0.70	4.60
Oxygen	19.40	4.30	16.80	12.90
Nitrogen	76.69	73.70	78.00	75.60
Hydrocarbons	2.84	20.80	4.50	6.90

*Shallow fields (350 to 700 ft) probably in Appalachia.[4]

temperatures above the pour point as the crude moves to the surface and to remove or inhibit the formation and deposition of organic solids such as paraffins and asphaltenes. Since the use of bottomhole heaters and equivalent hot-fluid circulation systems can affect only the production borehole and its immediate vicinity, the practice is associated with stimulation, remedial, and preventive treatments. Downhole heating by burners and electrical downhole heaters, by intermittent injection of hot fluids* into production wells, and by circulation of hot fluids still is practiced today in certain parts of the world. Its use appears to have diminished over the years, as implied by the emphasis on the early use of downhole heaters given in Nelson and McNiel's[2] excellent account of the development of thermal recovery processes and in API's *History of Petroleum Engineering*.[3]

In-situ combustion of reservoir crudes probably occurred during air injection projects used in the early 1900's to enhance oil recovery. In 1917 Lewis[4] reported gas analyses from several air injection projects in which the oxygen concentrations were deficient relative to those of nitrogen (Table 1.1). Although carbon dioxide concentrations also were low, possibly indicating an oxygenation of the crude rather than active combustion, such oxygenation methods do produce heat. Lewis noted the following.

> The excess of nitrogen in the air-gas is probably due to the extraction of oxygen from the air during its passage through the sand rather than to the picking up of nitrogen. ... the change of composition of the air is probably due in most part to chemical reactions between the oxygen and the oil, or with other substances underground, but it may also be partly caused by a greater percentage of oxygen being absorbed in the oil when the air is passed through the sand.

However, the generation of heat in those cases (if it in fact occurred) does not constitute an example of a thermal recovery process as defined here, for there was no apparent *intent* to generate and use heat in the reservoir.

Some key elements of underground combustion processes in oil reservoirs – including injection of air to burn part of the crude to generate heat and reduce the crude viscosity while providing a driving force to displace the oil – were recognized as early as 1920 by Wolcott[5] and Howard[6] and culminated in patents issued in 1923. The first published large-scale field operations of the underground combustion process were carried out in the USSR in 1933.[7] Those tests, however, were carried out in coal seams in what is now known as an in-situ coal gasification process. It will be argued by some that these are not thermal recovery operations, but they certainly fit the definition of thermal recovery given in the opening paragraph. The first attempt at an application to oil reservoirs is also Soviet in origin and occurred in 1934.[8] Beyond the initial attempt, however, the process does not appear to have been pursued further until about a decade later.

In the U.S. the first reported field applications of the combustion process are probably those instigated by E.W. Hartman starting in 1942.[9] Although the intent was to radiate heat at 900°F from a bottomhole heat exchanger into a watered-out reservoir north of Bartlesville, OK, rather than to inject the heated air into the formation, the early (10-day) response at distant wells (660 ft) strongly suggests that hot air was injected into the formation. Increased production at several nearby wells was accompanied by increased gas flow rates, increased crude gravities, and increased temperatures, all typical of a combustion process. In the second test in the Ardmore district, which took place within the next 2 years, Hartman knew he was injecting heated air into the formation.

In-situ combustion, as we know it today, was developed very rapidly in the U.S., starting with the laboratory research of Kuhn and Koch,[10] published in 1953, and that of Grant and Szasz,[11] published the following year. Those researchers visualized a traveling heat wave – i.e., the heat left behind in the burnt zone was carried downstream by the injected cool air. A succession of technical papers soon followed those early publications. Of the later papers, the one by Wilson *et al.*[12] introduced the concept of sequential oil and steam zones, and the one by Dietz and Weijdema[13] showed how the heat recuperation aspects of in-situ combustion recognized by Grant and Szasz could be improved significantly by injecting water as well as air.

Currently, we recognize three principal forms of the combustion process: dry forward and reverse combustion and wet forward combustion. The words forward and reverse are applied when the combustion front moves with or against the air stream, respectively. With few and notable exceptions, most of the early field applications of the combustion process to oil reservoirs have been in the U.S.

Steam drives date from at least as early as 1931-32, when steam was injected for 235 days into an 18-ft sand at a depth of 380 ft in the Wilson and Swain

*We shall discuss elsewhere the massive injection of hot fluids such as steam to heat the wellbore and significant portions of the reservoir.

lease near Woodson, TX. According to Stovall,[14] injection pressures ranged from 150 to 200 psi and injection rates from 35,000 to 52,500 lbm/D (100 to 150 BWPD), although steam was injected for only 4 hours daily during the last 70 days. Fig. 1.1 shows the well locations. There is no apparent record of any use of steam drives in the next 20 years, when a steam drive pilot was operated at Yorba Linda.[15] The first large-scale steam drives were those in Schoonebeek[16] and Tia Juana.[17]

An early steam drive pilot test at Mene Grande[18] played a major role in the development of the cyclic steam injection process:

> In October 1959, CSV suspended a steam drive pilot test in the tar sands of the Mene Grande field. ... During steam injection into the sands at a depth of some 550 ft, the overburden pressure was exceeded. Cratering accompanied by steam, water, and oil eruptions occurred around the injection wells. ... When the test was discontinued it was decided to relieve the pressure in the injectors. Surprisingly, they produced small amounts of steam and considerable amounts of oil (100 to 200 B/D) although they had never before produced any oil.

Today, cyclic steam injection (also known as steam soak, push-pull, and huff 'n' puff) consists of injecting steam for a period of days or weeks and waiting a few days before putting the well on production. Under the right conditions the procedure, with little delay, can allow the operator to convert the injection well to a producing well that yields significant volumes of stimulated oil. Steam injection cycles can be repeated, usually with diminishing success. A number of older cyclic steam injection operations now are being converted to steam drive to recover the remaining oil.

The use of scaling rules[19] and scaled physical models[20] played an important role in the development of both steam drive and cyclic steam injection processes.

The earliest record of a hot-gas injection process in an oil reservoir is a proposal by Lindsly[21] in 1928. He actually recognized that the crude would pyrolyze, that light ends in the crude would be preferentially stripped off, and that these would condense to raise the API gravity and lower the viscosity of the crudes. The first recorded field test was in the Chusov Township formation in the USSR in 1935.[22] A mixture of at least 20% air and hot gases at 842 to 932°F was injected into a shallow, heterogeneous limestone having an average thickness of about 340 ft. Although the operators did not want to rely on combustion because of the nonuniformity of the oil distribution, it is likely that combustion of the viscous crude occurred.

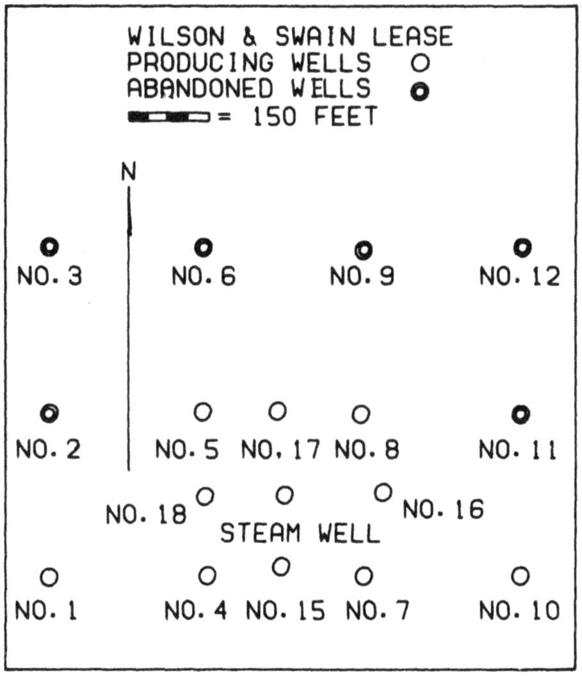

Fig. 1.1 – Plat of well locations from the earliest article on steam drive (1934).[14]

1.4 Potential Importance

An example of how thermal recovery can increase production markedly is shown in Fig. 15.1, which gives data for California from 1964 through 1978. Most of the thermal oil produced in California is in the range of 10 to 20°API. The incremental production rate in 1979 was 199,000 B/D, which amounted to 20% of the state's total production.[23]

Historically, thermal recovery processes have been aimed at viscous crudes. This has been a natural direction to follow, since viscous crudes have not been recoverable by any other practical in-situ method. Thermal recovery methods, however, usually can be applied to any crude; they merely have to be technically feasible in the field being considered and be competitive economically with alternative methods, especially in the case of low-viscosity crudes.

Along with viscous crudes, coal and oil shale are the most likely in-situ targets for thermal recovery processes. There are vast deposits of very heavy oils, tars, coal, and oil shale throughout the world. In the U.S. there is a concentration of heavy oil reservoirs in California, where the impact of thermal recovery already has been felt. Vast deposits of heavy oil sands exist in Canada and Venezuela, and these to a large extent also will be amenable to recovery by thermal processes. In the western part of the U.S. there are vast accumulations of oil shale and coal, much of which probably will have to be recovered through some form of in-situ thermal recovery process. In-situ processes for coal and oil shale have yet to come of age, although the potential is there. Table 1.2 gives selected examples of the various types of resources that currently appear to require some form of in-situ thermal recovery process for proper development. Considering that these are only selected examples and, thus, that they fall short of being representative of the worldwide energy resources, it seems clear that the potential of in-situ thermal recovery processes is extremely great.

TABLE 1.2 – RESOURCES FOR POSSIBLE USE OF IN-SITU THERMAL RECOVERY PROCESSES

	Amount of Resource*	
Type of Resource		Equivalent Oil (10^9 bbl)
Alberta oil sands[24]		700
Orinoco heavy oil belt[25]		1,000
Powder River basin, coal seams >50 ft thick and >1,000 ft deep [26]	1.9×10^{11} tons	900**
Piceance Creek basin, oil shale assaying >25 gal/ton[27]		450

*The amount of resource is generally much larger than the amount recoverable.
**The assumed fuel value is 14,000 Btu/lbm coal.
Note: The total energy consumed in the world in 1980 was equivalent to about 49×10^9 bbl of oil.

1.5 Organization of the Monograph

For those who want to know what to expect in the following chapters, some comments on the organization of the monograph are in order. The next chapter describes the types of thermal recovery processes. Chap. 3 presents the physical, chemical, and mathematical basis for heat and fluid flow in reservoirs. It is not essential for the reader to master Chap. 3 to be able to understand and apply much of the material in the remainder of the monograph. However, it is an important chapter, and it is suggested that readers – even those who are not particularly well versed in science or mathematics – should at least browse through it to enhance their basic understanding of the various elements making up thermal recovery processes. Throughout Chap. 3, particular efforts have been made to present information in a descriptive rather than in a highly mathematical manner. Chap. 4 is devoted to selected reservoir engineering topics common to thermal recovery processes. Chap. 5 discusses the evaluation of the size of the equivalent heated zone in the reservoir and gives background pertinent to the estimation of temperature distributions in the reservoir and its adjacent formations. Chaps. 6 through 9 describe in some detail the characteristics, performance, and design of thermal recovery operations. In Chap. 10, methods are given for determining heat losses from surface and subsurface lines and for calculating bottomhole temperatures in wells and steam qualities in injection wells. Chap. 11 discusses well completions, facilities, operational problems, and surveillance methods. Chap. 12 discusses the impact of economics and the reservoir and fluid properties conducive to thermal operations. Chap. 13 covers pilot testing. Chap. 14 considers other uses of thermal recovery processes, such as in-situ coal gasification and in-situ retorting of oil shale. The last chapter summarizes where we are and speculates on the future of thermal recovery operations. Chaps. 3 and 5 through 10 include example calculations. A complete nomenclature is provided. Separate appendices present conversion factors and properties of materials. A bibliography completes the monograph. It is intended that the manner of presentation will make the monograph both a readable book and a practical guide for day-to-day use.

1.6 Symbols and Units

In the application-oriented chapters (Chaps. 4 through 13), equations are presented in such a way that readers can use them either with the practical oilfield units prevalent in many sections of the world, including the U.S., or with any consistent set of units of their choice. So that the equations will be applicable to either type of unit system, we have introduced the appropriate conversion factor in front of an equation and identified it by enclosing it in $\langle\ \rangle$. For example, the familiar expression for the steady-state pressure drop Δp between two concentric radii r_w and r_e resulting from a rate of flow q of an incompressible liquid of viscosity μ through a reservoir having a permeability k and thickness h would be expressed by

$$\Delta p = \left\langle 2\pi(141.2) \right\rangle \frac{q\mu}{2\pi kh} \ln(r_e/r_w), \qquad (1.1)$$

when using pounds per square inch, reservoir barrels per day, centipoise, millidarcies, and feet.

Those who wish to work in a consistent set of units are to omit the factor within $\langle\rangle$. The remaining expression,

$$\Delta p = \frac{q\mu}{2\pi kh} \ln(r_e/r_w), \qquad (1.2)$$

identifies the basic equation, including natural factors such as 2π. Since the basic equation would hold in any consistent set of units, it would remain unchanged in any consistent unit system currently in use or that might come into use. Specifically, the equation would remain unchanged in any consistent metric system that may be adopted in the U.S.

Those who wish to use oilfield units would read the equation as

$$\Delta p = 141.2 \frac{q\mu}{kh} \ln(r_e/r_w), \qquad (1.3)$$

where the 2π (or any other natural factor that might arise in some other equation) is included in the conversion factor.

Of course, the factor in $\langle\ \rangle$ is 1 where the equations are consistent in the customary oilfield units (such as in most of Chap. 10), in which case the $\langle\ \rangle$ is not used. If in doubt, use the units indicated and omit (where provided) the factor within $\langle\ \rangle$.

The oilfield units used in the monograph are as follows.

Permeability	millidarcies (md)
Viscosity	centipoise (cp)
Time	days (D)
Distance	feet (ft)
Volume	
Reservoir	acre-feet
Liquids	barrels (bbl or B)
Gases	10^3 standard cubic feet (Mscf)
Pressure	pounds of force per square inch (psi)
Area	
Reservoir	acres
cross section	square feet (sq ft)
Flow rate	
Liquids	barrels per day (B/D)
Gases (except steam)	10^3 standard cubic feet per day (Mscf/D)
Steam	barrels of equivalent water per day (BWPD)
Volumetric flux	
Liquids	cubic feet per square foot per day (cu ft/sq ft-D)
Gases	standard cubic feet per square foot per day (scf/sq ft-D)
Velocity	feet per day (ft/D)
Temperature	degrees Fahrenheit (°F)
Mass	pounds (lbm)
Heat, energy	British thermal units (Btu)

The Society of Petroleum Engineers, publisher of this monograph, has developed a set of standard symbols for use in petroleum engineering and in related disciplines.[28-30] A sincere effort has been made to adhere to those standards, although some deviations have been found to be necessary.

The Nomenclature defines the quantities the symbols represent and gives their oilfield units and the dimensions of the symbols in terms of the five basic units: length (L), time (t), mass (m), temperature (T), and electric charge (q). Constants and conversion factors are presented in Appendix A.

References

1. Perry, G.T. and Warner, W.S.: "Heating Oil Wells by Electricity," U.S. Patent No. 45,584 (July 4, 1865).
2. Nelson, T.W. and McNiel, J.S. Jr.: "Past, Present, and Future Development in Oil Recovery by Thermal Methods," *Pet. Eng.*, Part I, (Feb. 1959) B27, Part II (March 1959) B75.
3. *History of Petroleum Engineering,* API Div. of Production, Dallas (1961).
4. Lewis, J.O.: "Methods of Increasing the Recovery from Oil Sands," *Bull. 148, Petroleum Technology,* USBM (1971) **37**.
5. Wolcott, E.R.: "Method of Increasing the Yield of Oil Wells," U.S. Patent No. 1,457,479 (filed Jan. 12, 1920; issued June 5, 1923).
6. Howard, F.A.: "Method of Operating Oil Wells," U.S. Patent No. 1,473;348 (filed Aug. 9, 1920; issued Nov. 6, 1923).
7. Elder, J.L.: "The Underground Gasification of Coal," *Chemistry of Coal Utilization,* Supplementary Volume, H.H. Lowry (ed.), John Wiley and Sons Inc., New York City (1963) 1023-1040.
8. Sheinman, A.B., Malofeev, G.E., and Sergeev, A.I.: "The Effect of Heat on Underground Formations for the Recovery of Crude Oil – Thermal Recovery Methods of Oil Production," Nedra, Moscow (1969); Marathon Oil Co. translation (1973) 166.
9. Gibbon, A.: "Thermal Principle Applied to Secondary Oil Recovery," *Oil Weekly,* (Nov. 6, 1944) 170-172.
10. Khun, C.S. and Koch, R.L.: "In-Situ Combustion – Newest Method of Increasing Oil Recovery," *Oil and Gas J.,* (Aug. 10, 1953) **52**, 92-96, 113, 114.
11. Grant, B.R. and Szasz, S.E.: "Development of Underground Heat Wave for Oil Recovery," *Trans.,* AIME (1954) **201**, 108-118.
12. Wilson, L.A., Wygal, R.J., Reed, D.W., Gergins, R.L., and Henderson, J.H.: "Fluid Dynamics During an Underground Combustion Process," *Trans.,* AIME (1958) **213**, 146-154.
13. Dietz, D.N. and Weijdema, J.: "Wet and Partially Quenched Combustion," *J. Pet. Tech.* (April 1968) 411-413; *Trans.,* AIME, **243**.
14. Stovall, S.L.: "Recovery of Oil from Depleted Sands by Means of Dry Steam," *Oil Weekly* (Aug. 13, 1934) 17-24.
15. Stokes, D.D. and Doscher, T.M.: "Shell Makes a Success of Steam Flood at Yorba Linda," *Oil and Gas J.* (Sept. 2, 1974) 71-76.
16. van Dijk, C.: "Steam-Drive Project in the Schoonebeek Field, The Netherlands," *J. Pet. Tech.* (March 1968) 295-302; *Trans.,* AIME, **243**.
17. de Haan, H.J. and Schenk, L.: "Performance and Analysis of a Major Steam Drive Project in the Tia Juana Field, Western Venezuela," *J. Pet. Tech.* (Jan. 1969) 111-119; *Trans.,* AIME, **246**.
18. Giusti, L.E.: "CSV Makes Steam Soak Work in Venezuela Field," *Oil and Gas J.* (Nov. 4, 1974) 88-93.
19. Geertsma, J., Croes, G.A., and Schwartz, N.: "Theory of Dimensionally Scaled Models of Petroleum Reservoirs," *Trans.,* AIME (1956) **207**, 118-123.
20. Stegemeier, G.L., Laumbach, D.D., and Volek, C.W.: "Representing Steam Processes With Vacuum Models," *Soc. Pet. Eng. J.* (June 1980) 151-174.
21. Lindsly, B.E.: "Recovery by Use of Heated Gas," *Oil and Gas J.* (Dec. 20, 1928) 27.
22. Dubrovai, K.K., Sheinman, A.B., Sorokin, N.A., Sacks, C.L., Pronin, V.I., and Charuigin, M.M.: "Experiments on Thermal Recovery in the Chusovsk Town," *Pet. Economy* (Nov. 5, 1936).
23. "Annual Review of California Oil and Gas Production," Conservation Committee of California Oil Producers, Los Angeles (1979).
24. Nicholls, J.H. and Luhning, R.W.: "Heavy Oil Sand In-Situ Pilot Plans in Alberta (Past and Present)," *J. Cdn. Pet. Tech.* (July-Sept. 1977) 50-61.
25. Fiorillo, G.: "Exploration of the Orinoco Oil Belt – Review and General Strategy," paper UNITAR/CF10/V/3 presented at the 11th Intl. Conference on Heavy Crude and Tar Sands, Caracas, Feb. 7-17, 1982.
26. Roupert, R.C., Choate, R., Cohen, S., Lee, A.A., Lent, J., and Spraul, J.R.: "Energy Extraction from Coal In-Situ – A Five-Year Plan," T1D27203, prepared by TRW Systems, available from Natl. Technical Information Service (1976).
27. Duncan, D.C. and Swanson, V.E.: "Organic-Rich Shale of the U.S. and World Land Areas," USGS Circular 523 (1965).
28. "Letter Symbols for Petroleum Reservoir Engineering, Natural Gas Engineering, and Well Logging Quantities," *Trans.,* AIME (1965) **234**, 1463-1496.
29. "Supplement to Letter Symbols and Computer Symbols for Petroleum Reservoir Engineering, Natural Gas Engineering, and Well Logging Quantities," *Trans.,* AIME (1972) **253**, 556-574.
30. "Supplements to Letter and Computer Symbols for Petroleum Reservoir Engineering, Natural Gas Engineering, and Well Logging Quantities," *J. Pet. Tech.* (Oct. 1975)1244-1264; *Trans.,* AIME, **259**.

Chapter 2

Thermal Recovery Methods

All thermal recovery processes tend to reduce the reservoir flow resistance by reducing the viscosity of the crude. The thermal recovery processes used today fall into two classes: those in which a hot fluid is injected into the reservoir and those in which heat is generated within the reservoir itself. The latter are known as in-situ processes, an example of which is in-situ combustion or fireflooding. Processes combining injection and in-situ generation of heat have been tested[1] but currently are not practiced to any great extent. Thermal recovery processes also can be classified as thermal drives or stimulation treatments. In thermal drives, fluid is injected continuously into a number of injection wells to displace oil and obtain production from other wells. The pressure required to maintain the fluid injection also increases the driving forces in the reservoir, increasing the flow of crude. Thus, a thermal drive not only reduces the flow resistance but also provides a force that increases flow rates. In thermal stimulation treatments, only the reservoir near production wells is heated. Driving forces present in the reservoir – such as gravity, solution gas, and natural water drive – effect the improved recovery rates once the flow resistance is reduced. Stimulation treatments also can be combined with thermal drives, in which case the driving forces are both natural and imposed. In thermal stimulation treatments, the reduction in flow resistance also may result from the removal of organic or other solids from openings in the casing, the liner, the screen, and even from the pores of the reservoir rock.

This chapter describes the various types of thermal recovery processes in common use today (later chapters provide details about each process). But first, the next section points out the need for thermal recovery processes by discussing pertinent features of a conventional waterflood applied to a reservoir containing a very viscous crude.

2.1 Need for Thermal Recovery Processes

Consider two reservoirs that are entirely similar except for the viscosity of the crude. Linear displacement of oil by a specified volume of water will result in an oil saturation in the water-swept zone that is larger for the one having the higher viscosity. This is illustrated schematically in Fig. 2.1. At sufficiently low crude viscosities, the oil saturation upstream of the displacement front is essentially irreducible, leading to what is known as piston-like displacement. Notice that in this example the oil saturation drops across the displacement front by an amount equal to $1 - S_{or} - S_{iw}$. For a crude of high viscosity, the drop in oil saturation at the displacement front is much smaller, and a substantial amount of mobile crude is bypassed by the leading edge of the displacement front. Because of its high viscosity, the crude has a much higher flow resistance than does the water. The water simply moves faster than the crude, leaving some mobile crude behind. For the same volume of water injected, then, the oil displacement front would be closer to a producer in a heavy-oil reservoir.

This means that for the more viscous crudes, at the same water injection rate and injected volume, water will break through at the producer earlier; more water must be handled earlier in the life of the project; and less oil will be produced. Oil will continue to be produced after water breakthrough but the average oil cut (the fraction of oil in the produced stream) will be lower. With high-viscosity crudes, many pore volumes of injected water are required to reduce the remaining oil to near its irreducible value.

Not only is the displacement efficiency poor, but the volumetric sweep efficiency is decreased at the higher crude viscosity. This decrease in volumetric sweep efficiency is the result of two different mechanisms. First, during the displacement of a viscous crude by water, the macroscopic displacement front is never planar but varies according to the velocity of the flow lines. Since the flow resistance of a flow channel is least where the displacement front is most advanced, the flow velocities in those channels are the highest. The result is that the displacement front arrives at the producer soonest along the flow line of highest velocity and that breakthrough occurs earlier (for a given injection rate) for the more viscous oil. Examples of the effect of increasing oil viscosity (which is proportional to mobility ratio) on the sweep efficiency at the time of water breakthrough and later are given in Fig. 2.2, which applies to a homogeneous repeated five-spot well pattern. The range of mobility ratios encountered in heavy-oil reservoirs is significantly larger than those shown in this figure, but corresponding results are not available. Layers of different permeabilities[3] and gravity tongues[4] are examples of other conditions leading to macroscopic bypassing.

The second mechanism reducing the volumetric sweep efficiency (a mechanism also reflected in Fig. 2.2) is associated with small-scale viscous instabilities. Instabilities in the displacement front will develop in the most homogeneous porous media; they arise from small perturbations in an otherwise uniform flow pattern. These pertubations may be caused, for example, by small differences in the shape of grains or pores. According to Chuoke *et al.*,[5] the distance between the viscous fingers that are most likely to form decreases with increasing crude viscosity, and the fingers are likely to be formed at relatively low flow rates. An example of viscous instabilities in a "homogeneous" laboratory pack is shown in Fig. 2.3. It is conjectured that some of the oil between adjacent fingers may be bypassed and trapped more easily when, because of high viscosities, the fingers are close together – i.e., the displacement efficiency is lower the higher the oil viscosity. (The rather poor displacement efficiency of oil by water is discussed in more detail in Section 6.2.) Thus, the presence of inhomogeneities in any reservoir not only ensures the presence of microscopic viscous fingering of the type shown in Fig. 2.3 but also results in bypassing and trapping of oil on quite a large scale (see Fig. 4.5 for a field example), the effect being more pronounced the more viscous the oil.

Although this discussion has been based on the displacement of oil by water, the same would be true of any other noncondensable and immiscible fluid having a lower mobility than the oil. The effects are even more pronounced for the displacement of oil by gas, where the viscosity ratio of the displaced fluid to the displacing fluid is even larger. The assessment is that the displacement of a high-viscosity crude by water or by an immiscible, noncondensing gas is inefficient. This has been known for a long time.

What can be done to improve the recovery from reservoirs containing viscous crudes? Let us examine some of the dimensionless groups that affect flow

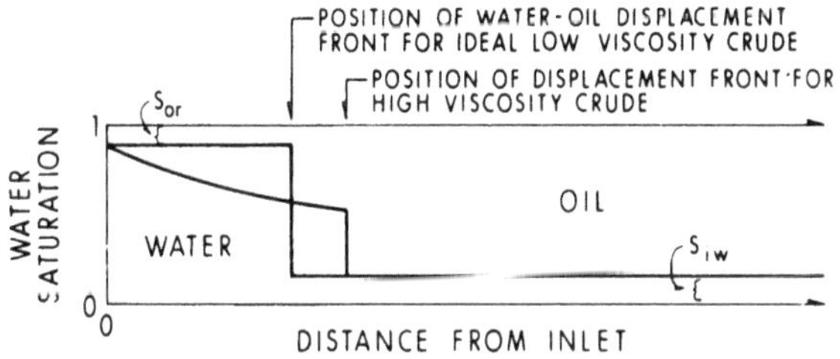

Fig. 2.1 – Qualitative effect of crude viscosity on the oil saturation distribution.

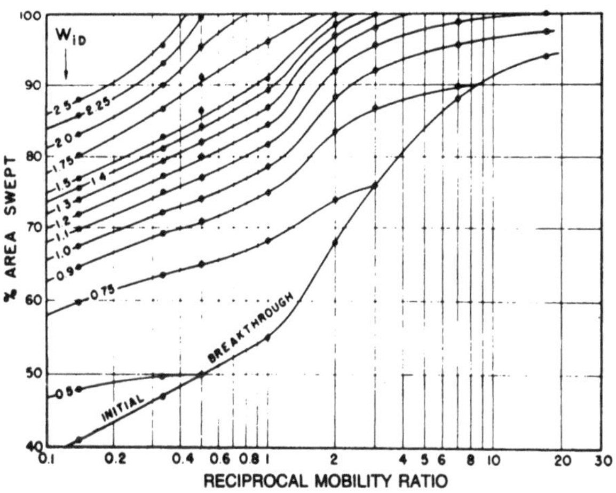

Fig. 2.2—Effect of mobility ratio on the area swept by waterflooding a repeated direct line-drive square well pattern,[2] as corrected in Ref. 10. (W_{iD} is the cumulative injected water expressed as a fraction of the floodable pore volume).

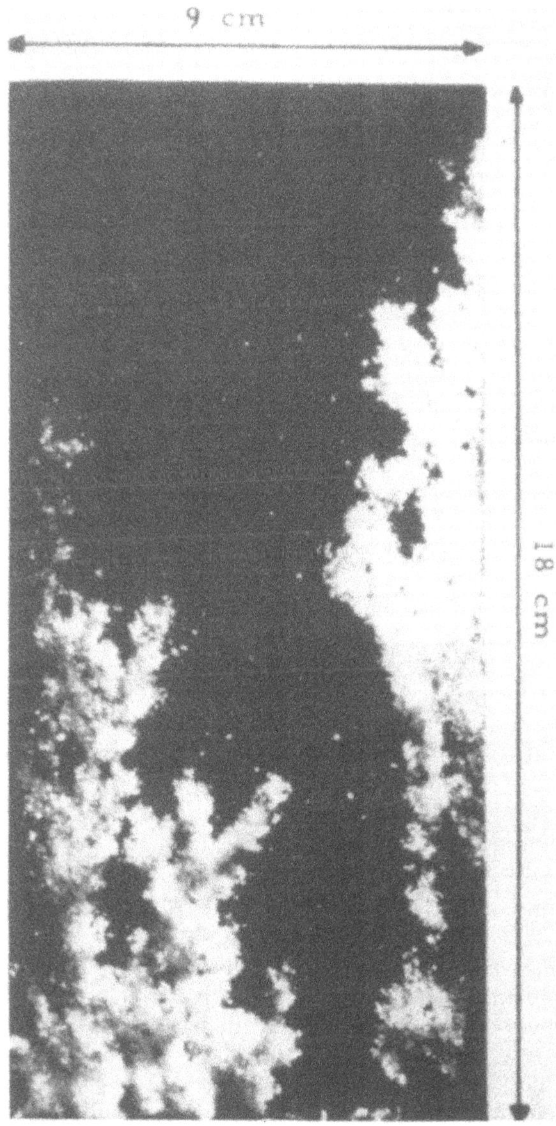

Fig. 2.3 – Formation of fingers in a porous oil-bearing medium containing 0.15 PV of connate water.[5] Displacing water appears white, $\mu_o = 200$ cp, $\mu_w = 1$ cp, $\sigma = 25$ dyne/cm, $k = 200$ darcies, and $u = 0.115$ cm/s.

mechanisms in the reservoir. There are essentially three independent dimensionless parameters that affect the flow rate of crude:

$$\frac{Lg\ \Delta\rho \cos\Theta}{\Delta p},\ \frac{\sigma \cos\theta_c}{\sqrt{k}\Delta p},\ \text{and}\ \frac{k\Delta p}{\mu L u}.$$

In these ratios,

k = permeability of the reservoir,
Δp = pressure drop between injectors and producers,
μ = crude viscosity,
L = distance between injection and production wells,
u = volumetric flow velocity, also known as the Darcy velocity and volumetric flux,
σ = interfacial tension between crude and water,
θ_c = wetting contact angle,
g = acceleration constant due to gravity,
$\Delta\rho$ = density difference between water and crude, and
Θ = formation dip.

The units of these quantities are unimportant for the following discussion. These ratios represent, respectively, the ratio of gravity to applied pressure forces, the ratio of capillary pressure forces to applied pressure forces, and the ratio of applied pressure forces to viscous forces. Only the factors related to crude properties appearing in these dimensionless ratios are considered in the following paragraphs.

In the ratio of the gravity forces to applied pressure forces, the only potentially controllable crude property is the density difference between the crude and the displacing fluids. But the density of the crude (or of the oil phase) does not change significantly during the course of any recovery process.

The factor $\sigma \cos\theta_c$ can be varied in the ratio of capillary pressure forces to applied pressure forces. Since capillary forces tend to hold the crude in the pore network, it would be desirable to make the interfacial tension essentially zero. This has the potential of allowing the crude to be dissolved or emulsified in the aqueous phase. In addition, the crude-containing aqueous phase, if formed, may have a much lower viscosity than the original crude. These dissolution and emulsification processes constitute what is known as chemical flooding, a subject that has been under active research and development for more than 2 decades. Chemical flooding research has been aimed primarily at low-viscosity crudes. It is expected that their general application to viscous crudes will await the resolution of problems associated with the recovery of low-viscosity crudes. This is not to say that no work has been done on the use of chemicals to emulsify and produce heavy oils. Doscher *et al.*,[6] for example, reported in 1958 on the successful use of heated caustic solutions to produce 30 wt% tar-in-water emulsions from Athabasca. But the technical and economic merits of such processes remain to be proved in projects of commercial size.

In the ratio of applied pressure forces to viscous forces, the viscosity is the only crude property. Everything else being the same, the crude flux u (and its associated flow rate) increases as the viscosity μ decreases. There are two basic physical ways to reduce the viscosity of the crude. One is to mix the crude with a low-viscosity solvent, and the other is to increase its temperature (which, by the way, also reduces the capillary forces). Solvents have been tried as displacement fluids both in drives and in stimulation treatments. However, the use of solvents remains to be proved commercially competitive with other processes.

The other way to alter crude viscosity is chemically by thermal decomposition (visbreaking, thermal cracking, or pyrolysis). The crude itself is destroyed by pyrolysis, the products of which are small amounts of solid carbonaceous residue (popularly but erroneously called coke) and gas and a relatively large fraction of a less-viscous liquid. Visbreaking is a mild form of pyrolysis that decreases viscosity without significantly changing other crude-oil properties of interest. Some crude alteration occurs in all thermal recovery processes, but a significant amount of conversion by pyrolysis occurs only at high temperatures.

As it has happened, thermal processes have been the only practical means of improving the oil recovery performance of reservoirs containing viscous crudes. The use of heat in wellbores to decrease the crude viscosity and increase crude production rates has been accepted as viable for many decades. Cyclic steam injection has been commercially successful since it was discovered in the early 1960's. Hot water, hot gas, steam, and various kinds of combustion drives have gained varying degrees of acceptance. Because of its success, relative simplicity, and low cost, the use of heat to improve oil recovery has gained wide acceptance by the oil industry.

Although the development of thermal recovery processes was prompted initially by difficulties in displacing and producing heavy crudes, the drive processes technically can be applied also to low-viscosity crudes. Steam- and combustion-drive processes have recovered low-viscosity crudes in several fields.[7,8] Although the amount of oil recovered may be higher, these processes seldom can compete commercially with cold-water flooding. Hot-water flooding and thermal stimulation methods are somewhat ineffective for low-viscosity crudes.

2.2 Hot-Fluid Injection

As the name implies, hot-fluid injection processes involve the injection of preheated fluids into a relatively cold reservoir. Generally, the injected fluids are heated at the surface, although wellbore heaters have been used on occasion. Fluids range from the common ones such as water (both liquid and vapor) and air, to others such as natural gas, carbon dioxide, exhaust gases, and even solvents. The choice is controlled by cost, expected effect on crude production response, and availability of fluids.

Sometimes, the choice of injection fluid is governed by the nature of the reservoir. Hot gas or even steam (but not hot water) may be considered for injection into the crest of a reservoir with a small, low-pressure gas cap – making sure to maintain gravitational stability by not exceeding the critical rate.[4] Injection of hot water near zones swept by an encroaching natural water drive may prevent oil resaturation of the zones already swept. But injection of steam under similar circumstances should be considered carefully, as the cooling by any encroaching water could have a significantly adverse effect on the rate of growth of the steam zone. Also, steam or fresh water should not be injected, without due consideration of the possible consequences, into reservoirs containing significant amounts of swelling clays.

The effective mobility ratio associated with hot-fluid injection is very unfavorable for noncondensable gases, less so for hot water, and the least unfavorable (or even favorable) for condensable gases such as steam. The effectiveness of steam displacement processes is related to the fact that the ratio of the pressure gradients upstream and downstream of the condensation front, evaluated at the condensation front, is affected greatly by the condensation mechanism itself. This effect, which will be discussed in more detail in Chap. 7, does not occur with hot-water and hot-gas injection.

Since thermal drives involve moving hot fluids from one well to another, they require reservoir continuity over distances somewhat greater than the well spacing.

All processes in which hot fluid is injected through the wellhead suffer from heat losses from the injection wellbore to the formations overlying the reservoir (the overburden). Such heat losses can be a significant fraction of the injected heat when the wells are deep or poorly insulated and the injection rates are low. Under such conditions the temperature of an injected noncondensable fluid entering the formation may be significantly lower than at the wellhead. When the injected fluid is condensable, as in the case of steam, the heat losses cause some of the vapors to be condensed. However, the temperature remains approximately constant as long as there is vapor present, unless there is a large friction pressure drop or a large gravity head – two effects that generally tend to offset each other and are negligible in most current operations.

Surface lines generally are insulated, and injection wells can be completed in such a way as to reduce heat losses. Because of the thermal stresses on the casing, it is usual to use new wells, cemented to surface, for injection. Existing wells in good condition often are used for production in drive projects.

Hot-Water Drives

Hot-water flooding is said to be almost as old as conventional waterflooding,[9] although early operations have not been documented adequately. In this process, the water is filtered, treated to control corrosion and scale, heated, and if necessary, treated to minimize the swelling of clays in the reservoir. The primary role of the heated water is to reduce the oil viscosity and, thereby, improve the displacing efficiency over that obtainable from a conventional waterflood. The design and operation of hot-water drives have many elements in common with conventional waterflooding, and the reader is referred to Craig's monograph[10] for general information.

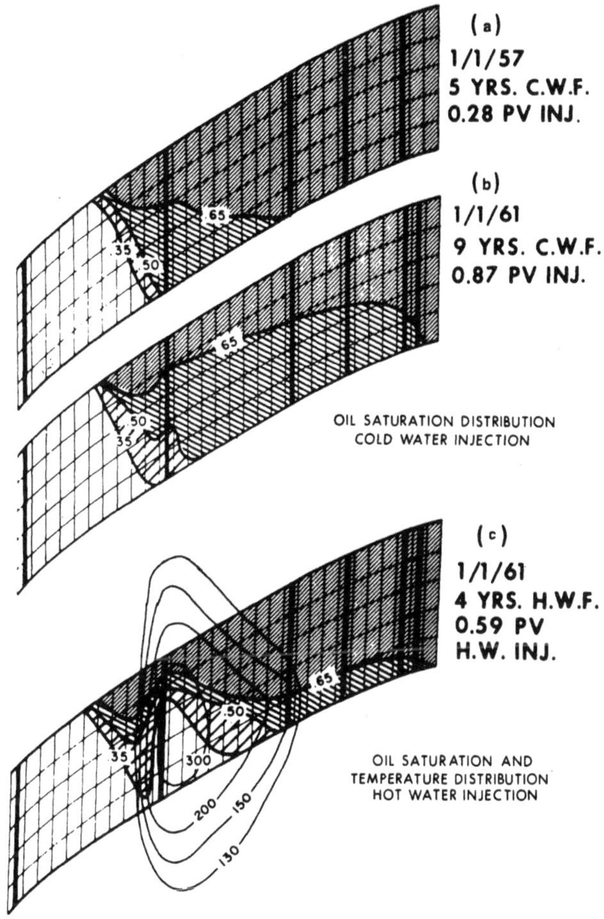

Fig. 2.4 – Calculated saturation and temperature distributions.[11]

Fig. 2.4 serves to make a number of points about hot-water floods. It compares Spillette and Nielsen's[11] calculations for cold-water and hot-water flooding. Fig. 2.4a shows the calculated saturation profiles resulting from a 5-year injection of cold water into the well (shown by the heavy line) at the left. Fig. 2.4b shows the calculated effect of an additional 4 years of conventional waterflooding, the injection now being into the second well from the left. Fig. 2.4c shows the calculated effect of injecting 380°F water for 4 years at the same rate and into the same well as in Fig. 2.4b, starting from conditions existing at the end of the 5-year waterflood shown in Fig. 2.4a. The total amount of water injected (measured in pore volumes) is the same for the hot-water flood as for the conventional waterflood.

One important thing to notice is that the transition zone, the distance between the 0.35 and 0.65 oil-saturation contours, is reduced in the hot-water flood. This is evidence of improved displacement

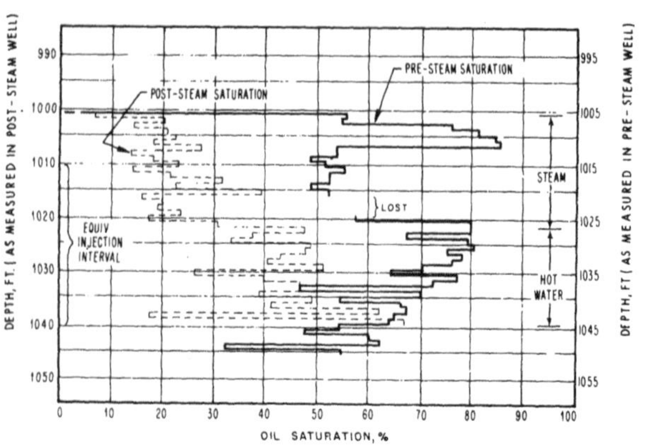

Fig. 2.5 – Comparison of core analyses before and after steam drive.[14]

efficiency – i.e., in the context of the discussions related to Fig. 2.1, the displacement front tends to become more piston-like as temperature increases. Thermal expansion of oil also contributes to the improved displacement efficiency of thermal projects. Another item to notice is the underrunning of the water near the base of the sand in Fig. 2.4a, even in the conventional waterflood. This underrunning is the result of buoyancy forces between the water and the oil and is obviously significant, even though the authors report using a permeability five times greater at the top than at the bottom of the sand. In hot-water floods, the gravity separation usually is accentuated by the increased density difference between water and oil as temperatures increase. Because of this buoyancy and other factors, the contours of equal temperature and saturation are certainly not vertical within the reservoir sand. It is worthwhile to point out that after injection of 0.59 PV of hot water, only about 30% of the reservoir shown in the cross section has been heated, and that the average temperature rise in the heated zone is well below the temperature rise at the injection well. It is also important to recognize that most of the reservoir heat is in the zones from which most of the oil already has been displaced. Indeed, all thermal drives are characterized by the presence of large amounts of heat in oil-depleted portions of the reservoir, which has prompted a number of modifications aimed at scavenging or recycling the heat to improve the efficiency of the process. For hot-water drives, some of this heat can be scavenged by injecting unheated water near the end of the project.

These observations are applicable to all thermal recovery processes. Accordingly, only additional comments will be made in discussing the other methods.

Steam Drives

Once-through boilers customarily are used to produce steam and varying amounts of hot water. The hot water usually is injected into the reservoir along with the steam, although sometimes the amount of hot water injected is reduced by separating the vapor and liquid phases upstream of the wellhead.[12] The separated hot water usually is recycled through the boiler feed system (if the water is sufficiently clean) or used to preheat the boiler feedwater. Because heating promotes deposition of dissolved solids, it is especially important in steam injection operations to treat the boiler feedwater properly. Precautions for reducing heat losses in surface and injection lines are similar to those to be taken in hot-water flooding. Because of the condensation resulting from heat losses, the water content of the steam entering a formation is generally higher than that at the wellhead. Injected or condensed fresh (hot) water may create problems in reservoirs containing swelling clays.

In addition to possessing the displacement mechanisms and thermal phenomena present in hot-water floods (which lead to improvements in displacement efficiency resulting from a decrease in viscosity, a reduction in capillary forces, and thermal expansion of the crude), steam drives also (1) cause boiling and steam stripping of the light components in the crude, (2) promote the formation of a low-viscosity oil bank near the condensation front, (3) yield unusually low residual oil saturations, and (4) improve the effective mobility ratio of the displacing process. Most of these effects have been discussed by Willman *et al.*[13]

In the steam-stripping process, steam removes relatively light components from the crude, much as a dry current of air passing over a pool of water lowers the vapor pressure and picks up moisture. Steam stripping removes a larger fraction of the crude than would be suggested by its boiling point distribution (e.g., the boiling point distribution of a particular crude might indicate that 20% of the crude boils at 400°F, but when the vapor pressure is maintained at a low level by the flowing vapor, more than 20% of the crude would be removed). As the steam condenses, most of the stripped components in the steam also condense, and these light ends condensing at the steam condensation front help to improve the displacement efficiency. The mixing of the condensing light ends with crude reduces the viscosity of the crude contacted by a subsequent advance of the condensation front. Eventually, the oil bypassed by the advancing condensation front is so light and distillable that it shrinks significantly as steam continues to pass by. Such shrinkage of the bypassed oil can result in very low ultimate values for the residual oil in steam drives. (These effects would be more pronounced with crudes containing a high percentage of light fractions but would occur to some extent with any crude.) In addition, the condensation phenomenon itself, quite apart from any distillation effects, results in more favorable effective mobility ratios than would have been expected on the basis of viscosity ratios alone (see Section 7.2).

The effectiveness of steam displacement can be seen from Fig. 2.5, taken from Blevins *et al.*[14] Initial saturations were obtained at the injection well, which is 40 ft from the well cored after the steam drive was terminated. The oil saturation in the upper 20 ft of the interval at the post-test core hole averages 20%, and that in the lower 18 ft averages 42%. This indicates that steam had swept the top 20 ft and hot water had swept the lower 18 ft. Note that the steam rose some 10 ft above the injection interval, to the top of the reservoir, in 40 ft of travel from the injection well.

Hot-Gas Drives

Injection of hot noncondensable gases has been attempted only infrequently. Some cases are reported by Sheinman *et al.*[15] The low viscosity of noncondensing gases results in poor mobility ratios and, consequently, poor displacement efficiencies (Fig. 2.2). Stripping of light ends also may be important in hot-gas drives. Here condensation of the stripped components results from cooling of the gas stream (which is analogous to meteorological precipitation) rather than, as in the case of steam, from the condensation of the carrier vapor phase. Because of the low density of gases, high volumetric gas injection rates (referred to standard conditions) are required to provide heat injection rates comparable with those obtainable with water or steam. At per-well injection rates normally encountered in field operations, wellbore heat losses reduce the temperature of hot dry gases significantly.

2.3 In-Situ Combustion

In in-situ combustion, oxygen is injected into a reservoir, the crude in the reservoir is ignited, and part of that crude is burned in the formation to generate heat. Air injection is by far the most common way to introduce oxygen into a reservoir. Since the injected air normally is cool (except as compression would warm it), the surface lines need be designed only as would be required to conform with prudent practice.

The wellbore near the pay zone, or for that matter any part of the injection well that might come in contact with free oxygen and fuel (crude), should be designed for high thermal stresses. Crude is likely to enter the wellbore by gravity drainage where the air enters the formation preferentially over a short segment of a large open interval having adequate vertical permeability. This crude inflow may be increased as the reservoir temperature near the wellbore increases as a result of the heat generated either by the ignition system used in the wellbore or by the combustion process itself, including reverse combustion following spontaneous ignition a short distance into the reservoir. In designing injection wells, precautions should be taken against any likelihood of combustion in the wellbore.

Injection wells generally require special design considerations only where they are specifically vulnerable. Production wells, on the other hand, can be expected to be assaulted to varying degrees by corrosion, erosion, and high temperature from the producing interval to the wellhead, the most severe conditions being at the producing interval. The standard well equipment, therefore, must be modified accordingly. The degree of modification will depend on the crude and water in the reservoir, the friability of the sand, the steel used in the tubing and casing, the method of well completion, the amount of heat and free oxygen, and the type of in-

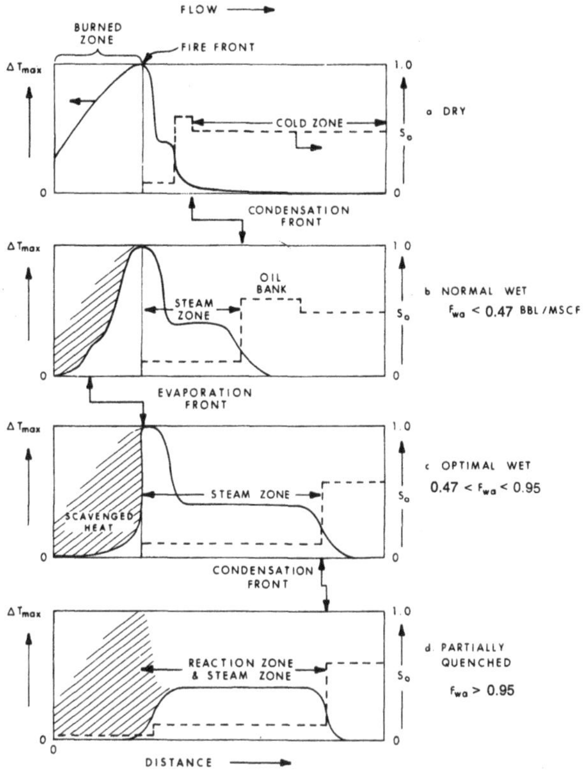

Fig. 2.6—Schematic representation of combustion processes.[17,19]

situ combustion process or control measures used.

Injection of air alone is known as dry underground combustion, in-situ combustion, or fireflooding.[16] For our purposes, unless otherwise specified, underground or in-situ combustion means dry forward combustion. In dry forward combustion, the combustion front moves out from the injection well as air injection is continued. The combustion front moves in the same direction as the air. Reverse combustion occurs when the combustion front moves in a direction opposite that of the injected air. Reverse combustion is achieved by igniting the crude near a production well while temporarily injecting air into it. Upon resumption of the normal air injection program, the combustion front will move toward the injection wells.

Wet and partially quenched combustion,[17] also known by the acronym COFCAW[18] (combination of forward combustion and waterflooding), uses water injection during the combustion process to recuperate the heat from the burned zone and adjacent strata. In this process, the ratio of injected water to air is used to control the rate of advance of the combustion front, the size of the steam zone, and the temperature distribution.

Fig. 2.6 schematically indicates some of the characteristics of the combustion process for different water/air ratios, F_{wa}. These concepts are developed on the basis of idealized horizontal flow and vertical combustion fronts, and they are only approximately representative of actual operations in the field, where the flow behavior is much more complicated.

Dry Forward Combustion

Temperature levels in dry forward combustion are affected by the amount of fuel burned per unit bulk volume of reservoir rock. The temperature levels in turn affect the displacement, distillation, stripping, cracking, boiling of the crude, and formation of "solid" fuel downstream of the combustion front. Temperatures in the range of 650 to 1,500°F have been observed frequently both in the laboratory and in the field. At high temperatures, the combustion zone is very thin. At moderate temperatures, the combustion reaction proceeds slowly enough to allow significant leakage of free oxygen in the direction of flow, thus increasing the thickness of the reaction zones. At lower temperatures, a smoldering reaction with the bypassed air may occur over distances of several feet. This usually happens when air is injected into previously unheated crude-containing reservoirs; the ensuing smoldering reaction generates heat and ultimately causes spontaneous ignition. Air bypassing may also occur in any part of the reservoir if the local air flux is very great, even when fuel is present and temperatures are high.

Typical equivalent oil saturations burned within the burned region are in the range of 0.06 to 0.12; the rest of the crude is displaced.

A characteristic of the dry forward combustion process is that the temperature of the burned zone remains quite high because the heat capacity of the air injected is too low to transfer a significant amount of heat. For this reason, water sometimes is used during or after the combustion process to help transfer the heat from the burned zone and to use it efficiently downstream, where the oil is.

Wet Forward Combustion

The addition of water during the combustion process has several interesting consequences. One is that heat is transferred more effectively than with air alone. A second one is that the steam zone ahead of the combustion front is larger and, thus, the reservoir is swept more efficiently than with air alone. The improved displacement from the steam zone results in lower fuel availability and consumption in the combustion zone, so that a greater volume of the reservoir is burned for a given volume of air injected.

Water must be injected in the wet combustion process. In low-permeability reservoirs, it may be difficult to inject both air and water simultaneously at the desired rates. In that case, the water and air may be injected alternately, each phase being injected for days. The duration of the air and water injection periods is controlled to achieve the desired average water/air ratio.

The water/air ratio also is controlled to obtain desired improvements in combustion-front velocities or temperature levels (Fig. 2.6). At low water/air ratios, all the water that reaches the combustion front already has been converted to steam. If the water/air ratios are kept sufficiently high, most of the water reaching the combustion front still will be in the liquid phase. This reduces the maximum temperature level, in some cases to that corresponding to the partial pressure of steam in the steam/gas mixture. Usually, such temperatures are adequate for thermal drives.

Gravity segregation between water and air does influence the wet combustion process. In extreme cases, water may not reach the upper part of the sand intervals, so that only dry combustion (with its relatively poor heat recuperation) takes place. In the lower parts of the interval, combustion may not be sustained because of the presence of too much water and too little air; in the center section, wet combustion may occur at some unknown water/air ratio. Personal experience with scaled laboratory experiments in gravity-dominated systems indicates that the average performance of wet combustion has the claimed advantages (although to a lesser degree) even under such adverse conditions of gravity segregation. The process has been applied successfully in several fields, as discussed at the end of Section 8.5.

Reverse Combustion

In reverse combustion the combustion front is initiated at the production well and moves backward against the air flow. Where there is no channeling – and typically in the laboratory there is none – the crude upstream of the receding combustion front tends to flow through the front. As the crude and the high-temperature combustion front come together, the crude is cracked severely and a relatively large amount of solid fuel is formed. This is in contrast with the flow of heated crude in cocurrent combustion processes, in which a substantial fraction of the heated crude moves away from the high-temperature combustion zone and only a small amount of fuel is formed. The generous amount of solid fuel formed in reverse combustion operations is responsible for the relatively large amount of equivalent oil saturation burned in the process. Also, a substantial equivalent oil saturation is found in the burned zone (including discernible amounts of unburned solid fuel). Thus, the amount of displaceable oil is somewhat lower than in cocurrent processes. On the other hand, the API gravity of the recovered product is increased significantly by extensive cracking within the combustion and burned zones. Because the more volatile pyrolysis products flow through the hot burned zone, no oil bank builds up and the resistance to flow is fairly low.

A major difficulty with the process is that of keeping it going. As discussed in more detail in Chap. 8, sustained air injection into an unheated reservoir generally leads to spontaneous ignition near the injection well.[20] Since the oxygen then is depleted not far from the injection well, reverse combustion cannot be maintained. In most cases, a reverse combustion operation cannot be expected to last for more than a few weeks, and usually for only a matter of days.[21] The time required for spontaneous ignition to occur is governed primarily by two factors: the reactivity of the crude and the initial reservoir temperature. Self-ignition times have been discussed by Burger[22] and Tadema and Weijdema.[23]

The only known operations with the reverse combustion process are the field test in the Bellamy

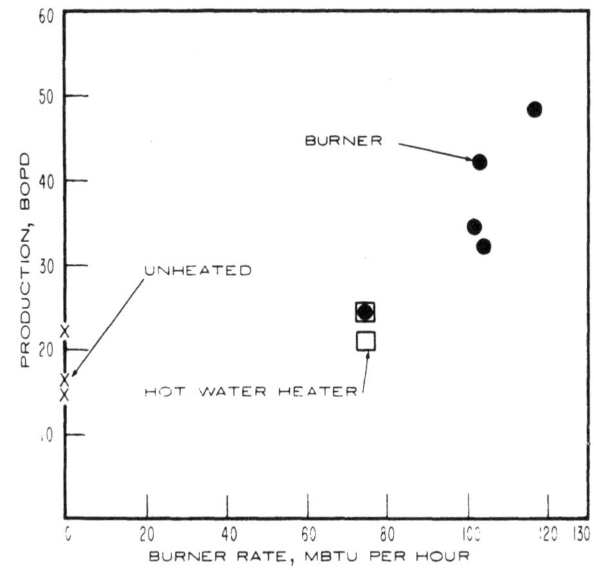

Fig. 2.7 – Production increase with burner rate.[27]

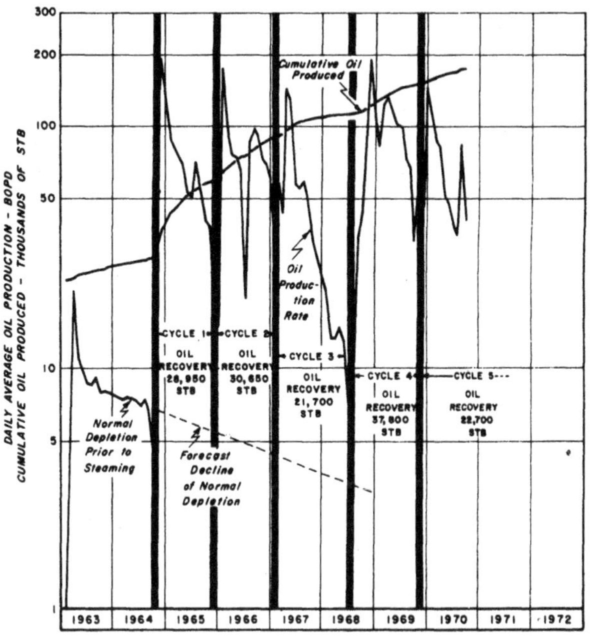

Fig. 2.8 – Typical response to cyclic steam injection.[31]

field reported by Trantham and Marx,[24] those at Athabasca reported by Giguere,[25] and that at the Carlyle field reported by Elkins *et al.*[20] The spacing between wells was so close in the Bellamy test (and the reservoir temperature was so low at the depth of 50 ft) that the operations had ended before self-ignition could occur. At Athabasca, where the reservoir temperature can be about 40°F, self-ignition probably would not occur for a matter of years.

2.4 Thermal Stimulation

By definition, stimulation processes decrease resistance to flow and, thus, allow the driving forces present in the reservoir to increase crude production. In thermal stimulation, the reduction in flow resistance is achieved by heating the wellbore and the reservoir near it. One mechanism that is always in force in thermal stimulation is a reduction in the viscosity of the crude and of the water; reducing the viscosities tends to reduce the flow resistance. A second mechanism is wellbore cleanup, in which the following might occur: organic solids near a wellbore may be melted or dissolved; clays may be stabilized; the absolute permeability may be increased by the high temperatures; or fines that could be inhibiting flow in gravel packs may be flushed away. Wellbore cleanup usually has a rather minor effect after the first stimulation cycle.

Thermal stimulation currently is the only effective treatment for viscous-oil reservoirs with poor lateral continuity. Because the effects are confined to the neighborhood of the wellbore, thermal stimulation improves oil production rates rather quickly. In drives, on the other hand, no significant sustained increase in production rates can be expected until an oil bank or heat (or both) reaches a producing well.

Wellbore Heating

In wellbore heating, the wellbore is heated at the producing interval. Generally, production and heating are performed concurrently, but sometimes they are alternated. The reservoir is heated by thermal conduction countercurrent to the transfer of heat out of the reservoir by the produced fluids.

The wellbore normally is heated either by using a gas-fired downhole burner or a downhole electric heater hanging on an electric power cable or by circulating fluids heated at the surface. Heating by passing electric currents through a reservoir also has been proposed.[26]

Typical increases in crude production rates obtained with downhole gas burners have been reported by DePriester and Pantaleo,[27] the source of Fig. 2.7.

Wellbore heating largely has been supplanted by cyclic steam injection, which is generally much more effective.

Hot-Fluid Injection

The current common practice is to inject steam into a formation for a few weeks, wait a few days to let the heat "soak in" and allow the steam to condense, and then put the well on production.[28,29] This process is known by a number of names: steam soak, steam stimulation, huff'n'puff, and cyclic steam injection.

Other fluids can be used instead of steam, but none have been found to be as effective. Hot water introduces a larger volume of water per unit of heat injected, and its use results in higher water saturations. This can affect the producing oil cut adversely. One of the few documented hot-water soaks is that reported by Dietrich and Willhite.[30]

Cyclic steam injection is popular because the production response is obtained earlier and the amount of recovered oil per amount of steam injected is often higher than in thermal drives. Also, relatively small steam boilers can be used, and they can be moved from well to well. Wells can be steam-soaked several times, the main requirements being natural driving forces, such as solution gas or gravity drainage, and sufficient oil near the wells. An example of the oil production response obtained from a five-cycle steam-soak operation is given in Fig. 2.8.

Combustion

Combustion as a stimulation treatment appears to be used primarily to burn solid organic matter, stabilize clays, and increase the absolute permeability of the formation near the wellbore of a production well. Combustion stimulation treatments have been carried out (White and Moss[32]) but only infrequently.

2.5 Comparison of Recovery Methods

Thermal recovery processes are used for either stimulation or flooding. The choice frequently is governed by the properties of the reservoir. In reservoirs that are small or that have relatively poor continuity, it may not be feasible to drill enough wells to implement a flooding operation. Indeed, if it is prohibitively costly to drill additional wells to ensure adequate communication over close spacing, there may be no choice but to consider a stimulation treatment to increase both recovery rates and ultimate economic recovery. Where the crude viscosity is high, cyclic steam injection has proved to be a successful stimulation technique as long as natural driving forces are available to produce the crude once its viscosity is reduced. Combustion stimulation has been found to be successful in burning solid organic particles, in stabilizing clays, and in increasing absolute permeabilities near the treated well. These effects are particularly attractive in low-permeability reservoirs.

Cyclic steam injection also is used as a precursor to steam drives. In reservoirs containing very viscous crudes, the flow resistance between wells may be so great that steam injection rates are severely limited, making steam drives both technically inefficient and uneconomic. Cyclic steam injection reduces the flow resistance near wells, where the resistance is most pronounced; this alone improves the injection rate attainable during steam drives by reducing the resistance to flow between wells. Repeated cyclic steam injection reduces the flow resistance still farther from the wells and may lead to connecting the heated zones of adjacent wells and further improving the operability of steam drives.

In addition to its sometimes being a necessary precursor to a steam drive, cyclic steam injection also is preferred for economic reasons. Because it is a stimulation treatment, it accelerates oil production. There is no long delay in obtaining a production response, as is the case for steam drives. Furthermore, since steam injection into any well lasts only a few weeks and the resulting production lasts several months, portable steam generators have been developed. A single portable steam generator can be moved from well to well to apply the cyclic steam injection process to several wells at a reasonable capital cost. A drawback to cyclic steam injection, however, is that the ultimate recovery may be low relative to the total oil in place in the reservoir. Ultimate recoveries from steam drives are generally much larger than those from cyclic steam injection. Thus, cyclic steam injection followed by a steam drive is an attractive combination in that crude production is accelerated quickly and the ultimate recovery is quite high.

Combustion displacement processes may be more attractive than steam drives. Conditions that might call for combustion rather than steam injection include: (1) high sustained injection pressures (above 1,500 psi), (2) excessive heat losses from the injection well (in reservoirs more than 4,000 ft deep), (3) a lack of a supply of fresh water or treatment costs that make the use of steam prohibitively expensive, (4) serious clay swelling problems due to fresh condensate, (5) undesirable or prohibited use of fuel to fire steam generators, and (6) thin or low-porosity sands, where the heat recuperation in wet combustion processes tends to make them more efficient than steam drives. Wilson and Root[33] have discussed the comparative costs of reservoir heating using steam and air injection. The successful development of downhole steam generators[34] would affect Conditions 1 and 2 and probably the comparative costs of the processes.

Wet combustion would be considered instead of dry combustion where there is ample available water and where water/air injectivity is favorable. Wet combustion would not be used where there is little likelihood that the water would move through the burned zone to recuperate heat effectively, as in gravity-dominated operations such as the ones at Moco[35] and West Newport Beach.[36]

The effective mobility ratio of steam drives is more favorable than that of combustion processes, as discussed in connection with the sweep efficiency results presented in Fig. 2.2. Also, for combustion processes there is a rather high reported incidence of well failure associated with high temperatures, corrosion, and erosion. These factors prompt us to recommend that *when the economics are the same (laying aside considerations of risk)*, steam injection is to be preferred to a combustion drive. However, each process has its limitations, and sometimes the conditions may point to only one thermal process. That process may well be a combustion one.

In the final analysis, the prudent operator chooses the process that is most cost-effective for his conditions. This choice requires a proper assessment not only of costs and risks but also of the expected production schedule.

Calculation of the production schedules – i.e., the amount of oil estimated to be produced as a function of time (for any producing mechanism) – is one of the main goals of reservoir engineering. The approach followed in this monograph regarding the estimation of recovery resulting from the several thermal recovery processes is to present the calculation methods that are applicable to simple geometries and that can be used to explain the various mechanisms pertinent to field projects. The geometry most often considered is the uniform parallel (linear) flow system, and emphasis is given to those methods for which the use of hand-held calculators is sufficient. As has been discussed in connection with hot water and steam (Figs. 2.4 and 2.5, respectively), buoyancy and permeability layering effects often prevent the flow from being uniform over a vertical section of the reservoir. The same is true of combustion. In most cases, there are no simple methods for predicting the crude recovery under such conditions. Indeed, numerical simulation may be the only way to estimate crude production response where there is buoyancy or vertical nonuniformity. Although, in terms of accuracy, numerical simulation methods of predicting recovery are far superior to desk-top methods, the latter can be very useful because of their simplicity. Ways of roughly estimating recoveries by using the desk-top approach are discussed in this book.

References

1. Walter, H.: "Application of Heat for Recovery of Oil: Field Test Results and Possibility of Profitable Operation," *J. Pet. Tech.* (Feb. 1957) 16-22.
2. Dyes, A.B., Caudle, B.H., and Erickson, R.A.: "Oil Production After Breakthrough – As Influenced by Mobility Ratio," *Trans.*, AIME (1954) **201**, 81-86.
3. Gaucher, D.H. and Lindley, D.C.: "Waterflood Performance in a Stratified Five-Spot Reservoir – A Scaled-Model Study," *Trans.*, AIME (1960) **218**, 208-215.
4. Dietz, D.N.: "A Theoretical Approach to the Problem of Encroaching and Bypassing Edge Water," *Proc.*, Koninkl. Ned. Akad, Wetenschap (1953) B56, 38.
5. Chuoke, R.L., Van Meurs, P., and Van Der Poel, C.: "The Instability of Slow, Immiscible, Viscous Liquid-Liquid Displacements in Porous Media," *Trans.*, AIME (1959) **216**,188-194.
6. Doscher, T.M., Labelle, R.W., Sawatsky, L.H., and Zwicky, R.W.: "Steam Drive Successful in Canada's Oil Sands," *Pet. Eng.* (Jan 1964) 71-78.
7. Konopnicki, D.T., Traverse, E.F., Brown, A., and Deibert, A.D.: "Design and Evaluation of the Shiells Canyon Field Steam-Distillation Drive Pilot Projects," *J. Pet. Tech.* (May 1979) 546-560.
8. Poettmann, F.H., Schilson, R.E., and Surkalo, H.: "Philosophy and Technology of In-Situ Combustion in Light Oil Reservoirs," *Proc.*, Seventh World Pet. Cong., Mexico City (1967) **III**, 487.
9. Nelson, T.W. and McNiel, J.S. Jr.: "Oil Recovery by Thermal Methods, Part I," *Pet. Eng.* (Feb. 1959) B27-B32.
10. Craig, F.F. Jr.: *The Reservoir Engineering Aspects of Waterflooding*, Monograph Series, Society of Petroleum Engineers of AIME, Dallas (1971) **3**.
11. Spillette, A.G. and Nielsen, R.L.: "Two-Dimensional Method for Predicting Hot Waterflood Recovery Behavior," *J. Pet. Tech.* (June 1968) 627-638 and discussion (July 1968) 770; *Trans.* AIME, **243**.
12. Herrera L., A.J.: "The M6 Steam Drive Project Design and Implementation," *The Oil Sands of Canada, Venezuela, 1977* (1977) CIM Special Volume 17, 551-560.
13. Willman, B.T., Valleroy, V.V., Runberg, G.W., Cornelius, A.J., and Powers, L.W.: "Laboratory Studies of Oil Recovery by Steam Injection," *J. Pet. Tech.* (July 1961) 681-690; *Trans.*, AIME, **222**.
14. Blevins, T.R., Aseltine, R.J., and Kirk, R.S.: "Analysis of a Steam Drive Project, Inglewood Field, California," *J. Pet. Tech.* (Sept. 1969) 1141-1150.
15. Sheinman, A.B., Malofeev, G.E., and Sergeev, A.I.: "The Effect of Heat on Underground Formations for the Recovery of Crude Oil – Thermal Recovery Methods of Oil Production," Nedra, Moscow (1969); Marathon Oil Co. translation (1973).
16. Kuhn, C.S. and Koch, R.L.: "In-Situ Combustion – Newest Method of Increasing Oil Recovery," *Oil and Gas J.* (Aug. 10, 1953) **52**, No. 14, 92.
17. Dietz, D.N. and Weijdema, J.: "Wet and Partially Quenched Combustion," *J. Pet. Tech.* (April 1968) 411-413; *Trans.*, AIME, **243**.
18. Parrish, D.R. and Craig, F.F. Jr.: "Laboratory Study of a Combination of Forward Combustion and Waterflooding – The COFCAW Process," *J. Pet. Tech.* (June 1969) 753-761; *Trans.*, AIME, **246**.
19. Buxton, T.S. and Craig, F.F. Jr.: "Effect of Injected Air-Water Ratio and Reservoir Oil Saturation on the Performance of a Combination of Forward Combustion and Waterflooding," AIChE Symposium Series (1973) **69**, No. 127, 27-30.
20. Elkins, L.F., Skov, A.M., Martin, P.J., and Lutton, D.R.: "Experimental Fireflood—Carlyle Field, Kansas," paper SPE 5014 presented at the 1974 SPE Annual Meeting, Houston, Oct. 6-9.
21. Dietz, D.N. and Weijdema, J.: "Reverse Combustion Seldom Feasible," *Prod. Monthly* (May 1968) **32**, No. 5, 10.
22. Burger, J.G.: "Spontaneous Ignition in Oil Reservoirs," *Soc. Pet. Eng. J.* (April 1976) 73-81.
23. Tadema, H.J. and Weijdema, J.: "Spontaneous Ignition in Oil Sands," *Oil and Gas J.* (Dec. 14, 1970) 77-80.
24. Trantham, J.C. and Marx, J.W.: "Bellamy Field Tests: Oil from Tar by Counterflow Underground Burning," *J. Pet. Tech.* (Jan. 1966) 109-115; *Trans.*, AIME, **237**.
25. Giguere, R.J.: "An In-Situ Recovery Process for the Oil Sands of Alberta," paper presented 26th Canadian Chemical Engineering Conf., Toronto, Oct. 3-6, 1976.
26. Abernethy, E.R.: "Production Increase of Heavy Oil by Electromagnetic Heating," *J. Cdn. Pet. Tech.* (July-Sept. 1976) 91-97.
27. DePriester, C.L. and Pantaleo, A.J.: "Well Stimulation by Downhole Gas-Air Burner," *J. Pet. Tech.* (Dec. 1963) 1297-1302.
28. Hong, K.C. and Jensen, R.B.: "Optimization of Multicycle Steam Stimulation," *Soc. Pet. Eng. J.* (Sept. 1969) 357-367; *Trans.*, AIME, **246**.
29. Rivero, R.T. and Heintz, R.C.: "Resteaming Time Determination – Case History of a Steam Soak Well in Midway Sunset," *J. Pet. Tech.* (June 1975) 665-671.
30. Dietrich, W.K. and Willhite, G.P.: "Steam Soak Results, Sisquoc Pool – Cat Canyon Field, Santa Barbara County," paper presented at Petroleum Industry Conf. on Thermal Oil Recovery, Los Angeles (June 1966).
31. Yoelin, S.D.: "The TM Sand Steam Stimulation Project," *J. Pet. Tech.* (Aug. 1971) 987-994; *Trans.*, AIME, **251**.
32. White, P.D. and Moss, J.T.: "High-Temperature Thermal Techniques for Stimulating Oil Recovery," *J. Pet. Tech.* (Sept. 1965) 1007-1011.
33. Wilson, L.A. and Root, P.J.: "Cost Comparison of Reservoir Heating Using Steam or Air," *J. Pet. Tech.* (Feb. 1966) 233-239; *Trans.*, AIME, **237**.
34. Bader, B.E., Fox, R.L., Johnson, D.R., and Donaldson, A.B.: *Deep Steam Project, Quarterly Report for April 1-June 30, 1979*, Sand 79-2091, Sandia Laboratories, Albuquerque, NM (1979).
35. Gates, C.F. and Sklar, I.: "Combustion as a Primary Recovery Process – Midway Sunset Field," *J. Pet. Tech.* (Aug. 1971) 981-986; *Trans.*, AIME, **251**.
36. Koch, R.L.: "Practical Use of Combustion Drive at West Newport Field," *Pet. Eng.* (Jan. 1956) 72.

Chapter 3

Physical and Mathematical Description of Heat and Mass Transfer in Porous Media

Sections 3.1 and 3.2 of this chapter, which contain definitions and a discussion of some of the important assumptions normally made in discussing thermal recovery processes, would benefit all readers. Other parts of this chapter are *not* intended for those who have little or no mathematical background or facility. Also, the chapter is not intended for the expert, for whom the presentation would not be sufficiently advanced. Rather, this chapter is intended as a quick reference to the physical and mathematical descriptions of heat and mass transfer in porous media. Since most of this chapter is *not* directed toward applications, but merely provides support of a general nature, any consistent set of units (rather than field units) applies unless otherwise indicated.

The transfer of heat and the flow of fluids in permeable media are amenable to physical description and mathematical analysis. An understanding of the various physical and chemical laws governing heat transfer is necessary to properly design, implement, control, and interpret thermal projects. Numerical solutions of the mathematical problems representing physical reality are essential, even if they are only estimates. Often one can simplify complex problems without losing much significance and still obtain meaningful solutions. In practice, calculations are necessary to size equipment and to estimate the duration and economic attractiveness of projects.

This chapter discusses the underlying physical principles related to energy and mass transfer and how their mathematical representations are used to develop the differential equations representing the energy and mass balances. In the next chapters, simplified versions of these differential equations are used to provide insight into the evaluation of heat transfer phenomena applied to reservoir problems and to estimate oil recoveries.

A condition generally assumed to prevail in all reservoir processes is that every point within the reservoir is in thermodynamic equilibrium.[1-3] For example, the partial pressure of steam is related thermodynamically to the steam temperature through the relation known as the Clausius-Clapeyron equation, given symbolically as

$$p_s = p_s(T_s). \quad \ldots\ldots (3.1)$$

The partial pressure of steam is also equal to the pressure of the gas phase multiplied by the mole fraction of the gas phase occupied by steam. Even though the pressure and temperature vary from location to location within the reservoir (so that on a global basis there is neither mechanical nor thermal equilibrium), it is assumed that local equilibrium exists so that wherever water vapor and water are present the water vapor's partial pressure and temperature are related by the Clausius-Clapeyron equation. It may appear that thermodynamic equilibrium, though generally assumed to exist in the reservoir, does not always occur. An often-cited example is the supersaturation of solution gas, which sometimes appears to occur during primary production. This and other apparent departures from thermodynamic equilibrium are probably due to the application of global behavior (bulk balances) to essentially microscopic phenomena. On a sufficiently local scale, the assumption of thermodynamic equilibrium can be used with confidence.

Another condition generally assumed to prevail is that the fluids and the reservoir-rock minerals in any small element of volume are at the same temperature. This implies that there is essentially no time lag between the temperature of the fluids in a pore and the average temperature of the surrounding minerals and implies negligible contact resistance to heat flow between the fluids and the minerals. Obviously this assumption can be a close approximation only in cases where the size of the mineral grains is relatively small.[3] That the temperatures of a fluid and of its adjacent grains are the same is a good working assumption in most applications of practical importance, one well-known exception being in the application of thermal recovery processes to rubbled oil shale. Special studies sometimes require that both temperatures be considered.

3.1 Concepts and Definitions

Before discussing heat transfer mechanisms it is important to define the physical significance of thermal properties, quantities, and concepts to be used. (For additional background and examples, refer to Jakob,[4] Bird *et al.*,[5] Carslaw and Jaeger,[6] and Zemanski.[7])

Heat is a form of energy. The heat content of a material, also known as its enthalpy, is the amount of thermal energy in a given mass of the material above a prescribed reference temperature and reference pressure. The *enthalpy* content per unit mass of material (h), also known as the specific enthalpy, is equal to its internal energy per unit mass (e) plus a flow energy term proportional to the ratio of the pressure p to density ρ:

$$h=e+p/(\rho J)-[e_r+p_r/(J\rho_r)], \quad \ldots\ldots\ldots (3.2)$$

where J is the mechanical equivalent of heat and the r subscripts identify quantities evaluated at the reference state. Since all forms of energy are referenced to an arbitrary state, the reference condition usually is omitted from an expression for energies. For example, when the internal energy per unit mass (e) of a fluid is to be evaluated, it is understood that the quantity sought is $e(T,p)-e(T_r,p_r)$. But the reference conditions should *always* be given when numerical values are presented or discussed.

Enthalpies for crude oil fractions, water, and steam are given in Section B.5 of Appendix B.

Temperature (T) is a manifestation of the average kinetic energy of the molecules of a material due to thermal agitation. Temperature is not energy; rather, it is a measure of the thermal energy content of a material. The temperature of a material usually is measured by contacting it with another system (such as a thermometer) whose degree of thermal agitation can be related to a measurable physical change (such as the expansion of mercury in a capillary), after the two systems have come to thermal equilibrium.

Heat capacity at constant pressure (C), also known as isobaric specific heat, is the quantity of heat (i.e., the change in enthalpy) required to increase the temperature of a unit mass of the material by one degree of temperature, while maintaining a constant pressure:

$$C=\left(\frac{\partial h}{\partial T}\right)_p. \quad \ldots\ldots\ldots (3.3)$$

The heat capacity at constant volume is defined as

$$C_V=\left(\frac{\partial e}{\partial T}\right)_V. \quad \ldots\ldots\ldots (3.4)$$

Since the heat capacity of a phase is not a strong function of temperature (except near the critical temperature), it is often convenient to express the specific enthalpy and internal energies as

$$h=C(T-T_r) \quad \ldots\ldots\ldots (3.5)$$

and

$$e=C_V(T-T_r), \quad \ldots\ldots\ldots (3.6)$$

where the heat capacities are now understood to represent *average* values over the temperature range of interest.

Sometimes it is more convenient to express the heat capacity of a substance on the basis of unit volume instead of unit mass. It then is called the volumetric heat capacity M, which is equal to the product ρC, where ρ is the bulk density of the material.

Heat capacities per unit mass for crude oils and fractions, water and steam, gases, reservoir rocks, and minerals are given in Section B.5 of Appendix B.

Thermal conductivity (λ) is a material property that indicates the quantity of heat transferred in unit time through the material per unit cross-sectional area normal to a unit temperature gradient, under steady-state conditions and in the absence of any movement of fluid or particles. Thermal conductivity is defined essentially by Eq. 3.12. Materials having high thermal conductivities are called conductors, and materials having low thermal conductivities are called insulators. In general, the thermal conductivity of any material varies with pressure and temperature. In many reservoir engineering calculations, average values over the expected conditions are adequate, unless there is a phase change. Thermal conductivities of reservoir rock and minerals, liquids, and gases are given in Section B.6 of Appendix B.

Thermal diffusivity (α) is defined as the ratio of the thermal conductivity to the volumetric heat capacity:

$$\alpha=\lambda/\rho C. \quad \ldots\ldots\ldots (3.7)$$

Thermal diffusivities can be determined from the values of densities, heat capacities, and thermal conductivities given in Sections B.1, B.5, and B.6, respectively, of Appendix B. Some specific values are given in Section B.7.

Latent heat of vaporization (L_v) is the amount of heat necessary to change a unit mass of liquid into its vapor without a change in temperature and is numerically equal to the latent heat of condensation. For every fluid component, there is a certain amount of heat associated with its phase change, its magnitude being independent of the direction of the phase change. For mixtures, the latent heat of vaporization depends on temperature, pressure, and composition. For single components, it depends only on either temperature or pressure.

Values of latent heat of vaporization of reservoir fluids may be obtained from Section B.5.

Heat of reaction (Δh_r) is the amount of heat released or absorbed during a chemical reaction per unit mass of reactant. For example, the reaction of oxygen and fuel during combustion releases heat. Such a reaction is called exothermic. An example of an endothermic reaction, one in which addition of heat is required before it can proceed, is the thermal decomposition of limestone and dolomite. Heats of combustion of crudes are given in Figs. 8.12 and 8.13.

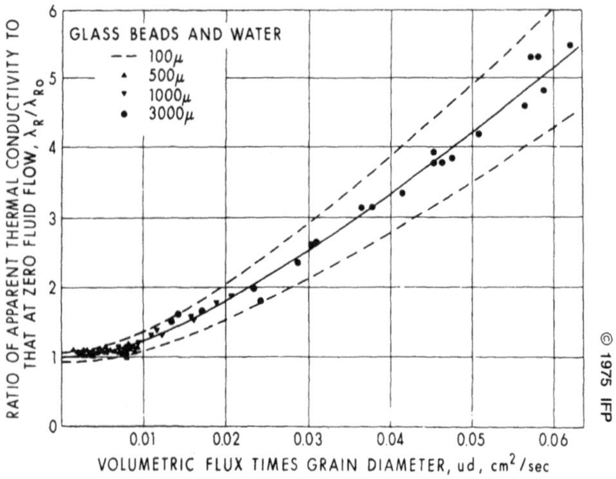

Fig. 3.1 – Influence of fluid velocity on the longitudinal thermal dispersivity of a porous medium.[18]

Gravitational potential energy is the energy per unit mass resulting from the position of an element of mass in the earth's gravitational field above a reference plane. It is given by

$$\frac{gz^*}{Jg_c}, \qquad (3.8)$$

where z is the distance above some arbitrarily chosen reference datum, g is the acceleration constant due to gravity, and g_c is a conversion factor in Newton's law of motion.

The contribution of the potential energy to the total energy in a typical thermal recovery project is small except where the change in z between the top and the bottom of a well can be large. This is not to say that the contribution due to gravity on the potential gradient in Darcy's law is negligible.

Kinetic energy is the energy of an element of mass associated with its motion. On a unit mass basis, it is usually approximated by

$$\frac{|u|^2}{2\phi^2 Jg_c}, \qquad (3.9)$$

where $|u|$ is the magnitude of the volumetric flux, and ϕ is the porosity. Kinetic energy contributions are usually greatest near the wellbore where the fluid velocities are largest. But the contribution of kinetic energy to the energy balance of a reservoir is, for practical purposes, negligible.

Total energy is the sum of the contributions previously discussed. In thermal recovery processes, the total energy per unit mass (e_t) can be considered to be composed of enthalpic and potential components:

$$e_t = h + \frac{gz}{Jg_c}. \qquad (3.10)$$

Since the heat capacity is not a strong function of temperature, it is often convenient to express the total energy as

$$e_t = C(T - T_r) + \frac{gz}{Jg_c}, \qquad (3.11)$$

which follows from the definition of heat capacity at constant pressure given by Eq. 3.3 and its approximation given by Eq. 3.5. Completeness would require the inclusion of the kinetic energy given by Eq. 3.9 in Eqs. 3.10 and 3.11.

3.2 Mechanisms of Heat Transfer

There are only three mechanisms for transferring heat: conduction, convection, and radiation. Each of these can be described in both physical and mathematical terms.

Heat conduction is the process by which heat is transferred through nonflowing materials by molecular collisions from a region of high temperature to a region of lower temperature. The physical law describing heat conduction, known as Fourier's first law, usually is expressed as

$$u_{\lambda x} = -\lambda \frac{\partial T}{\partial x}, \qquad (3.12)$$

where $u_{\lambda x}$ is the rate of heat transfer by conduction in the positive x direction per unit cross-sectional area normal to the x direction (also known as the conductive heat flux in the x direction), λ is the thermal conductivity of the material, and $\partial T/\partial x$ is the temperature gradient in the x direction. The minus sign shows that the transfer is in the direction of decreasing temperature. Similar expressions apply for conductive heat transfer in the y and z directions. Where conduction and convection occur simultaneously, dispersion of a flowing fluid as it moves through a porous medium (discussed in Section 3.4) increases the apparent or effective thermal conductivity of the porous medium, as shown in Fig. 3.1.*

Heat convection is the name commonly used to describe the process by which energy is transferred by a flowing fluid. Consider a heated fluid flowing at a volumetric flux u, the direction of the flow being unspecified. The associated convective heat flux, whose direction always parallels that of the fluid flow, usually is written as

$$u_T = u\rho C(T - T_r), \qquad (3.13)$$

where T_r is the reference temperature. This expression is an approximation to the convective energy flux given by

$$u_T = u\rho\left(h + \frac{gz}{Jg_c} + \frac{|u|^2}{2\phi^2 Jg_c}\right), \qquad (3.14)$$

where the expression in parenthesis represents the total energy of the system per unit mass. Since the potential and kinetic energy contributions can be converted to heat, but are not heat in themselves, u_T is a convective energy flux rather than a convective

*Unless otherwise specified, the coordinate systems used in this book have the z ordinate oriented vertically upward.

*For reservoirs that are anisotropic in thermal conductivity, permeability, and dispersion coefficients, see Refs. 6, 14, and 18, respectively.

heat flux. Convective heat transfer has components in the x, y, and z directions that are proportional to the x, y, and z components of the fluid flow.

The rate of advance of a convective heat front (v_T) resulting from the flow of a hot liquid advancing at an average velocity v through a cooler rock of a given porosity ϕ is expressed by

$$v_T = \frac{\phi v M_f}{M_R} \quad \text{......................} \quad (3.15)$$

Here M_R is the volumetric heat capacity of the fluid-filled reservoir, and M_f is the volumetric heat capacity of the fluid. The direction of the flow again is unspecified, and the direction of the velocity of the convective heat front always parallels that of the fluid flow. As can be ascertained from the values of ϕ, M_R, and M_f, the velocity ratio v_T/v normally would be about one-third in reservoir processes. Thus, even where there are no heat losses, injection of 1 PV of hot water would bring only about one-third of the flooded reservoir to the injection temperature. This is illustrated schematically in Fig. 3.2, which shows that in the absence of conduction there can be no heat transferred beyond the convectively heated zone and that the entire heated zone is at the injection temperature. Since there would be an infinite temperature gradient at the leading surface of the heated zone, inclusion of conduction heat transfer would smooth the temperature profile, as indicated schematically by the dashed line.

Radiation is the process by which heat is transferred by means of electromagnetic waves. The rate of radiation heat transfer from a heated surface per unit surface area is given by the Stefan-Boltzmann law as

$$u_r = \sigma\epsilon(460 + T)^4, \quad \text{....................} \quad (3.16)$$

where σ is the Stefan-Boltzmann constant (1.713×10^{-9} Btu/sq ft -hr-°R^4), the temperature T is in degrees Fahrenheit, and ϵ is the emissivity of the surface. Emissivities are dimensionless, are equal to or less than one, and depend strongly on the nature of the surface. Table B.17 gives emissivities for some metals. There is little thermal radiation through opaque materials such as rocks; therefore, it is not considered to be an important heat transfer mechanism in porous media. Exceptions would be where the radiating surfaces are widely separated and the space between them is filled with fluid (particularly gas). Within the reservoir such situations seldom arise, except perhaps in the in-situ retorting of rubbled oil shale and the gasification of coal.

Radiation is an important heat transfer mechanism in fired heaters and boilers, and it contributes to heat losses from some surface and subsurface flow systems. The effect of heat losses is discussed in Chap. 10.

With the foregoing definitions and neglecting radiation effects, the total energy flux due to flow of a fluid in the x direction is the sum of the conductive and convective components:

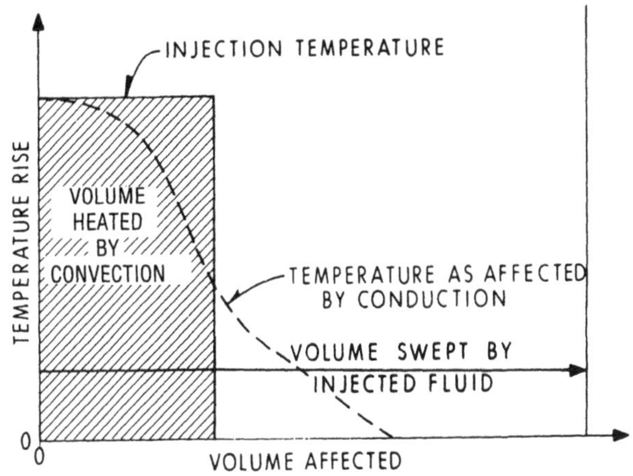

Fig. 3.2 – Temperature profiles resulting from convection and conduction.

$$u_{e,x} = u_{\lambda,x} + u_{T,x}$$
$$= -\lambda \frac{\partial T}{\partial x} + u_x \rho_f h_f, \quad \text{................} \quad (3.17)$$

where $u_{e,x}$ is the total energy flux in the x direction, $u_{T,x}$ is the component of the convective energy flux in the x direction, and u_x is the component of the volumetric flux in the x direction. Similar expressions for the total heat flux in the y and z directions are

$$u_{e,y} = -\lambda \frac{\partial T}{\partial y} + u_y \rho_f h_f \quad \text{................} \quad (3.18)$$

and

$$u_{e,z} = -\lambda \frac{\partial T}{\partial z} + u_z \rho_f \left(h_f + \frac{gz}{Jg_c} \right). \quad \text{.......} \quad (3.19)$$

3.3 General Energy Balance

The first law of thermodynamics states that energy can be neither created nor destroyed. Since kinetic energy and mechanical work done by the thermal expansion of the reservoir on its surroundings are usually negligible, the first law can be stated, on a unit volume basis, as

net energy transfer + energy input

from sources = gain in internal energy. .... (3.20)

In developing the differential equation describing the conservation of energy (or energy balance), the stationary rectangular parallelpiped shown in Fig. 3.3 is considered. A Cartesian coordinate system is used, with the z axis oriented vertically upward. The sides of this infinitesimal volume element are of lengths Δx, Δy, and Δz. Energy is transferred across each of the six faces. The total energy transferred into the volume element across the area $\Delta y \Delta z$ over a period of time Δt is $u_{e,x} \Delta y \Delta z \Delta t$, and the total energy transferred out of the volume element across the opposite face is $(u_{e,x} + \Delta u_{e,x}) \Delta y \Delta z \Delta t$. The net energy transferred to the volume element is obtained by adding the contributions parallel to three coordinate axes. Thus,

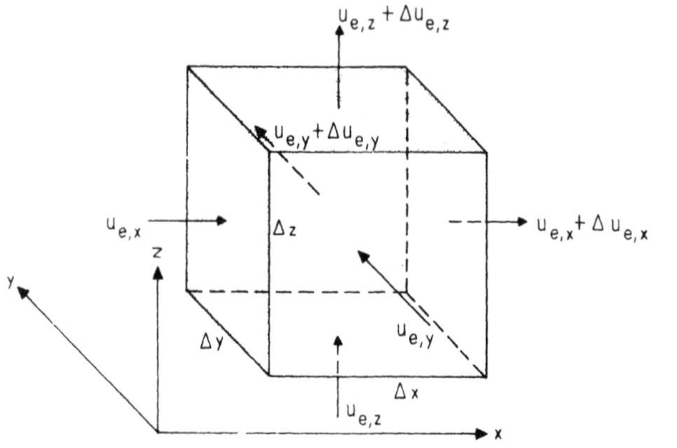

Fig. 3.3 – Volume element for derivation of energy balance.

$$\text{net energy transfer} = (u_{e,x}\Delta y\Delta z + u_{e,y}\Delta x\Delta z + u_{e,z}\Delta x\Delta y)\,\Delta t - [(u_{e,x}+\Delta u_{e,x})\Delta y\Delta z + (u_{e,y} + \Delta u_{e,y})\Delta x\Delta z + (u_{e,z}+\Delta u_{e,z})\Delta x\Delta y]\Delta t \quad \text{.......... (3.21)}$$

$$= -(\Delta u_{e,x}\Delta y\Delta z + \Delta u_{e,y}\Delta x\Delta z + \Delta u_{e,z}\Delta x\Delta y)\,\Delta t. \quad \text{.......... (3.22)}$$

The rate of energy input from sources, per unit volume, is $\dot{Q}$. Over the time period Δt and the volume element $\Delta x\Delta y\Delta z$, the amount of energy from sources is expressed as

$$\text{energy input from sources} = \dot{Q}\Delta x\Delta y\Delta z\Delta t. \quad \text{.. (3.23)}$$

The internal energy of the volume element at any time t is given by $\rho e\Delta x\Delta y\Delta z$. Since the volume element is stationary, the gain of internal energy within it is independent of the space variables and is only a function of time. The internal energy at a time $t+\Delta t$ is $[\rho e + \Delta(\rho e)]\Delta x\Delta y\Delta z$, and over the time period Δt,

$$\text{gain in internal energy} = \Delta(\rho e)\,\Delta x\Delta y\Delta z. \quad \text{... (3.24)}$$

Substitution of Eqs. 3.22, 3.23, and 3.24 into Eq. 3.20 results in

$$-(\Delta u_{e,x}\Delta y\Delta z + \Delta u_{e,y}\Delta x\Delta z + \Delta u_{e,z}\Delta x\Delta y)\cdot\Delta t + \dot{Q}\Delta x\Delta y\Delta z\Delta t = \Delta(\rho e)\,\Delta x\Delta y\Delta z. \quad \text{......... (3.25)}$$

Dividing by $\Delta x\Delta y\Delta z\Delta t$ gives

$$-\left(\frac{\Delta u_{e,x}}{\Delta x} + \frac{\Delta u_{e,y}}{\Delta y} + \frac{\Delta u_{e,z}}{\Delta z}\right) + \dot{Q} = \frac{\Delta(\rho e)}{\Delta t}. \quad \text{................. (3.26)}$$

And, finally, taking the limit as Δx, Δy, Δz, and Δt approach zero results in the differential-equation form of the energy balance:

$$\frac{\partial u_{e,x}}{\partial x} + \frac{\partial u_{e,y}}{\partial y} + \frac{\partial u_{e,z}}{\partial z} = -\frac{\partial(\rho e)}{\partial t} + \dot{Q}. \quad \text{... (3.27)}$$

When n_P phases are present, the internal energy per unit bulk volume (ρe) is given by

$$\rho e = (1-\phi)M_\sigma\Delta T + \phi\sum_{i=1}^{n_p} S_i\rho_i e_i, \quad \text{......... (3.28)}$$

where M_σ is the volumetric heat capacity of the reservoir solids, S_i is the saturation of the ith phase, ρ_i is its density, and e_i is its internal energy per unit mass. (It should be understood that reference conditions – T_r, p_r, z_r, etc. – are to be the same for all the terms in equations such as Eq. 3.28.) The total energy flux components in the x, y, and z directions are the sum of a conductive heat flux and the convective energy fluxes for each flowing phase. For example, the energy flux component in the z direction is

$$u_{e,z} = -\lambda\frac{\partial T}{\partial z} + \sum_{i=1}^{n_p} u_{i,z}\rho_i[h_i + gz/(Jg_c)]. \quad \text{........................ (3.29)}$$

Similar expressions represent the total energy flux in the x and y directions. In Eq. 3.27, $\dot{Q}$ (the total rate of heat input from sources, per unit volume) is made up of contributions such as heat from injection and production wells, heats of combustion and reaction, and endothermic heats of mineral decomposition. Heat sources in principle can be functions of space and time, such as a moving combustion front. The strength of any source also may be affected by the dependent variables of the system, such as temperature and concentration, and may vary with time (as does, for example, a heat injection rate).

Substitution of Eqs. 3.28 and 3.29 into Eq. 3.27 provides the energy balance equation. Although not written in full, later examples will show how these substitutions are made to arrive at energy balances applicable to specific problems.

3.4 Mechanisms of Mass Transport

A particular fluid component can be transported by molecular diffusion and by bulk flow. Although diffusion alone can cause component movement in a no-flow system, usually both diffusion and bulk flow occur. The concept of fluid or component movement by means of diffusion and bulk flow parallels heat transfer by conduction and convection, respectively, in that diffusion and conduction result from molecular interactions while bulk flow and convection are macroscopic phenomena. Although it is well known that bulk fluid flow in porous permeable media is governed by Darcy's law, it is beneficial to discuss the effect of temperature gradients.

For nonisothermal systems, a potential has yet to be defined (if it indeed exists) for which the volume rate of flow per unit cross-sectional area (or volumetric flux, also known as the Darcy velocity) is proportional to the potential gradient. Still, to the best of our knowledge, the generalized form of Darcy's law,

$$u_{i,x} = -\frac{k_i}{\mu_i}\frac{\partial p_i}{\partial x}, \quad \text{.................... (3.30)}$$

$$u_{i,y} = -\frac{k_i}{\mu_i}\frac{\partial p_i}{\partial y}, \qquad (3.31)$$

and

$$u_{i,z} = -\frac{k_i}{\mu_i}\left(\frac{\partial p_i}{\partial z} + \rho_i \frac{g}{g_c}\right), \qquad (3.32)$$

holds in the presence of a temperature gradient even when the effective permeability k, the viscosity μ, and the density ρ of phase i are functions of temperature. Here, $u_{i,x}$ is the component of the volumetric flux in the x direction for phase i (and similarly for $u_{i,y}$ and $u_{i,z}$), g is the acceleration constant due to gravity, g_c is the conversion factor in Newton's law of motion, and p_i is the pressure of phase i.

Modifications of Darcy's law to account for turbulence and other inertial effects also are considered to be valid in the presence of temperature variations.

In addition to bulk flow, mass transfer by diffusion within a phase is important in some recovery processes. The diffusive mass flux, or rate of mass flow per unit cross-sectional area, is given by Fick's law:

$$u_{\eta,x} = -D\frac{\partial c}{\partial x}, \qquad (3.33)$$

which is considered valid in the presence of temperature variations. Here, $u_{\eta,x}$ is the component of the diffusive mass flux in the x direction, D is the diffusion coefficient, and c is the concentration of the diffusing component. For example, at or near a burning interface the oxygen concentration gradient is usually large. Diffusion may provide a significant oxygen flux to the combustion zone. This would be in addition to a contribution arising from any bulk flow of an oxygen-containing gas phase. The total oxygen mass flux in the gas phase in the x direction is

$$(u_{m,x})_{O_2} = -\frac{k_g}{\mu_g}\frac{\partial p_g}{\partial x} c_{O_2} - D\frac{\partial c_{O_2}}{\partial x}, \qquad (3.34)$$

where the oxygen concentration is given in mass of oxygen per unit volume of gas phase, and $(u_{m,x})_{O_2}$ represents the x component of the mass flux of oxygen. Similar expressions hold for each fluid component within any phase. Because diffusion coefficients for gases are of the order of 100 to 1,000 times as great as those for liquids, only diffusion mass transport in gases is considered in this monograph. Expressions similar to Eq. 3.34 can be developed for diffusion mass fluxes in the y and z directions.

The effective diffusion coefficient D appearing in Eqs. 3.33 and 3.34 is not the molecular diffusivity. Rather, it is an effective diffusivity found to be affected by the porosity, tortuosity, average pore or grain size, degree of heterogeneity and cementation of the rock, and fluid velocity. It more commonly is called the dispersion coefficient to reflect the dispersive effects of the porous medium in enhancing the mixing of miscible components during bulk flow. Data from Perkins and Johnson,[8] who present an excellent discussion on dispersion, show that the

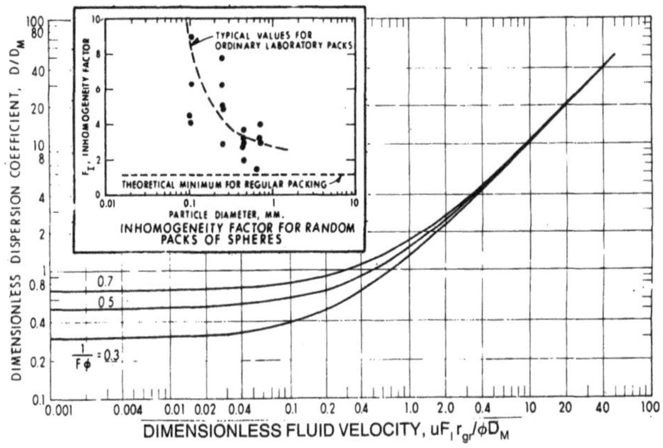

Fig. 3.4 – Dispersion coefficient in the flow direction.[8]

dispersion coefficient in the direction of flow, graphed in Fig. 3.4, is given by

$$D = \frac{D_M}{F_R \phi} + 0.5\frac{u r_{gr} F_I}{\phi}, \qquad (3.35)$$

where F_R is the formation electrical resistivity factor, D_M is the molecular diffusivity for the component in square centimeters per second, u is the phase volumetric flux in centimeters per second, and r_{gr} is the grain radius in centimeters; the inhomogeneity factor F_I is given in the insert to Fig. 3.4.

The diffusion-dispersion process just described occurs within a phase. Diffusion-dispersion processes also play a significant role in mass transport (transfer) between different phases, even in isothermal systems. One of the most common examples of interphase mass transport affected by diffusion is the recycling of a condensate gas reservoir by dry gas. Here the lighter components in the condensate diffuse through it toward the interface with the injected lean gas phase. From there they diffuse into the gas phase. Then the enriched gas flows according to Darcy's law, but the transport of the components in the condensate toward the gas phase is strictly a diffusion process. In this isothermal stripping process, the condensate is denuded of its lightest components when it comes in contact with a lean gas. The stripping process also is influenced by the vapor pressure of the components; the lightest components (which have the highest vapor pressure) are stripped off preferentially; the very heaviest often remain in the reservoir. Since vapor pressures are increased markedly by increases in temperature, stripping can be pronounced in thermal operations, especially in those directed at light crudes. Stripping effects are particularly significant in thermal operations involving a hot gas phase, such as in steam, combustion, or hot-gas drives.

Injected or evolved gas reduces the partial pressure of other hydrocarbon vapors in equilibrium with the crude. These reduced vapor pressures, in turn, reduce the boiling temperature of crude components below their normal boiling points. Fig. 3.5, taken from Nelson,[9] gives a correlation for the effect of vapor pressure on the boiling point of normal paraffin

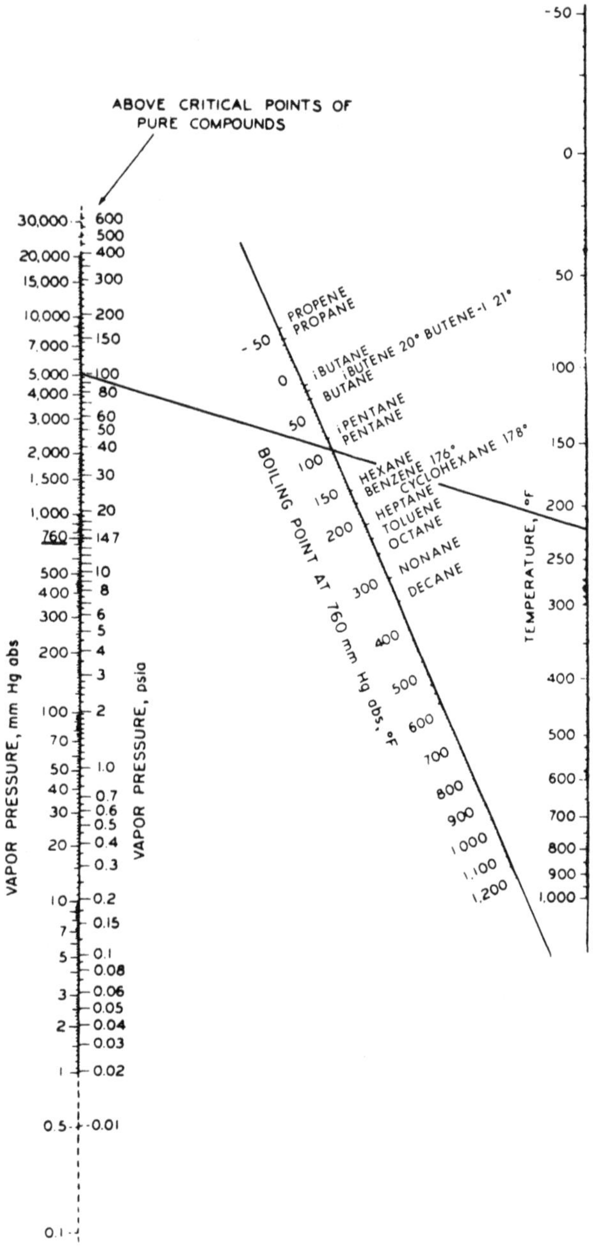

Fig. 3.5 – Vapor pressure of normal paraffin hydrocarbons.[9]

hydrocarbons. For example, at a vapor pressure of 100 psia, the boiling point of normal pentane is about 200°F, whereas it is 97°F at the reference pressure. On reaching cooler portions of the reservoir, some or all of the components in the vapor phase may condense to liquids. For example, in a steam process, the condensation of steam necessarily will be accompanied by condensation of some hydrocarbon components that had been distilled from the crude. The condensed hydrocarbon components mix miscibly with the crude present at the point of condensation, a process controlled by diffusion and dispersion. What is important is that at the steam displacement front, the oil would be (after a while) much lighter and less viscous than crude originally in the reservoir. At an advanced stage in a steam drive the residual oil trapped by the steam front, when rich in light ends, may be stripped to a very low ultimate residual oil saturation. Such a phenomenon has been alluded to by Hagoort *et al.*[10]

The partition of any component among the several phases present is governed by local thermodynamic conditions. Partition factors (known as equilibrium constants, K ratios, and K-values and defined by Eqs. 3.37 through 3.39) are functions of the local pressure, temperature, and composition.

Solubility and adsorption effects can delay the rate of advance of a component through the reservoir. For example, in a combustion process, the appearance of carbon dioxide at distant production wells may be delayed because of its solubility in both water and oil phases. The carbon dioxide will show up in the produced gases when the partitioning between phases has been satisfied. This delay, which is different for different components, sometimes is referred to as the reservoir chromatographic effect.

Thus, a number of mechanisms play a role in moving fluids from one point to another within a reservoir. (See Collins[2] for further discussion of bulk and interphase mass transport in porous media.) Also, mass transport phenomena have been studied in great detail in other disciplines (e.g., see Bird *et al.*[5] and Nelson[9]).

One last mechanism should be mentioned – compaction. Although it is not common, it is important in many thermal projects in the Bolivar coast of Venezuela in forcing fluids from reservoir rocks. An example is the Tia Juana steam-drive project[11] in which compaction significantly reduced the pore volume, in effect squeezing the oil out of the reservoir.

3.5 Continuity Equation

A mathematical description of fluid flow in a permeable medium is obtained from three laws or principles: (1) the law of conservation of mass, (2) laws describing fluid transport, such as Darcy's and Fick's laws, and (3) equations of state. The law of conservation of mass, when applied to an arbitrary volume, requires that

net mass transfer + mass input from sources

= accumulation of mass. . . . . (3.36)

This equation must be true for each component in the reservoir fluids. Most reservoirs contain water, oil, and gas phases, each displaying limited solubility or miscibility in the others. Each of these phases may contain many miscible components. For instance, analysis may indicate that the gas phase contains methane, carbon dioxide, water vapor, hydrogen sulfide, propane, butane, and several other hydrocarbon and nonhydrocarbon components. To some degree, all components will be present in all three phases. That is, the total number of molecules of any component in a small reservoir volume will be divided among those in the gas, oil, and water phases. The mole fraction of any component j in the gas, oil, and water phases will be denoted by y_j, x_{oj}, and x_{wj}, respectively, so that the following equilibrium conditions can be written as follows:

$$y_j = x_{oj} K_{oj}\,(p, T, \text{other components}) \quad \ldots\ldots (3.37)$$

and

$$y_j = x_{wj} K_{wj}\,(P, T, \text{other components}). \quad \ldots\ldots (3.38)$$

K_{oj} is called the equilibrium ratio or K-factor and represents the ratio of the mole fraction of component j in the gas phase to the mole fraction of component j in the oil phase. K_{wj} is the equilibrium ratio representing the ratio of the mole fraction of component j in the gas phase to the mole fraction of component j in the water phase. Sometimes it is convenient to eliminate y_j from Eqs. 3.37 and 3.38, which gives

$$x_{oj} = x_{wj} \frac{K_{wj}}{K_{oj}}\,(p, T, \text{other components}). \quad \ldots (3.39)$$

Only two of the three equilibrium ratios are independent.

Groups of components normally are treated as a pseudocomponent[12] to reduce the number of equilibrium factors required for analysis.

An example of K-values for hexanes in an Oklahoma City crude, taken from Katz and Hachmuth,[13] is given in Fig. 3.6.

For water vapor, it is normal to use an equilibrium ratio derived from Dalton's law of partial pressure,

$$y_w = p_s(T)/p_g, \quad \ldots\ldots (3.40)$$

where $p_s(T)$ is the saturation pressure of steam (which is also the *partial* pressure of steam at temperature T), and p_g is the pressure of the gas phase. A second equilibrium condition for the water component is

$$y_w = x_{ow} K_{ow}. \quad \ldots\ldots (3.41)$$

Since there is always a finite amount of water in the vapor phase, it follows that $(0 < y_w \leq 1)$. This is true even when the solubility of water in oil is assumed to be zero $(x_{ow} = 0)$.

The total mass flux of component j in the x direction is given by

$$(u_{m,x})_j = M_j \left(u_{g,x} \frac{\rho_g}{M_g} y_j + u_{o,x} \frac{\rho_o}{M_o} x_{oj} + u_{w,x} \frac{\rho_w}{M_w} x_{wj} - D_j \frac{\partial}{\partial x} \frac{\rho_g y_j}{M_g} \right), \quad \ldots (3.42)$$

which follows directly from Eq. 3.34 and the definitions of the y_j, x_{oj}, and x_{wj} and the concentration of the component in the gas phase, c_j. Similar expressions hold for fluxes of component j in the y and z directions. In this equation, $u_{i,x}$ is the x component of the volumetric flux of phase i, D_j is the dispersion coefficient for component j in the gas phase, ρ_i is the density of phase i ($i = g$, o, or w), and M_j and M_g are the molecular weights of component j and the gas, respectively. The c_j and y_j are related by

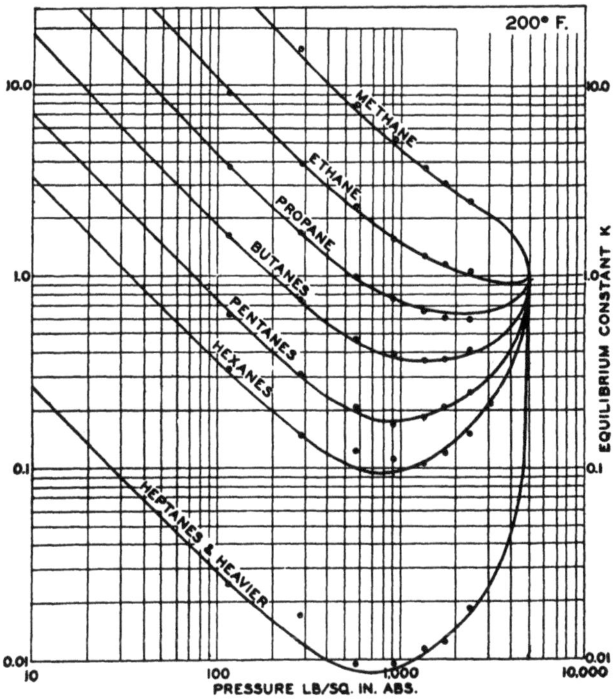

Fig. 3.6 – Variations of equilibrium constants with pressure at 200°F (after Katz and Hachmuth[13]).

$$c_j = M_j \rho_g y_j / M_g. \quad \ldots\ldots (3.43)$$

An approach entirely analogous to that used in obtaining the energy balance (Eqs. 3.20 through 3.27) gives

$$-\left[\frac{\partial}{\partial x}(u_{m,x})_j + \frac{\partial}{\partial y}(u_{m,y})_j + \frac{\partial}{\partial z}(u_{m,z})_j \right] \quad \ldots\ldots (3.44)$$

as the terms of the differential equation arising from the first term of Eq. 3.36.

The contribution due to mass sources of component j per unit volume of reservoir, corresponding to the second term of Eq. 3.36, is expressed as

$$w_j. \quad \ldots\ldots (3.45)$$

And the contribution representing the accumulation of mass of component j, the third term in Eq. 3.36, is

$$\frac{\partial}{\partial t}\left[M_j \phi (\rho_g S_g y_j / M_g + \rho_o S_o x_{oj} / M_o + \rho_w S_w x_{wj} / M_w) \right], \quad \ldots\ldots (3.46)$$

where ϕ is the porosity; ρ_i is the density of phase i ($i = g, o$, or w); M_i is the average molecular weight of phase i; S_i is the saturation of phase i; x_{oj}, y_j, and x_{wj} are the mole fractions of component j in the oil, gas, and water phases; and M_j is the molecular weight of component j.

Substitution of Eqs. 3.30 through 3.32 and Eqs. 3.44 through 3.46 into Eq. 3.36 gives the differential equations arising from the law of conservation of mass for each component j:

$$\frac{\partial}{\partial x}\left(x_{oj}\frac{k_o}{\mu_o}\frac{\rho_o}{M_o}\frac{\partial p_o}{\partial x}+D_j\frac{\partial}{\partial x}\frac{\rho_g y_j}{M_g}+y_j\frac{k_g}{\mu_g}\frac{\rho_g}{M_g}\cdot\frac{\partial p_g}{\partial x}+x_{wj}\frac{k_w}{\mu_w}\frac{\rho_w}{M_w}\frac{\partial p_w}{\partial x}\right)+\frac{\partial}{\partial y}\left(x_{oj}\frac{k_o}{\mu_o}\cdot\frac{\rho_o}{M_o}\frac{\partial p_o}{\partial y}+D_j\frac{\partial}{\partial y}\frac{\rho_g y_j}{M_g}+y_j\frac{k_g}{\mu_g}\frac{\rho_g}{M_g}\frac{\partial p_g}{\partial y}+x_{wj}\frac{k_w}{\mu_w}\frac{\rho_w}{M_w}\frac{\partial p_w}{\partial y}\right)+\frac{\partial}{\partial z}\left[x_{oj}\frac{k_o}{\mu_o}\frac{\rho_o}{M_o}\cdot\left(\frac{\partial p_o}{\partial z}+\frac{g\rho_o}{g_c}\right)+D_j\frac{\partial}{\partial z}\frac{\rho_g y_j}{M_g}+y_j\frac{k_g}{\mu_g}\frac{\rho_g}{M_g}\cdot\left(\frac{\partial p_g}{\partial z}+\frac{g\rho_g}{g_c}\right)+x_{wj}\frac{k_w}{\mu_w}\frac{\rho_w}{M_w}\left(\frac{\partial p_w}{\partial z}+\frac{g\rho_w}{g_c}\right)\right]=\frac{\partial}{\partial t}\left[\phi\left(x_{oj}\frac{\rho_o}{M_o}S_o+y_j\frac{\rho_g}{M_g}S_g+x_{wj}\frac{\rho_w}{M_w}S_w\right)\right]-w_j/M_j. \quad \text{(3.47)}$$

Notice that diffusion of components has been considered only in the gas phase. Diffusion in liquids is to be added to Eq. 3.47 as warranted. The source term w_j may be a function of position, time, temperature, pressure, and compositions. It may represent the injection of a component at a fixed well location or the generation of the component throughout any part of the reservoir resulting from in-situ reactions. In combustion operations, for example, the disappearance of oxygen from the reservoir in the combustion zone is treated as a moving negative source (called a sink), while the injection of oxygen may be treated as a fixed source.

For a reservoir problem with n_C components, there are n_C differential equations of the type given by Eq. 3.47, one for each component. In addition, we have the one energy balance obtained by substituting Eqs. 3.28 and 3.29 into Eq. 3.27. Thus, we have a total of n_C+1 differential equations. We also have $2n_C$ independent equations representing "equilibrium ratios" of the type given by Eqs. 3.37 through 3.39. This gives a total of $3n_C+1$ equations. On the other hand each of the continuity equations has three unknowns: y_j, x_{oj}, and x_{wj}. Also, the temperature, and the pressures and saturations of each phase are unknowns. For three phases, the total number of unknowns is thus $3n_C+7$. The six additional relations required to allow the determination of all the unknowns are

$$S_o+S_g+S_w=1, \quad \text{(3.48)}$$

$$P_{c,wo}=p_o-p_w, \quad \text{(3.49)}$$

$$P_{c,og}=p_g-p_o, \quad \text{(3.50)}$$

$$\sum_{j=1}^{n_C} y_j=1, \quad \text{(3.51)}$$

$$\sum_{j=1}^{n_C} x_{oj}=1, \quad \text{(3.52)}$$

and

$$\sum_{j=1}^{n_C} x_{wj}=1. \quad \text{(3.53)}$$

Here, $P_{c,wo}$ and $P_{c,og}$ are the capillary pressures between the oil and water phases and between the gas and oil phases, respectively. Capillary pressures generally are considered to be a function of the saturations but also may depend on temperature and pressure.

The development of the continuity equation followed here is known as compositional, since it is applied to each component of interest in the reservoir process being considered. It is a very general and powerful approach and reduces to the more familiar forms of the continuity equations used in reservoir engineering known as phase continuity equations. Some examples will show the generality of Eq. 3.47 and of its supportive relationships.

Although the phases usually found in thermal recovery processes are gas, oil, and water, other phases may be considered in special applications. For example, a solid fuel may be considered to be an additional phase in combustion operations. Also, because not all liquid hydrocarbons are miscible with each other, there may be more than one oil phase. Table 3.1 gives the symbol for the mole fraction of any component j in any phase i. The last line reflects the conditions imposed by the constraints similar to those given by Eqs. 3.51 through 3.53. That is, for any phase, the sum of the mole fractions of its components must equal unity. Note that the subscript w is used to denote the water component, whether it is in the liquid or vapor form. Because of usage, the subscript w also is used to represent the water phase, even though the water phase may contain large concentrations of dissolved matter.

The system of equations presented thus far, though idealized to represent isotropic properties and diffusion in the gas phase only, is formidable. However, in many instances of practical interest, the equations can be simplified considerably by reducing

TABLE 3.1 – MOLE FRACTION OF COMPONENT j IN PHASE i

	Phase				
Component	$i=g$ (gas)	$i=o$ (oil)	$i=w$ (water)	ith Phase	n_pth Phase
$j=1=w$ (water)	y_w	x_{ow}	x_{ww}	x_{iw}	x_{n_pw}
2	y_2	x_{o2}	x_{w2}	x_{i2}	x_{n_p2}
jth	y_j	x_{oj}	x_{wj}	x_{ij}	x_{n_pj}
n_cth	y_{n_c}	x_{on_c}	x_{wn_c}	x_{in_c}	$x_{n_pn_c}$
$\sum_{j=1}^{n_c}$	1	1	1	1	1

either the number of pertinent mass and heat transfer mechanisms or the number of components and phases or both. Except in the simplest cases, they usually can be solved only with the aid of high-speed digital computers. At the end of the chapter three examples are given of how these general heat and mass transfer equations have been reduced in practical applications. But first the phase mass balances in general use in reservoir engineering are obtained from the more general compositional continuity equations.

Continuity Equations for Oil, Gas, and Water Phases

In the "phase" continuity equations, the gas and water phases are treated as components and the oil phase is considered to have gas dissolved in it. The gas is considered to have a fixed composition, and the amount of gas dissolved in the oil is considered to be a known function of pressure and temperature. The basis for all mass balances is the volume of the oil phase at stock-tank conditions. That is, the balances are made on the volumes of oil, gas, and water referred to what is known as stock-tank conditions of pressure and temperature. Any reference conditions can be used as stock-tank conditions. A reference temperature of 60°F is standard in the U.S., but the reference pressure varies from one state regulatory agency to another. Once the stock-tank conditions are specified, the volume balances on the oil, gas, and water can be converted to mass balances through the densities of the phases at stock-tank conditions. Since the oil phase is composed of stock-tank oil and gas, we can write

$$\frac{\rho_{o,r}}{M_{o,r}}+\frac{R_s\rho_{g,r}}{M_{g,r}}=B_o\frac{\rho_o}{M_o}, \quad \ldots\ldots\ldots\ldots\ldots (3.54)$$

where

$\rho_{o,r}$ = mass density of the oil phase at stock-tank conditions,
$M_{o,r}$ = molecular weight of the oil phase at stock-tank conditions,
R_s = volume of gas dissolved in the oil phase per unit volume of oil (both volumes referred to stock-tank conditions),
$\rho_{g,r}$ = mass density of the gas phase at stock-tank conditions,
$M_{g,r}$ = molecular weight of the gas phase at stock-tank conditions,
B_o = oil formation volume factor, defined as the ratio of the volume of the oil phase at any pressure and temperature to the volume that it would occupy at stock-tank conditions,
ρ_o = mass density of the oil phase, and
M_o = molecular weight of the oil phase.

In Eq. 3.54 the first two terms are the number of moles of stock-tank oil and gas, respectively, present in a unit volume of stock-tank oil. The third term is the number of moles of oil phase per unit volume of stock-tank oil. The equation states that the number of moles of oil phase is the sum of the number of moles of the stock-tank oil and gas in it. It can be rearranged as

$$\frac{1}{B_o}\frac{M_o}{M_{o,r}}\frac{\rho_{o,r}}{\rho_o}+\frac{R_s}{B_o}\frac{M_o}{M_{g,r}}\frac{\rho_{g,r}}{\rho_o}=1, \quad \ldots\ldots (3.55)$$

where now the first term is the mole fraction of stock-tank oil in the oil phase, which has been defined as x_{oo}:

$$x_{oo}=\frac{1}{B_o}\frac{M_o}{M_{o,r}}\frac{\rho_{o,r}}{\rho_o}. \quad \ldots\ldots\ldots\ldots\ldots\ldots (3.56)$$

Similarly,

$$x_{og}=\frac{R_s}{B_o}\frac{M_o}{M_{g,r}}\frac{\rho_{g,r}}{\rho_o}, \quad \ldots\ldots\ldots\ldots\ldots\ldots (3.57)$$

and $x_{ow}=0$.

Substitution of these expressions for x_{oo}, x_{ow}, and x_{og}, together with $y_o=y_w=x_{wo}=x_{wg}=0$ and $y_g=x_{ww}=1$, into Eq. 3.47 gives

$$\frac{\partial}{\partial x}\left(\frac{1}{B_o}\frac{k_o}{\mu_o}\frac{\partial p_o}{\partial x}\right)+\frac{\partial}{\partial y}\left(\frac{1}{B_o}\frac{k_o}{\mu_o}\frac{\partial p_o}{\partial y}\right)$$

$$+\frac{\partial}{\partial z}\left[\frac{1}{B_o}\frac{k_o}{\mu_o}\left(\frac{\partial p_o}{\partial z}+\frac{g\rho_o}{g_c}\right)\right]$$

$$=\frac{\partial}{\partial t}\left(\frac{\phi}{B_o}S_o\right)-\frac{w_o}{\rho_{o,r}} \quad \text{(3.58)}$$

for the oil phase,

$$\frac{\partial}{\partial x}\left(\frac{1}{B_w}\frac{k_w}{\mu_w}\frac{\partial p_w}{\partial x}\right)+\frac{\partial}{\partial y}\left(\frac{1}{B_w}\frac{k_w}{\mu_w}\frac{\partial p_w}{\partial y}\right)$$

$$+\frac{\partial}{\partial z}\left[\frac{1}{B_w}\frac{k_w}{\mu_w}\left(\frac{\partial p_w}{\partial z}+\frac{g\rho_w}{g_c}\right)\right]$$

$$=\frac{\partial}{\partial t}\left(\frac{\phi}{B_w}S_w\right)-\frac{w_w}{\rho_{w,r}} \quad \text{(3.59)}$$

for the water phase, and

$$\frac{\partial}{\partial x}\left(\frac{R_s}{B_o}\frac{k_o}{\mu_o}\frac{\partial p_o}{\partial x}+\frac{1}{B_g}\frac{k_g}{\mu_g}\frac{\partial p_g}{\partial x}\right)$$

$$+\frac{\partial}{\partial y}\left(\frac{R_s}{B_o}\frac{k_o}{\mu_o}\frac{\partial p_o}{\partial y}+\frac{1}{B_g}\frac{k_g}{\mu_g}\frac{\partial p_g}{\partial y}\right)$$

$$+\frac{\partial}{\partial z}\left[\frac{R_s}{B_o}\frac{k_o}{\mu_o}\left(\frac{\partial p_o}{\partial z}+\frac{g\rho_o}{g_c}\right)\right.$$

$$\left.+\frac{1}{B_g}\frac{k_g}{\mu_g}\left(\frac{\partial p_g}{\partial z}+\frac{g\rho_g}{g_c}\right)\right]$$

$$=\frac{\partial}{\partial t}\left(\frac{\phi R_s}{B_o}S_o+\frac{\phi}{B_g}S_g\right)-\frac{w_g}{\rho_{g,r}} \quad \text{(3.60)}$$

for the gas phase.

In developing these "phase" continuity equations, the assumption is that the gas composition is constant, which means that the molecular weight of the gas is independent of pressure and temperature and that the formation volume factor for gas is given by

$$B_g=\rho_{g,r}/\rho_g. \quad \text{(3.61)}$$

Also, note that the M_j appearing in the denominator of the last term in Eq. 3.47 must be interpreted as the molecular weight of the phases at stock-tank conditions ($M_{i,r}$). The phase continuity equations (Eqs. 3.58 through 3.60) are the same ones developed by Muskat,[14] except that we have considered the case in which gas does not dissolve in the water phase. Solubilities of additional components can be included by procedures similar to those used in considering the gas dissolved in the oil phase.

Example 3.1 – Steam Injection With Three Phases and Two Components

Consider a steam injection process with gas, dead oil, and water phases, and with hydrocarbon and water components. Diffusion is neglected.

We have $n_P=3$ (three phases) and $n_C=2$ (two components). Accordingly, the values of the molar fractions are, following the manner of Table 3.1,

	Phase		
Component	$i=g$ (gas)	$i=o$ (oil)	$i=w$ (water)
$j=1=w$ (water)	$y_w=1$	$x_{ow}=0$	$x_{ww}=1$
$j=2=o$ (oil)	$y_o=0$	$x_{oo}=1$	$x_{wo}=0$
$\sum_{j=1}^{2}$	1	1	1

Since the only component in the gas phase is steam (the oil is dead), it follows that $y_w=1$ and $y_o=0$. Because the oil is dead, there can be no other components dissolved in it, so that $x_{ow}=0$ and $x_{oo}=1$. The water phase has no other components in it, so that $x_{ww}=1$ and $x_{wo}=0$.

For the oil component ($j=o$), Eq. 3.47 reduces to

$$\frac{\partial}{\partial x}\left(\frac{k_o}{\mu_o}\frac{\rho_o}{M_o}\frac{\partial p_o}{\partial x}\right)+\frac{\partial}{\partial y}\left(\frac{k_o}{\mu_o}\frac{\rho_o}{M_o}\frac{\partial p_o}{\partial y}\right)$$

$$+\frac{\partial}{\partial z}\left[\frac{k_o}{\mu_o}\frac{\rho_o}{M_o}\left(\frac{\partial p_o}{\partial z}+\frac{g\rho_o}{g_c}\right)\right]$$

$$=\frac{\partial}{\partial t}\left(\phi\frac{\rho_o}{M_o}S_o\right)-w_o/M_o,$$

and for the water component ($j=w$), it reduces to

$$\frac{\partial}{\partial x}\left(\frac{k_g}{\mu_g}\frac{\rho_g}{M_g}\frac{\partial p_g}{\partial x}+\frac{k_w}{\mu_w}\frac{\rho_w}{M_w}\frac{\partial p_w}{\partial x}\right)+\frac{\partial}{\partial y}\left(\frac{k_g}{\mu_g}\right.$$

$$\left.\cdot\frac{\rho_g}{M_g}\frac{\partial p_g}{\partial y}+\frac{k_w}{\mu_w}\frac{\rho_w}{M_w}\frac{\partial p_w}{\partial y}\right)+\frac{\partial}{\partial z}\left[\frac{k_g}{\mu_g}\frac{\rho_g}{M_g}\right.$$

$$\left.\cdot\left(\frac{\partial p_g}{\partial z}+\frac{g\rho_g}{g_c}\right)+\frac{k_w}{\mu_w}\frac{\rho_w}{M_w}\left(\frac{\partial p_w}{\partial z}+\frac{g\rho_w}{g_c}\right)\right]$$

$$=\frac{\partial}{\partial t}\left[\phi\left(\frac{\rho_g}{M_g}S_g+\frac{\rho_w}{M_w}S_w\right)\right]-w_w/M_w.$$

Note that $M_g=M_w$ and that M_o and M_w are constant.

We also have

$$S_o+S_g+S_w=1,$$

$$P_{c,wo}=p_o-p_w,$$

and

$$P_{c,og}=p_g-p_o.$$

These equations, together with the energy balance, constitute a system of six equations in the seven unknowns S_o, S_w, S_g, p_o, p_g, p_w, and T. Where there is no steam present, $S_g=0$ and the six unknowns can be solved with the six equations. Where there is steam, Eq. 3.40 is used with $y_w=1$ to obtain

$$p_g=p_s(T),$$

which provides the seventh equation. In the superheated-steam case, $S_w=0$ and the number of equations also reduces to six.

When contributions due to the potential energy are negligible, Eqs. 3.27 through 3.29 combine to become

$$\frac{\partial}{\partial x}\left(-\lambda\frac{\partial T}{\partial x}+u_{o,x}\rho_o C_o\Delta T+u_{w,x}\rho_w C_w\Delta T+u_{g,x}\rho_s h_s\right)+\frac{\partial}{\partial y}\left(-\lambda\frac{\partial T}{\partial y}+u_{o,y}\rho_o C_o\Delta T+u_{w,y}\rho_w C_w\Delta T+u_{g,y}\rho_s h_s\right)+\frac{\partial}{\partial z}\left(-\lambda\frac{\partial T}{\partial z}+u_{o,z}\rho_o C_o\Delta T+u_{w,z}\rho_w C_w\Delta T+u_{g,z}\rho_s h_s\right)=-\frac{\partial}{\partial t}(\rho e)+\dot{Q},$$

where $\Delta T=T-T_r$, and T_r is the reference temperature.

These are the seven equations solved by Coats *et al.*[15] in their numerical simulation of a steam drive displacing a dead oil.

Example 3.2 – Steam Injection With Three Phases and Four Components

Consider a steam injection process with gas, oil, and water phases and four components. The components are (1) a hydrocarbon solution gas that can exist in both the gas phase and the oil phase, (2) a distillable portion of the oil that can exist in both the gas phase and the oil phase, (3) a nonvolatile hydrocarbon, and (4) water. Diffusion is neglected.

We have $n_P=3$ and $n_C=4$. Accordingly, the values of the molar fractions are shown in the following table, in the manner of Table 3.1.

Component	Phase $i=g$ (gas)	$i=o$ (oil)	$i=w$ (water)
$j=4=w$ (water)	$y_w=\frac{p_s}{p_g}$	$x_{ow}=0$	$x_{ww}=1$
$j=1$ (solution gas)	y_1	x_{o1}	$x_{w1}=0$
$j=2$ (distillable liquid hydrocarbon)	y_2	x_{o2}	$x_{w2}=0$
$j=3$ (nonvolatile hydrocarbon)	$y_3=0$	x_{o3}	$x_{w3}=0$
$\sum_{j=1}^{4}$	1	1	1

These molar concentrations give rise to three nontrivial equilibrium conditions:

$$y_1=x_{o1}K_{o1}(p_g,T),$$
$$y_2=x_{o2}K_{o2}(p_g,T),$$

and

$$y_w=p_s(T)/p_g.$$

For Component 1, Eq. 3.47 reduces to

$$\frac{\partial}{\partial x}\left(x_{o1}\frac{k_o}{\mu_o}\frac{\rho_o}{M_o}\frac{\partial p_o}{\partial x}+y_1\frac{k_g}{\mu_g}\frac{\rho_g}{M_g}\frac{\partial p_g}{\partial x}\right)+\frac{\partial}{\partial y}\left(x_{o1}\frac{k_o}{\mu_o}\frac{\rho_o}{M_o}\frac{\partial p_o}{\partial y}+y_1\frac{k_g}{\mu_g}\frac{\rho_g}{M_g}\frac{\partial p_g}{\partial y}\right)+\frac{\partial}{\partial z}\left[x_{o1}\frac{k_o}{\mu_o}\frac{\rho_o}{M_o}\left(\frac{\partial p_o}{\partial z}+\frac{g\rho_o}{g_c}\right)+y_1\frac{k_g}{\mu_g}\cdot\frac{\rho_g}{M_g}\left(\frac{\partial p_g}{\partial z}+\frac{g\rho_g}{g_c}\right)\right]=\frac{\partial}{\partial t}\left[\phi\left(x_{o1}\frac{\rho_o}{M_o}S_o+y_1\frac{\rho_g}{M_g}S_g\right)\right]-w_1/M_1.$$

For Component 2, an equation in all other respects similar to the last one is obtained by writing (x_{o2}, y_2) for (x_{o1}, y_1) and w_2/M_2 for w_1/M_1. For Component 3,

$$\frac{\partial}{\partial x}\left(x_{o3}\frac{k_o}{\mu_o}\frac{\rho_o}{M_o}\frac{\partial p_o}{\partial x}\right)+\frac{\partial}{\partial y}\left(x_{o3}\frac{k_o}{\mu_o}\frac{\rho_o}{M_o}\cdot\frac{\partial p_o}{\partial y}\right)+\frac{\partial}{\partial z}\left[x_{o3}\frac{k_o}{\mu_o}\frac{\rho_o}{M_o}\left(\frac{\partial p_o}{\partial z}+\frac{g\rho_o}{g_c}\right)\right]=\frac{\partial}{\partial t}\left(\phi x_{o3}\frac{\rho_o}{M_o}S_o\right)-w_3/M_3.$$

For Component 4,

$$\frac{\partial}{\partial x}\left(\frac{k_w}{M_w}\frac{\rho_w}{M_w}\frac{\partial p_w}{\partial x}+y_w\frac{k_g}{\mu_g}\frac{k_g}{M_g}\frac{\partial p_g}{\partial x}\right)$$

$$+\frac{\partial}{\partial y}\left(\frac{k_w}{\mu_w}\frac{\rho_w}{M_w}\frac{\partial p_w}{\partial y}+y_w\frac{k_g}{\mu_g}\frac{\rho_g}{M_g}\frac{\partial p_g}{\partial y}\right)$$

$$+\frac{\partial}{\partial z}\left[\frac{k_w}{\mu_w}\frac{\rho_w}{M_w}\left(\frac{\partial p_w}{\partial z}+\frac{g\rho_w}{g_c}\right)+y_w\frac{k_g}{\mu_g}\frac{\rho_g}{M_g}\right.$$

$$\left.\cdot\left(\frac{\partial p_g}{\partial z}+\frac{g\rho_g}{g_c}\right)\right]=\frac{\partial}{\partial t}\left(\phi\frac{\rho_w}{M_w}S_w\right.$$

$$\left.+y_w\phi\frac{\rho_g}{M_g}S_g\right)-w_w/M_w\,.$$

In addition we have

$$x_{o1}+x_{o2}+x_{o3}=1,$$
$$y_1+y_2+y_w=1,$$
$$S_o+S_g+S_w=1,$$
$$P_{c,wo}=p_o-p_w,$$

and

$$P_{c,og}=p_g-p_o.$$

The energy balance, neglecting potential contributions, is identical in form with that obtained in Example 3.1.

The energy equation, the preceding nine equations, and the three nontrivial equilibrium conditions constitute a system of 13 equations in the unknowns x_{o1}, x_{o2}, x_{o3}, y_1, y_2, y_w, S_g, S_w, S_o, p_g, p_w, p_o, and T. The equilibrium conditions are used to eliminate y_1, y_2, and y_w. Capillary pressures are used to eliminate p_g and p_w. S_o can be eliminated using $S_g+S_o+S_w=1$. The resultant seven equations and seven unknowns x_{o1}, x_{o2}, x_{o3}, S_g, S_w, T, and p_o represent the system of equations solved by Coats[16] in simulating a steamflood with distillation and solution gas.

Example 3.3—Reverse Combustion With One Phase and Two Components

An example from an in-situ combustion study brings out additional features of the continuity equations and energy balance. Warren *et al.*[17] considered an adiabatic linear system, in which the gas saturation and gas velocity were constant in space and time, and an oxygen sink (a negative source) resulting from a temperature-dependent reaction between the oxygen and a low concentration of residual fuel, so low that its burning affected neither the effective porosity nor the gas velocity.

In this problem there is only one phase ($n_P=1$, the gas phase), and two components ($n_C=2$, oxygen and reaction products). Accordingly, the values of the molar fractions are as follows, in the manner of Table 3.1.

Component	Phase $i=g$ (gas)
$j=1$ (oxygen)	y_{g1}
$j=2$ (combustion products)	y_{g2}
$\sum_{j=1}^{2}$	1

There can be no equilibrium values in a one-phase system.

Oxygen is the only component followed, since the concentration of the other gas component is $y_2=1-y_1$, and its determination is trivial. Since

$$u_g=\frac{k_g}{\mu_g}\frac{\partial p_g}{\partial x},$$

and ϕ and S_g are constants, the continuity equation (Eq. 3.47) reduces to

$$\frac{\partial}{\partial x}\left(D_1\frac{\rho_g}{M_g}\frac{\partial y_1}{\partial x}-y_1\frac{\rho_g}{M_g}u_g\right)$$

$$=\frac{\partial}{\partial t}\phi S_g\left(y_1\frac{\rho_g}{M_g}\right)-w_1/M_1.$$

Note that here we do consider the diffusion of the oxygen in the gas phase. The quantity $y_1\rho_g/M_g$ is oxygen concentration in moles of O_2 per unit volume of gas phase, c_1, which leads to

$$\frac{\partial}{\partial x}\left(D_1\frac{\partial c_1}{\partial x}-u_gc_1\right)=\phi S_g\frac{\partial c_1}{\partial t}-w_1/M_1.$$

Since oxygen is disappearing from the system, the source w_1 then is called a mass sink. Thus, the term w_1/M_1 represents the rate of oxygen consumption in moles per days per unit volume.

The energy balance obtained from Eqs. 3.27 through 3.29 becomes

$$\frac{\partial}{\partial x}\left(\lambda\frac{\partial T}{\partial x}-u_g\rho_gh_g\right)=\left[(1-\phi)\rho_\sigma C_\sigma\right.$$

$$\left.+\phi S_g\rho_gC_{Vg}\right]\frac{\partial T}{\partial t}-\Delta h_{o2}w_1/M_1.$$

The last term represents the heat source as the product of the mass rate of oxygen consumption per unit reservoir volume (w_1/M_1) and the quantity of heat liberated per unit mole of oxygen consumed (Δh_{o2}). These last two equations are the same as those developed by Warren *et al.*[17] in their investigation of reverse combustion.

References

1. Spillette, A.G.: "Heat Transfer During Hot Fluid Injection into an Oil Reservoir," *J. Cdn. Pet. Tech.* (Oct.-Dec. 1965) 213-217.

2. Collins, R.E.: *Flow of Fluids Through Porous Materials,* Reinhold Publishing Corp., New York City (1961); reprinted by Petroleum Publishing Co., Tulsa, OK (1976).
3. Jenkins, R. and Aronofsky, J.S.: "Analysis of Heat Transfer Processes in Porous Media – New Concepts in Reservoir Heat Engineering," *Proc.,* 18th Technical Conf. on Petroleum Production, Penssylvania State U., University Park, PA (1954) 69.
4. Jakob, M.: *Heat Transfer,* John Wiley and Sons Inc., New York City (1949) **1**.
5. Bird, R.B., Stewart, W.E., and Lightfood, E.N.: *Transport Phenomena,* John Wiley and Sons Inc., New York City (1960).
6. Carslaw, H.J. and Jaeger, J.C.: *Conduction of Heat in Solids,* Oxford at the Clarendon Press, Oxford, England (1959).
7. Zemanski, M.W.: *Heat and Thermodynamics,* McGraw-Hill Book Co. Inc., New York City (1943).
8. Perkins, T.K. and Johnson, D.C.: "A Review of Diffusion and Dispersion in Porous Media," *Soc. Pet. Eng. J.* (March 1963) 70-84; *Trans.,* AIME, **228**.
9. Nelson, W.L.: *Petroleum Refinery Engineering,* fourth edition, McGraw-Hill Book Co. Inc., New York City (1958).
10. Hagoort, H., Leijnse, A., and van Poelgeest, F.: "Steam-Strip Drive: A Potential Tertiary Recovery Process," *J. Pet. Tech.* (Dec. 1976) 1409–1419.
11. de Haan, H.J. and Schenk, L.: "Performance Analysis of a Major Steam Drive Project in the Tia Juana Field, Western Venezuela," *J. Pet. Tech.* (Jan. 1969) 111–119; *Trans.*, AIME, **246**.
12. Lee, S.T., Jacoby, R.H., Cheu, W.H., and Culham, W.E.: "Experimental and Theoretical Studies on the Fluid Properties Required for Simulation of Thermal Processes," *Soc. Pet. Eng. J.* (Oct. 1981) 535-550.
13. Katz, D.L. and Hachmuth, K.H.: "Vaporization Equilibrium Constants in a Crude Oil-Natural Gas System," *Ind. Eng. Chem.* (1937) **29**, No. 9.
14. Muskat, M.: *Physical Principles of Oil Production,* McGraw-Hill Book Co. Inc., New York City (1949) **302**.
15. Coats, K.H., George, W.D., Chu, C., and Marcum, B.E.: "Three-Dimensional Simulation of Steamflooding," *Soc. Pet. Eng. J.* (Dec. 1974) 573-592; *Trans.*, AIME, **274**.
16. Coats, K.H.: "Stimulation of Steamflooding with Distillation and Solution Gas," *Soc. Pet. Eng. J.* (Oct. 1976) 235-247.
17. Warren, J.E., Reed, R.L., and Price, H.S.: "Theoretical Considerations of Reverse Combustion in Tar Sands," *Trans.*, AIME (1960) **219**, 109-123.
18. Bia, P. and Combarnous, M.: "Transfert de Chaleur et de Masse," Revue de l'Institut Francais du Pétrole (June 1975) 361-376.

Chapter 4

Some Reservoir Engineering Concepts

Several topics related to those aspects of reservoir engineering that are applicable to all thermal recovery processes are discussed in this chapter. The topics include reservoir modeling, the effects of channeling or bypassing of injected fluids, the bases for predicting project performance, the estimation of the flow resistance between wells, design considerations, and a definition of thermal oil.

4.1 Reservoir Modeling

Descriptions of reservoirs, and the processes used to recover hydrocarbons from them, always are idealized. These idealizations lead first to conceptual models. When quantitative answers are needed, the conceptual models are translated into physical models (e.g., models using sandpacks and cores to study displacement, and multidimensional scaled physical models) or mathematical models. With the latter models, mathematical analysis and numerical methods, aided by a wide range of computers and calculators, are used to obtain results. Thus, theoretical concepts, physical models, mathematical techniques, and computers and calculators are the tools available for solving reservoir problems. The degree of idealization varies with the needs of the study and with the capabilities and tools available to the engineer. Even with the most sophisticated tools available to us, we can represent the reservoir and its contents and flow mechanisms only by gross idealizations. It probably will remain so, despite great improvements and advances in the tools used to carry out such investigations.

The effect of detailed descriptions of the reservoir and its flow processes on oil recovery can be studied only by means of numerical simulators. For the rather simple numerical examples presented in this monograph, simplified descriptions (models) of the processes are necessary. Evaluation of the resultant mathematical expressions requires no more than a desk-top calculator. Simplicity has merits, and the usefulness of the simpler models has been proved over the past decades in several ways. In waterflooding,[1] for example, an extensive body of literature was developed to obtain production responses in horizontal, isothermal, homogeneous reservoirs of uniform and constant properties (including reservoir thickness, permeability, porosity, and initial saturation) containing incompressible fluids having constant and uniform properties throughout the reservoir. These assumptions were used for many years to study the effect of well patterns and mobility ratio on vertical-front (i.e., neglecting gravity effects) piston-like (i.e., neglecting capillary and saturation gradients) displacement and have provided great insight into flooding processes. In well test analysis,[2,3] similar assumptions often are made of the reservoir and its initial conditions. The reservoir usually is assumed to be filled with a single fluid having uniform properties and a constant compressibility. Interpretations of well test data, which at best have been validated for relatively uniform phase distributions, are extremely useful in determining not only the average reservoir pressure but also the reservoir transmissivity[4] (kh/μ) and the ratio of hydraulic diffusivity ($k/\phi\mu c$) to the square of the effective well radius (r_{wa}).

Frontal Advance and Bypass Models of Displacement Processes

In thermal recovery, which developed on the heels of waterflooding, similar assumptions and models were used during the initial development of the process. Displacement fronts, such as combustion and steam condensation fronts, initially were assumed to be vertical and to extend throughout the full thickness of horizontal reservoirs. Where reservoirs had some dip, the fronts were assumed to be normal to the plane of the reservoir. Similar assumptions also were made about saturation and temperature profiles – i.e., isosurfaces were considered to be normal to the bedding and to extend through the full thickness of the reservoir. In this monograph, models having these characteristics are called frontal advance (FA) models. This name is used to distinguish them from the bypass models, which are considered more realistic but also more complicated to describe, both physically and mathematically. In the bypass models, the fronts are not vertical. The causes of bypassing include inhomogeneities of reservoir properties (such as absolute and relative permeabilities, capillarity,

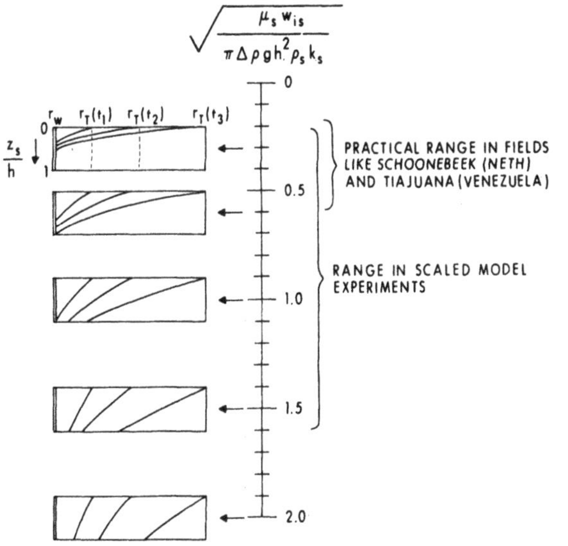

Fig. 4.1 – Effect of the ratio of viscous to gravity forces on the location of steam displacement fronts.[7]

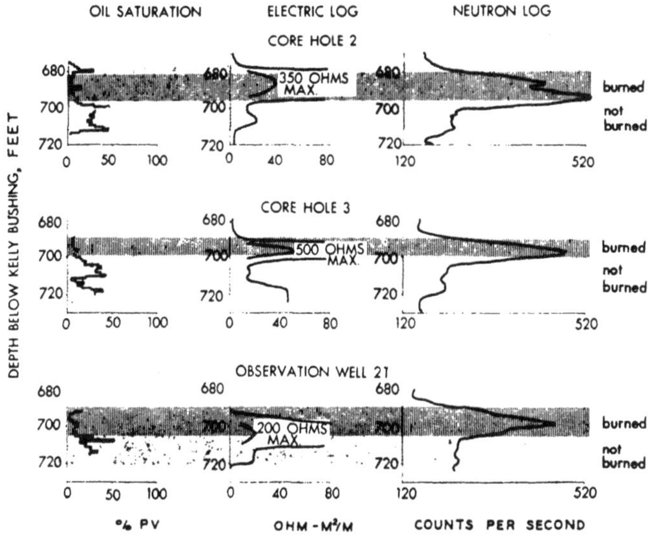

Fig. 4.2 – Postthermal recovery coring and logging results from Wells CH-2, CH-3, and OW-21, South Belridge field.[8]

and porosity) and initial saturation and pressure distributions, anisotropies in the reservoir properties, partial well completions, and buoyancy forces.

Although bypass models of reservoir displacement processes are obviously more realistic than FA models, it is often much more difficult to describe them properly with anything but numerical simulators. A few publications[5-7] have considered highly simplified thermal bypass models and have shown significant differences in the behavior of results from the two types of models. Generally, however, simulators are used when bypassing is considered to be important.

Of the several causes of bypassing, the buoyancy forces are the only ones that act consistently. Denser fluids always tend to flow downward and stay at the bottom of a reservoir, allowing lighter fluids to occupy the shallower, upper parts. The degree to which this happens will be affected by the other factors that influence bypassing. In extreme cases, for example, where there is a high-permeability (thief) zone in the lower part of a reservoir, it may be difficult to discern the effect of buoyancy because of the dramatic channeling through the thief zone. Also, the effect of buoyancy will not be pronounced where the time available for the fluids to separate under the influence of gravity is small compared with the time it takes the fluids to move from an injector to a producer. Thus, buoyancy effects will be slight where the horizontal flow velocities are high compared with the vertical.

Bypassing due solely to buoyancy is illustrated in Fig. 4.1. This figure, taken from van Lookeren,[7] shows the effect of the ratio of viscous to gravity forces on the shape of a radial steam displacement front near an injection well. The equations given in that reference are some of the few available for calculating the shape of steam displacement fronts for radial flow, as well as for linear flow systems in dipping formations, from operational and reservoir parameters. Note the pronounced effect that a small change in the buoyancy parameter has on the shape of the steam displacement fronts.

Buoyancy effects are more pronounced in combustion and steam drives than in waterfloods because of the larger density difference between vapors and liquids than between water and oil. Increasing the spacing between wells tends to increase the time available for separation and increases the buoyancy effects. Low injection pressures, which are economically advantageous in any project, result in low gas-phase densities and high density differences; these also promote buoyancy. Buoyancy is important. Its potential significance in thermal operations cannot be overemphasized. In fact, it can control the type of displacement prevailing in a thermal project. Gravity override in combustion and steam injection operations has been observed and reported frequently,[7,8] and underride of injected water due to buoyancy forces also has been observed, both in conventional and hot-water floods.[9,10] Examples of the effects buoyancy can have in thermal recovery operations are given in Figs. 4.2, 4.3, and 7.11. In Fig. 4.3, the horizontal lines indicate the location of the top and bottom of the reservoir. Since factors other than buoyancy also would yield effects similar to those shown in the figures, these examples of fluid segregation are not necessarily due *solely* to buoyancy.

An example of a combustion project where bypassing was controlled primarily by reservoir layering rather than by buoyancy is shown in Fig. 4.4.

Some fundamental differences inherent in the FA and bypass models (bypass not necessarily resulting from buoyancy forces) are discussed in the following sections.

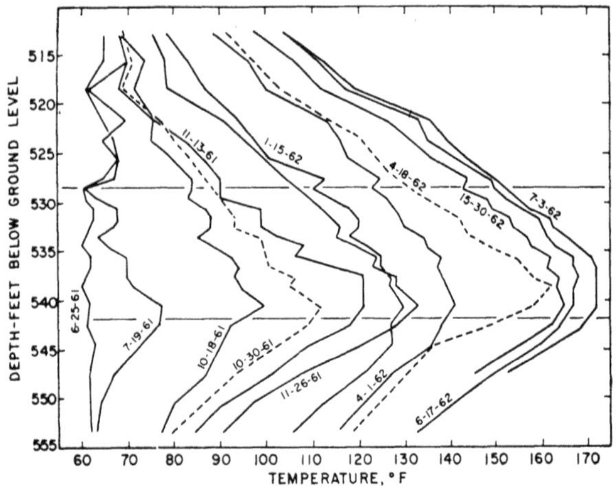

Fig. 4.3 – Temperature profiles showing underrunning of injected water.[10]

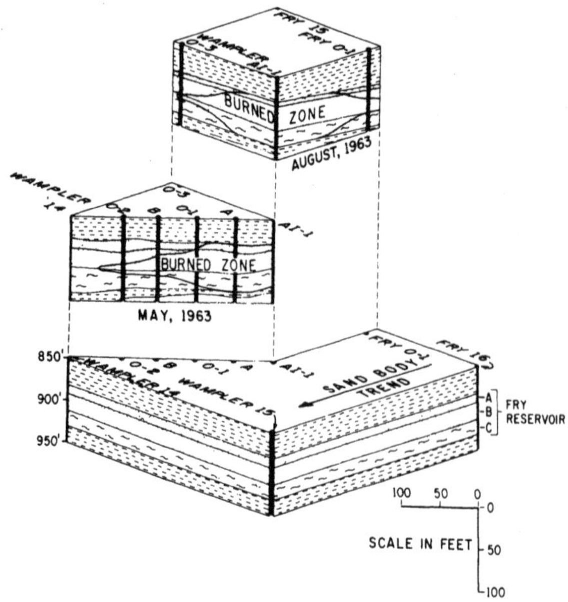

Fig. 4.4 – Block diagram of test site showing bypassing in a burned reservoir.[11]

4.2 Differences in FA and Bypass Models

There is only one reality in physical processes, but there are many ways of interpreting and modeling this reality. Two models used to describe the reality of reservoir displacement processes are the FA (frontal advance) and bypass models. Only the bypass model is capable of representing all aspects of reality, and even it can be woefully inadequate. Since so much literature based on the FA model has been published, the two models and the many differences that arise from their basic assumptions should be discussed. The simpler FA model has its uses, although the bypass model is more general. The following sections examine selected aspects of heat transfer, fluid displacement, flow resistance, fillup, and breakthrough times; the discussion should be helpful in determining the better model to use in a given situation.

The main way to make use of the following discussion is by recognizing that the severity of bypassing can range from none to completely segregated flow – two extremes not included in Fig. 4.1. The discussion on bypass effects is more applicable the more severe the degree of bypassing; the discussion on FA models is more applicable the less severe the bypassing. It is important to understand at least the qualitative difference in behavior resulting from extremes in the severity of bypassing. Figs. 4.5 and 4.6 are used in the discussion.

Heat Transfer

For any given rate of heat injection or generation in the reservoir, the rate of advance of the leading edge of the heated zone always will be faster where there is bypassing. This is a direct consequence of the bypassing. Although the heat front penetrates faster, the average temperature over the entire thickness of the formation and upstream of the leading edge of the displacement front is lower than in the FA case. Since the vertical extent of the heated zone is smaller than the formation thickness, some of the heat lost from the heated zone is in the formation of interest. Thus, the fraction of the injected heat remaining in the formation is larger under bypass conditions than under FA conditions. In thin reservoirs, the bypassed portions of the reservoir are relatively close to the heated zone. These bypassed portions are heated by conduction and convection from the heated zone. Thus, for a given rate of heat injection, these factors taken together lead, under bypass conditions, to a larger, unevenly heated reservoir volume of lower average temperature but higher heat content than under FA conditions. Heat breakthrough will occur somewhat earlier under bypass conditions. Fig. 4.5 portrays these differences in the temperature distribution where the bypass zone is at the top of the reservoir.

Flow Resistance

Thermal displacement processes usually are associated with the formation of an oil bank (and other zones) downstream of a characteristic heat front. This characteristic heat front is different for different processes, generally being taken as the combustion front in combustion processes and the steam condensation front in steam displacement processes. In Fig. 4.6, well-defined oil banks are indicated immediately downstream of the characteristic heat front, although in some thermal displacement processes there is some separation between the oil bank and the characteristic heat front. For this assumed case, a material balance on the oil gives an oil bank volume V_{ob} as

$$V_{ob} = \frac{S_{oi} - S_{or} - S_{oF}}{S_{o,ob} - S_{oi}} V_d, \quad \ldots\ldots\ldots\ldots\ldots (4.1)$$

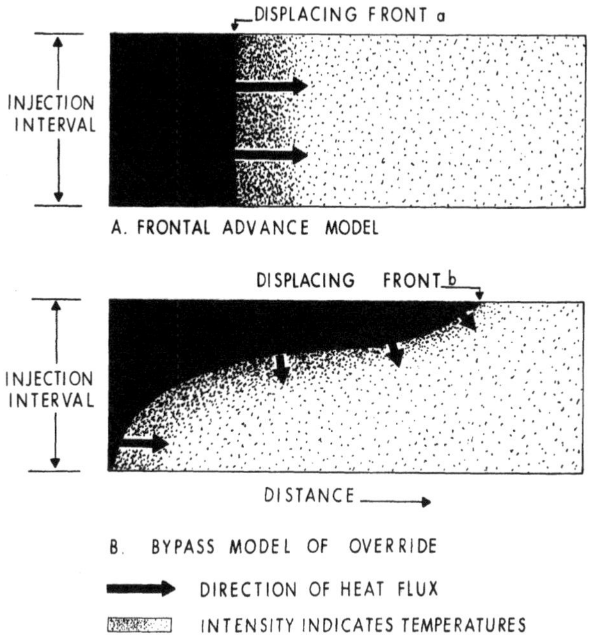

Fig. 4.5—Idealization of temperature distribution in frontal advance and bypass models. (Refer to Fig. 4.6 for corresponding oil bank distribution.)

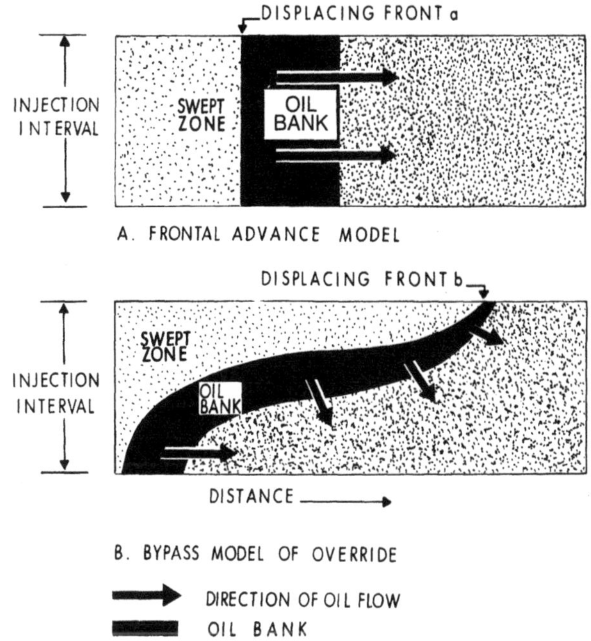

Fig. 4.6—Idealization of oil bank distribution in frontal advance and bypass models. (Refer to Fig. 4.5 for corresponding temperature distribution.)

where V_d is the volume displaced or swept by the characteristic heat front, $S_{o,ob}$ is the oil saturation in the oil bank (which is generally large), S_{oi} is the initial oil saturation, S_{or} is the residual oil saturation in the swept volume, and S_{oF} is the equivalent oil saturation burned in the swept volume. It generally is assumed that one of the quantities S_{or} or S_{oF} is zero. That is, there is no residual oil saturation in combustion processes (this is not quite true; see Section 8.1) and there is no oil burned in steam processes. When the oil bank occupies all the unswept part of the reservoir, oil bank fillup is said to occur. In the FA model, the greatest flow resistance generally occurs at the time of fillup, because the oil near the producing interval is essentially unheated (see Fig. 4.5). For very viscous crudes, the pressure drop required in an FA model to increase the volume of the swept zone at practical rates could preclude an economically attractive venture. In other cases the required rate of growth could not be achieved without injecting at pressures above fracture-initiation and fracture-extension pressures.

Where bypassing occurs, there are three clear effects on the flow resistance.

1. The distance from the leading surface of the swept zone to the producer is smaller than under FA conditions (see Fig. 4.6), which tends to reduce the flow resistance.

2. The average temperature in the unswept part of the reservoir is higher where bypassing occurs, tending to reduce the viscosity of the crude and, thus, the flow resistance between wells.

3. The surface of the interface between the swept and unswept zones is likely, under bypassing conditions, to be larger, which would lead to thinner oil banks – i.e., for a given volume of oil bank, a larger cross-sectional area will lead to a thinner zone in the direction of flow. And the thinner the oil bank, the lower the associated flow resistance.

When it is recognized that most of the heat downstream of the swept zone would be relatively close to the displacement front, it can be seen that the average temperature in a thin oil bank would be higher than in a thick one; hence, low flow resistances would be more common under bypassing than under FA conditions. There do not appear to be any significant factors to counter these conclusions.

Fluid Displacement

Consider how a liquid (or a component in a liquid) can be transported from Point A to Point B in the reservoir. Fig. 4.7 illustrates the mechanisms. Phases (such as oil, water, and gas) are transported according to Darcy's law. Thus, a liquid can be moved from Point A to Point B by flow of that liquid phase. Components within the liquid also can be transported by diffusion (intraphase transport). But a liquid (or a component in a liquid, such as heptane in crude) also can move from Point A to Point B by two other routes. First, the liquid may form a vapor phase by evaporation and then be transported between two points, A and B, as illustrated by the Path AA′B′B in Fig. 4.7. Second, the liquid may be dissolved partially in a second liquid phase, and this second liquid phase then may be transported between the points. Re-equilibrium between the two liquid phases may result in an increase in the liquid content of interest at Point B, via the Path AA″B″B. Of the alternative mechanisms, transfer to the gas phase is typically more important for recovering oil by thermal methods than the transfer to another liquid, even in very viscous oils. Analogous mechanisms

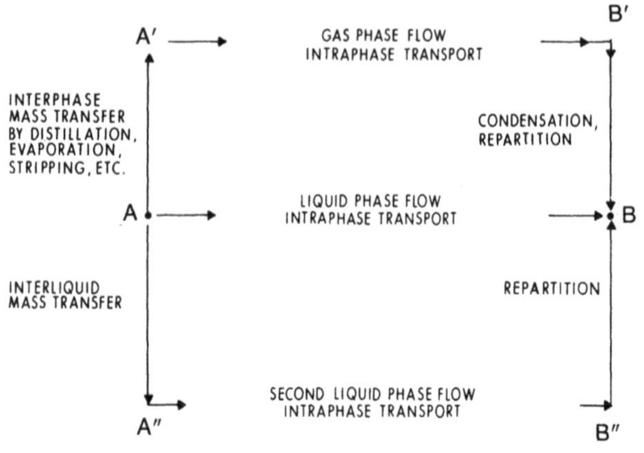

Fig. 4.7 – Schematic representation of how a component may be transported from Point A to Point B in a reservoir.

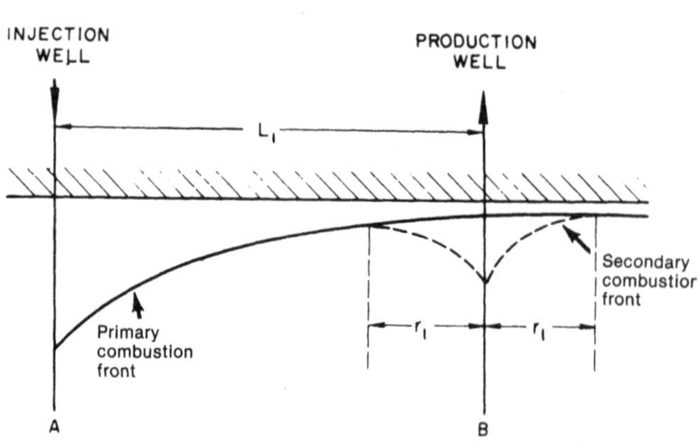

Fig. 4.8 – Combustion-front behavior near production wells.[5]

exist for multiple flow paths for the transport of a gas (or a gas component) between two points.

Now, under bypass conditions there is always more surface area between the swept and unswept regions than there is under FA conditions. In those thermal recovery processes where the swept zone contains a significant amount of flowing gas or vapor (which excepts only hot-water floods), distillation mass transfer plays a significant role within the swept zone and in a small heated zone immediately downstream of the swept zone. Under bypass conditions there is some mass transfer from the relatively large heated volumes immediately adjacent to the swept zones. In other words, bypassed oil – whether arising from a viscous displacement in the swept zone and present as remaining oil or present as heated but unswept oil adjacent to a swept zone – may be transported from Point A to Point B by a combination of mass transfer to the gas phase and bulk flow of the gas phase. This stripping effect is relatively important in combustion and steam-drive processes. Similar descriptions would apply to the transport of combustion products dissolved in the liquid phases, and one liquid dissolved in another. These contributions are not of general practical interest or importance, though their study may provide valuable insight and information on fundamental aspects of fluid displacement.

The term "drag process" sometimes is used to characterize a project or process in which there is pronounced bypassing and in which the injected fluid is dragging rather than *pushing* the reservoir oil to the production wells. Although the words "drag" and "displacement" evoke different configurations of multiphase flow, Darcy's law describes both flow processes. But where bypassing is severe, the attainable pressure gradients may be too low to displace the oil at high rates.

Breakthrough

One of the characteristics of bypass models is early breakthrough of the heat and displacement fronts at production wells. One beneficial consequence of early breakthrough is that a heated path – and, thus, one of relatively low flow resistance – connects an injection and a production well. Breakthrough generally causes pressures to drop and heats the vicinity of the producer (assuming it had not been already by cyclic steam injection or other stimulation), thus increasing the oil mobility. Sometimes, however, breakthrough is followed by operational difficulties.

Under bypass conditions, breakthrough occurs over only a limited portion of the producing interval. Heating may be localized and have limited impact on the oil production until a substantial portion of the underlying or adjacent oil column is heated. Also, a reduction in the pressure gradient in the vicinity of a producer offsets the beneficial effects of increased oil mobility. After breakthrough, the injected fluid tends to circulate or blow through the swept reservoir zone. Inefficient use of the injected fluid – which is characterized by the production of large quantities of steam, heated water, or unused oxygen – should be avoided. This generally is done by reducing the injection rate or increasing the backpressure of the offending well, either of which is likely to offset the beneficial effects of the increased oil mobility.

Abrasion of wellbore hardware by formation sand or fines carried in high-velocity produced-gas streams is accentuated whenever breakthrough occurs over a limited interval – i.e., under bypass conditions. High gas velocities, associated with a given mass flow rate over a limited interval, also may lead to a slightly lower combustion efficiency in the reservoir and may result in early oxygen breakthrough in combustion operations.

In combustion operations under bypass conditions, increasing fluxes of (bypassing) oxygen resulting from convergent flow into a producer may lead to the formation of an active combustion front near the producer that is separate from the primary combustion front near the injector.[5] This is illustrated in Fig. 4.8, where the two active combustion zones are connected by an inefficient combustion zone, allowing the passage of unused oxygen. The secondary combustion front formed near the producer behaves in much the same manner

TABLE 4.1 – TYPICAL INPUT DATA REQUIRED BY THERMAL RESERVOIR SIMULATORS

Group	Property	Requirements
Reservoir*	Principal values of the anisotropic absolute permeability and thermal conductivity, assigned to the directions *x*, *y*, and *z*	Three values of permeability and conductivity
	Porosity and heat capacity of reservoir rock	Two values for each block
	Relative permeabilities to each phase	One relation for each phase at each grid block; each relation is a function of saturations and temperature
	Capillary pressure	Two relations as a function of saturations; several pairs allowed
	Reservoir geometry	Specify coordinate system to be used and locations of wells and boundaries
Overburden and underburden	Thermal conductivity and heat capacity	At least one of each for both caprock and base rock
Initial values	Saturations, pressure, temperature, and composition	One value for each variable at each grid block
Fluids	Density and viscosity of each phase; compressibility of the reservoir matrix	One relation for each phase; each relation is a function of temperature, pressure, and possibly composition
	Component properties and *K* values	Relations as a function of pressure and temperature
	Latent heat of vaporization and saturation pressure	Latent heat of vaporization and pressure/temperature relation at saturation for each component that undergoes a phase change
	Enthalpy and internal energy of each phase	A relation for each quantity for each phase as a function of temperature, pressure, and possibly composition
Well and boundary conditions	Rates, pressures, and temperatures	Maximum and minimum values, constraints, and penalties

*Properties of grid blocks, including dimensions and locations, are used to define the reservoir.

as the main front at breakthrough, except possibly for the fact that it may be observed unusually early in the life of the project. Thus, some careful interpretation may be warranted. The reason a secondary combustion front develops where the oxygen flux is high is that the local rates of heat generation exceed the local rates of heat loss, leading to higher temperatures and more efficient oxygen consumption – hence, to the formation of a secondary active combustion front.

The time at which breakthrough occurs, the degree of blowthrough of the injected fluid, the amount of abrasion, the oxygen in the produced gas, and the formation of secondary combustion fronts (the latter two in combustion projects only) all are affected by the degree of bypassing existing in a project.

4.3 Bases for Predicting Project Performance

The philosophies regarding the thoroughness and detail required of a performance-prediction study depend on the background and knowledge of the engineer making the study; on the demands of the management; on the time, manpower, tools, and money available for the study; and on the information available about the reservoir and its fluids.

The trend is toward predictions based on the use of numerical reservoir simulators, especially among the young engineers. Many engineers who acquired experience before numerical reservoir simulation came of age prefer relatively simple models for predicting performance and emphasize analysis of the reservoir's previous performance and the known response of thermal projects in similar reservoirs under comparable conditions. Either approach may be valid, but more important than the approach is the exercise of good engineering judgment.

A number of cyclic steam injection projects have been based solely on data input consisting of the production response in a similar, neighboring operation rather than on a formally calculated prediction. This is an extreme approach, but it appears warranted where the risk of failure (either in money or lost production) is low and success is deemed likely on the basis of local experience. For large commercial projects involving the commitment of a great deal of money, where the development cost per barrel of daily oil production can exceed 3,000 times the price of a barrel of crude, it would be prudent to base performance predictions on extensive analysis and physical and numerical reservoir simulation of field pilot-testing results. Obviously, this approach requires enormous effort. (Pilot testing is discussed in detail in Chap. 13).

Relatively simple models and procedures for predicting the reservoir performance of thermal proceses are emphasized in this monograph. These models, which provide insight into the characteristics of each thermal recovery process, often have been found adequate for estimating reservoir performance, and the calculations are described easily since they require no more than a pocket-size calculator. References to numerical reservoir studies are given for each process, and some uses of such studies are discussed at appropriate places in the text.

There are procedures for solving differential equations describing heat and mass transport that are intermediate between using the simple models discussed in the text and using the numerical reservoir simulators available in the industry. Such procedures typically involve computer programs that simulate particular process phenomena. One good example of such a process simulator was developed by Gottfried[12] to study thermal recovery of oil in a linear system. Reservoir simulators differ from process simulators in that they (1) accommodate a greater number of grid blocks, dimensions (three), and phases (three), (2) can refer pressures to a common datum, (3) provide a variety of options descriptive of flow from the reservoir into the well (called well models), (4) can operate in three dimensions and in a cross-sectional mode (two dimensions) and a plan mode (two dimensions), and (5) provide easy input and output information. In short, a numerical reservoir simulator is a highly flexible, general-purpose, user-oriented process simulator. Usually, because it limits the number of grid blocks and the time-step sizes, it sacrifices some accuracy for the sake of computation speed.

These reservoir simulators require a large amount of data input about the reservoir (geometry and distribution of properties), its fluids (saturation, pressures, properties, and initial conditions), wells (location, intervals open, skin effect, and well model to be used), and operational variables (rates and pressures and the constraints of both). Table 4.1 suggests the amount of information required. Specifically, reservoir properties such as permeability and porosity are required inputs for each grid block, and several sets of relative permeabilities and capillary pressures usually can be used to describe the reservoir. Transmissivities between grid blocks can be used to represent leaky or fully sealing faults. In other words, the simulator requires more information about the distribution of properties in the reservoir than is normally available. Assumptions must be made about the homogeneity and trends in reservoir properties. How sensitive the results are to these and other assumptions can be determined by rerunning the simulator, but this can be expensive and is not done except for those variables considered to have the most bearing on the results. Simulations normally are not run with all reservoir properties being different in every grid block. In fact, it is doubtful that reservoir simulations ever have been run that way. But a conscious decision must be made by the user to specify the properties describing the reservoir and its fluids. In some simulators, properties such as those of saturated steam are built in. Some reservoir simulators automatically assign the same absolute permeability value to all grid blocks when the reservoir is identified as being both homogeneous and isotropic. Thus, the job of putting in data for a number of properties can be reduced considerably. However, the fact that a simulator has been programmed to reassign data when a reservoir is simple enough does not mean that fewer data are required to describe the reservoir. A permeability value (at least three values, if anisotropy can be described in the simulator) still must be provided at each grid block. The user assigns these data, either directly or through assumptions or programmed (default) instructions.

By contrast, the simple models generally require the entering of few but critical data. They often are limited to linear or radial flow, and if not, the total volume swept is assumed to be independent of flow directions. Frequently, displacement is assumed to be piston-like. This means that there is a sharp drop in the oil saturation across the displacement front, leaving a uniformly low amount of oil in the swept zone. In some cases, however, saturation distributions are considered and used in predicting performance. The oil displaced from the swept zone usually is assumed to be produced, although some recovery-efficiency factor based on experience, correlations, or laboratory studies often is used to account for the effects of bypassing (channeling), early breakthrough, nonuniform displacement, delays in production resulting from fillup requirements, and other factors. It is, in fact, in the choice of appropriate recovery-efficiency factors that engineering judgment is to be exercised most carefully. The simple models are basically a combination of mass and energy balances relating potential recovery to cumulative injected fluid or heat or both. The recovery efficiency factor adjusts the maximum values given by the various models to what may be expected realistically. These factors are typically in the range of 0.5 to 0.7.

One of the most difficult results to obtain with the simple models is the production rate schedule. Often these models can yield only average rates. Even then the production rates are calculated on the basis of assumed injection rates and pressures. Being based on energy and mass balances, however, they yield results that are often adequate for screening analysis.

Although coarser than results from the simpler models, other results that may be suitable for screening analysis are the correlations between reservoir and operational parameters and recovery. An example of such a correlation is given by Brigham *et al.*[13] for the combustion process.

4.4 Flow Resistance Between Wells

To make estimates of injection and production rates likely to be attained in fluid displacement processes, it is necessary to have information about the flow

TABLE 4.2 – GEOMETRIC FACTORS FOR SELECTED WELL PATTERNS

Well Pattern	Geometry	F_G
Staggered line drive	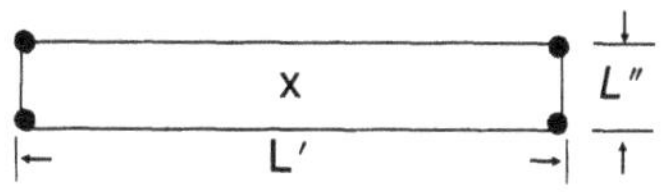	$2\left(\ln \frac{L'}{r_w} - \ln 2K - 0.5 \ln \sqrt{m(1-m)}\right)$
Five spot	as above with $L'' = L'$	$2\left(\ln \frac{L'}{r_w} - 0.9640\right)$
Direct line drive	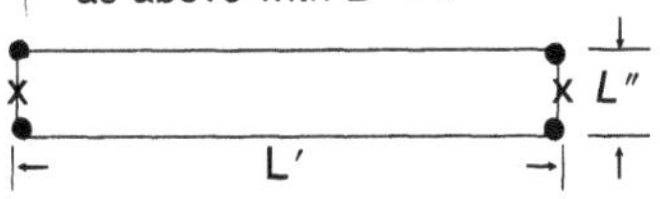	$2\left(\ln \frac{L'}{r_w} - \ln 2K - 0.25 \ln m\right)$
Seven spot	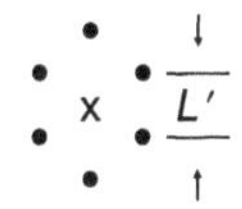	$\frac{3}{2}\left(\ln \frac{L'}{r_w} - 0.5691\right)$
Arbitrary		Eqs. 4.5 through 4.7

Notes:
1. The first four well patterns are repeated. Points and crosses indicate the locations of injection and production wells.
2. For the staggered and direct line-drive well patterns, the values of K and m may be found from the following table.

L''/L'	m	K	L''/L'	m	K
2.347	0.01	1.5757	0.4261	0.99	3.6956
1.994	0.03	1.5828	0.5015	0.97	3.1558
1.599	0.10	1.6124	0.6254	0.90	2.5781
1.360	0.20	1.6596	0.7352	0.80	2.2572
1.211	0.30	1.7139	0.8258	0.70	2.0754
1.097	0.40	1.7752	0.9117	0.60	1.9496
1.000	0.50	1.8541	1.000	0.50	1.8541

Additional values of the complete elliptic integral of the first kind, $K(m)$, may be found in a number of mathematical tables.[19]

resistance between wells of interest. Strictly speaking, flow resistance is defined for reservoir fluid systems in which the local fluid content per unit volume does not change with time – i.e., the reservoir fluid system must be either incompressible or at steady state. The flow resistance of a reservoir containing a single fluid of constant viscosity is discussed first. Displacement of fluids of different mobilities, and the effect of changes in local fluid density with time, are discussed later.

Steady-state flow resistances between wells generally are given only for highly idealized, thin, horizontal reservoirs of uniform and isotropic permeability, uniform thickness, and containing only one mobile fluid. Thus, the reservoir transmissivity $\mathfrak{T} = (kh/\mu)$ is considered constant and uniform throughout the reservoir.

For the highly idealized repeated well patterns such as the five-spot, the flow rates at reservoir conditions at all wells are considered to be equal in magnitude. When all well radii are equal, the pressure drops between nearby injection and production wells are also equal. The flow resistance between an injector and any one of its four nearby producers then is called the flow resistance of the five-spot well pattern and is given by

$$\frac{1}{I} = \frac{1}{J} = \frac{\Delta p}{q} = \frac{\Delta p}{i} = \langle 141.2(2\pi) \rangle \frac{F_G}{2\pi \mathfrak{T}}, \quad \ldots . (4.2)$$

where for a repeated five-spot well pattern[14] of side L',

$$F_G = 2\left(\ln \frac{L'}{r_w} - 0.9640\right). \quad \ldots\ldots\ldots\ldots\ldots (4.3)$$

Note that the flow resistance between two wells is equal to the reciprocal injectivity $1/I$ and reciprocal productivity $1/J$.

When all injection and production rates are equal at reservoir conditions, F_G is a geometric factor that depends only on the relative well locations and on the characteristic ratio of injection rate to production rate. Otherwise, F_G also depends on the flow rate at reservoir conditions at each well. Expressions for the geometric factor F_G corresponding to other idealized well patterns are given in Table 4.2. Muskat[15] calls the quantity

$$2\pi/F_G = \langle 141.2(2\pi) \rangle q/\mathfrak{T}\Delta p \quad \ldots\ldots\ldots\ldots (4.4)$$

the production capacity or flow capacity of a well pattern. Fig. 4.9 plots $10^4/141.2F_G = 70.82/F_G = 10^4\ q/\mathfrak{T}\Delta p$ for several of the more common well patterns. Morel-Seytoux[16] refers to $q/\mathfrak{T}\Delta p$ as the normalized injectivity.

The flow resistance between any injector J and any producer K in the presence of an arbitrary number of injectors (n_i) and producers (n_p) can be found by superposition of the pressure fields resulting from individual wells.[15-17] The pressure resulting from an

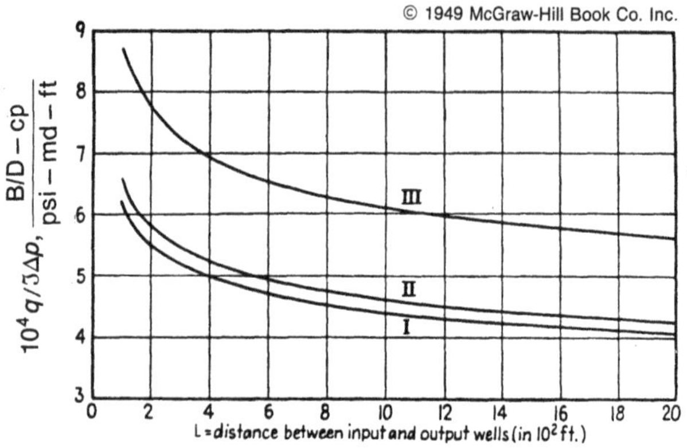

Fig. 4.9A – Values of flow capacities for selected well patterns.[15] Curve I is direct line drive with $L''=2L'$, Curve II is five-spot network, and Curve III is seven-spot network. (Refer to Table 4.2 for meaning of L' and L''. $r_w=0.25$ ft in all cases.)

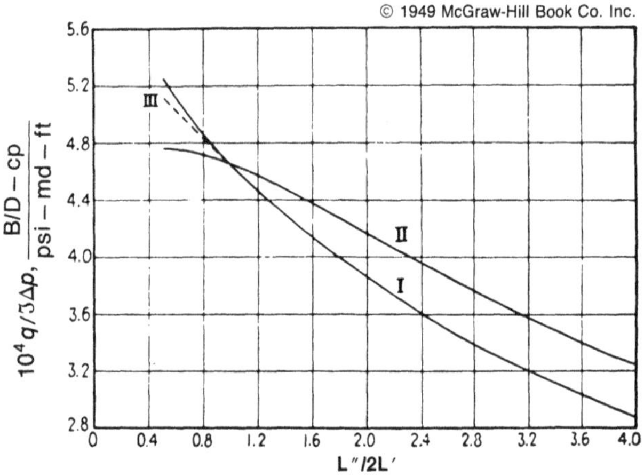

Fig. 4.9B – Values of flow capacities for selected well patterns.[15] Curve I is direct line drive with L' fixed at 660 ft, Curve II is direct line drive with L'' fixed at 330 ft, and Curve III is staggered line drive with L' fixed at 660 ft. (Refer to Table 4.2 for meaning of L' and L''. $r_w=0.25$ ft in all cases.)

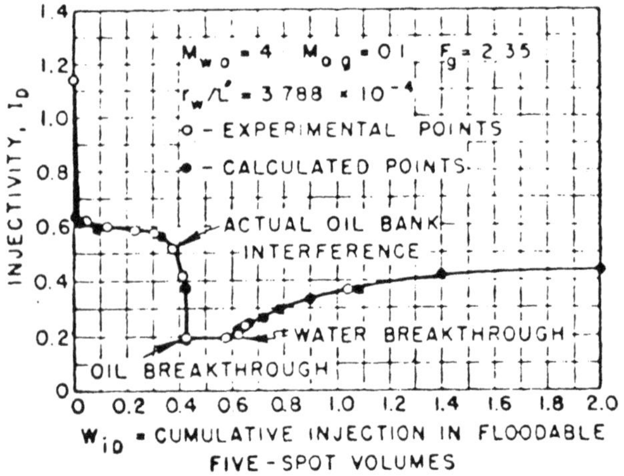

Fig. 4.10 – Variation in injectivity for a water/oil/gas system.[14]

individual well is given by Eq. 1.1. Since the flow rate i_J at the chosen injector J may not be equal in magnitude to the flow rate q_K at the chosen producer K, the flow resistance between them is based arbitrarily on the flow rate q at the producer. Again, all rates are at reservoir conditions. Thus,

$$\frac{1}{J_{J,K}}=\frac{\Delta p_{J,K}}{q_K}=\langle 141.2(2\pi)\rangle \frac{F_G}{2\pi\mathfrak{J}}. \quad \ldots\ldots\ (4.5)$$

For arbitrary well locations,

$$F_G=\frac{i_J}{q_K}\ln\left(\frac{L_{J,K}}{r_{w,J}}\right)-\sum_{j=1}^{n_i}{}' \frac{i_j}{q_K}\ln\left(\frac{L_{J,j}}{L_{K,j}}\right)$$

$$+\ln\left(\frac{L_{J,K}}{r_{w,K}}\right)+\sum_{k=1}^{n_p}{}' \frac{q_k}{q_K}\ln\left(\frac{L_{J,k}}{L_{K,k}}\right), \quad \ldots (4.6)$$

where

$$L_{j,k}=\sqrt{(x_j-x_k)^2+(y_j-y_k)^2}=L_{k,j}. \quad \ldots. (4.7)$$

is the distance between injector j located at coordinates (x_j,y_j) and producer k located at coordinates (x_k,y_k). The symbol Σ' is used to denote that the injector $(j=J)$ or producer $(k=K)$ being considered is to be omitted from the sum. Their contributions are represented by the two terms outside the summations and include the effect of their respective well radii. The injection rate at the jth well is indicated by i_j; the production rate at the kth producer is indicated by q_k. All i_j and q_k are positive quantities, as is F_G. Whenever the flow rates are the same at all wells, their ratios become unity and F_G is a purely geometric factor. Although Eqs. 4.5 through 4.7 provide the means for estimating the flow resistance between any two wells, it is considerably easier to use the expressions for F_G given by Eq. 4.3, Table 4.2, or Fig. 4.9 where the actual well pattern can be represented by the repeated well pattern indicated in Table 4.2. Eqs. 4.5 through 4.7 would be useful where there is no pattern to the well locations and the number of wells is not too large. As a practical matter, an approximate flow resistance between any two wells may be obtained from Eqs. 4.5 through 4.7 by considering only those wells that are within a characteristic radius from the midpoint of the two wells. The characteristic radius is about three times the distance between the two wells of interest.

It follows from Eq. 4.6 that doubling the rate of only one of several wells changes the flow resistances between all wells. On the other hand, the flow resistances are unaffected when all rates are multiplied by the same factor.

Since flow rates are expressed at reservoir conditions, Eqs. 4.2 through 4.7 also apply to gas. For gas rates expressed in thousands of standard cubic feet per day, Eq. 4.5 becomes

$$\frac{1}{J_{J,K}}=\frac{\Delta(p^2)_{J,K}T_{sc}}{2p_{sc}q_J z T_{ab}}=\langle 2.516\times10^4(2\pi)\rangle\frac{F_G}{2\pi\mathfrak{J}},$$

$$\ldots\ldots\ldots\ldots\ldots\ldots\ldots\ldots\ (4.8)$$

where p_{sc} is the pressure at standard conditions, $\Delta(p^2)_{J,K}$ is the difference in the square of the pressure between wells J and K, T_{sc} is the temperature at standard conditions, T_{ab} is the average temperature of the gas in the reservoir, and z is the gas supercompressibility factor. Pressures are in pounds absolute per square inch, and temperatures are in degrees Rankine.

The foregoing concepts of flow resistance for determining the relationships between flow rates and pressure differences between wells are based on the flow of a single fluid at steady state. During thermal recovery operations, the properties of the fluid being produced are generally very different from those of the fluid being injected. The flow resistance between two wells then depends also on the distribution of fluids within the reservoir. And since fluid displacement operations are dynamic, the flow resistance between wells changes with time, even when all fluids can be considered incompressible.

Simple models have been used in waterflooding to estimate how the flow resistance between wells varies during an operation. Factors affecting the injectivity (or productivity) of a waterflood include the relative size, shape, and fluid mobility of the flooded region, the oil bank, and the as-yet-unaffected portion of the reservoir. Fig. 4.10 shows an example (based on an FA model and fluids of constant density) of how the calculated injectivity of a waterflood changes with time in a reservoir initially containing oil, gas, and water. In the figure, the initial dimensionless injectivity is fairly high because the reservoir contains mobile gas. The injectivity decreases as the viscous oil bank grows with cumulative injection. After the oil bank reaches the producer and oil production starts, the oil bank becomes smaller. Then the injectivity rises again – slowly at first, then more rapidly after water breakthrough. The final dimensionless injectivity of 0.42 corresponds to the effective transmissivity of the water-filled reservoir, the steady-state value given by Eqs. 4.2 and 4.3. More pronounced variations in injectivity would be expected in thermal projects where FA conditions and more pronounced mobility contrasts prevail.

Deppe[17] shows how to include the effect of boundaries, and of fluid banks of different mobilities, in calculating flow resistances for arbitrary well locations. This and other references, though very useful, are based on FA models. Bypass models, which result in higher injectivities than FA models, are often more representative of thermal operations – especially for combustion and steam injection.

Another quantity of interest is the productivity index, which is based on the difference between the average pressure in the reservoir and the bottomhole producing pressure. The injectivity index is defined analogously. These indices, which are a measure of the flow resistance around a well, usually are obtained by means of well tests and are used to determine (1) trends in well performance in established operations and (2) potential rates in new operations and after stimulation treatments.

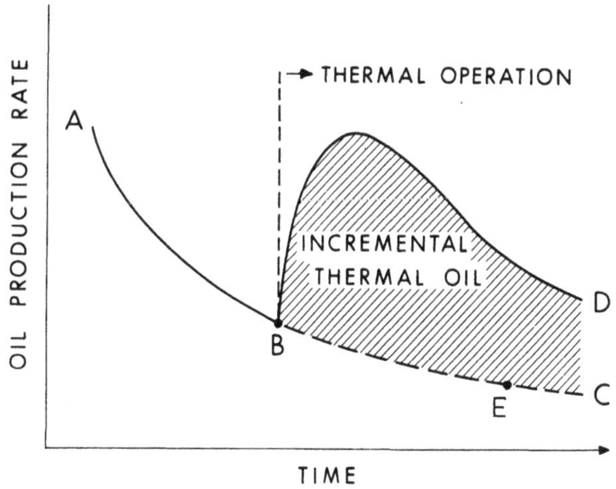

Fig. 4.11 – Incremental thermal oil.

4.5 Idealized Process Design

Some brief comments on design procedures are in order. Performance prediction, from a reservoir engineering point of view, emphasizes the determination of the amount of oil recovery and the rates at which the oil is produced. The production rates schedule with time is extremely important in the economics of a process, but it usually can be estimated only crudely with the simpler predictive models. Of course, any performance prediction must be justifiable and self-consistent.

Self-consistency refers to the internal balances within the approximations used to make the performance prediction. Examples of self-consistency include (1) using injection rates compatible with the pressure limitations imposed by fracture-initiation or fracture-extension pressures and with the flow resistance of the reservoir, (2) checking that the estimated production rates calculated from the cumulative recovery vs. time response can be achieved taking into account such factors as the pumps, well capacity, and flow resistance of the reservoir, and the viscosity of the oil at the conditions expected to prevail near the producers, and (3) checking that the characteristics of the reservoir (such as fractures, layering, saturation distributions, continuity, and geologic setting) and of well locations and methods of completion would lead to the desired performance.

The project designer must be able to support his choice of input data and calculations and to show consistency in the results. However, the degree of support and consistency will depend on the objectives of the design. A screening analysis should require much less effort and detail than the design of an expensive commercial expansion. But good engineering judgment is needed in all areas of design – whether one is dealing with reservoirs, wells, surface operations, or economics.

Although the reservoir aspects of thermal recovery are emphasized in this monograph, the design of wells and surface lines, function of surface facilities, sizing of selected major equipment, and methods for calculating the economic attractiveness of a contemplated venture also are discussed briefly. All of these factors must be evaluated to arrive at the overall project design and especially to determine if there is any economic incentive for undertaking the venture. Again, the detail required depends on the level or the objectives of the design.

4.6 Thermal Oil

Not all the oil produced in a thermal recovery project is necessarily the result of the thermal operation. Fig. 4.11 shows the oil production history of a field, with a thermal operation initiated at Point B. Some production would have been obtained beyond Point B even if no thermal operation had been implemented. This latter production, given by the area under Segment BE, usually is determined by extrapolation of the available history (Segment AB). The extrapolation may range from fitting a curve visually through the existing history to a full-scale numerical simulation study in which Segment AB is matched and BE is predicted. The total recovery from the thermal operation, given by the total area under Segment BD of the curve, may be appreciably greater than the extrapolated production in the absence of the thermal operation. Note that in this case the economic limit is reached later in the thermal operation (Point D) than it would be were the normal operation to continue (Point E).

The incremental production resulting from the thermal operation, called the incremental thermal oil, may be defined as the difference between the area under Segment BD and the area under Segment BC. Because of the uncertainties in the extrapolated curve segment, BC, there are uncertainties in the value of the incremental thermal production. Unfortunately, other ways of calculating the thermal oil also are used (including using a horizontal line from Point B, or equating the thermal oil with the total production beyond Point B), which leads to inconsistencies in what is meant by thermal oil. Reported values of thermal oil may be derived by one of several methods, and the method is not always stated.

Note that the incremental thermal oil may have been produced by mechanisms that were present in the reservoir before the thermal operations were initiated. For example, a reduction in the viscosity of the oil may result in improved gravity drainage and significant production increases in outlying wells. Estimates have been made of the contributions of several mechanisms to incremental thermal production.[18] Some of those mechanisms, such as compaction, are not in themselves thermal, but they may be promoted or enhanced by the thermal process.

References

1. Craig, F.F. Jr.: *The Reservoir Engineering Aspects of Waterflooding*, Monograph Series, Society of Petroleum Engineers, Dallas (1971) **3**.
2. Matthews, C.S. and Russell, D.G.: *Pressure Buildup and Flow Tests in Wells*, Monograph Series, Society of Petroleum Engineers, Dallas (1967) **1**.
3. Earlougher, R.C. Jr.: *Advances in Well Test Analysis*, Monograph Series, Society of Petroleum Engineers, Dallas (1977) **5**.
4. Ramey, H.J. Jr.: "Commentary on the Terms 'Transmissibility' and 'Storage'," *J. Pet. Tech.* (March 1975) 294-295.
5. Prats, M., Jones, R.F., and Truitt, N.E.: "In-Situ Combustion Away from Thin, Horizontal Gas Channels," *J. Pet. Tech.* (March 1968) 18-32.
6. Neuman, C.H.: "A Mathematical Model of the Steam Drive Process—Applications," paper SPE 4757 presented at the 1975 SPE California Regional Meeting, Ventura, April 2–4.
7. van Lookeren, J.: "Calculation Methods for Linear and Radial Steam Flow in Oil Reservoirs," *Soc. Pet. Eng. J.* (June 1983) 427–439.
8. Gates, C.F. and Ramey, H.J. Jr.: "Field Results of South Belridge Thermal Recovery Experiment," *Trans.*, AIME (1958) **213**, 236-244.
9. Spillette, A.G.: "Heat Transfer during Hot Fluid Injection into an Oil Reservoir," *J. Cdn. Pet. Tech.* (Oct.-Dec. 1965) 213-218.
10. Martin, W.L., Dew, J.N., Powers, M.L., and Steves, H.B.: "Results of a Tertiary Hot Waterflood in a Thin Sand Reservoir," *J. Pet. Tech.* (July 1968) 739-750; *Trans.*, AIME, **243**.
11. Clark, G.A., Jones, R.G., Kiney, W.L., Shilson, R.E., Surkalo, H., and Wilson, R.S.: "The Fry In Situ Combustion Test-Performance," *J. Pet. Tech.* (March 1965) 348-353.
12. Gottfried, B.S.: "A Mathematical Model of Thermal Oil Recovery in Linear Systems," *J. Pet. Tech.* (Sept. 1965) 196-210; *Trans.*, AIME, **234**.
13. Brigham, W.E., Satman, A., and Soliman, M.Y.: "Recovery Correlations for In-Situ Combustion Field Projects and Application to Combustion Pilots," *J. Pet. Tech.* (Dec. 1980) 2132–2138.
14. Prats, M., Matthews, C.S., Jewett, R.L., and Baker J.D.: "Prediction of Injection Rate and Production History for Multifluid Five-Spot Floods," *J. Pet. Tech.* (May 1959) 98-105; *Trans.*, AIME, **216**.
15. Muskat, M.: *Physical Principles of Oil Production*, McGraw-Hill Book Co. Inc., New York City (1949).
16. Morel-Seytoux, H.J.: "Unit Mobility Ratio Displacement Calculations for Pattern Floods in a Homogeneous Medium," *Soc. Pet. Eng. J.* (Sept. 1966) 217-227; *Trans.*, AIME, **237**.
17. Deppe, J.C.: "Injection Rates – The Effect of Mobility Ratio, Area Swept, and Pattern," *J. Pet. Tech.* (June 1961) 81-91; *Trans.*, AIME, **222**.
18. de Haan, H.J. and Schenk, L.: "Performance Analysis of a Major Steam Drive Project in the Tia Juana Field, Western Venezuela," *J. Pet. Tech.* (Jan. 1969) 111-119; *Trans.*, AIME, **246**.
19. Pierce, B.O. and Foster, R.M.: *A Short Table of Integrals*, fourth edition, Blaisdell Publishing Co., Waltham, MA (1957).

Chapter 5

Heating the Reservoir

The general purpose of thermal recovery is to raise the temperature of the crude in a reservoir so that it can be produced more readily. It is desirable to heat the reservoir efficiently, being aware that not all the heat injected into or generated in a reservoir stays in it. Some of the heat in the reservoir is lost through produced fluids, and some is lost to the adjacent overburden and underburden formations. When heat lost to the adjacent formations is controlled by conduction heat transfer (which is usually the case), it can be estimated readily. The amount of heat lost from the reservoir through produced fluids is generally difficult to forecast without the aid of numerical or physical simulators. This chapter presents some simplified methods for estimating the heat remaining in the reservoir and the heat lost through produced fluids.

The fraction of the heat injected into (or generated in) a formation that remains in the formation, known as the heat efficiency, is often independent of the thermal recovery process used, be it steam, hot water, or combustion. In any one reservoir, the heat efficiency before breakthrough is quite sensitive to the life of the project and somewhat insensitive to other operating conditions. This insensitivity makes the heat efficiency an excellent screening tool for ascertaining conditions under which a significant portion of the injected heat remains in the reservoir. The temperature distribution in the reservoir, on the other hand, is very much dependent on the thermal process and on the operating conditions. Except in the simplest cases, which will be discussed, temperature distributions within a reservoir can be described only with the aid of numerical or physical simulators. Although it is useful to know the amount and the distribution of heat in the reservoir, such information generally must be supplemented by recovery calculations to determine the attractiveness of thermal recovery projects.

The first section of the chapter examines the extent to which the volumetric heat capacity of the reservoir is sensitive to the variations in fluid content and porosity typically encountered in thermal recovery operations. Following sections discuss methods for estimating losses of heat from the reservoir to the adjacent formations and through the produced fluids.

Subsequent sections discuss the estimation of the size and growth rate of heated zones and the estimation of the temperature distribution between wells. Although injection temperatures and heat injection rates usually are assumed to be constant, variable rates also are considered.

5.1 Effective Volumetric Heat Capacity of Reservoir Formations

Some simple methods for estimating the heat losses to the formations adjacent to the reservoir, and the temperature distribution within the reservoir, are discussed in some detail in the following sections. In addition to considering homogeneous reservoirs having uniform properties (including thickness), all these methods consider that the bulk reservoir properties are independent of the variations in fluid saturations.

An important factor to determine, therefore, is how sensitive to variations in fluid content is the effective isobaric heat capacity of a fluid-filled porous formation. The amount of heat Q required to increase the temperature of a bulk volume of formation V_R by an amount ΔT, and at constant pressure, is

$$Q = \langle 43{,}560 \text{ cu ft/acre-ft} \rangle V_R M_R \Delta T, \quad \ldots\ldots (5.1)$$

where M_R is the isobaric volumetric heat capacity of the bulk, fluid-filled reservoir. That is, M_R is simply the amount of heat required to raise a unit bulk volume by one degree of temperature and is equal to the product of the effective density and the isobaric specific heat capacity of the bulk formation. The heat content of this bulk volume is also equal to the sum of the individual heat contents of its components. For a formation of porosity ϕ filled with a nonvolatile oil, water, and a gas phase containing steam and noncondensable gas, the heat content can be written

$$Q = \langle 43{,}560 \text{ cu ft/acre-ft} \rangle V_R \{ (1-\phi) M_\sigma \Delta T + \phi S_o M_o \Delta T + \phi S_w M_w \Delta T + \phi S_g [f M_g \Delta T + (1-f)(\rho_s L_v + \rho_s C_w \Delta T)] \}, \quad \ldots\ldots\ldots (5.2)$$

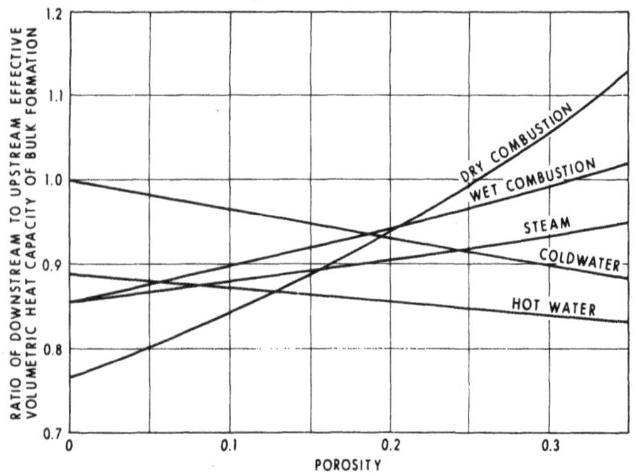

Fig. 5.1 – The ratio of downstream to upstream bulk formation volumetric heat capacities for various types of thermal recovery methods.

where f is the volume fraction of noncondensable gas in the vapor phase; M_σ, M_o, M_w, and M_g are the isobaric volumetric heat capacities of the solid, oil, water, and gas phases, respectively; S_o, S_w, and S_g are the saturation of the fluid phases; L_v is the latent heat of vaporization of water; ρ_s is the steam density, and C_w is the isobaric heat capacity of water per unit mass.

Equating the two expressions gives the effective volumetric heat capacity of the formation in terms of the rock and fluid properties as

$$M_R=(1-\phi)M_\sigma+\phi S_oM_o+\phi S_wM_w+\phi S_g \cdot\left[fM_g+(1-f)\left(\frac{\rho_sL_v}{\Delta T}+\rho_sC_w\right)\right]. \quad ..(5.3)$$

As can be seen, there is a contribution from each of the four phases present: solid, oil, water, and gas. The gas-phase contribution itself is made up of two parts: a contribution arising from the volumetric heat capacity M_g of the noncondensable gas and a contribution due to the steam of density ρ_s. The steam contribution has two terms, one due to the latent heat of vaporization L_v, and the other due to the sensible heat.

In a more general development, the enthalpy of all components would be considered in a manner similar to that used to consider the enthalpy of water. For example, if some components in the oil change phase over the temperature change, it would be proper to include the associated latent heats of vaporization in Eq. 5.2. Although the latent heat per unit mass of some hydrocarbon components may be as high as 20% of that of water, their contribution to the effective volumetric heat capacity of the formation would be small. The reason for this can be discerned from Fig. 5.1. If the steam case has only a 15% change on the downstream/upstream volumetric heat capacity ratio, the effect of a volatile oil would be appreciably less.

Fig. 5.1 demonstrates the variation in the effective volumetric heat capacity of the bulk formation resulting from rather large changes in fluid saturation. It shows this variation for hot water, steam, and dry and wet combustion drives for porosities up to 0.35. Table 5.1 gives the properties and conditions considered for these cases. The saturation values considered in the zones upstream and downstream of a displacement front may be somewhat extreme for the processes. This allows the effect of saturation differences between these two zones to have a noticeable effect on the corresponding volumetric heat capacities. Except for

TABLE 5.1 – CONDITIONS CONSIDERED IN PREPARING FIG. 5.1

	Process*			
Property	Hot Water	Steam	Dry Combustion	Wet Combustion
Upstream				
S_o	0.3	0.2	0	0
S_w	0.7	0.3	0	0.3
S_g	0	0.5	1.0	0.70
S_s	0	0.5	0	0.35
T, °F	350	550	900	550
M_σ, Btu/cu ft-°F	39.7	41.2	46.0	41.2
M_o, Btu/cu ft-°F	30.8	33.1	–	33.1
M_w, Btu/cu ft-°F	58.7	60.4	–	60.4
M_s, Btu/cu ft-°F	–	3.1	–	3.1
M_g**, Btu/cu ft-°F	–	0.71	0.55	0.51
$\rho_s L_v/\Delta_T$, Btu/cu ft-°F	–	4.32	–	4.32
Downstream				
S_o	0.7	0.7	0.5	0.45
S_w	0.3	0.3	0.3	0.35
S_g	0	0	0.2	0.2
T, °F	150	150	150	150
M_σ, Btu/cu ft-°F	35.2	35.2	35.2	35.2
M_o, Btu/cu ft-°F	27.6	27.6	27.6	27.6
M_w, Btu/cu ft-°F	61.3	61.3	61.3	61.3
M_g**, Btu/cu ft-°F	1.15	1.15	1.15	1.15

*The conditions for the cold-water flood case differ from those of the hot-water drive only in that the upstream temperature is 150°F.
**Evaluated at 1,000 psia.

wet combustion, only one component was considered in the gas phase in any one process, and typical values of the volumetric heat capacities for minerals and fluids were used. The upstream temperature depends on the process, and the reservoir temperature is 150°F. Volumetric heat capacities of the dry reservoir rock were taken from the sandstone values given in Fig. B.68, using a density of 165.4 lbm/cu ft under all conditions. For oil, heat capacities were calculated with Eq. B.38 and densities were read from Fig. B.31. For water, heat capacities were read from Fig. B.64 and densities from Table B.2. For air, heat capacities were read from Fig. B.62 and corrected for pressure using Fig. B.63. Air densities were assumed to behave ideally. Steam properties were taken from Tables B.2 and B.6. For steam drive and wet combustion, the effective volumetric heat capacities upstream and downstream of the displacement front are within ±10% of each other for all porosities above 0.185. For dry combustion, the same is true for porosities from 0.16 to 0.33. For hot water, the deviation from equality is greater, the value downstream being smaller than the one upstream by 11 to 17% for all saturations below 0.35. The deviation for hot water does decrease as the injection temperature is reduced, as indicated by the limiting curve for conventional waterflooding. Deviations are less than ±17% for all processes over the porosity range from 0.085 to 0.35.

These comments should not be interpreted to mean that an arithmetic average value necessarily should be used in calculating the distribution of heat within the reservoir. The value to be used should be selected with care after the problem has been considered. For example, if the model to be used assumes that all the heat in the reservoir is in the steam swept zone, the volumetric heat capacity to be used in the calculations should be that in the steam zone (upstream of the condensation or displacement front) rather than the average of values upstream and downstream of the condensation front.

Although the effects of saturation and porosity variations normally are taken into account when multiphase thermal problems are solved with high-speed computers, the influence of these variables on the volumetric heat capacity of the reservoir is slight enough that in analytical methods constant average values can be used to develop some insight into reservoir heating. Furthermore, the error introduced by using average values would be small where most of the heat stored in the reservoir is within a clearly defined region (e.g., upstream of a displacement front). The following section presents results on the equivalent size of the heated reservoir volume resulting from injection of a heated fluid at constant rate and temperature. It is based on the assumption that the volumetric heat capacity of the heated zone is a constant.

5.2 Growth of the Equivalent Heated Volume Under a Constant Rate of Heat Input

Since Lindsly's study in 1928,[1] a number of authors have been concerned with the effect of heat losses to overburden and underburden on the size of the heated zone. Marx and Langenheim's results,[2] in particular, are useful for bringing out some important features about reservoir heating. They considered hot-fluid injection into a well at both constant rate and constant temperature. Although they recognized the presence of a radial temperature transition zone (see Fig. 5.2), they assumed the temperature of the heated zone to be everywhere at the downhole temperature of the injected fluid T_{idh} and the reservoir temperature outside the hot zone to be at the initial and reference temperature T_i. In their development, there is no temperature variation within the heated volume of the reservoir, either vertically or horizontally. A bulk heat balance made on a horizontal reservoir of uniform and constant properties gives

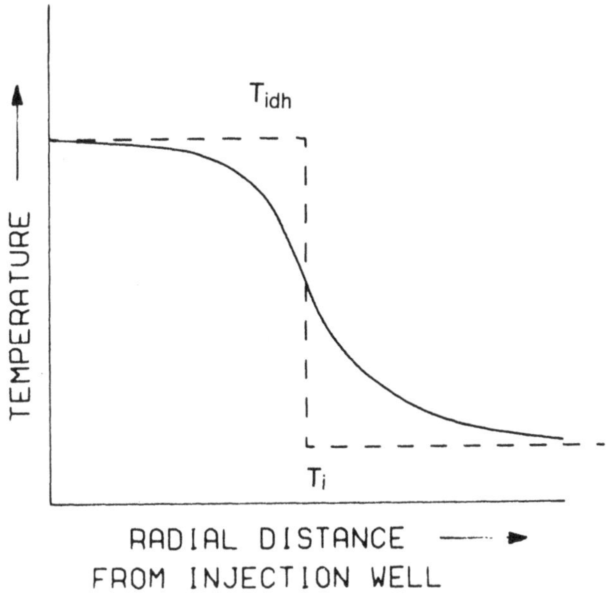

Fig. 5.2 – A qualitative comparison between the true reservoir temperature distribution and a step approximation.[2]

$$\dot{Q}_i = \langle 43{,}560 \text{ sq ft/acre} \rangle M_R h_t \Delta T_i \frac{dA}{dt} + \dot{Q}_l. \quad (5.4)$$

In Eq. 5.4, the quantity $M_R h_t A \Delta T_i$ represents the amount of heat required to bring a reservoir volume of a given area A, gross thickness h_t, and volumetric heat capacity M_R to a temperature ΔT_i above the initial (reference) reservoir temperature. Also, $\dot{Q}_i$ is the constant rate of heat injection, $\dot{Q}_l$ is the total rate of heat loss to the overburden and underburden, and dA/dt is the rate of growth of the areal extent of the heated zone.

As Marx and Langenheim[3] later pointed out, the shape of the heated area is general, though their initial illustrative example (see Fig. 5.2) discusses a symmetrically expanding cylindrical heated zone.

The total rate of heat loss $\dot{Q}_l$ increases with time as the heated zone grows. Marx and Langenheim assume that the heat losses from the horizontal

TABLE 5.2 – AUXILIARY FUNCTIONS

t_D	E_h	G	$e^{t_D}\,\mathrm{erfc}\sqrt{t_D}$	erfc t_D
0.0	1.0000	0	1.0000	1.0000
0.01	0.9290	0.0093	0.8965	0.9887
0.0144	0.9167	0.0132	0.8778	0.9837
0.0225	0.8959	0.0202	0.8509	0.9746
0.04	0.8765	0.0347	0.8090	0.9549
0.0625	0.8399	0.0524	0.7704	0.9295
0.09	0.8123	0.0731	0.7346	0.8987
0.16	0.7634	0.1221	0.6708	0.8210
0.25	0.7195	0.1799	0.6157	0.7237
0.36	0.6801	0.2448	0.5678	0.6107
0.49	0.6445	0.3158	0.5259	0.4883
0.64	0.6122	0.3918	0.4891	0.3654
0.81	0.5828	0.4721	0.4565	0.2520
1.00	0.5560	0.5560	0.4275	0.1573
1.44	0.5087	0.7326	0.3785	0.0417
2.25	0.4507	1.0143	0.3216	0.0015
4	0.3780	1.5122	0.2554	0.0000
6.25	0.3251	2.0318	0.2108	
9	0.2849	2.5641	0.1790	
16	0.2282	3.6505	0.1370	
25	0.1901	4.7526	0.1107	
36	0.1629	5.8630	0.0928	
49	0.1424	6.9784	0.0798	
64	0.1265	8.0907	0.0700	
81	0.1138	9.2177	0.0623	
100	0.1034	10.3399	0.0561	

reservoir are only by vertical conduction into the adjacent formations. Their results give a total rate of heat loss to the adjacent formations, which is

$$\dot{Q}_l=\dot{Q}_i\,(1-e^{t_D}\mathrm{erfc}\sqrt{t_D}), \quad \ldots\ldots\ldots\ldots\ldots (5.5)$$

where t_D is a dimensionless time given by

$$t_D=4\left(\frac{M_S}{M_R}\right)^2\frac{\alpha_S}{h_t^2}\,t, \quad \ldots\ldots\ldots\ldots\ldots (5.6)$$

and erfc(x) is the complementary error function:

$$\mathrm{erfc}(x)=\frac{2}{\sqrt{\pi}}\int_x^\infty e^{-s^2}ds$$

$$=1-\frac{2}{\sqrt{\pi}}\int_o^x e^{-s^2}ds$$

$$=1-\mathrm{erf}(x), \quad \ldots\ldots\ldots\ldots\ldots (5.7)$$

where erf(x) is the error function. In Eq. 5.6, α is the thermal diffusivity and the subscript S refers to the overburden and underburden (the surrounding formations), both of which are assumed to have the same thermal properties. Farouq Ali[4] has shown that when the thermal properties of the surrounding formations differ, average values should be used for $M_S\alpha_S$ in Eq. 5.6. Following are other results obtainable from the Marx and Langenheim development.

1. The areal extent of the equivalent heated zone is

$$A=\left(\frac{\text{acre/sq ft}}{43{,}560}\right)\frac{\dot{Q}_iM_Rh_t}{4\Delta T_i\alpha_SM_S^2}\,G\,. \quad \ldots\ldots\ldots (5.8)$$

2. The rate of growth of the equivalent heated zone is

$$\frac{dA}{dt}=\left(\frac{\text{acre/sq ft}}{43{,}560}\right)\frac{\dot{Q}_i}{\Delta T_iM_Rh_t}e^{t_D}\mathrm{erfc}\sqrt{t_D}. \quad \ldots\ldots\ldots (5.9)$$

3. The heat remaining in the reservoir is

$$Q=\frac{\dot{Q}_iM_R^2h_t^2}{4\alpha_SM_S^2}\,G\,. \quad \ldots\ldots\ldots\ldots (5.10)$$

4. The cumulative heat lost to the adjacent formations is

$$Q_l=\dot{Q}_it-Q. \quad \ldots\ldots\ldots\ldots (5.11)$$

5. The fraction of the injected heat remaining in the reservoir, a quantity that has been called the reservoir heat efficiency, is

$$E_h=Q/Q_i=G/t_D. \quad \ldots\ldots\ldots\ldots (5.12)$$

The function G appearing in some of these expressions is

$$G=2\sqrt{\frac{t_D}{\pi}}-1+e^{t_D}\mathrm{erfc}\sqrt{t_D}, \quad \ldots\ldots\ldots\ldots (5.13)$$

so that all five of the quantities given above vary with time. Values of E_h, G, $e^{t_D}\mathrm{erfc}\sqrt{t_D}$, and of the complementary error function erfc t_D are presented in Table 5.2 over the range of values of the dimensionless time t_D of practical interest.

The facts that (1) the temperature is uniform throughout the heated zone and is equal to the temperature of the injection fluid and (2) all the reservoir heat is in the heated zone indicate that the results given by Eqs. 5.5 and 5.8 through 5.13 are more applicable to steam drives than to any other thermal process. This is certainly true of the areal extent, rate of growth, and heat content of the heated zone.

Although not apparent from the results of Marx and Langenheim, the expressions for the rate of heat losses to the adjacent formation and the reservoir heat efficiency have a general applicability to any thermal recovery process.

Prats[5] showed that, for the idealized reservoir properties considered by Marx and Langenheim[2] and Lauwerier,[6] the distribution of the heat between the reservoir and the adjacent formations given by Eq. 5.12 should be an excellent approximation when there is no vertical temperature variation within the reservoir. This applies even if the injection temperature or mass rate varies, as long as the net rate of heat input is constant. By net rate of heat input is meant the difference between the total rate at which heat is generated in or injected into the reservoir through any number of wells (located anywhere) and the total rate at which heat is withdrawn from the reservoir through production of hot fluids from any number of wells (also located anywhere). Because conductive heat losses through the lateral boundaries of the reservoir were considered to be negligible in the development, the location of the injection and production wells does not affect this result as long as they are not too close to the lateral reservoir boundaries. It follows from these considerations that the distribution of heat between the reservoir and the adjacent formations is *not affected* by flow geometry prior to heat breakthrough or by the process used to introduce the heat into the reservoir. This explains why Marx and Langenheim,[2] Lauwerier,[6] Malofeev,[7] Rubinshtein,[8] and others arrived at identical results for the distribution of the injected heat between the reservoir and adjacent formations.

The work of Prats[5] also shows that Eqs. 5.8, 5.10, and 5.12 are valid for heat transfer conditions not so restrictive as those used by some of the authors cited. The same expressions are valid (1) when heat transfer in the plane of the reservoir is not only by convection but also by conduction and even radiation and (2) when heat transfer in the adjacent formations is by three-dimensional (rather than by solely vertical) conduction. Thus, the heat efficiency expression given by Eq. 5.12 is more generally applicable than indicated by the conditions under which it was first obtained. In particular, it is applicable to *any* type of thermal recovery process, provided the underlying assumptions referred to in the preceding paragraph are reasonably representative of the actual phenomena.

The heat efficiency defined by Eq. 5.12 is plotted in Fig. 5.3 as a function of dimensionless time t_D. The function G defined by Eq. 5.13, which is required to determine the area of the heated volume and its heat content, may be found from Fig. 5.3 by multiplying E_h by t_D.

The *fraction* of the injected heat remaining in the reservoir (the heat efficiency E_h) does not depend on the injection temperature, the rate of fluid injection, or the rate of heat injection. For constant rate of heat input, it is a function of only the dimensionless time. Cumulative heat losses to the adjacent formations increase as the injection temperature increases, but so does the heat remaining in the reservoir. At high fluid injection rates, the heated zone is quite large and the cumulative heat loss from the enlarged heated zone is great. But the amount of heat remaining in the reservoir as a result of the high fluid injection rate is also great. These effects again cancel, so that the heat efficiency, again, is a function of only the dimensionless time. From the definition of the dimensionless time given by Eq. 5.6, it follows that the major variable other than time is the gross thickness of the reservoir h_t. Of course, the thermal properties vary somewhat, depending on mineral and fluid content (see Fig. 5.1 and Appendix B), but the range of variation is rather limited compared with the variations in gross reservoir thickness. Notice that the gross reservoir thickness appears as a square. If one reservoir is twice the thickness of another, then an injection period four times as long is required to retain the same fraction of the injected heat. If the duration of the heating period is the same in both reservoirs, the dimensionless time of the thinner reservoir is four times that of the other one. As can be seen from Fig. 5.3, a factor of four in the dimensionless time has a significant effect on the value of the heat efficiency.

The ratio of the thermal properties of the reservoir to those of adjacent formations appears in the dimensionless time (Eq. 5.6). Although the range of variation of the factors involving these properties is small relative to the range in gross reservoir thickness, their contribution can be great enough to affect the operating life, or the heat efficiency, by amounts that are operationally important.[4]

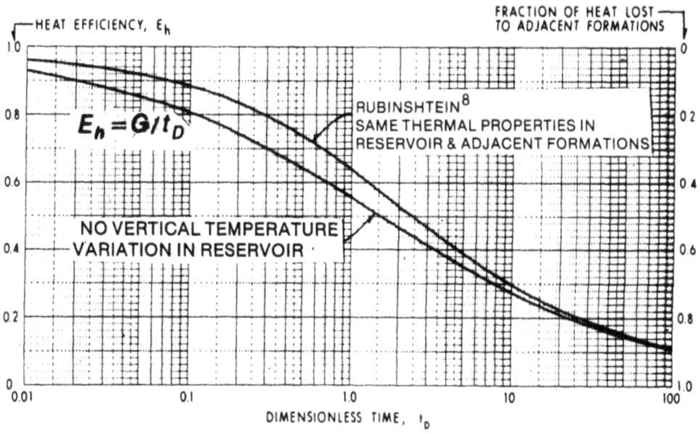

Fig. 5.3 – Distribution of heat between the reservoir and adjacent formations.

Rubinshtein[8] was the first investigator to consider how vertical temperature variations within the reservoir affect heat losses. His work dealt with radial hot-water injection, where the flow velocity was uniform within the vertical extent of the reservoir, implying a net-to-gross reservoir thickness ratio of one. His result for the reservoir heat efficiency is

$$E_h(t_{1D}) = 1 - a_2\left[\frac{2}{3}\sqrt{\frac{t_{1D}}{\pi}}(1 - a_2 f_3) + a_2 f_4\right], \quad \ldots\ldots\ldots\ldots (5.14a)$$

where

$$f_3 = \sum_{n=1}^{\infty} a_1^{n-1}\left(1 + \frac{n^2}{t_{1D}}\right)e^{-n^2/t_{1D}} \quad \ldots\ldots (5.14b)$$

and

$$f_4 = \sum_{n=1}^{\infty} n a_1^{n-1}\left(1 + \frac{2n^2}{3t_{1D}}\operatorname{erfc}\frac{n}{\sqrt{t_{1D}}}\right) \quad \ldots (5.14c)$$

depend on the dimensionless time t_{1D}, defined as

$$t_{1D} = \theta^2 t_D = \frac{\alpha_R M_R^2}{\alpha_S M_S^2} t_D. \quad \ldots\ldots\ldots\ldots (5.15)$$

The constant θ can be found from Eq. 5.15 and the constants a_1 and a_2 are defined as

$$a_1 = (1 - a_2) = (\theta - 1)/(\theta + 1). \quad \ldots\ldots\ldots\ldots (5.16)$$

The subscript R refers to properties of the reservoir. The reservoir heat efficiency given by the above expression was evaluated for equal thermal properties in the overburden and underburden. The effect of vertical temperature variations within the reservoir on E_h is compared with that given by Eq. 5.12 in Fig. 5.3. For the single comparison given (equal thermal properties everywhere, so that $\theta = 1$ and $a_1 = 0$), vertical temperature variations result in a maximum increase in heat efficiency of about 15%.

Example 5.1 – Calculation of Rate of Growth of Heated Zone

Steam is injected at 208 B/D of equivalent water into

a clean sandstone reservoir containing a 10°API crude and having a gross (and net) thickness of 25 ft, a porosity of 0.30, and a temperature of 103°F. Calculations indicate that the bottomhole injection pressure at the midpoint of the fully perforated injection interval is 995 psig and that the steam quality is 73%. It is assumed that there is no significant vertical temperature gradient within the reservoir, that the heated volume is at the steam injection temperature, and that the residual oil and water saturations in the heated zone are 0.2 and 0.4, respectively. Calculate the rate of growth of the equivalent heated zone after 7 months of continuous injection for the case where both adjacent formations are (1) shales and (2) dense ($\phi=0.02$) limestones. These thermal properties are given:

$$M_\sigma = 42.2 \text{ Btu/cu ft-°F},$$
$$\lambda_{sh} = 23.4 \text{ Btu/ft-D-°F},$$
$$\alpha_{sh} = 0.74 \text{ sq ft/D},$$
$$\rho_w = 64.3 \text{ lbm/cu ft at 60°F},$$
$$M_{\text{calcite}} = 33.8 \text{ Btu/cu ft-°F, and}$$
$$\alpha_{ls} = 1.2 \text{ sq ft/D}.$$

Note that to determine the rate of growth of the areal extent of the equivalent heated zone given by Eq. 5.9 requires that we determine the rate at which heat is injected into the reservoir, a number of thermal properties, and a function $e^{t_D}\text{erfc}\sqrt{t_D}$ that depends on the dimensionless time given by Eq. 5.6. First we obtain intermediate properties required for estimating the rate of growth of the heated zone. Since only steam is present in the vapor phase within the reservoir, $f=0$ (see Eq. 5.3). The absolute pressure is $995+14.7=1{,}010$ psi, and linear interpolation of the steam tables in Appendix B gives these data:

$$T_s = 546°\text{F},$$
$$L_v = 648 \text{ Btu/lbm},$$
$$h_s = 1{,}192 \text{ Btu/lbm},$$
$$h_w = 544 \text{ Btu/lbm},$$
$$\rho_s = 2.27 \text{ lbm/cu ft, and}$$
$$\rho_w = 46.2 \text{ lbm/cu ft}.$$

Also, the oil density corresponding to 10°API is 62.4 lbm/cu ft at 60°F.

The heat capacity of the oil is estimated as the arithmetic average of the heat capacities at the two temperatures using Eq. B.38:

$$C_o = [0.388+0.00045(103+546)/2]/\sqrt{1}$$
$$=0.534 \text{ Btu/lbm-°F}.$$

The heat capacity of water over the temperature range is estimated from

$$C_w = \frac{h_w(546°\text{F}) - h_w(103°\text{F})}{546°\text{F} - 103°\text{F}}$$
$$=(544 \text{ Btu/lbm} - 70.1 \text{ Btu/lbm})/(443°\text{F})$$
$$=1.07 \text{ Btu/lbm-°F}.$$

It is considered that the water initially in the reservoir is replaced effectively by fresh condensate so that the brine salinity at 60°F would have no bearing on the calculation of the reservoir heat capacity. The oil expands thermally. Its density in the heated zone is estimated with the aid of the specific gravities presented in Fig. B.31 to be

$$\rho_o(546°\text{F}) = (62.4 \text{ lbm/cu ft})(0.835)$$
$$=52.1 \text{ lbm/cu ft},$$

which is slightly denser than the water condensate at the same temperature.

From the available properties, the volumetric heat capacities of the reservoir and adjacent formations required for subsequent calculations are obtained as follows.

$$M_w = C_w\rho_w$$
$$= (1.07 \text{ Btu/lbm-°F})(46.2 \text{ lbm/cu ft})$$
$$= 49.4 \text{ Btu/cu ft-°F}$$
$$M_o = C_o\rho_o$$
$$= (0.534 \text{ Btu/lbm-°F})(52.1 \text{ lbm/cu ft})$$
$$= 27.8 \text{ Btu/cu ft-°F}.$$
$$M_{sh} = (\lambda/\alpha)_{sh}$$
$$= (23.4 \text{ Btu/ft-D-°F})/(0.74 \text{ sq ft/D})$$
$$= 31.6 \text{ Btu/cu ft-°F}.$$
$$M_{ls} = (1-0.02)(33.8 \text{ Btu/cu ft-°F}) + (0.02)(49.4 \text{ Btu/cu ft-°F})$$
$$= 34.4 \text{ Btu/cu ft-°F}.$$
$$M_R = (1-0.3)(42.2 \text{ Btu/cu ft-°F}) + 0.3(0.2)(27.8 \text{ Btu/cu ft-°F}) + 0.3(0.4)[(49.4 \text{ Btu/cu ft-°F}) + (648 \text{ Btu/lbm})/(546°\text{F}-103°\text{F}) + (1.07 \text{ Btu/lbm-°F})(2.27 \text{ lbm/cu ft})]$$
$$=37.6 \text{ Btu/cu ft-°F}.$$

The dimensionless time required to obtain the function G now can be calculated for the case where the adjacent formations are shales.

$$t_D = 4\left[\frac{31.6 \text{ Btu/cu ft-°F}}{37.6 \text{ Btu/cu ft-°F}}\right]^2 \cdot \frac{(0.74 \text{ sq ft/D})(7 \text{ months})(30.4 \text{ D/month})}{(25 \text{ ft})^2}$$
$$=0.712,$$

so that by interpolating in Table 5.2 we find

$$e^{0.712}\text{erfc}\sqrt{0.712} = 0.475.$$

In a similar manner, for the limestone we obtain

$$t_D = 4\left[\frac{34.4 \text{ Btu/cu ft-°F}}{37.6 \text{ Btu/cu ft-°F}}\right]^2 \cdot \frac{(1.2 \text{ sq ft/D})(212.8 \text{ days})}{(25 \text{ ft})^2}$$
$$=1.38,$$

and

$$e^{1.38}\text{erfc}\sqrt{1.38} = 0.385.$$

In calculating the rate of heat input into the reservoir, note that the total mass rate of injection is found by

$$w_t = (5.615 \text{ cu ft/bbl})(208 \text{ B/D})(62.4 \text{ lbm/cu ft}) = 72{,}900 \text{ lbm/D},$$

of which 73% is dry steam, or

$$w_s = (0.73)(72{,}900 \text{ lbm/D}) = 53{,}200 \text{ lbm/D}.$$

The rate of heat input into the reservoir then is obtained as follows:

$$\begin{aligned}\dot{Q}_i &= w_t\,[h_w(546°\text{F}) - h_w(103°\text{F})] + w_s L_v \\ &= (72{,}900 \text{ lbm/D})(544 \text{ Btu/lbm} - 70.1 \text{ Btu/lbm}) \\ &\quad + (53{,}200 \text{ lbm/D})(648 \text{ Btu/lbm}) \\ &= 6.90 \times 10^7 \text{ Btu/D}.\end{aligned}$$

The rate of growth of the heated area (in sq ft/D) now is obtained from Eq. 5.9 for the shale caprock and base rock:

$$\frac{dA}{dt} = \frac{(6.90 \times 10^7 \text{ Btu/D})(0.475)}{(443°\text{F})(37.6 \text{ Btu/cu ft-°F})(25 \text{ ft})} = 79 \text{ sq ft/D}.$$

In the case of the limestone caprock and base rock,

$$\frac{dA}{dt} = \frac{(6.90 \times 10^7 \text{ Btu/D})(0.385)}{(443°\text{F})(37.6 \text{ Btu/cu ft-°F})(25 \text{ ft})} - 64 \text{ sq ft/D}.$$

This problem illustrates (1) how to consider the initial reservoir temperature to calculate the rate of heat injection, (2) some uses of the steam tables, (3) how to evaluate thermal properties of formations, and (4) the importance of formation thermal properties on the rate of growth of the heated zone.

5.3 Growth of Equivalent Heated Volume Under Variable Rates of Heat Input

For a variable rate of heat input, the heat remaining in the reservoir can be estimated using the results given by Eq. 5.10. The estimation of the heat in the reservoir for variable rate conditions is based on the assumption that the variable rate of heat input can be represented by a series of N step changes occurring at a sequence of N times: $t_1, t_2, t_3, \ldots, t_n, \ldots, t_N$, with $t_1 = 0$.

For example, Fig. 5.4 shows a smooth curve of heat injection rate vs. time. This curve can be approximated by a series of N discrete step changes in $\dot{Q}_i$. The figure also shows how the smooth heat injection rate curve can be discretized. During the nth time interval, the heat injection rate is represented by

$$\dot{Q}_i(t) = \sum_{j=1}^{n} U(t-t_j)\Delta\dot{Q}_j, \quad \ldots\ldots\ldots\ldots (5.17)$$

where $\Delta\dot{Q}_j$ is the change in the heat injection rate that takes place at t_j. Note that $\Delta\dot{Q}_j$ is positive when the rate of heat injection increases at t_j and is

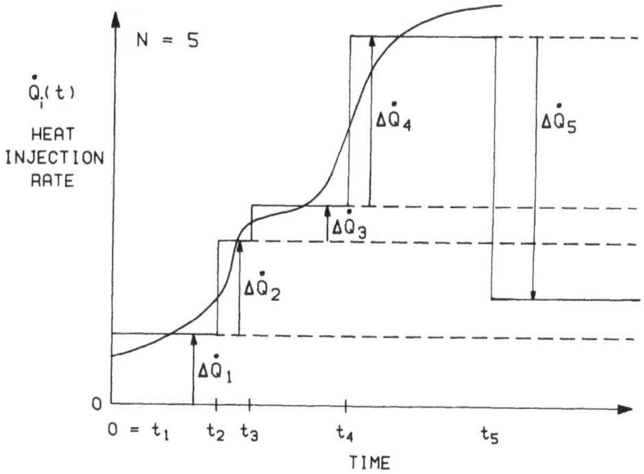

Fig. 5.4 – Variable rate of heat injection by step changes.

negative otherwise. The unit function $U(t-t_j)$ is either 0 or 1. It is 1 when $t > t_j$, and 0 when $t < t_j$. Also note that in the nth time interval,

$$t_n < t < t_{n+1}. \quad \ldots\ldots\ldots\ldots\ldots\ldots (5.18)$$

The cumulative heat injection at any time during the nth time interval is approximated from

$$Q_i(t) = \sum_{j=1}^{n} U(t-t_j)\,[(t-t_j)\Delta\dot{Q}_j]. \quad \ldots (5.19)$$

A plot of this quantity vs. time would have different slopes for times larger and smaller than the t_j values. A smooth curve may be drawn through the segmented sections of the $Q_i(t)$ vs. t graph, a procedure that is the inverse of the discretizing procedure used to convert the initial $\dot{Q}_i$ vs. t curve in Fig. 5.4 into a sequence of step changes.

As pointed out by Ramey,[9] the heat remaining in the reservoir at any time during the nth time interval can be obtained by superposition and is given by

$$Q(t) = \left(\frac{M_R}{M_S}\right)^2 \frac{h_t^2}{4\alpha_S} \sum_{j=1}^{n} U(t-t_j) \cdot \Delta\dot{Q}_j G(t_D - t_{Dj}). \quad \ldots\ldots\ldots (5.20)$$

Values of G required for the calculation may be obtained from Table 5.2, Fig. 5.3, or the definition of G (Eq. 5.13). Values of G are evaluated as a function of $t_D - t_{Dj}$, which is related to $t - t_j$ by means of the expression

$$t_D - t_{Dj} = 4\left(\frac{M_S}{M_R}\right)^2 \frac{\alpha_S}{h_t^2}(t-t_j). \quad \ldots\ldots (5.21)$$

The areal extent of the equivalent heated zone is found by replacing the product $\dot{Q}_i G$ in Eq. 5.8 by the sum in Eq. 5.20. (It is to be understood that the value of ΔT_i in the resultant equation may vary with time.) And the reservoir heat efficiency now is found by substituting the expressions for the heat in the reservoir equation (Eq. 5.20) and the cumulative heat injected equation (Eq. 5.19) into Eq. 5.12.

But the applicability of the results obtained or indicated for variable rate of heat injection needs additional discussion. The heat remaining in the reservoir, the cumulative heat losses, and the reservoir heat efficiency are generally applicable over a rather wide range of conditions and processes, as discussed in connection with the case of constant rate of heat injection. The areal extent of the heated area, on the other hand, may not be representative of that of a steam zone where the condensation front recedes with time. In cyclic steam injection, for example, the equivalent heated zone may be much larger than the actual steam zone. On the other hand, for growing steam zone volumes, the equivalence is expected to be fairly good, provided the amount of heat stored in the reservoir outside the steam zone is small. Conditions under which this is the case, which are valid for times smaller than a critical time defined by Mandl and Volek,[10] will be discussed in more detail in Chap. 7.

Results discussed in this section for variable heat injection rates are valid for any manner of introducing or withdrawing heat through wells. Note that the variability in the rate of heat injection can be due to changes in mass injection rate or in enthalpy or in any other means of introducing thermal energy into the reservoir. The rate of heat injection is, as for the constant rate case, the net rate of injection (injection minus production from all wells in the reservoir of interest). An example will bring out additional points.

Example 5.2 – Calculation of Heat Losses From the Reservoir to Adjacent Formations

As an example for estimating heat losses to overburden and underburden, consider a combustion project in which the air injection rate into one well was doubled to 2 MMscf/D after 6 months, while the air injection rate into each of two other wells remained at 1 MMscf/D. Oxygen consumption is complete. There are indications that combustion is occurring over the entire 10-ft, 0.30-porosity oil sand. Estimate the heat lost to the massive adjacent shales a year after ignition. There have been no indications of temperature increases at any of the seven producers. Injection pressure is 1,500 psig. The given properties are

$$M_\sigma = 49.5 \text{ Btu/cu ft-°F},$$
$$M_g = 2.12 \text{ Btu/cu ft-°F},$$
$$\alpha_S = 0.725 \text{ sq ft/D, and}$$
$$\lambda_S = 20.2 \text{ Btu/ft-D-°F}.$$

Notice from Eq. 5.11 that to determine the cumulative heat lost requires information on the cumulative heat injected and on the heat remaining in the reservoir. For variable rates, these quantities are given by Eqs. 5.19 and 5.20, respectively. The function G appearing in Eq. 5.20 is to be evaluated at dimensionless times defined by Eq. 5.21.

Since only two rate changes are indicated (one at the start and the other at 6 months), $N=2$. The value of t_1 is always zero, and t_2 is 6 months.

Since oxygen consumption is complete, we use the rule of thumb that approximately 100 Btu are generated for every standard cubic foot of air used. From this, the value of $\Delta\dot{Q}_1$ (at the start of the project) is

$$\Delta\dot{Q}_1 = (3 \text{ MMscf/D})(100 \text{ Btu/scf}) = 3\times10^8 \text{ Btu/D}.$$

Likewise the value of $\Delta\dot{Q}_2$ (at 6 months) corresponding to the increase in 1 MMscf/D in the injection rate of one of the three wells is

$$\Delta\dot{Q}_2 = (1 \text{ MMscf/D})(100 \text{ Btu/scf}) = 1\times10^8 \text{ Btu/D}.$$

Eq. 5.19 then is used to obtain the cumulative heat injection at 365 days:

$$Q_i(365) = (1 \text{ year} - 0 \text{ year})(365 \text{ D/yr})(3\times10^8 \text{ Btu/D}) + (1 \text{ year} - 0.5 \text{ year})(365 \text{ D/yr}) \cdot (1\times10^8 \text{ Btu/D}) = 1.28\times10^{11} \text{ Btu}.$$

In dry combustion operations, most of the heat in the reservoir is in the part that has been burned. There, the liquid saturations are negligible and the effective volumetric heat capacity is given from Eq. 5.3 as

$$M_R = (1-\phi)M_\sigma + \phi M_g = (1-0.30)(49.5 \text{ Btu/cu ft-°F}) + 0.30(2.12 \text{ Btu/cu ft-°F}) = 35.3 \text{ Btu/cu ft-°F}.$$

Also,

$$M_S = \lambda_S/\alpha_S = (20.2 \text{ Btu/ft-D-°F})/(0.725 \text{ sq ft/D}) = 27.9 \text{ Btu/cu ft-°F},$$

$$t_D = 4\left(\frac{27.9 \text{ Btu/cu ft-°F}}{35.3 \text{ Btu/cu ft-°F}}\right)^2 \cdot \frac{(0.725 \text{ sq ft/D})(365 \text{ days})}{(10 \text{ ft})^2} = 6.6,$$

$$t_{D2} = 3.3,$$

and

$$t_{D1} = 0.$$

The values of G corresponding to values of $t_D - t_{Dj}$ are found from the curve labeled $E_h = G/t_D$ in Fig. 5.3.

$$G(t_D - t_{D1}) = G(6.6) = 6.6\, E_h|_{t_D=6.6} = 6.6(0.32) = 2.1.$$

And, similarly,

$$G(t_D - t_{D2}) = 3.3(0.40) = 1.3.$$

The heat remaining in the reservoir at 365 days is calculated next from Eq. 5.20:

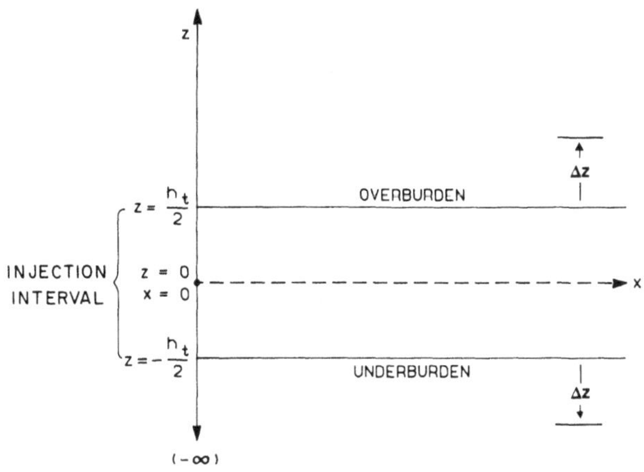

Fig. 5.5 – Coordinate system used by Lauwerier.[6]

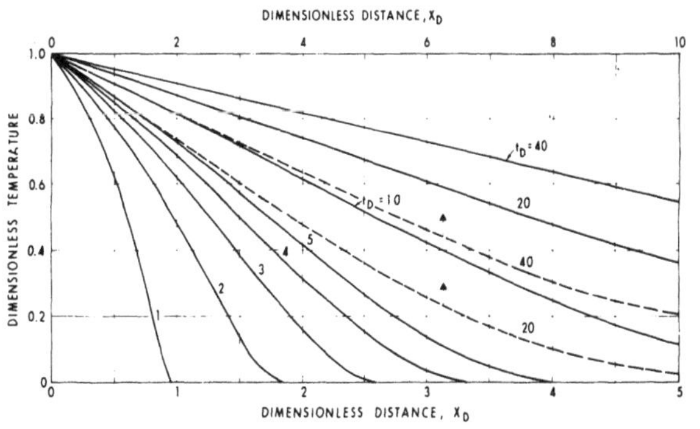

Fig. 5.6 – Reservoir temperature profiles resulting from injection of noncondensable hot fluid.

$$Q=\left(\frac{35.3 \text{ Btu/cu ft-°F}}{27.9 \text{ Btu/cu ft-°F}}\right)^2 \frac{(10 \text{ ft})^2}{(4)(0.725 \text{ sq ft/D})}$$

$$\cdot[(3\times10^8 \text{ Btu/D})(2.1)+(1\times10^8 \text{ Btu/D})(1.3)]$$

$$=4.2\times10^{10} \text{ Btu},$$

from which the cumulative heat lost to the adjacent formations is, from Eq. 5.11,

$$Q_l=12.8\times10^{10} \text{ Btu}-4.2\times10^{10} \text{ Btu}$$

$$=8.6\times10^{10} \text{ Btu}.$$

Thus, heat losses amount to 67% of the heat generated during the combustion process.

This example is intended to show (1) that the heat-loss calculations obtainable from the extended Marx and Langenheim results are also applicable to combustion processes, (2) that the results are applicable for any flow geometry involving any number of injection and production wells, (3) that it is unnecessary to know reservoir temperatures to calculate heat losses, except to estimate thermal properties, and (4) the use of superposition to account for variable rate of heat injection.

5.4 Heat Losses Through Produced Fluids

Currently, variations with time in the temperature of produced fluids can be predicted only crudely, unless a numerical or physical simulator is employed. Approximate methods for estimating the bottomhole temperature response at a particular distance from an injection well—methods amenable to calculation with the aid of graphs and pocket-size calculators—are limited to those of Lauwerier[6] in linear systems, Malofeev's[7] extension of this work to radial system, and Gringarten and Sauty's[11] results from a two-well pilot. It is emphasized that the temperature of produced fluids discussed in this chapter is at downhole conditions, which are the ones of interest in determining the rate at which heat is removed from the reservoir through produced fluids. Differences between bottomhole and wellhead production well temperatures can be estimated as shown in Chap. 10.

Lauwerier[6] is recognized as being the first to put on a firm basis approximate calculations of temperature distributions within the reservoir. Lauwerier considered hot-water injection at a constant flux u and temperature rise ΔT_i above the reservoir temperature into a *linear* horizontal reservoir of uniform and constant properties, shown schematically in Fig. 5.5. Heat transfer to the adjacent formation is by vertical conduction only. The temperature distribution within the reservoir is assumed to be independent of vertical position, and heat flow within the reservoir is by convection only. Lauwerier's temperature distribution in the reservoir and adjacent formations is given by

$$T_D=\frac{\Delta T}{\Delta T_i}=U(t_D-x_D)\operatorname{erfc}\left(\frac{x_D+\Delta z_D}{2\sqrt{t_D-x_D}}\right), \quad \text{.......... (5.22)}$$

where

$$x_D=\frac{4\alpha_S M_S^2 x}{u h_t^2 M_f M_R}, \quad \text{.......... (5.23)}$$

t_D is defined by Eq. 5.6, M_f is the volumetric heat capacity of the injected hot fluid,

$$\Delta z_D=\frac{2M_S}{h_t M_R}\Delta z, \quad \text{.......... (5.24)}$$

and Δz is the distance (always positive) into one of the adjacent formations measured from its common boundary with the reservoir. Thus, Δz is not defined within the reservoir. Accordingly,

$$\Delta z_D=0 \quad \text{.......... (5.25)}$$

when Eq. 5.22 is used to calculate temperatures within the reservoir.

Fig. 5.6 presents a plot of the dimensionless temperature profiles in the reservoir given by Eq. 5.22. Note that because of the unit function $U(t_D-x_D)$ there is no temperature rise for values of x_D larger than t_D, or $x>tuM_f/M_R$. We recognize uM_f/M_R as representing the velocity for convective heat transfer v_T (see Eq. 3.15 and its related discussion). Absence of a temperature rise for

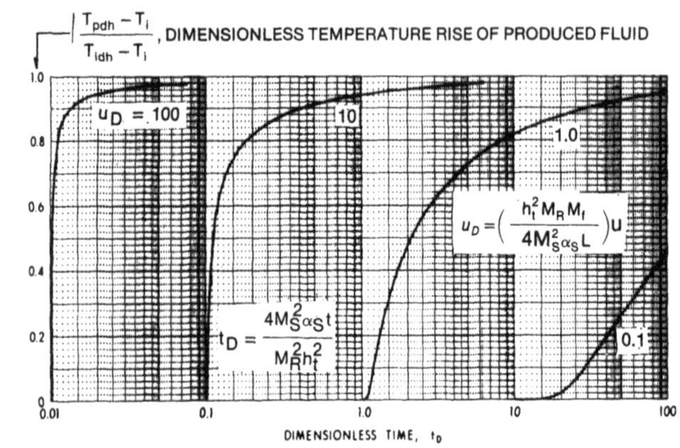

Fig. 5.7 – Temperature rise of produced fluid as a function of dimensionless time.

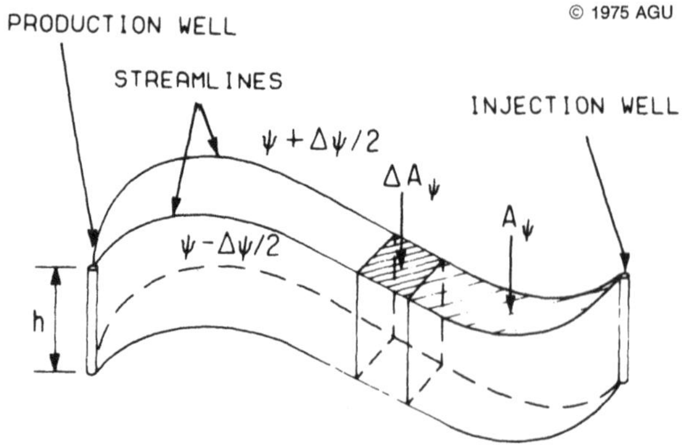

Fig. 5.8 – Visualization of a stream channel.[11]

$x_D > t_D$, a result that does not depend on the properties of the adjacent formations, shows that heat cannot be transferred *by convection* at a rate of frontal advance greater than

$$v_T = uM_f/M_R. \quad \ldots\ldots\ldots\ldots\ldots\ldots (5.26)$$

This expression holds for any convective flow system, where the flux u may vary in space and time. Of course, the velocity of convective heat transfer v_T is *always* in the same direction as the fluid flux u.

Malofeev[7] showed that Lauwerier's temperature expression (Eq. 5.22) was also valid for radial flow when x_D is defined by

$$x_D = \frac{4\pi r^2}{M_f h_t i}\,\frac{\alpha_S M_S^2}{M_R}, \quad \ldots\ldots\ldots\ldots\ldots\ldots (5.27)$$

where r is the radial distance from the injection well and i is the injection rate.

The heat remaining in the reservoir has been found[12] to be identical with that given by Eq. 5.10, both for the linear parallel flow considered by Lauwerier and for the radial flow considered by Malofeev.[7] Thus, the heat efficiency and the heat remaining in the formation are the same as those found from Marx and Langenheim's results.[2] This is so even though the flow geometries are different, and the reservoir temperature given by Eqs. 5.22 through 5.25 varies gradually with distance from the injection face.

The temperature in the reservoir given by Eqs. 5.22 through 5.25 can be used to evaluate the temperature change with time at a fixed distance from the injection face. For parallel linear flow, the temperature rise with time at a fixed distance x from the injection face can be considered as the temperature of the produced fluid when the value of $x=L$ represents the distance to the "producer." Fig. 5.7 shows the dimensionless temperature rise as a function of dimensionless time for several values of a dimensionless fluid flux obtained by evaluating $1/x_D$ from Eq. 5.23 at $x=L$. Heat breakthrough occurs earlier as the fluid velocity increases, as can be seen in Fig. 5.7. Therefore, the proportion of heat remaining in the reservoir is larger at higher velocities, so that the average reservoir temperature between "injector" and "producer" is also higher. Accordingly, at the higher flow velocities the temperature rises faster once heat breakthrough occurs (see Fig.5.7).

These comments are also formally applicable to radial flow from an injector, but here the presence of a producer concentric with the injector has no practical counterpart. The results are valid, however, for radial flow into a *producer* at $r=0$ when there is a constant temperature rise ΔT_i above the initial reservoir temperature at a cylinder a distance r_1 from the producer and x_D is redefined as

$$x_D = \frac{4\pi(r_1^2 - r^2)\alpha_S M_S^2}{M_R M_f h_t q}, \quad \ldots\ldots\ldots\ldots\ldots (5.28)$$

where q is the production rate.

In Lauwerier's results, both in the parallel linear and plane radial flow cases, the temperature rise is the same at the same distance from the injector along all flowlines, so that there cannot be heat transfer between flowlines. When the flowlines are not symmetrical with respect to the injector, however, there will be temperature gradients normal to the direction of flow, giving rise to conductive heat transfer. Gringarten and Sauty[11] assumed that (1) the flowlines would not change with time and (2) the conduction heat transfer normal to the flowlines could be neglected in calculating the temperature response between an injector and a producer in a homogeneous and uniform aquifer of infinite areal extent. Under the assumption that there is no heat transfer between streamlines, Gringarten and Sauty were able to make a heat balance along an arbitrary streamline connecting an injector and a producer, depicted in Fig. 5.8, and obtained a temperature distribution as a function of time along this arbitrary streamline in a manner completely analogous to that of Lauwerier.[6] Gringarten and Sauty's temperature equation along a streamline is identical in form with the one obtained by Lauwerier and given by Eq. 5.22. Of course, the definition of x_D is now different. Not only that, but x_D must now depend on the streamline, which is denoted by ψ. In the generalized results of Gringarten and Sauty, the factor $x/h_t u$ appearing in Eq. 5.23 is replaced by the quantity A_ψ/q_ψ:

$$\frac{x}{h_t u} = \frac{A_\psi}{q_\psi}, \quad \ldots\ldots\ldots\ldots\ldots\ldots\ldots\ldots (5.29)$$

where A_ψ is the cumulative area along a stream tube bounded between streamlines $\psi \pm \Delta\psi/2$, and q_ψ is the volume rate of flow within the stream tube.

For radial flow from an injector, for example, the injection rate per unit angle is $i/2\pi$. Since the flow is symmetrical,

$$q_\psi = \frac{\Delta\psi i}{2\pi}, \quad \text{(5.30)}$$

where, for radial flow, $\Delta\psi$ is equal to the increment of angle subtended between two flowlines. Also, the cumulative area from the injector between two flowlines subtending the angle $\Delta\psi$ is

$$A_\psi = \frac{r^2}{2}\Delta\psi. \quad \text{(5.31)}$$

Thus,

$$\frac{x}{h_t u} = \pi r^2/i, \quad \text{(5.32)}$$

independent of the flow channel. Substitution of $\pi r^2/i$ for $x/h_t u$ in Eq. 5.23 immediately yields Malofeev's extension for radial flow from an injector, given by Eq. 5.27. For other than linear or radial flow, however, A_ψ/q_ψ will depend on the streamlines.

Gringarten and Sauty evaluated the temperature of the produced fluid in an isolated two-spot resulting from the different contributions to the temperature along each stream tube. Results, given in Fig. 5.9, are functions of two parameters: a dimensionless heat injection rate and the dimensionless time given by Eq. 5.6. The dimensionless heat injection rate is defined as

$$\dot{Q}_{iD} = \frac{M_f M_R h_t i}{4\alpha_S M_S^2 L^2}, \quad \text{(5.33)}$$

where L is the distance between the injector and the producer. The dimensionless cumulative heat injected is then

$$Q_{iD} = \dot{Q}_{iD} t_D$$

$$= \frac{M_f it}{M_R L^2 h_t}. \quad \text{(5.34)}$$

Note that a value $Q_{iD} = 1$ means that the cumulative amount of heat injected is equal to the amount of heat required to bring a reservoir volume $L^2 h$ to the injection temperature. These results behave like those for the linear case shown in Fig. 5.7 in the sense that the producing temperature rises faster the higher the rate of heat injection.

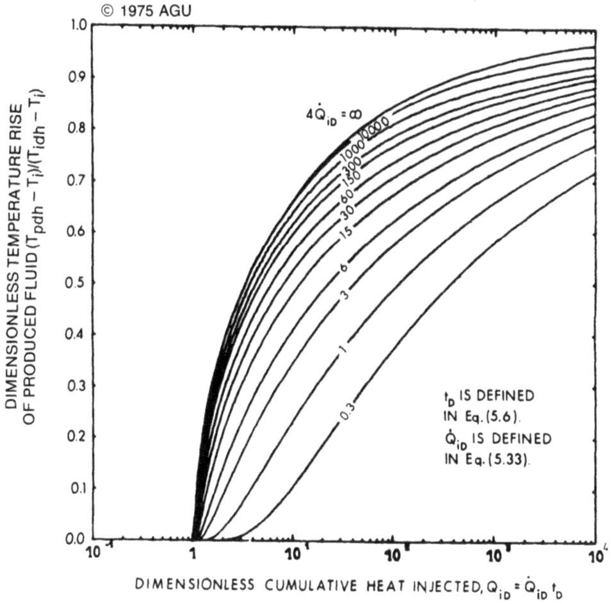

Fig. 5.9 – Dimensionless temperature at the production well of an isolated two-spot.[11]

5.5 Discussion

The procedures and results discussed thus far provide the means for *estimating* the heat lost from the reservoir in thermal recovery processes. The amount of heat lost to adjacent formations can be estimated from these procedures reasonably well when the vertical temperature gradients within the reservoir are small. The results discussed indicate that a vertical temperature distribution within the reservoir arising from heat losses gives rise to only a slight increase in the heat remaining in the reservoir. But it is possible for the heat in the reservoir to be appreciably different if the lack of vertical uniformity in temperature is a result of layers of different permeability or gravity override.

Prior to heat breakthrough, the bypassing of heat in practical situations will lead to higher reservoir heat efficiencies than would be estimated using the procedures discussed above. For example, in a thick sand with gravity override, there probably would be negligible heat losses to the underburden. Rough estimates of the heat in place may be made using the approximate thickness of the high-temperature zone, which by definition is smaller (perhaps appreciably so) than the reservoir thickness.

After heat breakthrough, the same comments apply when the rate of heat removal from the reservoir through the produced fluids either is known or is negligible compared with the rate of heat input. For it is the difference between the rate of heat input and the rate of production that enters the calculation of the heat in the reservoir. Since the foregoing methods for predicting the rate of heat withdrawal through produced fluids are rough approximations they will affect adversely the quality of the estimates of heat remaining in the reservoir. No such difficulty exists, of course, where bottomhole temperatures and the composition of the produced fluids are known. Although it could provide additional insight into thermal recovery processes, the amount of heat produced from field projects by means of produced fluids seldom is given.

Additional insight into the distribution of the heat within the reservoir can be gained through reported laboratory studies, numerical simulation, and analyses and measurements resulting from field projects.

Most laboratory studies of thermal processes have used tube experiments, thus modeling linear flow.[13] A few laboratory studies have been made of three-

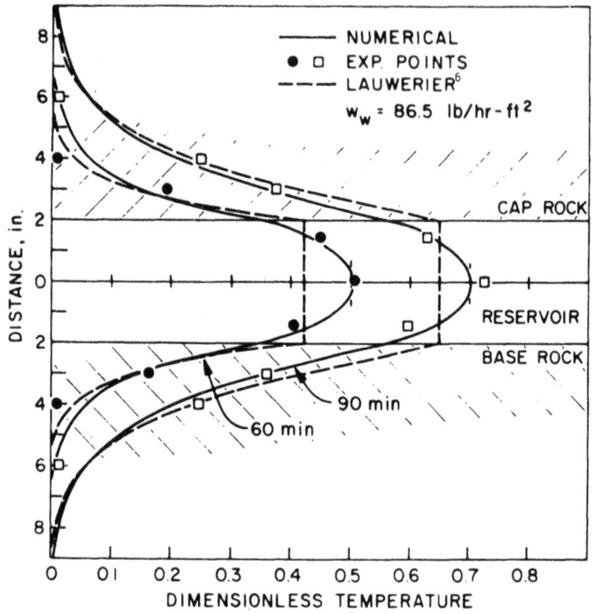

Fig. 5.10 – Cross-sectional hot-water temperature profiles 18 in. from inlet.[13]

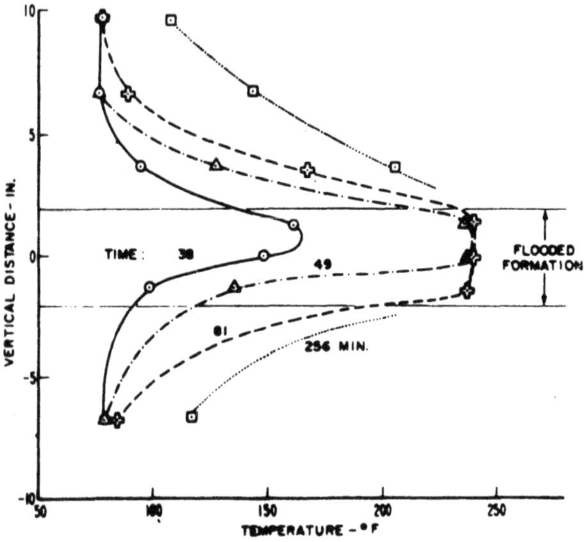

Fig. 5.11 – Temperature profiles on a vertical line 21 in. from the model injection well (steam injection rate 0.49 lbm/min.)[14]

dimensional processes, either in the neighborhood of injection wells[14] or modeling elements of symmetry representative of reservoir elements.[15] Figs. 5.10 through 5.12 show temperature distributions in both linear-flow and radial-flow laboratory systems with simulated caprocks and base rocks. Fig. 5.10 compares observed and numerically simulated temperatures resulting from the downward linear flow of 170°F hot water displacing unheated water with those obtained by Lauwerier's results. The observed and numerical temperatures agree quite well with each other and agree fairly well with the temperatures obtained by Lauwerier except near the leading edge of the heat front. These differences are due to thermal conduction, an important factor near the leading edge of a convection heat front.[15]

Figs. 5.11 and 5.12 give vertical temperature profiles for horizontal injection at two different injection rates. The effect of gravity override is noticeable in both cases, more so at the slower injection rate (Fig. 5.12). Both figures also show preheating ahead of the steam front, which is supported amply by theoretical[10] and numerical studies.[16] Isotherms after 40 minutes of injection are given in Fig. 5.13 for yet a third experiment.[17] The figure clearly shows the preheating within the reservoir ahead of and below the steam zone resulting from the flow of hot water through the condensation front (approximately given by the 340°F isotherm). In Fig. 5.14, temperatures at the center of the injection interval, obtained by numerical simulation, are plotted vs. radial distance and exhibit preheating downstream of the condensation front.

Typical isotherms for the combustion process in which the combustion front is assumed to be vertical are shown in Fig. 5.15. Note that the high temperature region is near the combustion front, with most of the reservoir heat stored upstream in the burnt region. With gravity override, calculated temperatures in a system with a radially symmetric front are relatively low at the combustion front, which may smolder rather than burn actively because the unused oxygen in the injected air bypasses along the top (see Fig. 5.16).

Alternative methods for heating an oil-bearing section in a reservoir with insufficient injectivity include heating through thief zones, gas caps, bottomwater zones,[18] or through fractures. If fractures are used, then horizontal ones are preferred.[19,20] Temperature distributions within the reservoir resulting from such uneven heating may be estimated roughly by using Lauwerier's results. In these cases the thickness of the heated zone (which must remain reasonably constant in time and distance) is used instead of the total formation thickness. The temperature distribution above and below the injection interval can then be estimated from Eq. 5.22, taking care to choose the properties of the caprock and base rock to represent those of the formations immediately above and below the injection interval.

When heating proceeds very rapidly through a horizontal high-permeability zone, the assumption may be made that the zone is at a reasonably uniform

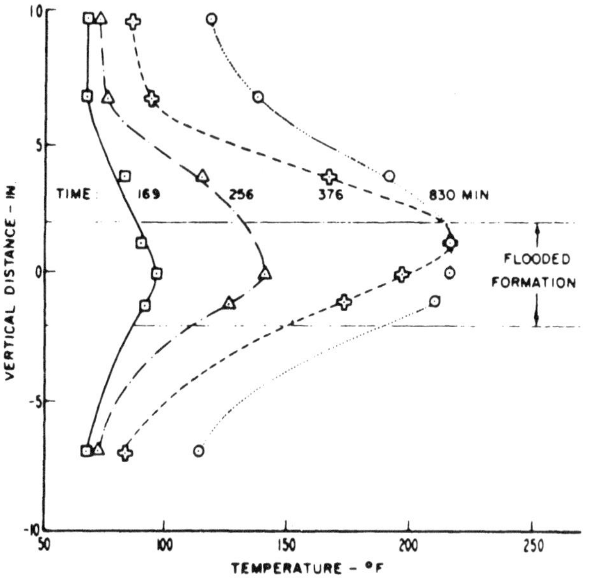

Fig. 5.12 – Temperature profiles on a vertical line 21 in. from the model injection well (steam injection rate 0.15 lbm/min.)[14]

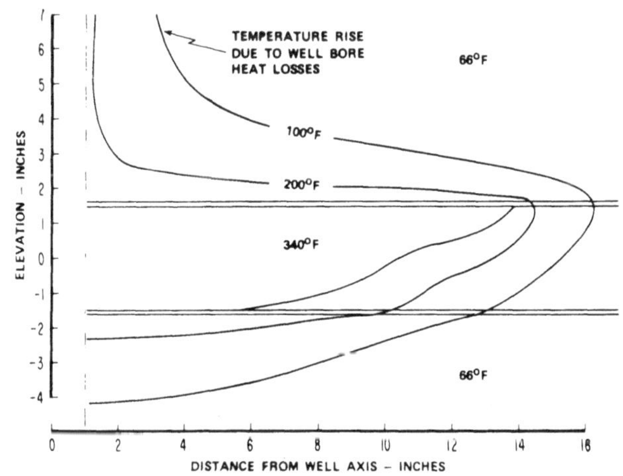

Fig. 5.13 – Temperature contours in model after 40 minutes of injection (injection pressure, 103 psig; steam injection rate, 0.44 lbm/min.)[17]

temperature for a considerable distance. Within this distance the temperature is essentially that at the injection face. Because (within limits) there is no dependence on position within the injection interval, x_D can be set to zero in Eq. 5.22 to estimate the temperature in the adjacent formations. The resultant equation,

$$\Delta T = \Delta T_i \operatorname{erfc}\left(\frac{\Delta z}{2\sqrt{\alpha_s t}}\right), \quad \text{.............. (5.35)}$$

gives the temperature distribution due to linear heat conduction from a plane maintained at a constant temperature ΔT_i and represents one of the fundamental solutions in conduction heat transfer.[23]

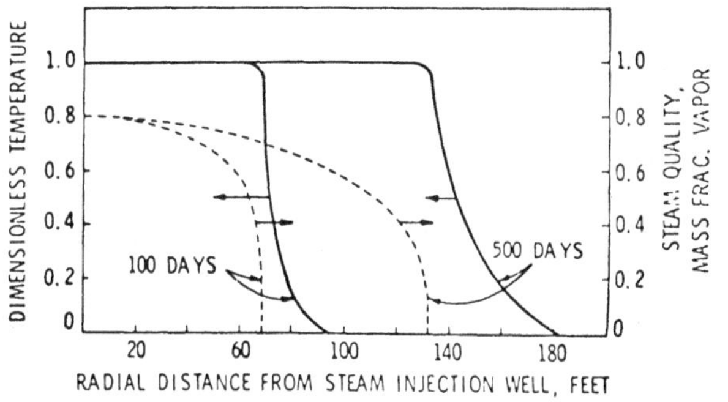

Fig. 5.14 – Temperature (at center of pay) and steam quality distribution.[16]

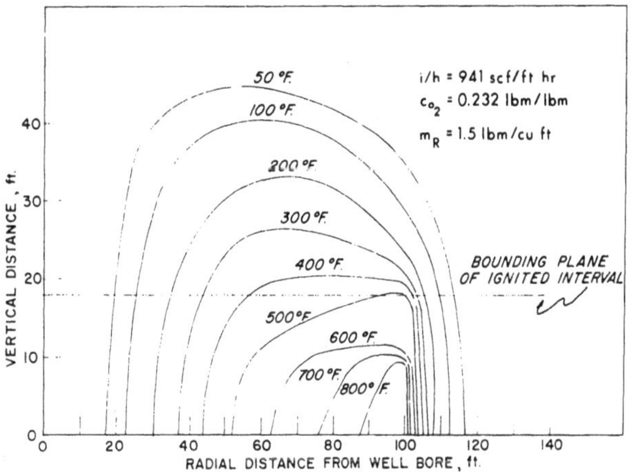

Fig. 5.15 – Typical isotherms above initial reservoir temperature in a combustion process.[22]

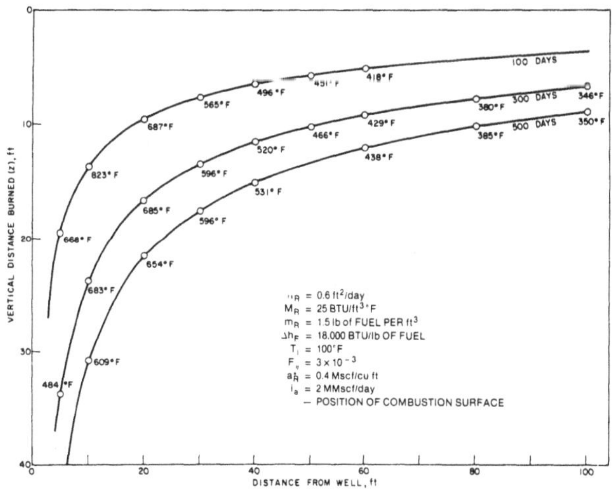

Fig. 5.16 – Temperatures on combustion surfaces under simulated gravity override conditions.[21]

References

1. Lindsly, B.E.: "Recovery by Use of Heated Gas," *Oil and Gas J.* (Dec. 20, 1928) 88-93.

2. Marx, J.W. and Langenheim, R.H.: "Reservoir Heating by Hot Fluid Injection," *Trans.*, AIME (1959) **216**, 312-315.
3. Marx, J.W. and Langenheim, R.H.: "Authors' Reply to H.J. Ramey, Jr.," in response to "Further Discussion of Paper Published in Transactions Volume 216," *Trans.*, AIME (1959) **216**, 365.
4. Farouq Ali, S.M.: *Oil Recovery by Steam Injection*, Producers Publishing Co. Inc., Bradford, PA (1970).
5. Prats, M.: "The Heat Efficiency of Thermal Recovery Processes," *J. Pet. Tech.* (March 1969) 323-332; *Trans.*, AIME, **246**.
6. Lauwerier, H.A.: "The Transport of Heat in an Oil Layer Caused by Injection of Hot Fluid," *Applied Scientific Research* (1955) **A5**, 145-50.
7. Malofeev, G.E.:"Calculation of the Temperature Distribution in a Formation When Pumping Hot Fluid into a Well," *Neft i Gaz* (1960) **3**, No. 7, 59-64.
8. Rubinshtein, L.L.: "The Total Heat Losses in Injection of a Hot Liquid into a Stratum," *Neft i Gaz* (1959) **2**, No. 9, 41-48.
9. Ramey, H.J. Jr.: "Discussion of Reservoir Heating by Hot Fluid Injection," *Trans.*, AIME (1959) **216**, 364-365.
10. Mandl, G. and Volek, C.W.: "Heat and Mass Transport in Steam-Drive Processes," *Soc. Pet. Eng. J.* (March 1969) 59-79; *Trans.*, AIME, **246**.
11. Gringarten, A.C. and Sauty, J.P.: "A Theoretical Study of Heat Extraction from Aquifers with Uniform Regional Flow," *J. Geophys. Research* (Dec. 1975) **80**, No. 35, 4956-4962.
12. Ramey, H.J. Jr.: "How to Calculate Heat Transmission in Hot Fluid Injection," *Pet. Eng.* (Nov. 1964) 110-120.
13. Chappelear, J.E. and Volek, C.W.: "The Injection of a Hot Liquid Into a Porous Medium," *Soc. Pet. Eng. J.* (March 1969) 100-114; *Trans.*, AIME, **246**.
14. Baker, P.E.: "An Experimental Study of Heat Flow in Steam Flooding," *Soc. Pet. Eng. J.* (March 1969) 89-99; *Trans.*, AIME, **246**.
15. Baker, P.E.: "Heat Wave Propagation and Losses in Thermal Oil Recovery Processes," *Proc.*, Seventh World Petroleum Cong., Mexico City (1967) **3**, 459-470.
16. Satter, A. and Parrish, D.R.: "A Two-Dimensional Analysis of Reservoir Heating by Steam Injection," *Soc. Pet. Eng. J.* (June 1971) 185-197; *Trans.*, AIME, **251**.
17. Baker, P.E.: "Effect of Pressure and Rate of Steam Zone Development in Steamflooding," *Soc. Pet. Eng. J.* (Oct. 1973) 274-284; *Trans.*, AIME, **255**.
18. Dillabough, J.A. and Prats, M.: "Proposed Pilot Test for the Recovery of Crude Bitumen from the Peace River Tar Sands Deposit – Alberta, Canada," *Proc.*, Simposio Sobre Crudos Pesados, Inst. de Investigaciones Petroleras, U. del Zulia, Maracaibo, Venezuela (1974).
19. Terwilliger, P.L.: "Fireflooding Shallow Tar Sands—A Case History," paper SPE 5568 presented at the 1975 SPE Annual Technical Conference and Exhibition, Dallas, Sept. 28–Oct. 1.
20. Giguere, R.J.: "An In-Situ Recovery Process for the Oil Sands of Alberta," paper presented at the 26th Canadian Chemical Engineering Conference, Toronto, Oct. 3-6, 1976.
21. Prats, M. Jones, R.F., and Truitt, N.E.: "In-Situ Combustion Away from Thin, Horizontal Gas Channels," *Soc. Pet. Eng. J.* (March 1968) 18-32; *Trans.*, AIME, **243**.
22. Thomas, G.W.: "A Study of Forward Combustion in a Radial System Bounded by Permeable Media," *J. Pet. Tech.* (Oct. 1963) 1145-1149; *Trans.*, AIME, **228**.
23. Closmann, P.J.: "Steam Zone Growth During Multiple Layer Steam Injection," *Soc. Pet. Eng. J.* (March 1967) 1-10.

Chapter 6

Hot-Water Drives

In its simplest form, a hot-water drive involves the flow of only two phases: water and oil. Steam and combustion processes, on the other hand, always involve a third phase: gas. Thus, the elements of hot-water flooding are relatively easy to describe, it being basically a displacement process in which oil is displaced immiscibly by both hot and cold water. Except for temperature effects and the fact that they generally are applied to relatively viscous crudes, hot-water floods have many elements in common with conventional waterfloods.

Because of the pervasive presence of water in all petroleum reservoirs, displacement by hot water must occur to some extent in all thermal recovery processes. It is known to contribute to the displacement of oil in the zones downstream of both steam drives and combustion drives. Thus, many elements of the discussion on hot-water drives presented in this chapter are applicable to appropriate regions in other processes.

The presence of a gas phase is known to affect the performance of waterfloods, and the same is thought to be true of hot-water drives. Only three of the effects that gas exerts on hot-water floods are discussed here. The first is that gases dissolved in the crudes, even heavy crudes, tend to come out of solution as the temperature increases. This results in an apparent initial expansion of the oil phase wherever gas bubbles form, but only until the gas bubbles coalesce into a (continuous) gas phase. Although this "foaming" effect may be significant (and occurs in addition to the expansion of the liquid itself), it appears to have received little attention thus far. The second effect, that of a trapped residual gas phase on hot-water flood recovery, is thought to be analogous to that found in waterflooding. But this effect is not discussed explicitly here, and the reader is referred to the pertinent waterflood literature.[1-3] The third effect is that the displaced oil tends to fill the space initially occupied by gas, which delays significant oil production. Aside from these comments, this chapter considers only the displacement of oil by water in the absence of gas.

The leading edge of the injected hot water loses heat so rapidly that it quickly reaches the initial reservoir temperature. Thus, at the leading edge of the displacement front, the oil mobility is that of the unheated oil. On the other hand, the viscosity of the injected hot water is lower than in conventional waterfloods. Thus, the mobility ratio of the oil ahead of the displacement front and the injected water near the injection well is more unfavorable in hot-water floods than in conventional waterfloods. This should result in somewhat earlier water breakthrough in hot-water floods. And there is some evidence for this. On the other hand, the mobility ratio of the fluids in the heated zones is more favorable in the hot-water floods than in conventional waterfloods. This results in better displacement efficiency from the heated zone and would improve the ultimate recovery even where residual oil saturation does not decrease with increasing temperature.

6.1 Mechanisms of Displacement

The experimental work of Willman *et al.*[4] shows that the improvement in recovery of viscous crudes by means of hot-water floods relative to normal (unheated) waterfloods is primarily due to (1) the improved oil mobility resulting from the reduction in the oil viscosity and (2) the reduction in residual oil at high temperatures.

For convenience, the reduction in residual oil with increasing temperature is discussed first. Thermal expansion of the crude obviously contributes to the reduction in residual oil at high temperatures. A residual oil saturation S_{or} at the initial reservoir temperature T_i occupies a volume $S_{or}[1+\beta_o(T-T_i)]$ at an elevated temperature T. (Here β_o is the volumetric thermal expansion coeffcient of the oil.) Let us say that the residual saturation measured at any temperature is constant. When the formation is allowed to return to its initial temperature T_i, the oil volume will decrease. If the same thermal expansion coefficient applies during the cooling as during the heating cycle (i.e., if the oil has not been altered), the apparent residual saturation would have been

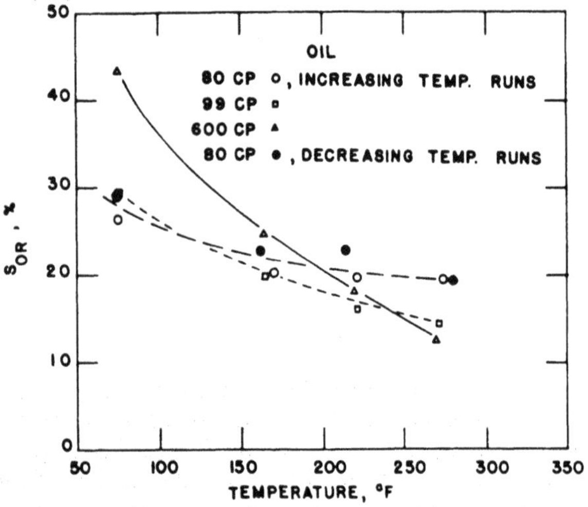

Fig. 6.1 – Residual oil saturation vs. temperature for Houston sand.[5]

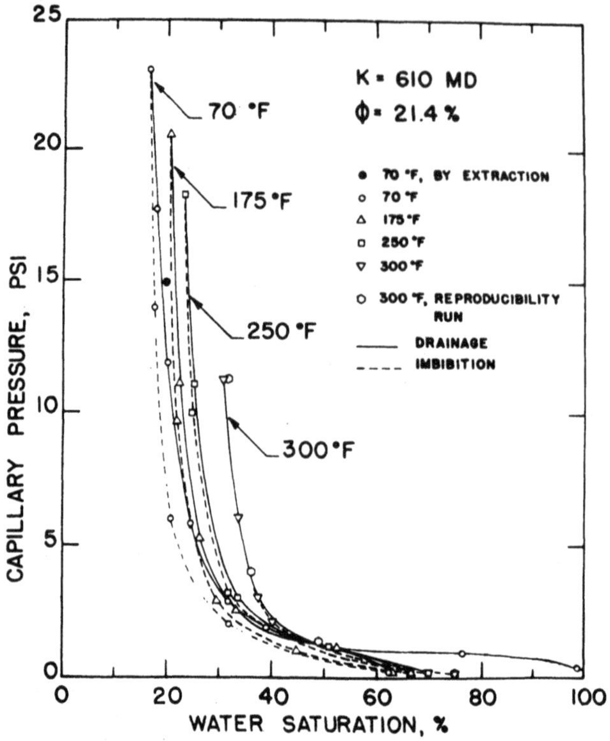

Fig. 6.2 – Capillary pressure vs. water saturation for a Berea sandstone core.[9]

reduced by an amount $S_{or}\beta_o(T-T_i)$. As can be seen from the values of the thermal expansion coefficient for crude oils given in Fig. B.30, a 300°F rise in temperature would reduce the residual oil saturation by 10 to 30% of the residual oil saturation obtained at the initial reservoir temperature.

In some cases, the reductions in residual oil are significantly more pronounced than can be explained by thermal expansion alone. Experimental values reported by Poston *et al.*,[5] Weinbrandt and Ramey,[6] and others[7,8] indicate reductions in residual oil saturations of 50% and higher. Fig. 6.1 shows some results reported by Poston *et al.*[5] Obviously, other phenomena are taking place. The most commonly held view is that the reductions in residual oil with increasing temperature above those explainable by thermal expansion are due to changes in surface forces at elevated temperatures. These surface forces include not only the interfacial ones, between the oil and water phases, but also the forces between the mineral surfaces and the liquids, especially those that may tend to hold complex organic compounds on the mineral surface.

These changes in surface forces do not necessarily reduce the capillary forces, since it appears that some of the rock/fluid systems studied become more water wet as the temperature increases. Shifts in capillary pressures and relative permeabilities in the direction of increasing water wetness with increasing temperature have been reported by Sinnokrot *et al.*[9] and Poston *et al.*,[5] respectively. Typical results are illustrated in Figs. 6.2 and 6.3, in which the increasing water wetness with increasing temperature is indicated by the increasing values of the irreducible water saturation. Because all water relative permeability curves tend to vanish at low saturations, it is difficult to show how the water residual saturations change with temperature from a conventional relative permeability plot such as the one in Fig. 6.3. However, these changes are easily discernible from the water saturation at an oil relative permeability of unity.

Fig. 6.4, adapted from Combarnous and Sourieau,[10] shows schematically how (1) thermal expansion, (2) viscosity reduction, (3) wettability, and (4) oil/water interfacial tension affect the displacement efficiency of crudes of different density. Qualitatively, thermal expansion is more important in light crudes, whereas in the heavy crudes the viscosity reduction and wettability changes are more important.

But much remains to be done to provide an adequate understanding of how increased temperatures affect both the magnitude and the direction of the change of those rock-fluid properties that affect the flow of fluids through a permeable rock or sandpack.

There are no published experimental results on the effect of increasing temperature on the displacement efficiency over a wide range of oil viscosities. The effect of oil/water viscosity ratio from 1 to 500 on the scaled displacement process in sandpacked tubes has been presented by Croes and Schwarz[11] for constant-temperature conditions. Their results, shown in Fig. 6.5, ignore the effects of temperature on both thermal expansion and changes in surface forces. At high oil viscosities, these results are useful in showing the early breakthrough of water and the

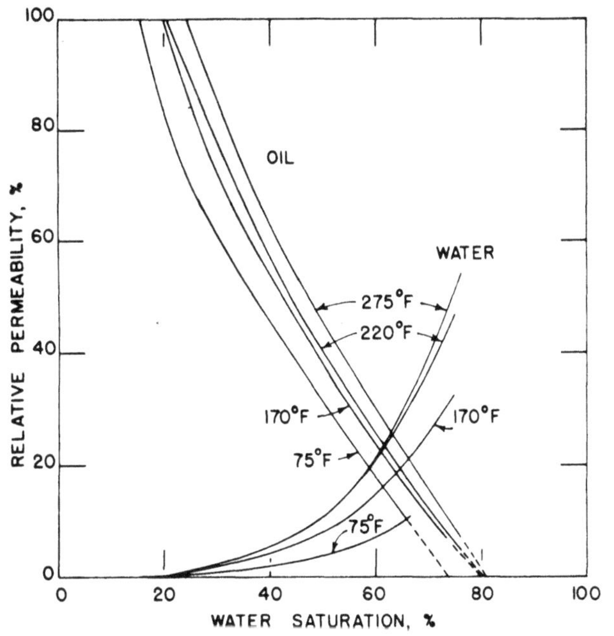

Fig. 6.3 – Water and oil relative permeability vs. water saturation, Houston sand, 80-cp oil.[5]

recovery of the bypassed oil by the water.

An informative way to acquire insight into the behavior of hot-water displacements is to compare results of hot and conventional waterfloods – e.g., those of Combarnous and Pavan.[8] In calculating oil/water displacements, Combarnous and Pavan assume that the temperature is known as a function of distance and time. They then solve the fractional flow equation in a linear system, considering that the fluid mobilities change not only with saturation but also with temperature. They take into account residual saturations and oil viscosity changes with temperature. Fig. 6.6 gives calculated results of three cases of water displacement. Temperature conditions differ from case to case. The saturation distribution after 90 minutes of displacing oil by water is given for all cases. These were calculated in the absence of capillary forces and, thus, yield sharp saturation fronts, as found through typical Buckley-Leverett[12] construction in conventional displacements. Case C is an isothermal displacement at 68.9°F. Case A is at 176°F, so that the oil/water viscosity ratio is lower than in the first case. Note that the water saturation attains higher values at the higher temperatures, reflecting a lower residual oil saturation. Note also that in Case A the larger amount of displaceable oil at the higher temperature results in a delay in the leading edge of the sharp displacement front.

In Case B in Fig. 6.6, oil initially at 68.9°F is displaced by water injected at 176°F. The lower curve (Curve D) shows that water cools considerably as it flows through the system, losing essentially all its heat near the displacement front. This lag between the fluid and heat fronts is due to the convective nature of the heat transport in hot-water floods. The phenomenon is described in some detail in Sec. 3.2 and is represented schematically in Fig. 3.2. Heat fronts generally cover only about one-third of the volume swept by the injected fluid. Because the

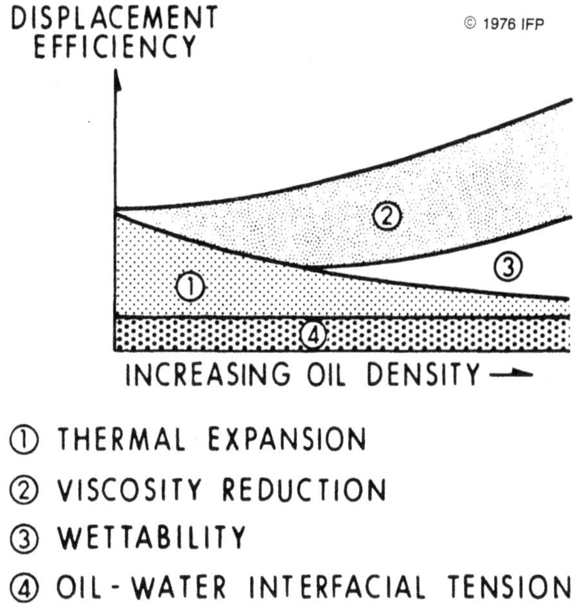

Fig. 6.4 – Relative contributions of mechanisms on the displacement efficiency of oil by hot water.[10]

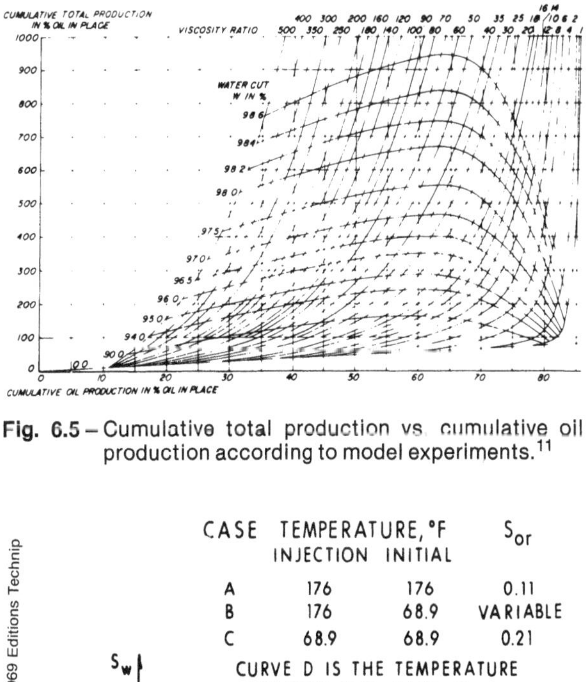

Fig. 6.5 – Cumulative total production vs. cumulative oil production according to model experiments.[11]

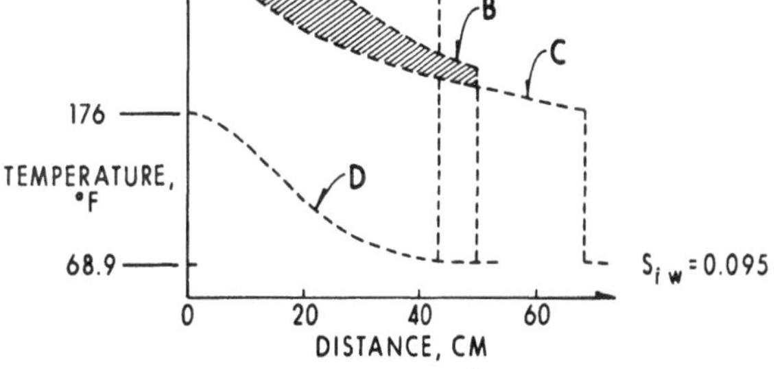

Fig. 6.6 – Calculated distributions of water saturations and temperatures resulting from displacement of oil by water at several injection and initial temperatures.[8]

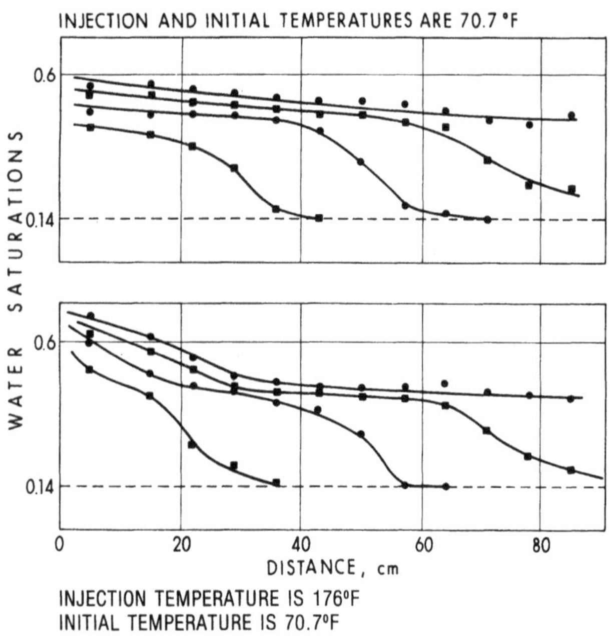

Fig. 6.7 – Observed saturation distributions at 30-minute intervals resulting from oil displacement by hot and cold water.[8]

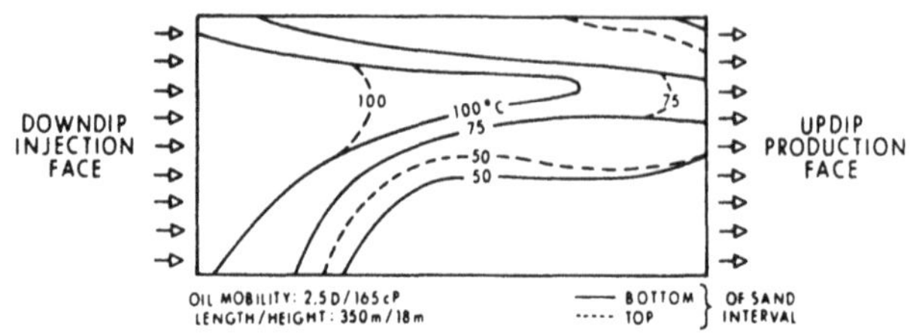

Fig. 6.8 – Plan view of temperatures after injection of 4 PV of hot water in a line-drive scaled laboratory experiment (properties are those of the prototype).[13]

temperature drops with increasing distance from the inlet face, the oil displacement is intermediate between those in Cases C and A. Thus, the location of the leading edge of the front is also intermediate. As is discussed later, experiments show that the oil saturation at the inlet face attains its residual value more slowly than in these calculations. Thus, the water front would be more advanced than indicated by the result shown for Case B.

In Case B, the water saturation profile has a rather pronounced inflection point – the point on the smooth curve at which the slope is the steepest – which has been described as a second displacement front. This second displacement front is associated with the location of the effective heat front and is referred to as the hot-water displacement front.[8] This name distinguishes it from the leading displacement front, which is always at essentially the initial reservoir temperature.

Experimental results corresponding to the temperature conditions of Cases C and B shown in Fig. 6.6 also have been provided by Combarnous and Pavan[8] and are shown in Fig. 6.7. As in the calculations, two displacement fronts can be recognized in the hot-water flood, but not in the cold-water flood. Note that the leading edge of the cold displacement front is not so sharp as the calculated ones shown in Fig. 6.6. The main reason for this dispersion is probably the presence of capillary forces in the experiments. Note also that the residual oil saturation is attained very slowly even near the injection face.

Other experimental results of Combarnous and Pavan (not presented here) show that the higher the temperature of the hot water, the earlier the water breakthrough. (The initial temperature was approximately the same in all the systems.) Although the differences in water breakthrough were small, they were consistent. Their results suggest that viscous instabilities actually may grow faster in hot-water floods than in conventional waterfloods. One possible explanation for this reported effect is that the portion of a water finger that is heated has less flow resistance than it would if it were cold, and the lowered flow resistance would accentuate the rate of growth of the more advanced fingers. As the oil is heated, however, its reduced viscosity and increased volume enhance the displacement of the bypassed oil. Thus, although the fraction of the reservoir swept at breakthrough appears to be slightly worse – at least in some experimental hot-water floods – improved displacement of the heated bypassed oil has the potential of yielding higher ultimate recoveries.

Any instabilities observed in linear tube experiments must be of small cross-sectional area, since tubes of small diameter generally have been used.[8,11] Where results of multidimensional scaled experiments of the hot-water process have been reported,[13] it appears that the hot water follows the paths created by the instabilities of the preceding cold-water flood. Since hot water cools faster in the smaller fingers, the higher temperatures are found in a small number of the larger channels, from which the intervening spaces are heated slowly. Reported[13] temperature distributions near the top and bottom of a line drive hot-water flood in a dipping reservoir are shown in plan view in Fig. 6.8. These results show the underrunning of the water (which is denser than the oil) and one large heated finger in the upper right quadrant.

These conclusions can be drawn about hot-water floods: (1) there are two recognizable displacement fronts, (2) the leading front is at the initial temperature of the reservoir, (3) the hot-water front substantially lags the cold-water front, (4) large volumes of injected hot water may be required to bring the oil saturation to its residual value even near an injection well, (5) oil is displaced throughout the entire zone swept by the injected water, and (6) the effect of instabilities appears to be quite important even in homogeneous formations. Items 4 through 6 are expected to be more pronounced the higher the oil viscosity. Also, they are not inconsistent with reported field observations.[14,15]

6.2 Performance Prediction

There are three essentially different approaches to estimating the performance of a hot-water drive. One approach, proposed by van Heiningen and Schwarz[16] makes use of the effect of oil viscosity on the isothermal recoveries as reflected, for example, in Figs. 6.5 and 6.9. The method calls for shifting from one viscosty ratio curve to another of lower value in a manner corresponding to the changes in the average temperature of the reservoir (which increases with time). In applying this procecure, the oil/water viscosity ratio as a function of temperature and the average reservoir temperature as a function of time are the principal items required. The procedure, which can be inferred from Fig. 6.9, clearly considers only viscosity effects, although the effect of thermal expansion of the fluids on the recovery can be included easily. Van Heiningen and Schwarz's procedure is easy to apply, but it is valid only where recovery curves such as the ones given in Fig. 6.5 are representative of the formation being considered. Furthermore, and this is true of all predictive methods, the recoveries must be reduced to account for variation in sweep efficiency resulting from well patterns and for the adverse effect of reservoir heterogeneities.

The second approach also is borrowed from waterflood technology and is based on the isothermal Buckley-Leverett displacement equations.[12] Modified forms of those equations for application to hot-water drives were first introduced by Willman *et al.*[4] in 1961; they have been used frequently as a fairly simple way of estimating the recovery performance of hot-water drives in linear and radial systems.[17,18] It is emphasized that estimates of recoveries from linear and radial flow systems must be reduced to allow for well-pattern and heterogeneity effects. A number of workers[19-21] have taken into account the effect of the well patterns by applying the Buckley-Leverett displacement along the stream channels characteristic of the well pattern, at least for isothermal waterfloods, and it appears that a similar approach could be followed with hot-water drives. The extension to general well patterns, though providing more significant results, requires the use of high-speed computers.

The third approach to estimating the performance of a hot-water drive is to use thermal numerical simulators. The simulators are capable of calculating more accurate recovery performances than can be achieved by the two simpler methods just discussed.

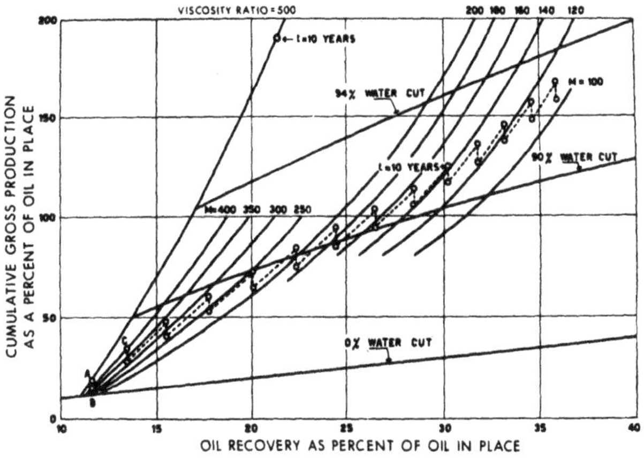

Fig. 6.9 – Cumulative gross production vs. oil recovery.[16]

However, they have two limitations: high cost (especially the cost of preparing the required data for input) and the quality of the input data (i.e., the results are no better than the data entered).

The Buckley-Leverett approach for estimating the oil displaced by hot water is discussed in the next section; the approach is useful as a screening procedure even where thermal numerical simulators are available.

Displacement by Hot Water

The approach proposed by Willman *et al.*[4] consists of discretizing the reservoir into n zones, each at a constant and uniform temperature T_j, different from zone to zone. The location and size of these zones vary with time in a manner consistent with energy balance considerations. Within each constant temperature zone, the isothermal Buckley-Leverett displacement equations apply and the rate of growth of the saturation front is a constant. But the rates of growth of the area encompassed by the saturation fronts change as each new temperature zone is entered. For linear or radial flows, the rate of growth of the saturation fronts, at a temperature T_j, is given by

$$\frac{dA}{dt} = \langle 1.289 \times 10^{-4} \text{ acre-ft/bbl} \rangle \cdot \frac{1}{\phi h_n} q(T_j) \frac{\partial f_w(S, T_j)}{\partial S}, \quad \text{.......... (6.1)}$$

where

$\frac{dA}{dt}$ = rate of growth of the area encompassed by the saturation front having a water saturation S, acres/D,
ϕ = porosity,
h_n = net sand thickness, ft,
f_w = fractional flow of water, which in hot-water floods depends on both saturation and temperature,
q = flow rate at ambient reservoir temperature and pressure, B/D and
T_j = is the temperature level of the jth zone, °F.

TABLE 6.1 – FORMATION PROPERTIES FOR EXAMPLE 6.1

Initial oil saturation S_{oi}	0.65
Initial water saturation S_{wi}	0.35
Irreducible water saturation S_{iw}	0.20
Waterflood residual oil saturation S_{or}	0.30
Porosity	0.243
Gross reservoir thickness, ft	33
Net reservoir thickness, ft	28
Absolute permeability, md	1300
Pattern area, acres	1.8
Initial reservoir temperature T_i, °F	117
Dip, degrees	0
Volumetric heat capacity of the reservoir M_R, (Btu/ft-°F)	40.2
Volumetric heat capacity of adjacent formations M_S, (Btu/cu ft-°F)	38.7
Thermal conductivity of overburden λ_S, (Btu/ft-D-°F)	24.2

For conventional waterfloods where the reservoir temperature is not considered to change with time, Eq. 6.1 reduces to the following familiar form.

$$\frac{dA}{dt} = \langle 1.289 \times 10^{-4} \text{ acre-ft/bbl} \rangle \cdot \frac{i}{\phi h_n} f_w'(S), \quad \ldots \ldots (6.2)$$

where

$$f_w'(S) = \frac{\partial f_w(S, T_i)}{\partial S}, \quad \ldots \ldots (6.3)$$

T_i being the original reservoir temperature. This is the form of the Buckley-Leverett displacement equation that describes the rate of advance of a saturation front. Since the Buckley-Leverett equation is not linear, its results cannot be superimposed.

Inherent in any application of the Buckley-Leverett displacement equation is the requirement that the fluids be of constant density – i.e., no thermal expansion or change of density is allowed. In thermal applications, this condition is applied within each constant-temperature zone, but the densities are allowed to differ from zone to zone.

For constant-density fluids and for negligible gravity and capillary effects, the fractional water flow is given by[12]

$$f_w(S, T) = \frac{1}{1 + [M(S, T)]^{-1}}, \quad \ldots \ldots (6.4)$$

where $M(S, T)$, the mobility ratio of the coflowing fluids, is given by

$$M(S, T) = \frac{k_{rw}}{\mu_w} \frac{\mu_o}{k_{ro}}. \quad \ldots \ldots (6.5)$$

The viscosity ratio of oil to water has a strong dependence on temperature, especially for viscous crudes. The relative permeability ratio, for convenience, usually is taken to be a function only of saturation, although, as shown in Section 6.1 and Fig. 6.3, it also depends on temperature.

Calculation of the area swept by the saturation fronts from Eq. 6.1 requires information about $\partial f_w / \partial S$ over a saturation range at each temperature T_j. Generally these slopes are determined graphically

TABLE 6.2 – RELATIVE PERMEABILITY DATA AND CALCULATED FRACTIONAL FLOW* FOR EXAMPLE 6.1

Water Saturation S	Reduced Saturation S_r**	k_{rw}/k_{ro}†	M‡	f_w	$(k_{rw}/k_{ro})'$	$(\log k_{rw}/k_{ro})'$	f_w'	k_{rw}
0.20	0	0	0	0	0	∞	0	0
0.21	0.02	2.00(–6)	8.23(–3)	8.16(–3)	6.04(–4)	695	2.45	2.40(–6)
0.25	0.10	2.57(–4)	1.06	0.514	1.61(–2)	145	15.7	3.00(–4)
0.28	0.16	1.10(–3)	4.53	0.819	4.48(–2)	94.0	6.05	1.23(–3)
0.30	0.20	2.33(–3)	9.59	0.906	7.49(–2)	77.4	2.98	2.40(–3)
0.325	0.25	4.63(–3)	19.1	0.950	1.30(–1)	64.5	1.33	4.69(–3)
0.35	0.30	8.61(–3)	35.4	0.973	0.211	56.4	0.654	8.10(–3)
0.36	0.32	1.08(–2)	44.5	0.978	0.253	53.9	0.503	9.83(–3)
0.37	0.34	1.34(–2)	55.2	0.982	0.302	51.8	0.394	1.18(–2)
0.38	0.36	1.66(–2)	68.3	0.986	0.361	50.2	0.310	1.40(–2)
0.39	0.38	2.03(–2)	83.6	0.988	0.428	48.6	0.247	1.65(–2)
0.40	0.40	2.47(–2)	102	0.990	0.508	47.4	0.199	1.92(–2)
0.42	0.44	3.61(–2)	149	0.993	0.712	45.4	0.131	2.56(–2)
0.44	0.46	5.22(–2)	215	0.995	1.00	44.2	8.85(–2)	3.32(–2)
0.45	0.50	6.25(–2)	257	0.996	1.19	43.8	7.33(–2)	3.75(–2)
0.50	0.60	1.53(–1)	630	0.998	2.92	44.0	3.02(–2)	6.48(–2)
0.55	0.70	3.97(–1)	1630	0.999	8.37	48.6	1.29(–2)	0.103
0.60	0.80	1.23	5060	1.00	32.9	61.5	5.27(–3)	0.154
0.61	0.82	1.61	6630	1.00	46.3	66.3	4.34(–3)	0.165
0.62	0.84	2.16	8890	1.00	67.8	72.3	3.53(–3)	0.178
0.63	0.86	2.98	12,300	1.00	104	80.1	2.84(–3)	0.191
0.64	0.88	4.29	17,700	1.00	169	90.7	2.23(–3)	0.204
0.65	0.90	6.51	26,800	1.00	299	106	1.72(–3)	0.219
0.70	1.00	∞	∞	1.00	∞	∞	0	0.300

*Calculated at the initial reservoir temperature of 117°F.

**$S_r = (S - S_{iw})/(1 - S_{or} - S_{iw})$.

†2.00(–6) represents 2.00×10^{-6}

‡Uses $\mu_w/\mu_o = 2.43 \times 10^{-4}$ from Table 6.4.

from a plot of the fractional flow f_w against saturation S, in which case one such plot is required for each T_j. A short cut is sometimes available. When the relative permeability ratio is considered to be independent of temperature, values of $f_w(S)$ and $\partial f_w/\partial S$ at different temperatures can be generated from those at the initial reservoir temperature by means of the relations

$$f_w(S,T)=\frac{f_w(S)}{f_w(S)+F_\mu[1-f_w(S)]} \quad \ldots\ldots\ldots (6.6)$$

and

$$\frac{\partial f_w(S,T)}{\partial S}=\frac{F_\mu f_w'(S)}{\{f_w(S)+F_\mu[1-f_w(S)]\}^2}, \quad \ldots\ldots (6.7)$$

where

$$F_\mu=\left(\frac{\mu_o}{\mu_w}\right)_{T_i}\bigg/\left(\frac{\mu_o}{\mu_w}\right)_T, \quad \ldots\ldots\ldots\ldots\ldots\ldots (6.8)$$

$$f_w(S)=f_w(S,T_i), \quad \ldots\ldots\ldots\ldots\ldots\ldots\ldots\ldots (6.9)$$

and f_w' is given by Eq. 6.3.

Example 6.1 – Calculation of Crude Displaced by Hot Water

Determine the oil displaced from the heated zone adjacent to a hot-water injection well, considering radial flow of the fluid and the temperature distribution given by Eqs. 5.22, 5.25, and 5.27, after a 2-year injection of 250 BWPD (measured at 60°F). The injection temperature is 392°F at the bottom of the well. Reservoir data, relative permeability, and crude properties are given in Tables 6.1 through 6.3. The reservoir temperature is 117°F. The relative permeability curves are dependent only on the reduced saturation S_r, which is defined in a footnote to Table 6.2. The residual water saturation increases and the residual oil saturation decreases linearly by 0.10 over the temperature range from 117 to 392°F. Thus, $1-S_{or}-S_{iw}$ remains constant. Expected values of crude and water properties are evaluated at selected temperature levels from 117 to 392°F. These are listed in Table 6.4.

The first step is to divide the temperature range from 117 to 392°F into a number of zones of constant temperature. Select five temperature levels: 144, 200, 254, 310, and 364°F. These temperatures were chosen to be the average values of the zones, which have increments of 55°F. The boundaries of these constant-temperature zones are the 117, 172, 227, 282, 337, and 392°F isotherms, whose position in time is calculated from Eq. 5.22. The areal locations of the isotherms as a function of time are plotted in Fig. 6.10. They are calculated using the definitions of the dimensionless radial distance and time given by Eqs. 5.27 and 5.6, the formation properties given in Table 6.1, and the injection parameters, in the following manner.

TABLE 6.3 – CRUDE PROPERTIES FOR EXAMPLE 6.1

API gravity, degrees	11.8
Viscosity, cp	
at 117°F	2,350
at 392°F	6.1
Specific heat, Btu/lbm·°F	0.48
Coefficient of thermal expansion, °F^{-1}	3.6×10^{-4}
Density is independent of pressure	

The relationship between area and dimensionless distance is

$$A=\pi r^2$$

$$=(1.289\times10^{-4}\text{ acre-ft/bbl})\cdot\frac{iM_wh_tM_R}{4\alpha_SM_S^2}x_D$$

$$=[(1.289\times10^{-4}\text{ acre-ft/bbl})\,(250\text{ B/D})$$

$$\cdot(62.4\text{ Btu/cu ft-°F})\,(33\text{ ft})$$

$$\cdot(40.2\text{ Btu/cu ft-°F})x_D]$$

$$\div\left[4\left(\frac{24.2\text{ Btu/ft-D}\cdot\text{°F}}{38.7\text{ Btu/cu ft}\cdot\text{°F}}\right)\right.$$

$$\left.\cdot(38.7\text{ Btu/cu ft-°F})^2\right].$$

$$A\text{ (in acres)}=0.7121x_D.$$

For simplicity in the calculations, this relationship between A and x_D will be considered to be independent of temperature. The relationship between real and dimensionless time is

$$t=\frac{h_t^2M_R^2t_D}{4\alpha_SM_S^2}$$

$$=\frac{(33\text{ ft})^2\,(40.2\text{ Btu/cu ft}\cdot\text{°F})^2\cdot t_D}{4\left(\dfrac{24.2\text{ Btu/ft-D-°F}}{38.7\text{ Btu/cu ft-°F}}\right)(38.7\text{ Btu/cu ft-°F})^2}.$$

$$t\text{ (in days)}=469.8\,t_D.$$

These expressions for A and t in terms of x_D and t_D then can be used to find the corresponding temperatures, using Eq. 5.22 with $\Delta z=0$:

$$T=117\text{°F}+(275\text{°F})$$

$$\cdot\text{erfc}\left(\frac{1.404A}{2\sqrt{2.129\times10^{-3}t-1.404A}}\right)$$

for values inside the radical ≥ 0.

TABLE 6.4 – CRUDE AND WATER PROPERTIES AS A FUNCTION OF TEMPERATURE FOR EXAMPLE 6.1

Temperature (°F)	Oil* Viscosity (cp)	Oil** Density (lbm/ft)	Water† Viscosity (cp)	Water† Density (lbm/ft)	μ_w/μ_o	F_μ	T_D	$Z=\text{erfc}^{-1}T_D$
117	2350	60.4	0.571	61.7	2.43×10^{-4}	1.00	0.0	∞
144	1030	59.8	0.448	61.3	4.35×10^{-4}	1.79	0.1	1.169
200	231	58.5	0.300	60.1	1.30×10^{-3}	5.34	0.3	0.7301
254	62.2	57.3	0.223	58.7	3.59×10^{-3}	14.8	0.5	0.4790
310	23.1	56.1	0.176	57.0	7.60×10^{-3}	31.3	0.7	0.2708
364	9.33	54.9	0.146	55.1	1.56×10^{-2}	64.2	0.9	0.0904
392	6.10	54.3	0.136	54.0	2.19×10^{-2}	90.3	1.0	0.0

*Intermediate values obtained by passing curve given by $\mu=Ae^{B/T}$ to points in Table 6.3. They also could have been obtained by linear interpolation in ASTM charts.

**$\rho_o(T)=\rho_o(60)|1-3.6\times10^{-4}(T-60)|$.

$$\rho_o(60)=\frac{141.5}{131.5+°\text{API}}(62.4)=61.6\text{ lbm/cu ft.}$$

†From Appendix B.

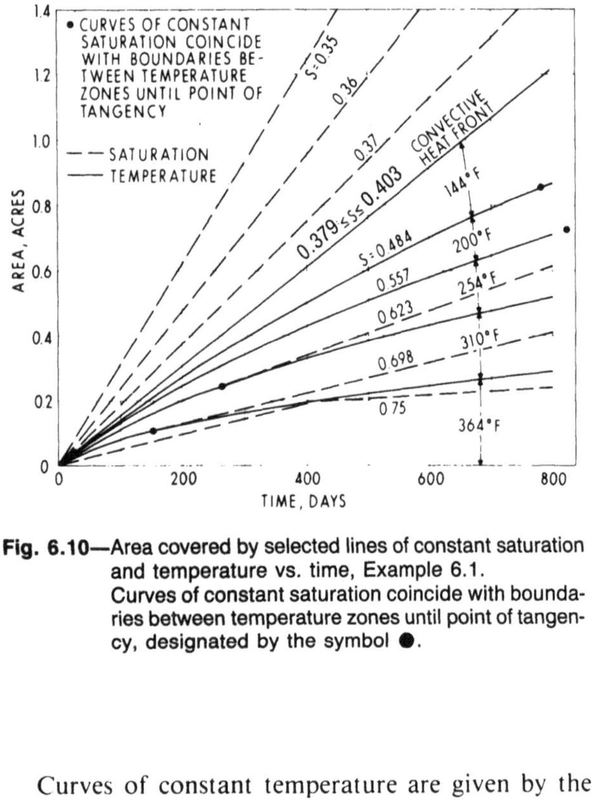

Fig. 6.10—Area covered by selected lines of constant saturation and temperature vs. time, Example 6.1.
Curves of constant saturation coincide with boundaries between temperature zones until point of tangency, designated by the symbol ●.

Curves of constant temperature are given by the combinations of A and t values that give the desired constant temperature. This can be done, for example, by choosing a value of A and then finding the value of t that gives the desired temperature. Such a trial-and-error procedure, in fact, could be used. The equivalent approach used here was to find the value of the quantity

$$Z=\frac{x_D}{2\sqrt{t_D-x_D}}$$

for which erfc Z gives the desired temperature, which is given in the last column of Table 6.4. With the value of Z given, values of t then can be found for any A from

$$t=659.7\,A(1+0.3511\,A/Z^2).$$

The boundaries of the temperature zones shown in Fig. 6.10 were calculated using this last equation. For example, for $T=117°$F it can be found from Table 6.4 that $Z=\infty$. Since $Z=\infty$, the position of the leading edge of the convective heat front at 117°F is found to be related to time by

$$A(117)=t/659.7\text{ acres.}$$

That is, the leading edge of the convective heat front at 117°F grows at a rate

$$\left.\frac{dA}{dt}\right|_{117}=1.516\times10^{-3}\text{acre/D.}$$

For the injection temperature of 392°F, the value of $Z=\text{erfc}^{-1}T_D$ found in Table 6.4 is 0. Thus, the equation relating t and A shows that it takes infinite time to move the 392°F isotherm any distance – i.e., only the injection well is at the injection temperature. The position of intermediate isotherms was obtained in a similar fashion.

The second step is to determine the values of $\partial f_w/\partial S$ required to evaluate Eq. 6.1. One way of doing this involves referring to Table 6.2. The mobility ratio $M(S,T_i)$ was calculated from its definition, given by Eq. 6.5, using values of the water/oil relative permeability ratio at selected saturations and the given fluid viscosities at the initial reservoir temperature. These mobility ratios at 117°F are given in the fourth column of Table 6.2. Corresponding values of the fractional water flow, calculated from Eq. 6.4, are given in the fifth column of Table 6.2.

In principle, values of f_w', given in the eighth column, could have been determined graphically from a plot of f_w vs. S. The fifth column, however, shows that values of f_w' obtained in such a manner would be zero for saturations above 0.55. Accordingly, values of f_w' were evaluated from an equivalent alternative representation,

$$f_w'=\frac{M}{(1+M)^2}\left(\frac{k_{ro}}{k_{rw}}\right)\left(\frac{k_{rw}}{k_{ro}}\right)',$$

where the prime indicates the derivative with respect to the water saturation S. Values of $(k_{rw}/k_{ro})'$ were obtained graphically from a plot of k_{rw}/k_{ro} vs. S. These are listed in the sixth column of Table 6.2. The plot of k_{rw}/k_{ro} vs. S required smoothing for obtaining derivatives. And some portions of the plot did not have enough significant digits for the slopes. Since

$$\left(\frac{k_{ro}}{k_{rw}}\right)\left(\frac{k_{rw}}{k_{ro}}\right)' \equiv 2.303\left(\log\frac{k_{rw}}{k_{ro}}\right)',$$

where again the primes indicate derivatives with respect to S, a second plot was made. This time, $\log k_{rw}/k_{ro}$ was plotted vs. S, and the derivatives were obtained graphically. Values of $(k_{rw}/k_{ro})'$ and $(\log k_{rw}/k_{ro})'$ were checked for consistencies using the last relationship. The accepted values of these quantities, obtained after consistently smoothing the data in both plots, are listed in Table 6.2 in the sixth and seventh columns.

Having values of $\partial f_w/\partial S$ at 117°F, it is now possible to determine the value of the saturation for which the frontal rate of advance at T_i is faster than that of the heat front. The value of $dA/dt|_{117}$ already has been calculated. Using this value and Eq. 6.2, the following is obtained.

$$(1.289\times10^{-4}\text{ acre-ft/bbl})\frac{i}{\phi h_n}f_w'(S)$$
$$=1.516\times10^{-3}\text{ acres/D,}$$

which gives

$$f_w'(S)=(1.516\times10^{-3}\text{ acres/D})(0.243)(28\text{ft})$$
$$\div\left[(1.289\times10^{-4}\text{ acre-ft/bbl})(250\text{B/D})\right.$$
$$\left.\cdot\frac{62.4\text{ lbm/cu ft}}{61.7\text{ lbm/cu ft}}\right]=0.3165.$$

This corresponds to $S=0.379$ (by linear interpolation between the entries in Table 6.2). Note that in evaluating f_w' the injection rate was adjusted to the equivalent volume rate of injection at the original reservoir temperature by multiplying it by the ratio of the densities at 60° and 117°F.

This result means that the isotherm of 117°F and that saturation front $S=0.379$ move together. Thus, the 117°F isotherm in Fig. 6.10 is also a line of constant saturation, the saturation being 0.379. All saturation fronts for which $S<0.379$ travel faster than the convective heat front. Their velocity is calculated from Eq. 6.2, from which their position with time has been calculated and plotted in Fig. 6.10 for saturation of 0.35, 0.36, and 0.37.

The next step is to consider the rate of growth of saturation fronts in regions of different temperatures. The calculation of the characteristic saturation for *one* temperature zone is shown in the following, the other saturations being obtained in similar fashion. The zone at 364°F is chosen to illustrate the procedure. Starting with the highest temperature level has merit, for it allows the early determination of the saturation (if indeed any exists) for which the growth rate exceeds that of the convective heat front. This is accomplished by evaluating Eq. 6.1 at 364°F and setting $dA/dt=1.516\times10^{-3}$ acre/D, which is the previously obtained value of the growth rate of the convective heat front. Then,

$$(1.289\times10^{-4}\text{ acre-ft/bbl})\frac{q(364)}{\phi h_n}\frac{\partial f_w(S,364)}{\partial S}$$
$$=1.516\times10^{-3}\text{ acres/D,}$$

which – upon substitution of the values of $q(364)=283$ B/D, $\phi=0.243$, and $h_n=28$ ft – gives

$$\frac{\partial f_w/\partial S(S,364)}{\partial S}=0.283.$$

At 364°F the characteristic saturation that grows at the same rate as the convective heat front is the value of S for which the last equation holds. This determination is essentially a trial-and-error procedure, which can be aided substantially by interpolation and extrapolation of intermediate guesses.

In this example, the residual oil and water saturations change by 0.10 over the temperature range from 117 to 392°F, whereas $1-S_{or}-S_{iw}$ remains at a constant 0.5. Since the residual oil saturations decrease and water saturations increase with increasing temperature, the saturation entries in Table 6.2 will increase by 0.09 at a temperature of 364°F, which is the proportionate amount of the maximum 0.10 change in residuals between 117 and 392°F. That is, the residual water saturation will be 0.29, and the last entry in the table will be 0.79 instead of 0.70. The important point for us is that at 264°F the values of k_{rw}/k_{ro} can be formed from those given in Table 6.2 by increasing those saturations by 0.09.

Eqs. 6.3 and 6.5 through 6.9 then are used to account for changes in the viscosity ratio of the crude. After some trial and error, it can be established that the desired value of S lies between 0.69 and 0.70. At $S=0.69$ it is found that $\partial f_w/\partial S=0.331$ and at $S=0.70$ it is 0.274. Interpolation for the desired value of 0.283 yields a saturation of 0.698.

Accordingly, at 364°F all water saturations lower than 0.698 tend to travel faster than the heat front. These saturation fronts, traveling faster than the heat front, enter cooler regions where the rate of their advance is reduced. This happens immediately upon injection. To determine the position of the $S=0.698$ saturation front, which is based on Eq. 6.1, then requires that $\partial f_w(0.698,T_j)/\partial S$ be calculated for the several temperature levels. Where the only temperature effect on the relative permeability ratio is a shift in the saturation scale, as is the case in the example, Eq. 6.7 can be used to calculate the values of

$$\frac{\partial f_w(0.698,\ T_j)}{\partial S}=\frac{F_\mu f_w'}{[f_w-F_\mu(1-f_w)]^2},$$

TABLE 6.5 – CALCULATION OF RATE OF ADVANCE OF S = 0.698 SATURATION FRONT – EXAMPLE 6.1

i	T_j °F	$0.698-0.1\left(\frac{T_j-117}{275}\right)$	f_w	f'_w	F_μ	$\left.\frac{\partial f_w}{\partial S}\right\vert_{T_j}$	$\left.\frac{62.4}{\rho_w}\right\vert_{T_j}$	$q(T_j)$ (B/D)	$\left.\frac{dA}{dt}\right\vert_{T_j}$ acre/D
1	364	0.608	1.00	4.53(−3)	64.2	2.83(−1)	1.13	283	1.56(−3)
2	310	0.625	1.00	2.97(−3)	31.3	9.30(−2)	1.09	272	4.79(−4)

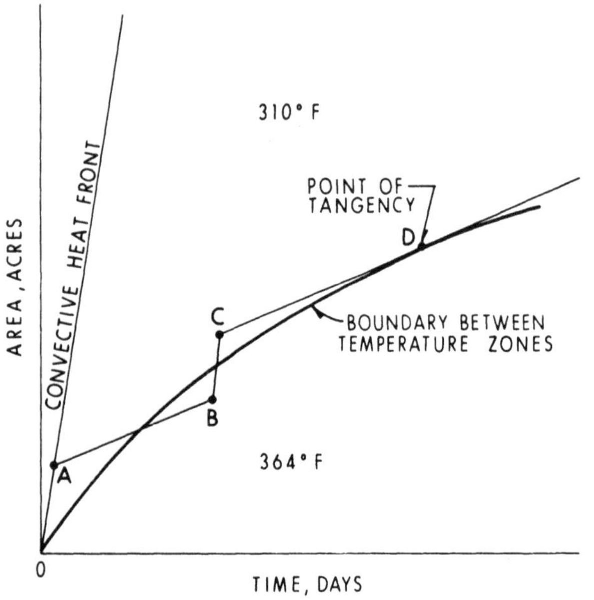

Fig. 6.11—Schematic construction of characteristic saturation Line OABCD at 364°F, Example 6.1.

where now the arguments of f_w and f'_w are

$$S=0.698-0.1\left(\frac{T_j-117}{275}\right), 0.2\leq S\leq 0.7 \text{ ... .}$$

Values of F_μ, $f_w(S)$, $f'_w(S)$, $S(T)$, $q(T)$, and dA/dt are given in Table 6.5. Values of F_μ previously were presented in Table 6.4. Values of $f_w(S)$ and $f'_w(S)$ were obtained by linear interpolation from the last two entries for these quantities in Table 6.2. Values of $q(T)$ were calculated by multiplying the injection rate of 250 B/D by the ratio of the water densities at 60°F and the corresponding temperatures found in Table 6.4.

The procedure for determining the position of the $S=0.698$ saturation front – which corresponds to 364°F – is to start by drawing a line from the origin of Fig. 6.10 having the slope of the rate of advance of the heat front. Details of the subsequent construction in the neighborhood of the origin are given in Fig. 6.11. As stated earlier, the front velocity slows to the velocity of the adjacent temperature zone (310°F) as soon as it moves from the injection well. This means that the advance of the front would tend to follow a Path OAB. As it reenters the 364°F zone it speeds, traveling along Path BC, repeating a zigzag path until it can travel within the 310°F zone without encountering the boundary with the 364°F zone. Since the Steps OA, AB, BC can be taken as small as desired, this means that the front will travel along the boundary between the two temperature zones until the rate of its advance equals the slope of the boundary at Point D. This construction leads to the saturation curve $S=0.698$ shown in Fig. 6.10.

Similar calculations are made for the saturations corresponding to the temperatures of each zone, the results being presented in Table 6.6. The determination of the position of the saturation fronts is also similar to that given for $S=0.698$. Their position is shown in Fig. 6.10. For each temperature zone there is a saturation that tends to move faster than the heat front but is slowed as it enters cooler regions. The saturation fronts are "trapped" on the boundary of two adjacent temperature zones until the rate of advance of the boundary between the two zones drops below the rate of advance of the saturation front at the temperature of the cooler zone. From this point, the point of tangency, the fronts move in the cooler zone at the rate of advance corresponding to the temperature of the cooler zone until it reaches the boundary of the next cooler zone, at which point a similar construction would be used. In the example, this situation does not arise. But it can be inferred from Fig. 6.10 that the position of the $S=0.698$ saturation front given by the dashed line eventually would intersect the boundary between the 254 and 310°F zones. For the zone at 144°F the rate of advance of the boundary with its cooler zone is that of the heat front, which is constant for the conditions in our example. This means that the

TABLE 6.6 – CHARACTERISTIC SATURATIONS AND RATES OF FRONTAL ADVANCE OF TEMPERATURE ZONES – EXAMPLE 6.1

Temperature, °F	144	200	254	310	364
Saturation	0.403	0.484	0.557	0.623	0.698
$\left.\frac{dA}{dt}\right\vert_{T_{i-1}}$, acre/D	1.52(−3)	4.53(−4)	5.47(−4)	6.93(−4)	4.79(−4)
$f_w(S,T_{j-1})$				1.0	1.00

corresponding saturation of 0.403 advances at the rate of the heat front. Since at the original reservoir temperature of 177°F the saturation of 0.379 also travels at the rate of advance of the heat front, all saturations between 0.379 and 0.403 (inclusive) travel at the same rate, giving rise to a shock front. However, that is an artifact of the discretization. For if a cooler temperature zone were considered—between 117 and 144°F—the corresponding saturation would be lower than 0.403. In the limit there would be neither a saturation jump nor a temperature jump for this example.

Saturations higher than 0.698, which correspond to the zone of highest average temperature, 364°F, are constructed in a slightly different manner. The procedure for determining the position of the $S=0.75$ saturation front is to draw a straight line from the origin of Fig. 6.10 having the slope of the rate of advance at 364°F, which is

$$\frac{dA}{dt}=\frac{(1.289\times10^{-4}\text{ acre-ft/bbl})}{(0.243)(28\text{ ft})}$$

$$\cdot(283\text{ B/D})(64.1)(1.72\times10^{-3})(0.8)$$

$$=4.74\times10^{-4}\text{ acres/D.}$$

This $S=0.75$ line, shown in Fig. 6.10, intersects the boundary of the 364°F zone at 410 days, at which point it would enter the zone at 310°F and be slowed to a rate of advance of 1.11×10^{-4} acre/D. A line having this slope is drawn from the point of intersection, even though the resulting line is in the 364°F zone. The two connected dashed lines, then, give the approximate position of the $S=0.75$ saturation front with time. This "bouncing" approach also could have been used in constructing the characteristic saturation of each temperature zone, as discussed in connection with Fig. 6.11.

For the initial saturation of 0.35, the saturation front is found from Eq. 6.2. This and all other fronts that travel faster than the heat front at 117°F are given by single straight lines, as shown in Fig. 6.10.

At the end of 2 years (730 days), the saturations and corresponding swept areas are read from Fig. 6.10 (ex-

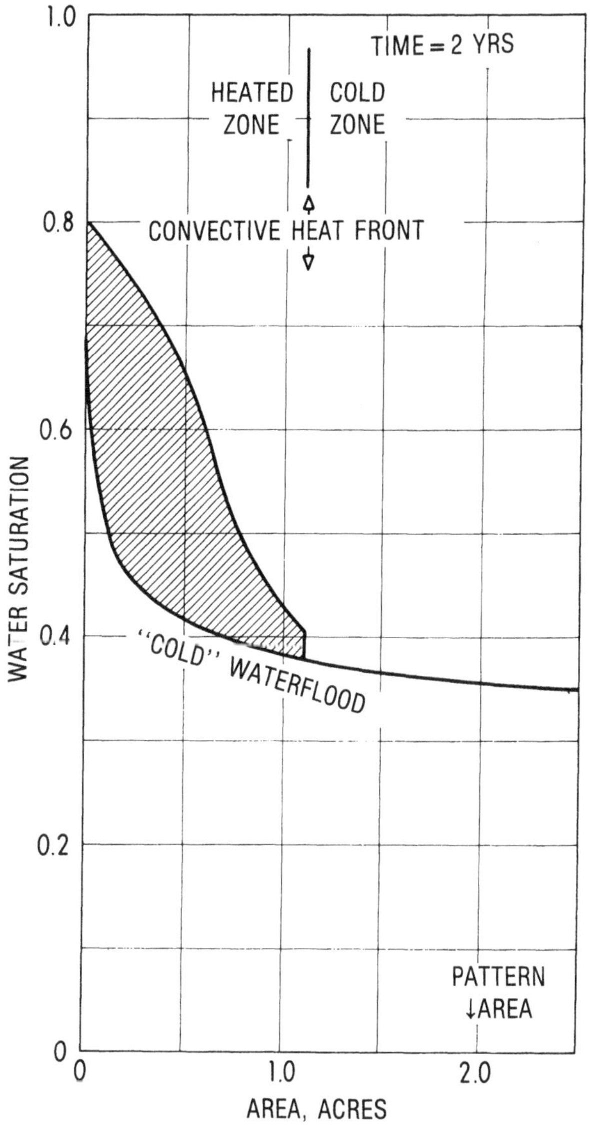

Fig. 6.12—Comparison of saturation distribution in cold- and hot-water floods, Example 6.1.

TABLE 6.7—ESTIMATION OF OIL DISPLACED—EXAMPLE 6.1

T	A (acres)	ΔA (acres)	$S_o(T)$	$\frac{\rho_o(T)}{\rho_o(117)}$	$S_o(117)$	$\Sigma S_o(117)\Delta A$	$\Sigma S_o(T)\Delta A$
364	0.275	0.275	0.232	0.909	0.211	0.058	0.064
310	0.490	0.215	0.303	0.929	0.281	0.118	0.129
254	0.665	0.175	0.385	0.949	0.365	0.182	0.196
200	0.812	0.147	0.480	0.969	0.465	0.251	0.267
144	1.11	0.298	0.560	0.990	0.554	0.416	0.434
117	1.80	0.69	0.633	1.000	0.633		

$$\bar{S}_o(T)=\frac{0.434}{1.11}=0.394 \qquad \bar{S}_o(117)=\frac{0.416}{1.11}=0.375$$

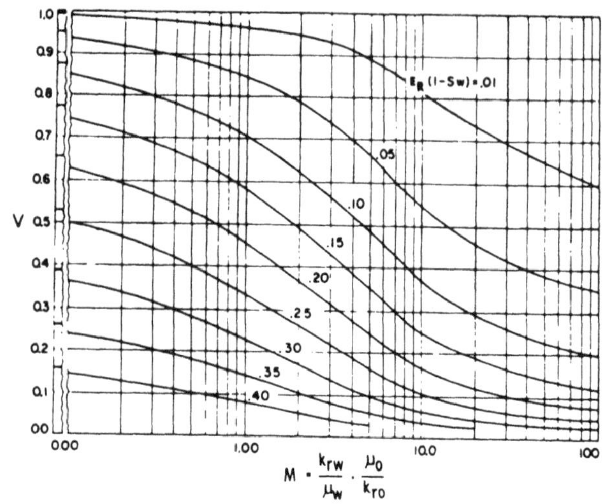

Fig. 6.13 – Permeability variation plotted against mobility ratio, showing lines of constant $E_R(1-S_w)$ for a producing WOR of 1.[1]

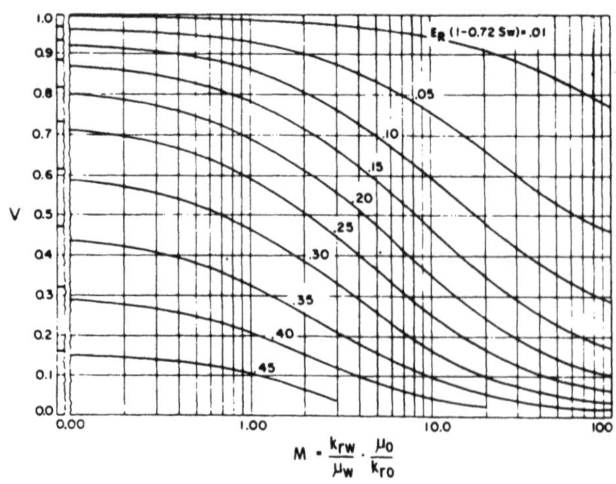

Fig. 6.14 – Permeability variation plotted against mobility ratio, showing lines of constant $E_R(1-0.72\,S_w)$ for a producing WOR of 5.[1]

trapolating or calculating results for low saturations where required) and the results are plotted in Fig. 6.12. Similar calculations give the saturation distribution due to waterflooding, which is also shown in Fig. 6.12. The shaded area represents the improved displacement resulting from the hot-water flood.

The estimate of oil remaining in the reservoir at 2 years is corrected for thermal expansion from the initial reservoir temperature using the oil densities listed in Table 6.4, the area of each temperature zone at 2 years being obtained from Fig. 6.10. Details of the calculations and the average oil saturation over the area of each temperature zone, determined from Fig. 6.12, are indicated in Table 6.7. The average oil saturation in the 1.11 acres of the heated zone is 0.394 and, when referred to the original reservoir temperature of 117°F, is estimated at 0.375. By analogous means, it can be determined that the average oil saturation (referred to 117°F) in the 1.8 acres is 0.474, whereas the one resulting from waterflooding is 0.595.

Since the initial oil saturation is 0.65, the oil displaced from the heated zone after 2 years is given by

$$V_o = (7758 \text{ bbl/acre-ft}) A \phi \Delta S_o h_n$$

$$= (7758 \text{ bbl/acre-ft})(1.11 \text{ acres})$$

$$(0.243)(0.65 - 0.375)(28 \text{ ft})$$

$$= 16{,}100 \text{ bbl at } 117°\text{F}.$$

This displacement from the heated region is supplemented by 619 bbl displaced from the 0.69 acres of the pattern volume swept by the cold water ahead of the heat front. The hot-water flood thus displaced 16,700 bbl of oil, compared with 5,200 bbl that would have been displaced by cold-water flood.

The foregoing example illustrates a procedure for estimating the oil displaced by a hot-water flood and discusses the effects of net/gross reservoir thickness, thermal expansion of the oil, temperature on residual saturations, and temperature on the fractional flow. Note that the effects of gas evolution from the oil as it is heated has been neglected.

Production Performance

The amount of oil displaced in a hot-water drive is invariably larger than the amount produced. As discussed in Chap. 4, the oil that is displaced but not produced is stored in unswept portions of the reservoir. In the case of viscous crudes especially, the mobility ratio between the advancing oil and any gas or water in the reservoir is very favorable. This means that crude would tend to fill regions of the reservoir initially filled with mobile gas and water before it is produced. Where an oil bank forms, consideration of these effects allow an estimation of the recovery history from estimates of the oil displacement history.

Although there are no simple methods for predicting oil recovery from hot-water floods, an approach is suggested that (1) is based on conventional waterflood technology, (2) appears to have some of the elements necessary to describe hot-water floods, and (3) considers only the effects of permeability variations and mobility ratio. It is by no means a proven method and is offered merely as an example of how it may be possible to adapt the existing technology from a related field when there is no other method for making the desired estimates of oil recovery.

As is generally done in applying simple methods to predict the performance of waterfloods,[1] it is assumed here that the crude recovery is not sensitive to the well pattern. And it is assumed that the method developed by Dykstra and Parsons[22] for waterflood performance predictions can be modified for hot-water drives.

The correlations developed by Dykstra and Parsons leading to Figs. 6.13 through 6.16 are based on calculations of layered linear flow models without

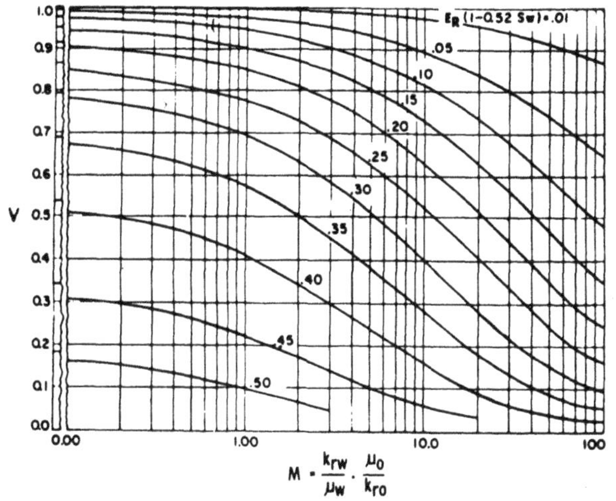

Fig. 6.15 – Permeability variation plotted against mobility ratio, showing lines of constant $E_R(1-0.52\,S_w)$ for a producing WOR of 25.[1]

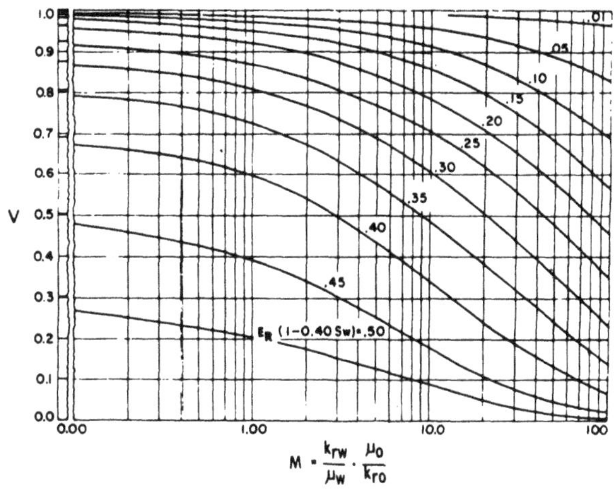

Fig. 6.16 – Permeability variation plotted against mobility ratio, showing lines of constant $E_R(1-0.40\,S_w)$ for a producing WOR of 100.[1]

crossflow. The basic information on the performance of a single layer was based on results of flood pot tests and included measured values of initial liquid saturations, mobility ratios based on displacement to an irreducible saturation, fractional oil recoveries, and produced water cuts. All experimental work was performed on core samples from fields in California, and permeability distributions were based on those samples. Nevertheless, the results have been used extensively to predict waterflood performance elsewhere. The method – which considers the adverse effects of permeability variations, mobility ratio, and initial mobile water saturations – is based on Figs. 6.13 through 6.16. These figures, reproduced from Craig's monograph on waterflooding[1] are based on Johnson's extensions[23] to the original work.

The value of E_R in these figures represents the fractional recovery of the initial oil in place calculated as previously outlined. To use these figures it is necessary to have values of the mobility ratio M and the permeability variations V.

The permeability variation (or coefficient of permeability variation[1,22]) is determined by ordering, in descending order, the permeability values for the reservoir of interest and calculating and plotting the percent of the total number of entries exceeding the value of each entry. An example of such a graph, which must be plotted on semilog probability paper, is shown in Fig. 6.17. The permeability variation V is calculated from

$$V = \frac{\bar{k} - k_\sigma}{\bar{k}}, \qquad (6.10)$$

where $\bar{k}$ is the permeability value exceeded by 50% of the samples and k_σ is the permeability value exceeded by 84.1% of the samples.

Craig[1] defined mobility ratio as the mobility of the displacing water at the average water saturation of the region upstream of the water displacement front divided by the mobility of the displaced oil just downstream of the water displacement front:

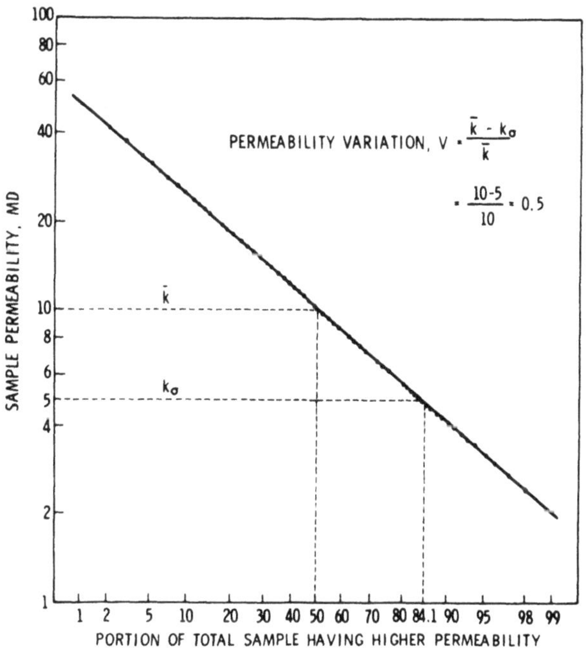

Fig. 6.17 – Log-normal permeability distribution.[1]

$$M_{w,o} = \frac{k_{rw}}{\mu_w}(\bar{S}) \Big/ \frac{k_{ro}}{\mu_o}(S_{wc}). \qquad (6.11)$$

With this definition of mobility ratio, the oil mobility is unaffected by temperature, whereas that of water is increased both because the water viscosity is reduced by the increase in temperature and because there is an increase in the water relative permeability associated with the higher average saturation within the swept zone. That is, the mobility ratio is larger (more unfavorable) for hot-water floods than for conventional waterfloods.

As illustrated by Craig,[1] the crude recovery at the abandonment water/oil ratio (WOR) is obtained by plotting the recovery values (as a fraction of the initial oil in place) calculated at values of WOR equal to 1, 5, 25, and 100 from Figs. 6.13 through 6.16. Because of the more unfavorable mobility ratio in hot-water drives, a recovery figure calculated in this manner would give poorer results than for waterfloods. The improved displacement efficiency of hot-water floods is taken into account (but only approximately) by multiplying the waterflood fractional recovery given in Figs. 6.13 through 6.16 by the ratio of the average oil saturation in the zone affected by the injected water under hot-water drive and waterflood conditions:

$$E_{Rhw}=E_R\frac{\bar{S}_{o,wf}}{\bar{S}_o(T_i)}, \quad \text{(6.12)}$$

where

E_{Rhw} = fractional recovery of oil in place by hot-water drive,
E_R = fractional recovery of oil in place by cold-water flooding given in Figs. 6.13 through 6.16,
$\bar{S}_{o,wf}$ = average oil saturation in the swept zone in a waterflood, and
$\bar{S}_o(T_i)$ = average oil saturation in the swept zone in a hot-water drive, referred to conditions existing in the reservoir prior to injection.

It is emphasized that E_R (and, thus, E_{Rhw}) does not include areal sweep efficiency effects, crossflow, heterogeneity effects other than layering, underrunning of water due to buoyancy, or capture factors. These factors, discussed briefly in Chap. 4, must be estimated separately as is done for waterflooding.

Values of $\bar{S}_o(T_i)$ are determined as in the preceding section. Values of $\bar{S}_{o,wf}$ can be determined using the Buckley-Leverett construction.[12] Craig illustrates how that is done in Section E.3 of his monograph.[1]

For mobility ratios above 100, Figs. 6.13 through 6.16 still can be used, since at large mobility ratios the fractional recovery at a given permeability variation V is approximately constant if WOR/M is constant. For example, the recovery efficiencies at a mobility ratio of 100 and a WOR of 100 are approximately the same as those at a mobility ratio and WOR of 5.

The procedure described above is unsubstantiated, but it appears reasonable. Also, there does not appear to be any other relatively simple method for estimating recovery that considers mobility ratio and nonuniform permeabilities. Although the definition of the mobility ratio given by Eq. 6.11 is not inconsistent with the laboratory observations reported by Combarnous and Pavan[8] nor with several reported field observations,[14,15,24,25] it must be remembered it was chosen only as an analogy to the one used by Craig. Also, the bypassing apparent in the field projects is at least partly due to heterogeneities or other adverse factors. An improved definition of the effective mobility ratio under hot-water flood conditions is required. Such an improved definition may be associated with the flow conditions of the hot-water displacement front, rather than with those of the leading edge of the injected-fluid front used in Eq. 6.11. The method discussed in this section obviously should be viewed as a tentative proposal, which should be tested against reported field experience.

Often, it is sufficient to consider only the amount of oil displaced. If the amount of oil displaced by a hot-water drive is unattractively low, no further calculations are necessary, because the recovery resulting from a hot-water (or any other) drive is always less than the amount displaced. Estimates of oil displaced can be converted to estimates of oil produced by using factors based on experience, the results of analogous operations in similar reservoirs, or on estimates of sweep or recovery efficiencies.[4,17,18]

A simple method proposed by Craig[1] estimates the ultimate recovery from conventional waterfloods by multiplying the maximum possible recovery by a factor $(1-V^2)/M$. Using the mobility ratio given by Eq. 6.11 appears to yield ultimate recoveries for hot-water floods that are lower than those obtained using Eq. 6.12. Ultimate recoveries from a hot-water flood may be governed more by the ultimate temperature in the reservoir (which may approach the injection temperature) than by the temperature at the start of the flood. Using the injection temperature to define the viscosity in the mobility ratio appears to yield estimates of ultimate recovery that are unrealistically high. If such results still look unattractive, further consideration of the project should be abandoned, because a more realistic model is only likely to provide a more pessimistic result. Fig. 6.9, discussed at the start of Section 6.2, also may be used to estimate the upper bounds of recovery from hot-water floods.

Example 6.2 – Calculation of Oil Recovery by Hot-Water Flood

Using the same parameters as in Example 6.1, estimate the crude recovery at a producing water cut of 0.98 for a flood pattern of 1.8 acres. The necessary additional relative permeability data are given in the last column of Table 6.2. The permeability variation is 0.7.

Calculation of the mobility ratio requires the value of the average water saturation in the part of the reservoir affected by the injected water. At the end of 2 years, the entire reservoir has been affected by the injected water, as can be seen in Fig. 6.12. The average oil saturation within the 1.1-acre heated zone, calculated in Table 6.7, is 0.394. The average oil saturation in the remaining 0.7 acres (1.8 acres minus 1.1 acres), which is at the original reservoir temperature of 117°F, is 0.633. This value also is given in Table 6.7. Thus, the volumetric average water saturation in the 1.8 acres is

TABLE 6.8 – ESTIMATION OF PRODUCTION RESPONSE FOR EXAMPLE 6.2

WOR	f_w	a	$(1-aS_{wc})E_R$	E_R	E_{Rhw}*
5	0.833	0.72	0.007	0.009	0.014
25	0.962	0.52	0.02	0.0024	0.038
100	0.990	0.40	0.04	0.047	0.074

*Underlined digits were used in plotting the results in Fig. 6.18 but are not significant. The results from that graph should be read to only *one* significant figure.

$$\bar{S}_w = [(1-0.394)(1.1)+(1-0.633)(0.7)]/1.8$$

$$=0.513.$$

This average saturation is found in the temperature zone of 200°F. The relative permeability to water at a temperature of 200°F and a saturation of 0.513 is found from Table 6.2 at $S=0.513-0.03=0.483$. The value of 0.03 corrects the saturation for the changes in residual saturation between 117 and 200°F as previously discussed in connection with Example 6.1. Linear interpolation of the relative permeability in Table 6.2 gives

$$k_w(0.483)=0.0555.$$

The relative permeability to oil at the initial reservoir saturations and temperature is

$$k_{ro}(0.35)=k_{rw}(0.350)\Big/\left.\frac{k_{rw}}{k_{ro}}\right|_{S=0.35}$$

$$=8.10\times10^{-3}/8.61\times10^{-3}$$

$$=0.941.$$

The mobility ratio calculated from Eq. 6.11 is then

$$M_{w,o}=\frac{(0.0555)}{(0.300\text{ cp})}\frac{(2350\text{ cp})}{(0.341)}=462,$$

where the corresponding viscosities have been taken from Table 6.4. The mobility ratio is highly unfavorable, and it is unwarranted to refine it to account for possible variations in its value during the course of a hot-water drive. The recovery at WOR = 100 and $M=500$ is about the same as that for WOR = 5, $M=25$. At a permeability variation of 0.7, the value of E_R $(1-0.40\ S_{wc})$ at WOR = 100 and $M=500$ is approximately 0.04. At WOR = 25 and $M=500$, E_R $(1-0.52\ S_{wc})\simeq 0.02$. At WOR = 5 and $M=500$, E_R $(1-0.72\ S_{wc})\simeq 0.007$.

The average oil saturation in the 1.8 acres due to a waterflood was reported (see Example 6.1) as 0.595, and the saturation (referred to T_i) due to a hot-water drive in the 1.11 acres covered by the convective heat front was reported at 0.375. Thus, the values of the fractional recovery E_R obtained from Figs. 6.12 through 6.15 is corrected by the ratio of 0.595/0.375 = 1.58. Results are listed in Table 6.8

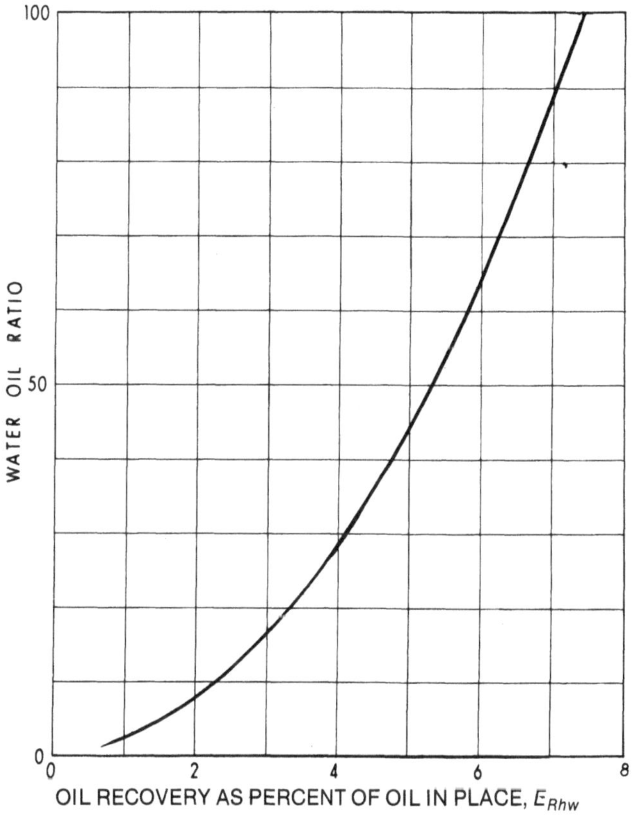

Fig. 6.18 – Estimates of predicted WOR vs. recovery – hot-water flood, Example 6.2.

and plotted in Fig. 6.18, from which it may be determined that the expected recovery at a fractional flow of water (water cut) of 0.98 (or a WOR of 49) is only about 5% of the initial oil in place.

The crude produced from the 1.8 acres after 2 years of hot-water injection is

$$N_p=(0.05)(1.8\text{ acres})(1{,}225\text{ bbl of oil/acre-ft})$$

$$\cdot(28\text{ ft})=3{,}700\text{ bbl}.$$

The recovery is very low. But this should be no surprise since all water cuts, even those at the highest temperature, are greater than 0.9.

This example illustrates (1) how to estimate the recovery from a hot-water drive at a predetermined abandonment water cut, (2) that reservoirs containing crudes of 2,000 cp and higher at initial reservoir conditions are not likely to be candidates for hot-water drive, (3) that the displacement and

TABLE 6.9 – HOT-WATER FLOOD OPERATIONS

Project	Location	(Chap. 6) Reference
Loco	Oklahoma	14
Kern River	California	24
Schoonebeek	Holland	15
N. E. Butterly	Oklahoma	25
Emlichheim	Germany	28
Arlansk	USSR	29

flow of viscous crudes occurs at relatively high water cuts, (4) that the displacement efficiency of hot-water drives is higher than that of waterfloods, (5) that the available simple methods can provide only a crude estimate of the recovery expected from a hot-water drive, and (6) that the mobility ratio of hot-water drives as defined here is more unfavorable than that of waterfloods.

6.3 Design

The same matters must be considered in designing hot-water floods as in designing any other fluid-injection project. These are illustrated for the steam drive process in Example 7.2. Some of the principal things to consider are the possibility of using existing wells as they become available or after they have been worked over and recompleted, the need for additional wells to reduce the spacing or to improve recovery, the effect of depth and average reservoir injectivity on the project life and economics, the type and location of surface facilities to be used, the availability and treatment of the water, and the environmental constraints on the use of fuel and the disposal of effluent. In thermal recovery projects, the principal additional considerations include the larger rates of fuel consumption, the need for cooling hot effluent, and the effect of temperature on the surface and downhole hardware.

From a reservoir point of view, any project design must consider how the performance is affected by the interaction between well spacing, reservoir depth, and formation injectivity. The reservoir depth generally determines the maximum injection pressure that can be used without fracturing the formation. This and the injectivity I of the formation determine the fluid injection rate according to the relation

$$i = I\Delta p. \qquad (6.13)$$

The injectivity (or injection rate per unit pressure drop) is discussed in some detail in Chap. 4.

It appears there is no published work specifically related to the determination of injectivity during a hot-water flood, but the concepts are the same for any frontal displacement process. Thus, where necessary, curves of approximate injectivity vs. time can be constructed using procedures entirely analogous to those discussed in Chap. 4. In heavy-oil reservoirs containing high crude saturations, the injectivity under piston-like displacement (frontal advance conditions) will be controlled primarily by the flow resistance of the crude. The injectivity generally would be too small to make any displacement process attractive, unless induced fractures, naturally existing fractures, or high-permeability water zones provide sufficient injectivity. Of course, such permeable paths then would promote bypassing of the crude by the injected (hot) fluids, but this sometimes can be used to advantage in the recovery process.[26,27] Under bypass conditions, the injectivity is determined primarily by the effective transmissibility of the injected fluid.

Project design, from the reservoir point of view, starts with a model of recovery performance as affected by spacing, heat injection rate, temperature of the injected fluid, pressure levels, and any other operating condition controllable by the engineer. Pressure differences and injectivity must be great enough to ensure injection rates that are high enough to effect a reasonably short project life and adequate thermal efficiency. Close spacing of wells tends to shorten the operating life and reduce heat losses to adjacent formations. The final choice of operating parameters is dictated by economics and often is influenced greatly by the number and location of wells available in the field for the project.

Large projects now are designed with the aid of thermal numerical simulators or (less and less often) with scaled physical models. The initial design of the large projects often is screened by using techniques such as the ones described in this chapter.

6.4 Examples of Field Applications

Hot-water flooding has not been a popular thermal recovery process. Only a few field pilots and commercial-size operations have been described or alluded to. Some of these field applications are listed in Table 6.9.

The first four of these projects are reported to be discontinued, and little information is readily available on the USSR and German operations. Severe channeling and high WOR's, which are indicative of poor sweep efficiencies, characterize the first four projects. Heat recuperation by cold-water follow-up has not been reported. At the Loco pilot,[14] total thermal recovery after the 1-year hot-water flood in a previously waterflooded thin sand (12.9 ft net, 1100 bbl/acre-ft) amounted to 156 bbl/acre-ft. Heat losses from this thin reservoir were reported to be about 60% of the injected heat. At the Northeast Butterly Oil Creek Unit, the hot-water drive phase of the project lasted about 4 years and produced less than 150,000 bbl of oil.[25] Most of the 375,000 bbl of thermal oil produced from the project resulted from cyclic hot-water stimulation, which included converting the injector in the original hot-water drive to production. At Kern River,[24] injection of 2.23×10^6 bbl of hot water in about a year at an average temperature of 300°F resulted in an oil recovery of 40,260 bbl. The pilot was terminated because of its poor performance.

The Schoonebeek hot-water drive[15] attained an

injection capacity of 95,000 B/D in 1966. In 1966, after about 10 years of operation, the oil recovery attributable to the hot-water drive was nearly 1.25×10^6 bbl, with substantial additional recovery still expected from any future cold-water follow-up to recuperate heat and continue the hot-water drive. Of the first four operations listed in Table 6.9, the crude viscosity at the initial reservoir temperature was lowest at Schoonebeek – 175 cp compared with 600, 2,000, and 4,060 cp at Loco, Northeast Butterly, and Kern River, respectively. The performance of the hot-water drive at Schoonebeek is shown in Fig. 6.19.

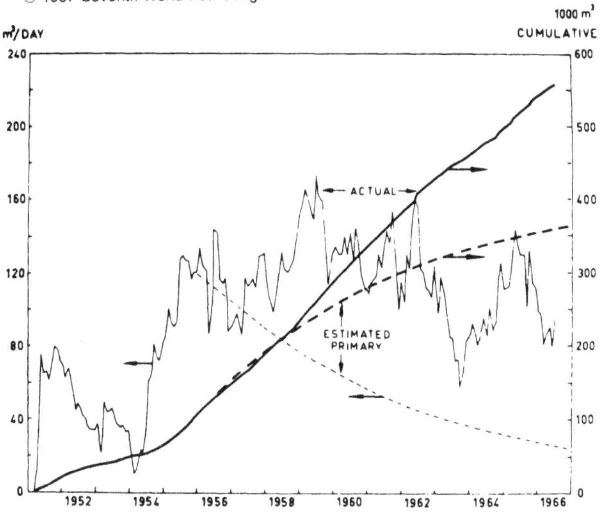

Fig. 6.19 – Reservoir performance of Schoonebeek hot-water flood.[15]

References

1. Craig, F.F. Jr.: *The Reservoir Engineering Aspects of Waterflooding,* Monograph series, Society of Petroleum Engineers, Dallas (1971) **3**.
2. Kyte, J.R., Stanclift, R.J. Jr., Stephan, S.C. Jr., and Rapoport, L.A.: "Mechanism of Water Flooding in the Presence of Free Gas," *Trans.,* AIME (1956) **207**, 215-221.
3. Dyes, A.B.: "Production of Water-Driven Reservoirs Below Their Bubble Point," *Trans.,* AIME (1954) **201**, 240-244.
4. Willman, B.T., Valleroy, V.V., Rumberg, G.W., Cornelius, A.J., and Powers, L.W.: "Laboratory Studies of Oil Recovery by Steam Injection," *J. Pet. Tech.* (July 1961) 681-690; *Trans.,* AIME, **222**.
5. Poston, S.W., Ysrael, S.C., Hossain, A.K.M.S., Montgomery, E.F.III, and Ramey, H.J. Jr.: "The Effect of Temperature on Irreducible Water Saturation and Relative Permeability of Unconsolidated Sands," *Soc. Pet. Eng. J.* (June 1970) 171-180; *Trans.,* AIME, **249**.
6. Weinbrandt, R.M., Ramey, H.J. Jr., and Cassé, F.J.: "The Effect of Temperature on Relative and Absolute Permeability of Sandstones," *Soc. Pet. Eng. J.* (Oct. 1975) 376-384.
7. Edmondson, T.A.: "Effect of Temperature on Waterflooding," *J. Cdn. Pet. Tech.* (Oct.-Dec. 1965) 236-242.
8. Combarnous, M. and Pavan, J.: "Deplacement Par L'Eau Chaude D'Huiles en Place Dans Un Milieu Poreux," *Comptes Rendus du Troisieme Colloque, Assn. de Recherche sur les Techniques de Forage et de Production,* Editions Technip, Paris (1969) 737-757.
9. Sinnokrot, A.A., Ramey, H.J. Jr., and Marsden, S.S. Jr.: "Effect of Temperature Level Upon Capillary Pressure Curves," *Soc. Pet. Eng. J.* (March 1971) 13-22.
10. Combarnous, M. and Sourieau, P: "Les Methodes Thermiques de Production des Hydrocarbures," *Revue,* Inst. Francais du Pétrole (July – Aug. 1976) Chap. 3, 543-577.
11. Croes, G.A. and Schwarz, N.: "Dimensionally Scaled Experiments and Theories and the Water-Drive Process," *Trans.,* AIME (1955) **204**, 35-42.
12. Buckley, S.E. and Leverett, M.C.: "Mechanism of Fluid Displacement in Sands," *Trans.,* AIME (1942) **146**, 107-116.
13. Harmsen, G.J.: "Oil Recovery by Hot-Water and Steam Injection," *Proc.,* Eighth World Pet. Cong., Moscow (1967) **3**, 243-251.
14. Martin, W.L., Dew, J.N., Powers, M.L., and Steves, H.B.: "Results of a Tertiary Hot Waterflood in a Thin Sand Reservoir," *J. Pet. Tech.* (July 1968) 739-750; *Trans.,* AIME, **243**.
15. Dietz, D.N.: "Hot Water Drive," *Proc.,* Seventh World Pet. Cong., Mexico City (1967) **3**, 451-457.
16. Van Heiningen, J. and Schwarz, N.: "Recovery Increase by 'Thermal Drive'," *Proc.,* Fourth World Pet. Cong., Rome (1955) Sec. II, 299.
17. Jordan, J.K., Rayne, J.R., and Marshall, S.W.III: "A Calculation Procedure for Estimating the Production History During Hot Water Injection in Linear Reservoirs," paper presented at the Twentieth Technical Conf. on Petroleum Production, Pennsylvania State U., University Park, May 9-10, 1957.
18. Farouq Ali, A.M.: *Oil Recovery by Steam Injection,* Producers Publishing Co. Inc., Bradford, PA (1970).
19. Higgins, R.V. and Leighton, A.J.: "A Computer Method to Calculate Two-Phase Flow in Any Irregularly Bounded Porous Medium," *J. Pet. Tech.* (June 1962) 679-683; *Trans.,* AIME, **225**.
20. Higgins, R.V. and Leighton, A.J.: "Computer Prediction of Water Drive of Oil and Gas Mixtures Through Irregularly Bounded Porous Media – Three-Phase Flow," *J. Pet. Tech.* (Sept. 1962), 1048-1054; *Trans.,* AIME, **228**.
21. Morel-Seytoux, H.J.: "Analytical-Numerical Method in Waterflooding Predictions," *Soc. Pet. Eng. J.* (Sept. 1965) 247-258; *Trans.,* AIME, **234**.
22. Dykstra, H. and Parsons, H.L.: "The Prediction of Oil Recovery by Waterflooding," *Secondary Recovery of Oil in the United States,* second edition, API, New York City (1950) 160-174.
23. Johnson, C.E. Jr.: "Prediction of Oil Recovery by Water Flood – A Simplified Graphical Treatment of the Dykstra-Parsons Method," *Trans.,* AIME (1956) **207**, 345-346.
24. Bursell, C.G., Taggert, H.J., and De Mirjian, H.A.: "Thermal Displacement Tests and Results, Kern River Field, California," *Prod. Monthly,* (Sept. 1966) **30**. No. 9, 18-21.
25. Holke, D.C. and Huebner, W.B.: "Thermal Stimulation and Mechanical Techniques Permit Increased Recovery from Unconsolidated Viscous Oil Reservoir," paper SPE 3671 presented at the 1971 SPE California Regional Meeting, Los Angeles, Nov. 4-5.
26. Terwilliger, P.L.: "Fireflooding Shallow Tar Sands—A Case History," paper SPE 5568 presented at the 1975 SPE Annual Technical Conference and Exhibition, Dallas, Sept. 28-Oct. 1.
27. Dillabough, J.A. and Prats, M.: "Proposed Pilot Test for the Recovery of Crude Bitumen from the Peace River Tar Sands Deposit – Alberta, Canada," *Proc.,* Simposio Sobre Crudos Pesados, Inst. de Investigaciones Petroleras, U. del Zulia, Maracaibo, Venezuela (1974).
28. Heymer, D.: "Sekundarverfahren unter Anwendung von Warme im Feld Emlichheim," *Z. deutsche geolog. Ges.,* (Aug. 1967) **119**, 570-573.
29. Viktorov, P.F. and Teterev, I.G.: "Characteristics of Operation and Possible Oil Recovery from the Sixth Formation of Arlansk Oil Field," *Neft i Gaz* (Jan. 1970) 31-4; Tulsa Abstract No. 131,602.

Chapter 7

Steam Drives

Steam drives differ markedly in performance from hot-water drives, the difference in performance being solely due to the presence and effects of the condensing vapor. The presence of the gas phase causes light components in the crude to be distilled and carried along as hydrocarbon components in the gas phase. Where the steam condenses, the condensable hydrocarbon components do likewise, thus reducing the viscosity of the crude at the condensation front. Moreover, the condensing steam makes the displacement process more efficient and improves the sweep efficiency. Thus, the net effect is that recovery from steam drives is significantly higher than from hot-water drives.

7.1 Mechanisms of Displacement

The experimental work of Willman *et al.*[1] shows that both steam and hot-water drives (1) improve oil mobility by reducing viscosity and (2) reduce residual oil at high temperatures. They achieve those effects to such a degree that the recovery of viscous crudes is better than could be attained by waterflooding.

Of course, all phenomena encountered in hot-water displacement (and discussed in Section 6.1) also are found in steam displacement. This is not only because the arrival of the condensation front is preceded by a hot-water drive – the steam condensate – but also because a hot-water phase is present along with steam in all cases of practical interest. Although the same phenomena are present – e.g., temperature still will affect surface forces – the magnitudes may be different as a result of the vapor phase. As in the case of hot-water drives, it is not well understood how the displacement process is affected by changes in surface forces resulting from temperature effects.

An important additional phenomenon affecting displacement in steam drives – a phenomenon first discerned by Willman *et al.*[1] – is the steam distillation of the relatively light fractions in the crude. Distillation causes the vapor phase to be composed not only of steam but also of condensable hydrocarbon vapors. Some hydrocarbon vapors will condense along with the steam, mixing with the original crude and increasing the amount of relatively light fractions in the residual oil trapped by the advancing condensate water ahead of the front. Dilution by light fractions causes some of the trapped oil to be displaced by the condensed water. The rest is stripped by the steam of all the remaining light ends, thus leaving less but heavier residue. The lighter components, stripped from the bypassed oil, help to regenerate and maintain a "solvent" bank just downstream of the condensation front. That this is the case may be seen from Fig. 7.1, which shows experimental results of Willman *et al.*[1] on the distillable content of the produced crude as a function of the cumulative oil produced. The produced crude did not change in composition until the steam zone was relatively near, at which point the volatile content increased markedly. Apparently, light ends in the vapor contributed to the high volatile content of the oil (including condensed hydrocarbons) produced after steam breakthrough.

The distillation of the crude bypassed by the advancing condensation front can result in very low ultimate residual oil saturations in the steam-swept zone. In principle, residual crude saturations can be essentially zero where the original crude has been diluted with large volumes of hydrocarbon condensate. Fig. 7.2, taken from Volek and Pryor,[2] shows the variation of residual oil in a tube steam-displacement laboratory experiment on a 24°API crude. The 20-in. sandpack initially was flooded to the cold-water flood residual oil saturation of 0.25. As can be seen, the oil saturation within most of the steam-swept zone is very low, in the range of 0.01 to 0.03. Near the condensation front, but still within the steam zone, the residual oil saturation has the much higher value of 0.16, suggesting that the steam is at least nearly "saturated" with hydrocarbon vapors. That is, distillation was not very pronounced under the conditions that prevailed in the interval between 24 and 30 in. from the injection end. Analysis of another experiment, shown in Fig. 7.3, confirms the expectation that the average boiling point (average carbon number) of the residual oil decreases with distance from the injection end. This experiment, which was conducted with 400-psi steam, was terminated at steam breakthrough. And yet, note that the average boiling point of the residual oil closest to the production end (Sample 4) is significantly higher than that of the original crude. This indicates that distillation had been important in the zone near the condensation front under the prevailing experimental conditions. Note also that the two samples closest to the steam inlet have essentially no components with less than 16 carbon atoms, these having been stripped off by the injected steam. Distillation, thus, is known to be of importance in the steam displacement of light crudes.

Injection of steam through granular packs containing crude oils, under conditions such that only vapor-phase recovery of the oil is obtained, shows that distillation is also important for heavy crudes, albeit to a lesser degree.[3] Fig. 7.4 shows that even a 9°API crude contains about 5 vol% of components distillable with 200-psia steam.

The concentration of the light components near the steam condensation front as a result of distillation and subsequent condensation would have a pronounced influence on the viscosity of the crude encountered by an advancing steam front. Only a small amount of diluent is required to reduce significantly the viscosity of heavy crudes, especially very heavy crudes. This relatively low-viscosity enriched crude is what is encountered by the advancing steam front. Just on the downstream side of the condensation front, then, the pressure gradients are significantly lower than if the original crude were present, a fact that will be used in the next section.

The rate of transport of the light components dissolved in the crude left within the steam zone and moving toward the vapor phase is controlled by diffusion and eddy flow within the oil phase and is reduced as the viscosity of the crude increases. Everything else being the same, then, the amount of a specific light component recovered over an interval of time would be larger the more volatile the crude. Here, of course, we are discussing only those components that are dissolved in the oil phase.

To develop very low (~0.0 to 0.05) residual oil saturations in steam-driven heavy crudes requires the generation of an adequate solvent bank. Such low residual saturations, therefore, should not be expected very near steam injectors. This effect raises questions about the determination of residual oil saturations in short-tube laboratory experiments.

Low values of residual oil saturations have been reported in field operations. Blevins[4] reported a 10-ft core interval in the Kern River field with an average saturation of about 0.03. Volek and Pryor[2] reported several 10-ft core intervals in the light-crude Brea field with "essentially zero oil saturations." It is clear that the displacement efficiency in steam drives can be very high.

7.2 Stability of Steam Fronts

Steam displacement is a stable process[5] – i.e., one that is not conducive to the formation and growth of viscous fingers. Small steam fingers, if formed, tend to lose heat at relatively high rates, ultimately resulting in condensation and disappearance of the

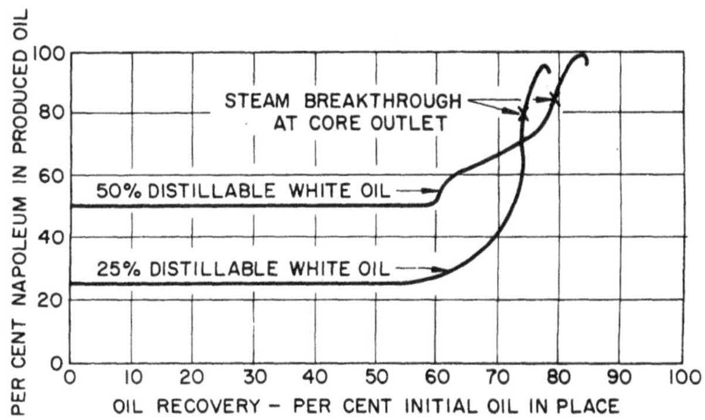

Fig. 7.1 – Composition of the oil produced during steam injection into a core initially containing two steam-distillable white oils and connate water.[1]

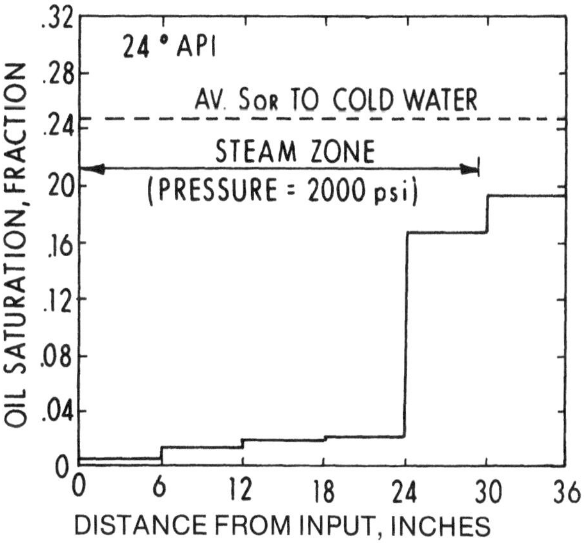

Fig. 7.2 – Residual oil saturation after steam injection.[2]

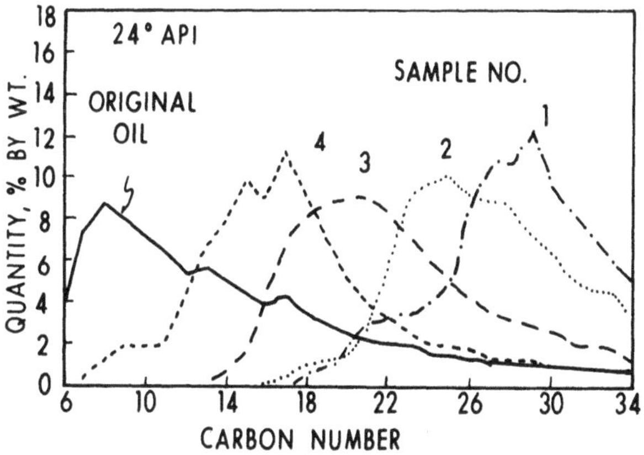

Fig. 7.3 – Hydrocarbon component distribution in residual oil after steamflood. (Sample *N* was obtained from the *N*th quarter of the tube residue closest to the inlet.)[2]

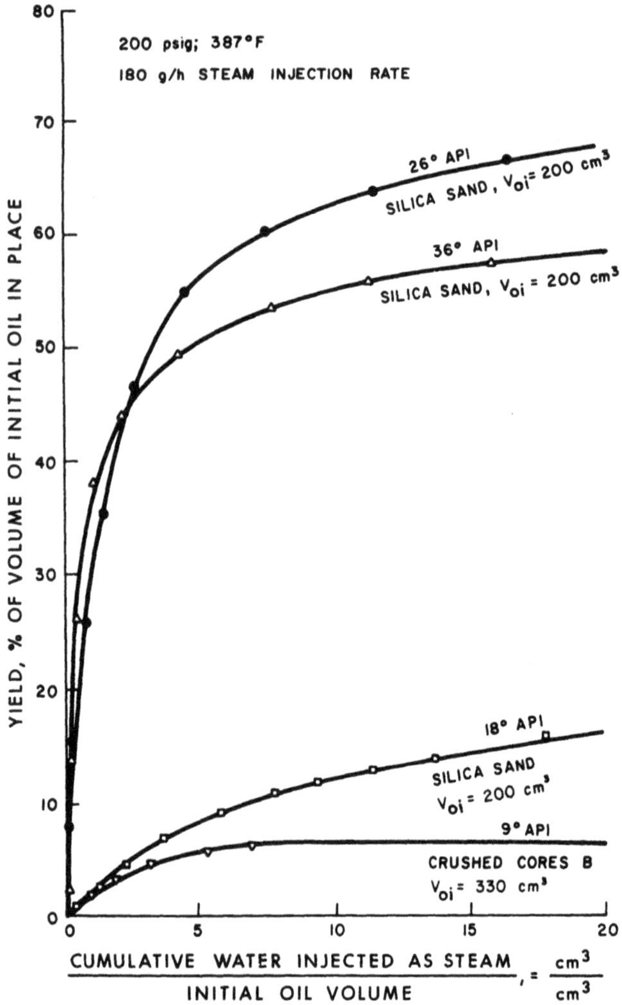

Fig. 7.4 – Crude oil steam distillation yield.[3]

steam and, thus, the finger. Although such simple physical arguments for the stability of steam displacement have been advanced for many years, in only the last few years has mathematical support been available. This was first provided by Miller[6] although he considered only steam displacing water. The stability analyses of oil-containing systems probably would not lead to radically different conclusions.

A parameter commonly used in reservoir engineering as a measure of the stability of a displacement front in the absence of capillary and gravity forces is the ratio of the pressure gradient $\partial p/\partial n$ normal to and on the downstream side of the displacement front to that on the upstream side. From Darcy's law, this pressure gradient ratio can be expressed as

$$(\partial p/\partial n)\big|_d \big/ (\partial p/\partial n)\big|_u = (u/\lambda)_d/(u/\lambda)_u. \quad \text{. . (7.1)}$$

Here, $(u/\lambda)_d$ is the volumetric flux u normal to the front divided by the mobility λ of *any* flowing phase present at the downstream side of the displacement front and, similarly, for the upstream side. Quantities upstream and downstream of the front are identified by the subscripts u and d, respectively. The choice of phases has no influence on the pressure gradient ratio, and different phases can (and sometimes must) be used on opposite sides of the displacement front. In solving for the pressure distribution in problems involving a swept and unswept zone separated by a displacement front, the ratio of the pressure gradients normal to the front must be specified. Examples are given by Muskat.[7] Because the pressure gradient ratio first was used in waterflooding, in which it was assumed that only water flows upstream of the displacement front and only oil downstream, it follows that $u_d = u_u$ and Eq. 7.1 reduces to

$$(\partial p/\partial n)\big|_d \big/ (\partial p/\partial n)\big|_u = \lambda_u/\lambda_d. \quad \text{. . . . . (7.2)}$$

The ratio of the pressure gradients then is given by the numerical value of the ratio of the mobility of the *only* fluid flowing in the swept zone (water) to that of the *only* fluid flowing in the unswept zone (oil). Because of the great importance of waterflooding in the early development of displacement processes, the ratio of the pressure gradients became synonymous with mobility ratio M, where

$$M = \frac{\lambda_u}{\lambda_d} = (k_r/\mu)_u/(k_r/\mu)_d. \quad \text{. . . . . (7.3)}$$

Over the years, the use of mobility ratio became so entrenched as being the number that determined the stability of a displacement process (as indeed it is for processes that are not characterized by condensation phenomena) that we lost sight of the fact that it is the value of the *pressure gradient ratio* that controls the behavior of the displacement front. Accordingly, in applying the customary use of mobility ratio at the steam displacement front, very large values of the mobility ratio generally are obtained because of the low viscosity of the steam vapor, which appears in the denominator of Eq. 7.3. Large values of the mobility ratio M correspond to unstable displacements. As stated earlier, all evidence indicates that steam displacement is a relatively stable process. And this has led to a paradox: the calculated unfavorable displacement by steam and the observed favorable displacement.

Let us consider Eq. 7.1 for the case of steam. If we write M_{eq} for the pressure gradient ratio and call this an equivalent mobility ratio to conform with past (and current) usage, the equivalent mobility ratio of steam drives may be expressed as

$$M_{eq} = (\partial p/\partial n)_d \big/ (\partial p/\partial n)_u \quad \text{. . . . . (7.4)}$$

$$= (u_w \mu_w/k_{rw})_d \, (k_{rs}/u_s \mu_s)_u \quad \text{. . . . . . . (7.5)}$$

$$= \frac{(u_w \rho_w)_d}{(u_s \rho_s)_u} \frac{\nu_{w,d}}{\nu_s} \frac{k_{rs}}{k_{rw,d}}, \quad \text{. . . . . (7.6)}$$

where ν is the kinematic viscosity,

$$\nu = \mu/\rho, \quad \text{. . . . . (7.7)}$$

μ is the fluid viscosity, k_r is the relative permeability, the subscripts s and w identify properties related to steam and water, and the subscript d is used to indicate that a property is evaluated at the downstream side of the displacement/condensation front. In developing Eq. 7.6, it has been assumed that the absolute permeability is the same on both sides of the displacement front. If this is not the case, then the right side of Eq. 7.6 should include a factor $k_{a,u}/k_{a,d}$.

Expressions for the equivalent mobility ratio identical with Eq. 7.6 have been presented by Hagoort[8] and Combarnous and Sourieau.[9] Neither study considers the effect of a moving oil phase, although Hagoort considers the presence of a residual oil saturation.

Combarnous and Sourieau evaluated the quantity

$$\frac{u_s \mu_s}{u_w \mu_w} = \frac{k_{rs}}{k_{rw}} \frac{1}{M_{eq}} \quad \text{.................. (7.8)}$$

for piston-like displacement of water by injection of (1) heated water (at the steam saturation temperature corresponding to the ambient pressure), (2) steam of 100% quality, and (3) superheated steam.

The calculations were based on a mass balance for the total water content (both liquid and vapor) across an advancing displacement front. Only steam *or* hot water is considered to exist in the upstream zone, and only liquid water exists downstream of the front. The energy balance across the advancing displacement front also is considered, with the temperature on the downstream side of the displacement front being the same as the initial reservoir temperature. The temperature at the upstream side of the displacement front is taken to be the temperature of the injected fluid. In the case of steam injection, the assumption of no liquid water in the steam zone is unrealistic, since it does not account for the condensate that forms as the steam front advances through the unheated reservoir. But, as Combarnous and Sourieau point out, the idealized model offers insight into the stability and equivalent mobility ratio of steam drives and hot-water displacement.

As indicated in Fig. 7.5, the initial temperature of the reservoir and the pressure at the displacement front have been related to representative reservoir depths by using average temperature and pressure gradients. Since the mathematical model considers the presence of only a single phase on each side of the displacement front, the relative permeability ratio appearing in Eq. 7.8 is unity. Thus, for steam displacement, the vertical axis in Fig. 7.5 is the reciprocal of the mobility ratio defined in Eq. 7.4. As can be seen from the results, heated water displacing formation water always has an unfavorable mobility ratio ($M_{eq} > 1$); values of $1/M_{eq}$ range approximately from 0.17 to 0.38. Displacement by saturated steam is stable ($M_{eq} < 1$) for temperature up to about 370°F, beyond which it is unstable ($M_{eq} > 1$). Inclusion of saturation distributions upstream and downstream of the displacement front would extend the temperature range over which the effective mobility ratio is favorable, which is consistent with laboratory observations.[5] But even with the very simplest of models, it is unmistakable that displacement by steam is more stable than by hot water. Likewise, superheated steam leads to more

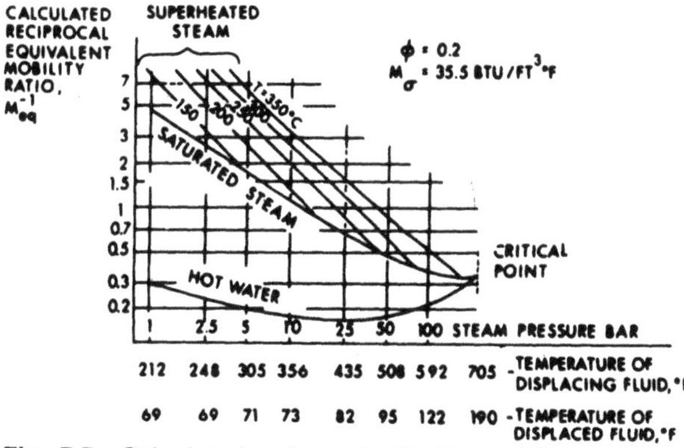

Fig. 7.5 – Calculated reciprocal effective mobility ratio for hot water and steam displacing cold water. (At a given pressure, the temperature of the displacing fluid is the steam saturation pressure, and that of the displaced fluid is related linearly to the pressure by means of typical and constant geothermal temperature and hydrostatic pressure gradients.[9])

stable displacements than does saturated steam. Except possibly very near the critical point of water, steam displacement is more stable as the temperatures decrease.

Although no published studies have considered the presence of a movable oil phase, the results shown in Fig. 7.5 indicate that steam displacement is more stable than displacement resulting from hot water under otherwise similar conditions. These theoretical calculations complement the experimental observations regarding the stability of the steam displacement process. For a better understanding of the steam displacement process, it would be highly desirable to extend the work cited here to consider the conditions at the steam/oil displacement front.

7.3 Prediction of Performance

For steam drives, because oil saturations in the steam-swept zones are low and effective mobility ratios are favorable, it has been easy to develop simple methods of predicting oil recovery. In these simple methods, variations in residual oil saturations within the steam zone, such as would result from distillation, generally are neglected. In other words, the values of the residual oil saturation used in the calculations are usually just those determined from simple steamflood experiments or assumed on the basis of experience. Experimentally obtained residual oil saturations do include distillation effects but generally do not include the full extent of solvent condensation effects. If it is thought that distillation and condensation phenomena reduce the steam drive residuals measured in the laboratory, corrections for the effect could be estimated. All the simple prediction methods use a single value of residual oil saturation in the steam-swept zone.

The first step in the calculations is to determine the volume of the steam zone and the amount of oil displaced from it. Where the gross volume of the steam zone is V_s, the amount of oil produced is

$$N_p = \langle 7{,}758 \text{ bbl/acre-ft} \rangle \phi \frac{h_n}{h_t} (S_{oi} - S_{or}) E_c V_s, \quad \text{.............................. (7.9)}$$

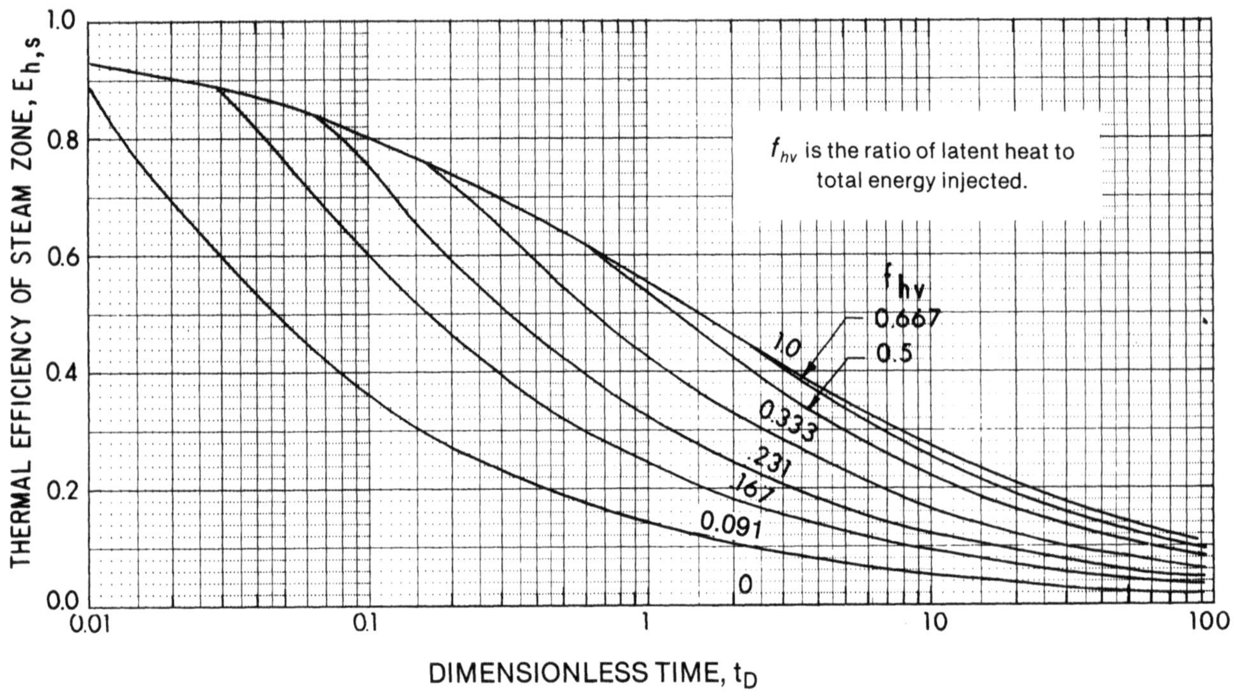

Fig. 7.6 – Fraction of heat injected in steamflood remaining in steam zone.[14] (f_{hv} is the ratio of latent energy to total energy injected).

where h_n is the net reservoir thickness, h_t is the gross reservoir thickness, S_{oi} is the initial oil saturation, S_{or} is the residual oil saturation, and E_c is the produced fraction of the oil displaced from the steam zone (sometimes called the capture factor).

The volume of the steam zone V_s can be found by methods such as the one proposed by Mandl and Volek.[10] Their method differs from those used by Farouq Ali,[11] Neuman,[12] Marx and Langenheim,[13] and Willman *et al.*[1] in that it identifies a critical time beyond which the zone downstream of the advancing condensation front is heated by hot water moving through the condensation front. Prior to the critical time t_c, all the heat in the reservoir is within the steam zone, and the results of Mandl and Volek[10] are the same as those of Marx and Langenheim.[13] The concept of a critical time is a significant contribution. Its derivation is based on three assumptions: (1) there is no gravity override of the steam zone, (2) all points on the condensation front advance at the same rate, and (3) the heat injection rate is constant. Departing from these assumptions – and in practice one *always* departs from them to some extent – will cause the critical time t_c to be smaller than that given by Mandl and Volek.[10] Other authors[1,11] consider the presence of hot water downstream of the vertical condensation front from the moment of injection. Although Willman *et al.*,[1] Farouq Ali,[11] and others consider the oil displacement by the hot-water condensate downstream of the condensation front, Myhill and Stegemeier[14] ignore the contribution as being small. Neuman[12] apparently considers the presence of a hot-water zone below steam zones affected by gravity override.

An approximate method developed by van Lookeren[15] and based on segregated flow principles is the only one that considers the effects of dip, the ratio of gravity to viscous forces, and the liquid level in the injection well on the production response. He provides results for both linear and radial flow and makes use of the equivalent mobility ratio given by Eq. 7.6. The method also yields approximate shapes for the steam zones under the influence of buoyant forces, which are discussed briefly in Chap. 4.

Myhill and Stegemeier[14] adopted and modified the steam zone volumes calculated by Mandl and Volek[10] so that the steam zone volume would vanish when no steam vapor is injected. The volume of the steam zone always can be related to the fraction of the injected heat present in the steam zone, $E_{h,s}$, by

$$V_s = \left\langle \frac{\text{acre-ft/cu ft}}{43{,}560} \right\rangle \frac{Q_i E_{h,s}}{M_R \Delta T}. \qquad (7.10)$$

M_R represents the total heat content of the steam zone per unit volume, and Q_i is the cumulative heat injected, calculated from the heat injection rate $\dot{Q}_i$:

$$\dot{Q}_i = w_i (C_w \Delta T + f_{sdh} L_{vdh}), \qquad (7.11)$$

where w_i is the mass rate of injection into the reservoir, ΔT is the temperature rise of the steam zone above the initial reservoir temperature (assumed to be the same as the temperature rise at downhole conditions, ΔT_i), (f_{sdh} and L_{vdh} are the steam quality and the latent heat of vaporization both at downhole conditions, respectively, and C_w is the average specific heat over the temperature range corresponding to ΔT. The function $E_{h,s}$ has been called the thermal efficiency of the steam zone and vanishes when no steam vapor is injected. For a constant rate of heat injection, it is given (after correcting a misplaced bracket in Eq. A-4 of Ref. 14) by

$$E_{h,s}=\frac{1}{t_D}\left\{G(t_D)+\left(1-f_{hv}\right)\frac{U(t_D-t_{cD})}{\sqrt{\pi}}\cdot\left[2\sqrt{t_D}-2\left(1-f_{hv}\right)\sqrt{t_D-t_{cD}}-\int_0^{t_{cD}}\frac{e^x\,\mathrm{erfc}\sqrt{x}\,dx}{\sqrt{t_D-x}}-\sqrt{\pi}G(t_D)\right]\right\}\quad\ldots(7.12)$$

The dimensionless time t_D is given by

$$t_D=4\left(\frac{M_S}{M_R}\right)^2\frac{\alpha_S}{h_t^2}t,\quad\ldots\ldots\ldots\ldots\ldots(5.6)$$

the function $G(t_D)$ is given by

$$G=2\sqrt{\frac{t_D}{\pi}}-1+e^{t_D}\,\mathrm{erfc}\sqrt{t_D},\quad\ldots\ldots\ldots(5.13)$$

the dimensionless critical time is defined by

$$e^{t_{cD}}\mathrm{erfc}\sqrt{t_{cD}}\equiv 1-f_{hv},\quad\ldots\ldots\ldots\ldots(7.13)$$

f_{hv}, the fraction of the heat injected in latent form, is given by

$$f_{hv}=(1+C_w\Delta T/f_{sdh}L_{vdh})^{-1},\quad\ldots\ldots(7.14)$$

and the unit function $U(x)$ is zero for $x<0$ and one for $x>0$.

$E_{h,s}$, the thermal efficiency of the steam zone, is plotted as a function of dimensionless time in Fig. 7.6 for several values of the parameter f_{hv}, the fraction of heat injected in latent form. Because of heat losses to the adjacent formations (see Chap. 5), the fraction of the injected heat present in useful form within the steam zone decreases with time. For a constant rate of heat injection (at constant steam quality), the volume of the steam zone calculated by Eq. 7.10 would continue to grow with time. But at long times, the growth rate of the steam zone would be very slow. When all the heat is injected as sensible heat ($f_{hv}=0$), the thermal efficiency of the steam zone is zero. The maximum thermal efficiency of the steam zone is obtained as the fraction of the heat injected in latent form approaches unity and is the same as that given by Marx and Langenheim. Note that because of the sensible heat of the steam vapor the value of f_{hv} can only approach unity in practical applications.

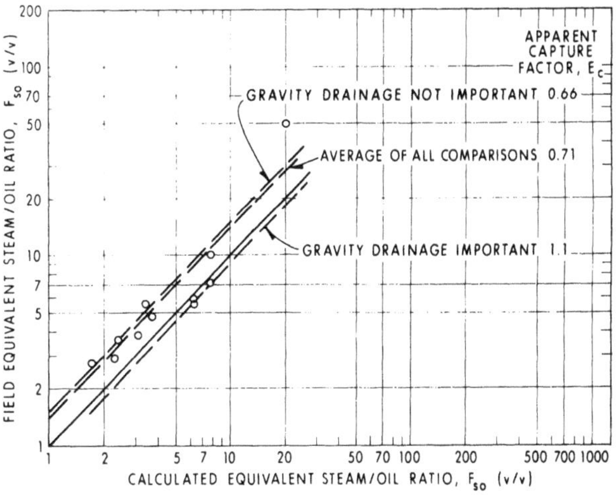

Fig. 7.7 – Comparison of calculated and observed steam/oil ratios for field projects. (Data are from Table 7.2.)

A quantity that is indicative of the success of a steam drive and changes slowly with time during a project is the ratio of the equivalent volume of steam injected (as condensate) per volume of oil produced. This quantity is known as the steam/oil ratio and is denoted by the symbol F_{so}. (The reciprocal quantity F_{os}, the oil/steam ratio, also is used extensively.) Accordingly, the steam/oil ratio F_{so} is given by

$$F_{so}=\frac{W_{s,eq}}{N_p},\quad\ldots\ldots\ldots\ldots\ldots\ldots(7.15)$$

where $W_{s,eq}$ is a measure of the steam used, expressed as equivalent volumes of water.

Bases for determining $W_{s,eq}$ include the steam injected, the steam generated, the feedwater passed through the generator, and the equivalent energy content. In some cases, the actual heat injected is reported without regard to the amounts in sensible

TABLE 7.1 – STEAM DRIVE FIELD PROJECTS – SUMMARY OF CONDITIONS[14]

			Thermal Properties*	Petrophysical Properties				Steam Parameters					
Field	Reference	Number of Injectors	T_i (°F)	h_t (ft)	h_n/h_t	ϕ	ΔS**	f_{sb}	f_{sdh}	p_s (psig)	i_s (B/D)	A (acres/well)	t (years)
Brea ("B" sand)	2	4	175	300	0.63	0.22	0.40	0.75	0.54	2000	500	10	8
Coalinga (Section 27, Zone 1)	16	40	96	35	1.0	0.31	0.37	0.7	0.55	400	500	9.2	4
El Dorado (northwest pattern)	17	4	70	20	0.85	0.26	0.20	0.75	0.45	500	200	1.6	1
Inglewood	18	1	100	43	1.0	0.37	0.40	0.75	0.7	400	1100	2.6	1
Kern River	19	85	90	55	1.0	0.32	0.40	0.7	0.5	100	360	2.5	5
Schoonebeek	20	4	100	83	1.0	0.30	0.70	0.85	0.7	600	1250	15	6
Slocum (Phase 1)	21	7	75	40	1.0	0.37	0.34	0.8	0.7	200	1000	5.65	2.5
Smackover	22	1	110	50	0.5	0.36	0.55	1.0	0.8	390	2500	10	1
Tatums (Hefner steam drive)	23	4	70	66	0.56	0.28	0.55	0.7	0.6	1300	685	10	5
Tia Juana	24	7	113	200	1.0	0.33	0.50	1.0	0.8	300	1400	12	5.3
Yorba Linda ("F" sand)		2	110	32	1.0	0.30	0.31	0.8	0.7	200	850	35	4.5

*M_R = 35 Btu/cu ft-°F, M_S = 42 Btu/cu ft-°F, λ_S = 1.2 Btu/ft-hr-°F.
**ΔS = Oil saturation at start of steaming minus the change in oil saturation from estimated primary during steam-drive period minus 0.15. The term 0.15 represents an average value of the residual oil saturation after steam drive.

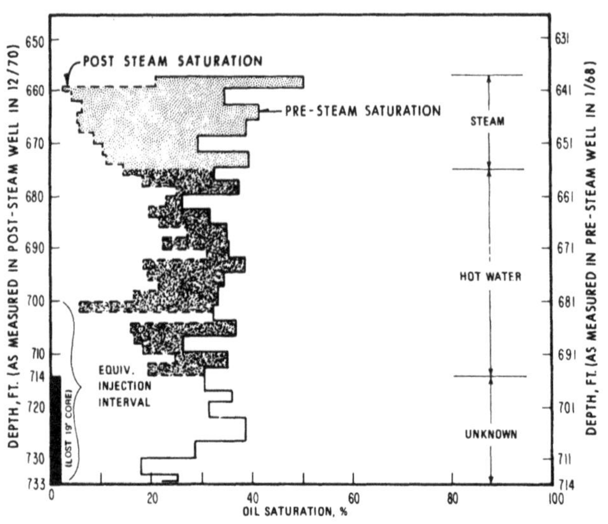

Fig. 7.8 – Comparison of core analysis of two wells 150 ft apart.[4]

TABLE 7.2 – COMPARISON OF FIELD RESULTS[14]

Field	Quantity of Steam Injected* W_{sD}	Steam-Zone Size** V_{sD}	Calculated Additional Equivalent Oil/Steam Ratio (vol/vol)	Field Additional Equivalent Oil/Steam Ratio (vol/vol)
Brea	0.5	0.15	0.13	0.14
Coalinga	0.94	0.45	0.16	0.18
El Dorado	1.6	0.315	0.05	0.02
Inglewood	1.26	1.256	0.41	0.28
Kern River	1.92	1.139	0.32	0.26
Schoonebeek	0.95	0.617	0.43	0.35
Slocum	1.41	1.202	0.29	0.18
Smackover	1.23	0.756	0.27	0.21
Tatums	1.54	0.397	0.13	0.10
Tia Juana	0.47	0.551	0.59	0.37
Yorba Linda "F"	0.54	0.280	0.16	0.17

*$W_{sD}=W_{s,eq}/(7758Ah_t)$.
**$V_{sD}=V_s/(7758Ah_t\phi)$.

and latent form. In others, barrels of feedwater are reported at a specified steam quality. Thus, there is no standard within the industry for reporting the steam/oil ratio given by Eq. 7.15. Since we propose using the recovery prediction method presented by Myhill and Stegemeier,[14] we shall use their definition of the equivalent volume of steam used.

Myhill and Stegemeier[14] converted the enthalpy of the steam at the generator outlet (relative to the ambient temperature of the feedwater) to the equivalent volume of injected steam $W_{s,eq}$ based on a reference enthalpy of 10^3 Btu/lbm. When expressed in terms of the heat injected Q_i, their relation becomes

$$W_{s,eq}=(2.853\times10^{-6}\text{ bbl/Btu})\cdot\left[\frac{C_w(T_{sb}-T_A)+f_{sb}L_{vb}}{C_w(T_{idh}-T_i)+f_{sdh}L_{vdh}}\right]Q_i. \quad (7.16)$$

Here, T_A is the ambient temperature of the boiler feedwater, and the subscripts b and dh represent boiler outlet and downhole conditions, respectively. Note that since the equivalent volume of steam injected is a measure of the thermal energy injected into the reservoir, it would not become zero even when the steam quality is zero.

Since only constant rates of heat injection have been considered in this chapter, the cumulative heat injected appearing in Eqs. 7.10 and 7.16 is equal to the product of the heat injection rate times the injection period. In actuality, constant rates of heat injection are never attained, so that in principle (1) the cumulative steam injected must be based on the average steam injection rate, (2) the cumulative equivalent volume of steam injected also must consider variations in temperature at the boiler inlet and outlet and at the bottom of the injection well, and (3) the thermal efficiency of the steam zone given by Eq. 7.12 should be obtained by superposition or equivalent procedures. Chap. 5 discusses procedures for taking into account these variable factors.

Fig. 7.7 compares steam/oil ratios calculated using Eqs. 7.9 through 7.16 with those obtained in field projects. The calculations were based on the assumption that the heat injection rates were constant over the life of the projects and used a capture factor of 100%. The capture factor is the fraction of the displaced oil that is produced. Since a number of these field projects already had primary production at the time steam injection started, the comparison is made with the field ratios associated with the oil production resulting from steam injection.

The additional equivalent steam/oil ratio for the field projects was based on the difference between the total recovery and the recovery from the primary production (extrapolated where necessary). That is, it is based on the quantity called "incremental thermal oil" in Fig. 4.11.

Tables 7.1 and 7.2 summarize the conditions on which the field and calculated additional equivalent steam/oil ratios are based.

Three fields – Yorba Linda, Coalinga, and Brea – have values close to those calculated. Those three fields have a high potential for gravity drainage. They have either relatively large dips or massive vertically continuous sand bodies and fairly high permeability. Of the remaining eight fields, only one (Tatums) has a significant dip, but its permeability is low. The average ratio of the field steam/oil ratio to the calculated steam/oil ratio for the latter group of fields is approximately 1.5 and is 0.9 for the three fields with good gravity drainage. Thus, the correlation shown in Fig. 7.7 indicates that the average steam/oil ratios from the field projects range from 1.5 to 0.9 of the calculated values, the latter value prevailing in reservoirs with good gravity drainage. Since the calculated steam/oil ratios are based on a capture factor of 100%, these results indicate that the capture factor E_c required to bring the results into agreement lies between 0.66 and 1.1, the latter value prevailing in reservoirs with good gravity drainage. Capture factors larger than 1.0 mean that more oil is produced than is displaced from the steam zone. This is possible where natural forces, such as gravity, have a significant influence on the production of the oil outside the steam zone.

Considering the large variation in reservoir properties and operating variables associated with these projects, it is surprising to find such a relatively small spread in the effective capture factor E_c. The significance is that Eq. 7.9 can be used, at least for screening purposes, to predict the recovery from steam drives, even though the authors only proposed the method to predict oil/steam ratios at a mature stage of steam injection. The authors have advised that this simple procedure be used with caution, at least until improved yet simple alternatives are developed.

Recovery prediction, although the most important aspect, is not sufficient to describe the total performance of a field project. The distribution of the steam within the reservoir is important for estimating injectivity, heat and steam breakthrough (which may have an impact on well spacing, project life, and poststeam breakthrough production performancc), and the overall recovery efficiency. One of the more important aspects regarding the distribution of the steam within the reservoir is the location of the steam zone within the vertical extent of the reservoir. It is well documented in both field and laboratory studies that steam rises to the top of the injection interval where adequate vertical communication is present.[4,12,15,26] Fig. 7.8, taken from Blevins *et al.*,[4] provides an example of the rise of steam considerably above the injection interval. The rise is due simply to the effect of buoyant forces, and the degree of the gravity override would be expected to be correlatable with the ratio of gravity forces to horizontal viscous forces, which may be expressed as[15]

$$\frac{Lg\Delta\rho}{\Delta p g_c} = \frac{Ak\rho g\Delta\rho}{\mu w_i g_c} \quad \text{. . . . . (7.17)}$$

In this expression, $g\Delta\rho/g_c$ represents the buoyant force gradient between two fluids having a density difference $\Delta\rho$, $\Delta p/L$ represents the pressure gradient, w_i/A is the mass injection rate w_i per cross-sectional area A, and ρ and k/μ are the density and the mobility of the injected fluid. Indeed, experimental results reported by Baker[26] and reproduced here as Fig. 7.9 show that the steam front tends to remain vertical over a longer period of time as the ratio of gravity to viscous forces decreases. In the two runs shown in Fig. 7.9, all the parameters appearing in Eq. 7.17 are constant except for the mass injection rate. Values of the vertical sweep efficiency at a time given by

$$\alpha t/h^2 = 0.08 \quad \text{. . . . . (7.18)}$$

have been calculated vs. relative values of the ratio of gravity forces to viscous forces defined by Eq. 7.17 from the results reported by Baker.[26] Here α is the thermal diffusivity of the water-saturated sand, h the sand thickness, and t is time. Results, plotted as Fig. 7.10, again show that vertical sweep efficiency increases as the ratio of gravity forces to viscous forces decreases. Although there is some scatter, there does appear to be a correlation between the ratio of gravity forces to viscous forces and the vertical sweep efficiency of the steam zone. The two isolated points on the middle left side of the plot correspond to the two runs using the smallest steam injection rates and may reflect experimental difficulties in controlling

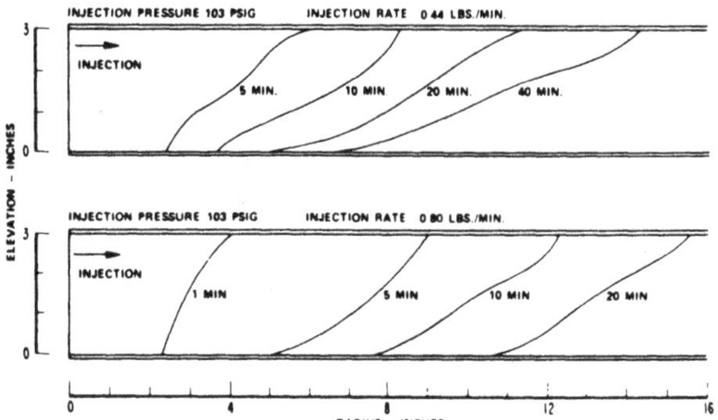

Fig. 7.9 – Steam front at different times in two model runs at high injection pressure and different injection rates.[26]

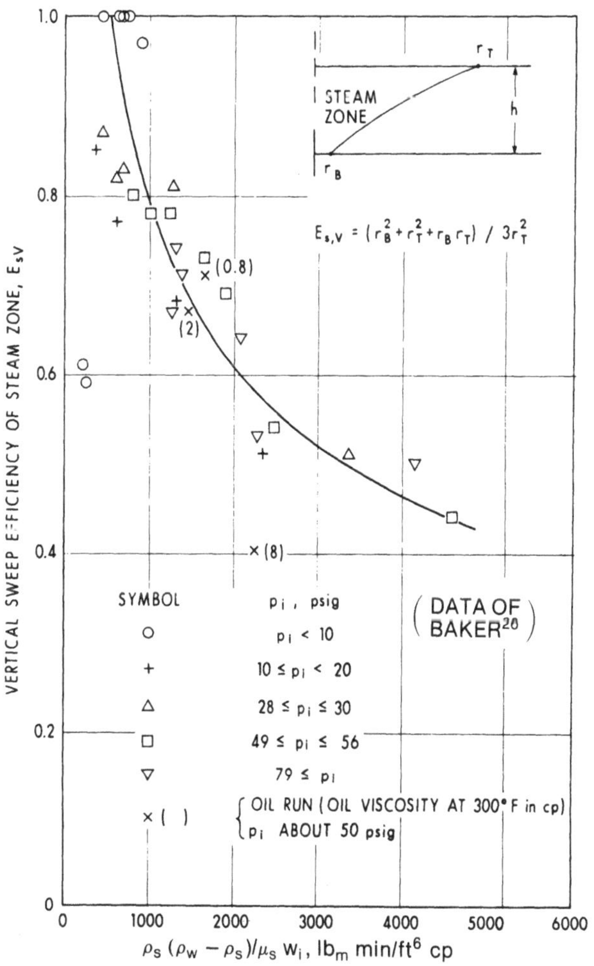

Fig. 7.10—Vertical sweep efficiency correlation for steam drives. In this figure, p_i represents the injection pressure.

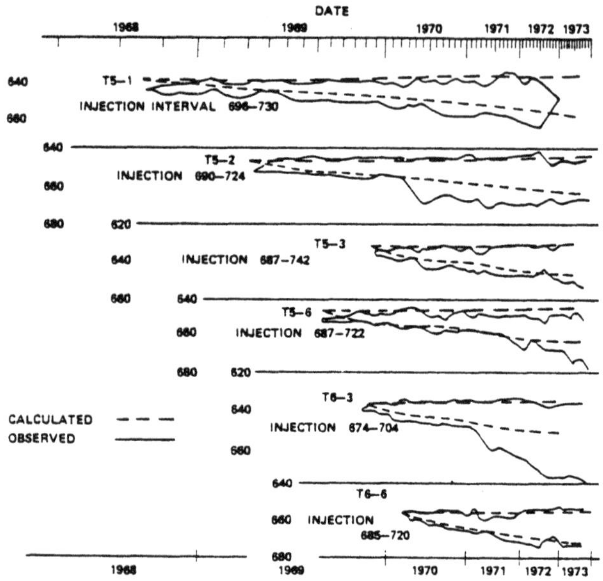

Fig. 7.11 – Calculated and observed steam zone thickness deduced from temperature profiles – 10-pattern steamflood, Kern River field, California.[12]

the rate of steam injection. These two points have been ignored in drawing the solid curve.

Although these results were obtained in the absence of oil, similar results were obtained when oils with viscosities less than about 2 cp at the steam temperature were used in the experiments. But an oil of 8-cp viscosity at steam temperature (300°F) gave a significantly lower vertical sweep efficiency. Thus, it appears that the presence of a heavy oil may promote gravity override,[15] perhaps by resisting the formation of the steam zone across the entire vertical extent of the sand. Data from the oil runs also were ignored in drawing the curve.

It is emphasized that these observations are valid in homogeneous and uniform sands; heterogeneities such as are caused by "thief" (or highly permeable) zones and poor vertical communication could lead to completely different behavior. The mobility ratio of a steam displacement may be favorable with water and low-viscosity oils but would be less favorable (and even unfavorable) as the viscosity of the crude increases.

Other than the work of van Lookeren,[15] the only published method for estimating the vertical cross section of the steam zone is that of Neuman.[12] Both assume the steam zone is always at the top of the sand. In addition to being independent of reservoir fluid flow properties and of the ratio of gravity forces to viscous forces, the volume of the steam zone calculated from Neuman's equation is proportional to the cumulative net heat injected into the reservoir and appears to be independent of the heat losses from the steam zone. Since Neuman did not provide the derivation of the equations, it is not possible to assess the substance of the method. However, the area is given as increasing linearly with the square root of steam injection time when the net heat injection rate is constant, and the thickness of the steam zone at any distance from the injection well is given as proportional to the square root of the time

TABLE 7.3 – RESERVOIR PARAMETERS FOR EXAMPLE 7.1

Porosity	0.18
Net thickness, ft	34.0
Gross thickness, ft	41.0
Average oil saturation	0.66
Residual oil saturation	0.17
Reservoir temperature, °F	93
Volumetric heat capacity of* heated reservoir, Btu/cu ft-°F	38.1
Volumetric heat capacity of adjacent formations, Btu/cu ft-°F	41.8
Thermal diffusivity of adjacent formations, sq ft/D	0.70

*Calculated from Eq. 5.3 to include the latent heat of condensation in the steam zone and the heat capacity of the nonpay interval:

$M_R = [h_n M_{R,\mathrm{pay}} + (h_t - h_n) M_{R,\mathrm{nonpay}}]/h_t$

elapsed since the leading edge of the front reached the point. Comparison between observed steam zone thicknesses and those calculated by Neuman's equation,

$$\Delta z_S = \frac{4(M\sqrt{\alpha})_S C_w \Delta T}{L_v M_{Rse}} \sqrt{\frac{\Delta t}{\pi}}, \qquad (7.19)$$

is given in Fig. 7.11. In this equation, $(M\sqrt{\alpha})_S$ is an average value of the properties in the overburden and in the reservoir beneath the steam zone, M_{Rse} is the volumetric heat capacity of the steam zone neglecting all contributions due to the steam itself, and Δt "is the time elapsed since steam first arrived where the steam zone thickness (Δz_s) is to be calculated." The agreement is reasonably good in most cases. Neuman also reports that approximately "half as much oil is displaced from the region below the steam zone as from the steam zone itself." This observation is based on the experience at Kern River[4] and Inglewood[18] and may differ markedly in steam drives in different reservoirs.

Other than the method of van Lookeren, there are no simple procedures for predicting the degree of gravity override on the shape of the steam zone. And no simple methods are known for predicting the effect of heterogeneities. This generally is done on the basis of experience unless numerical simulation is used to study the process. Even then the results may be particularly sensitive to the degree of vertical communication and homogeneity assumed in the study. The predicted recovery does not appear to be very sensitive to the shape of the steam zone, at least until steam breakthrough.[14] In principle, steps taken to curtail steam production after breakthrough should affect the subsequent recovery. With a large degree of override, steam breakthrough would occur relatively early and would reduce the amount of oil produced up to breakthrough.

For viscous crudes, breakthrough of steam under bypassing conditions is likely to lead to significant oil production after breakthrough. Under frontal drive conditions, postbreakthrough recovery would be expected to be lower than under bypass conditions.

Jones[25] recently has provided a program for a hand-held calculator to predict the performance of steam drives based on the models of van Lookeren[15] and of Myhill-Stegemeier[14] (the latter with some modifications).

Example 7.1 – Estimating the Recovery History by Steam Drive

The feedwater rate into a steam generator is 1,000 B/D, referred to a temperature of 60°F. The inlet feedwater temperature averages 80°F, and the output temperature and steam quality average 565°F and 0.78. The quality of the steam entering the reservoir is 0.73, and the bottomhole temperature is also 565°F. Pertinent reservoir properties are given in Table 7.3. Estimate the approximate recovery history due to the steam injection using the Myhill-Stegemeier[14] method.

The fraction of the heat injected in latent form f_{hv} is calculated from Eq. 7.14 to be

$$f_{hv}=\left[1+\frac{(1.08\text{ Btu/lbm-°F})(565-93)\text{°F}}{(0.73)(615.4\text{ Btu/lbm})}\right]^{-1}$$

$$=0.47.$$

Values of the fluid properties appearing in this equation are given in Table 7.4. The value $f_{hv}=0.47$ is used to enter Fig. 7.6 to calculate the heat efficiency of the steam zone as a function of time. For values of dimensionless time smaller than the critical value, the heat efficiency of the steam zone is given by the upper curve of Fig. 7.6. The value of the dimensionless critical time defined by Eq. 7.13 can be found from Table 5.2 as the value of t_D for which

$$e^{t_D}\text{erfc}\sqrt{t_D}=0.53,$$

or $t_D=t_{cD}=0.49$, as obtained by linear interpolation. After the critical time, the heat efficiency of the steam zone would follow the curve corresponding to $f_{hv}=0.47$ in Fig. 7.6. Rather than interpolate between the curves for $f_{hv}=0.5$ and 0.333 in Fig. 7.6, the curve for $f_{hv}=0.5$ is used for these calculations.

The relationship between real and dimensionless time is found from Eq. 5.6 and is

$$t=\frac{(38.1\text{ Btu/cu ft-°F})^2(41.0\text{ ft})^2 t_D}{(4)(0.70\text{ sq ft/D})(41.8\text{ Btu/cu ft-°F})^2}$$

$$t\text{ (in days)}=499\,t_D$$

$$t\text{ (in years)}=1.37\,t_D.$$

Values of the heat efficiency of the steam zone have been calculated at a number of selected times and are listed in Table 7.5.

The heat injection rate into the reservoir, $\dot{Q}_i$, which follows from Eq. 7.11, is

$$\dot{Q}_i=(1{,}000\text{ B/D})(62.4\text{ lbm/cu ft})(5.615\text{ cu ft/bbl})\cdot[(1.08\text{ Btu/lbm-°F})(565-93)\text{°F}+(0.73)(615.4\text{ Btu/lbm})]$$

$$=3.36\times10^8\text{ Btu/D}.$$

Cumulative heat injected also is listed in Table 7.5. The volume of the steam zone then is calculated as a function of time from Eq. 7.10:

$$V_s=\left[(3.36\times10^8\text{ Btu/D})\,t\,E_{h,s}\right]\div\left[(43{,}560\text{ cu ft/acre-ft})\cdot(38.1\text{ Btu/cu ft-°F})(565-93)\text{°F}\right]$$

$$=0.429\,t\,E_{h,s}\,.$$

Values of V_s, which also have been listed in Table 7.5, represent the gross volume of the steam zone in the reservoir, and are expressed in acre-ft when t is in days.

The areal extent of the steam zone then is given by

$$A\text{(in acres)}=V_s/(41.0\text{ ft}),$$

values of which are listed in Table 7.5.

The cumulative oil production, which is proportional to the steam zone volume, is calculated from Eq. 7.9 and is given by

$$N_p=(7{,}758\text{ bbl/acre-ft})(0.18)(34.0\text{ ft}/41.0\text{ ft})\cdot(0.66-0.17)(0.7)V_s$$

$$N_p\text{ (in bbl)}=397\,V_s\,.$$

TABLE 7.4 – DOWNHOLE FLUID PROPERTIES FOR EXAMPLE 7.1

Injection temperature, °F	565
Latent heat of condensation, Btu/lbm	615.4
Steam quality	0.73
Specific heat of water,* Btu/lbm-°F	1.08
Density of water, lbm/cu ft	45.0

*Calculated from the difference in the enthalpy of saturated liquid water at 575 and 93°F divided by (565 – 93)°F.

TABLE 7.5 – ESTIMATED PERFORMANCE – EXAMPLE 7.1

Dimensionless Time	Time (years)	Thermal Efficiency of Steam Zone (fraction)	Cumulative Heat Injected Q_i (Btu)	Gross Volume of Steam Zone V_s (acre-ft)	Areal Extent of Steam Zone A (acres)	Cumulative Oil Production N_p (bbl)	Equivalent Steam Injected $W_{s,eq}$ (bbl)	Steam/Oil Ratio F_{so}
0.05	0.068	0.855	8.34×10^9	9.10	0.222	3.62×10^3	2.50×10^4	6.8
0.10	0.137	0.805	1.68×10^{10}	17.3	0.422	6.87×10^3	5.00×10^4	7.2
0.20	0.273	0.742	3.35×10^{10}	31.7	0.775	1.26×10^4	1.00×10^5	7.9
0.50	0.683	0.642	8.38×10^{10}	68.7	1.68	2.73×10^4	2.50×10^5	9.4
1.00	1.37	0.537	1.68×10^{11}	115	2.82	4.58×10^4	5.00×10^5	11
2.0	2.73	0.427	3.35×10^{11}	183	4.46	7.26×10^4	1.00×10^6	14
5.0	6.83	0.300	8.38×10^{11}	321	7.84	1.27×10^5	2.50×10^6	19
10	13.7	0.225	1.68×10^{12}	483	11.8	1.92×10^5	5.00×10^6	26
20	27.3	0.165	3.35×10^{12}	706	17.2	2.81×10^5	1.00×10^7	35

TABLE 7.6 – RESERVOIR PARAMETERS FOR EXAMPLE 7.2

Porosity	0.32
Net thickness, ft	28.2
Gross thickness, ft*	35.0
Oil initially in place, bbl/acre-ft	1520
Current water cut	0.21
Residual oil to steam, bbl/acre-ft	372
Volumetric heat capacity of heated reservoir, Btu/cu ft-°F	38.8
Volumetric heat capacity of adjacent formations	41.9
Thermal diffusivity of adjacent formations, sq ft /D	0.81
Well radius, ft	0.25
Maximum allowed injection pressure, psig	780
Current reservoir pressure, psig	215
Formation volume factor at 215 psi	1.00
Permeability, md	2920
Initial reservoir temperature, °F	80
Oil viscosity at 80°F, cp	185

*Includes 5-ft clay streak between the two sands.

Finally, the cumulative steam/oil ratio is calculated from the cumulative production and cumulative equivalent steam injected (expressed as condensate), all of which are given in Table 7.5. The cumulative equivalent steam injection is calculated using Eq. 7.15,

$$\begin{aligned} W_{s,eq} &= (2.853\times10^{-6}\ \text{bbl/Btu}) \\ &\cdot\left[(1.08\ \text{Btu/lbm-°F})(565-80)\text{°F}\right. \\ &\left.+(0.78)(615.4\ \text{Btu/lbm})\right] \\ &\div\left[(1.08\ \text{Btu/lbm-°F})(565-93)\text{°F}\right. \\ &\left.+(0.73)(615.4\ \text{Btu/lbm})\right]Q_i \\ &=2.99\times10^{-6}\ Q_i\ , \end{aligned}$$

where $W_{s,eq}$ is in barrels. The steam/oil ratio is calculated using Eq. 7.15.

Results shown in Table 7.5 indicate that after 1 year the predicted steam/oil ratios would be greater than 10 bbl of equivalent steam per barrel of produced oil. If a capture factor of 100% could be realized instead of the 70% used here, the predicted steam/oil ratios would be lower by 43%. Despite the high initial oil saturation, the steam/oil ratio is high in this example, mainly because of the relatively low porosity of 0.18 and the fact that 17% of the gross pay is shale, which also is heated to steam temperature without contributing to the oil production. Other factors that would have resulted in lower steam/oil ratios include a higher quality of the injected steam and lower steam temperature. As Eq. 7.10 shows, the steam zone volume is inversely proportional to the difference between the steam temperature and the initial reservoir temperature. For the same amount of steam injected, low-temperature steam occupies a larger volume than a high-temperature steam and results in a larger steam zone volume and a greater amount of oil displaced and produced.

This example illustrates that (1) a high initial oil content does not guarantee an attractive steam/oil ratio, (2) a fairly low porosity and a low ratio of net/gross sand thickness can increase the steam/oil ratio significantly, (3) higher steam temperatures (or injection pressures) result in lower steam zone volumes and oil recoveries and high steam/oil ratios, (4) the steam/oil ratio increases with time between fillup and steam breakthrough, and (5) because of heat losses to adjacent formations, the rate of growth of the steam decreases with time. Point 4 probably could not be made where gravity override is significant.

The effect of higher steam temperatures resulting in smaller steam zone volumes, oil recoveries, and oil/steam ratios sometimes can be circumvented. It is generally desirable to inject at a high rate to shorten both the injection phase (for a given amount of heat injected) and the entire operation. This reduces not only operating costs but also the fractional heat losses, both from the wellbore and from the reservoir. But higher injection rates result in higher injection pressures, which for steam are translated into higher injection temperatures and densities. And higher steam pressures result in smaller steam zone volumes. More heat has been injected faster, but this has not improved the recovery. Quite the opposite. Sometimes, however, it is possible to inject at a high rate at the beginning and then reduce the injection pressure (and temperature) to allow the steam zone to expand more rapidly. In the case of variable heat injection rates, the steam zone volume can be calculated by using superposition, as illustrated in Chap. 5.

A method intermediate in complexity between the one discussed here and complete thermal numerical simulation is described by Davies *et al.*[27] In that method, streamtubes are idealized by converging and diverging straight-sided channels, to which the steam displacement calculations first made by Willman *et al.*[1] are applied. The method was developed for the repeated five-spot well pattern but could be extended for other cases. As done by Willman *et al.*,[1] Farouq Ali,[11] and others, Davies *et al.* also considers how displacement by the liquid water phase downstream of the condensation front contributes to oil recovery.

7.4 Design of Steam Drives

The design of any field project requires a correlation of its economics and its requirements for technical success. Chaps. 4, 11, and 12 provide some information bearing on design parameters, and other chapters also may be found useful.

The following should be considered in designing any field project:

1. Is the *reservoir* description adequate?
2. Is there enough *oil in place* to justify the effort?
3. Can the *old* wells be used for thermal operations?
4. Are there adequate sources of fresh water and fuel?
5. Is there sufficient information to estimate the likely range of *operating variables* (such as pressures and rates at injectors and producers) and *production performance*?

The reservoir description includes not only the vertical and horizontal distribution of porosity, permeability, oil in place, and gas and water saturations, but also the amount of clays sensitive to fresh water, the lateral continuity of the sands, and any other factor that may influence the operation of the steam drive. The work of Kendall[28] is an example of the geologic study that should be available before any supplemental recovery process is installed. Such studies may be reduced in scope or dispensed with when information about similar nearby reservoirs is available. But the importance of a good reservoir description cannot be overemphasized. Inadequate information often leads to delays[29] and technical disasters, even in pilot projects (see Chap. 13).

The elements that must be considered in designing a project are too varied and numerous for complete description in this work. But some of the considerations that are pertinent to any design are identified and discussed in this section by means of an example.

Example 7.2 – Designing a Steam Drive

A reservoir at a depth of 1,300 ft has produced by depletion only 4.9% of the initial oil in place, which was 1,520 bbl/acre-ft. The crude viscosity at reservoir temperature is 185 cp. Remaining primary reserves are estimated at 100 bbl/acre-ft. The reservoir is composed of an upper and a lower sand separated by a rather thin (5.0-ft) but continuous clay streak. The gross thicknesses of the upper and lower sands are 17 and 13 ft, and each has a net/gross thickness ratio of 94%. The properties of the sands (which are considered to be common) are given in Table 7.6. The reservoir is being produced through gravel-packed wells completed over the two sands on 2-acre spacing, the well pattern approximating a five-spot. There is no dip to speak of in the accumulation and no bottom water. Consider the requirements for a steam drive in this reservoir.

The well completion and repair records indicate that essentially all steam injection wells will need to be new. The existing wells, with a few exceptions, are serviceable as producers. The well patterns considered for the steam drive approximate the following options.

1. Plug and abandon every other producer and drill and complete new injectors at their locations to obtain a 4-acre five-spot pattern.
2. Infill drill new injectors to obtain a 2-acre five-spot pattern.
3. Plug and abandon every fourth producer and drill and complete new injectors at their location to obtain an 8-acre nine-spot pattern. These three well patterns are illustrated in Fig. 7.12.

If the reservoir and crude properties listed in Tables 7.6 and 7.7 and Eqs. 4.2 and 4.3 are used, it is found that *if* the only mobile phase in the reservoir is oil, the steady-state five-spot injection rate i and production rate q would be related to the downhole pressure drop between injector and producer by

$$i=\left(\frac{7.082\times10^{-3}}{2\pi}\right)\left[\frac{\pi kh/\mu}{\ln(L'/r_w)-0.964}\right]\Delta p.$$

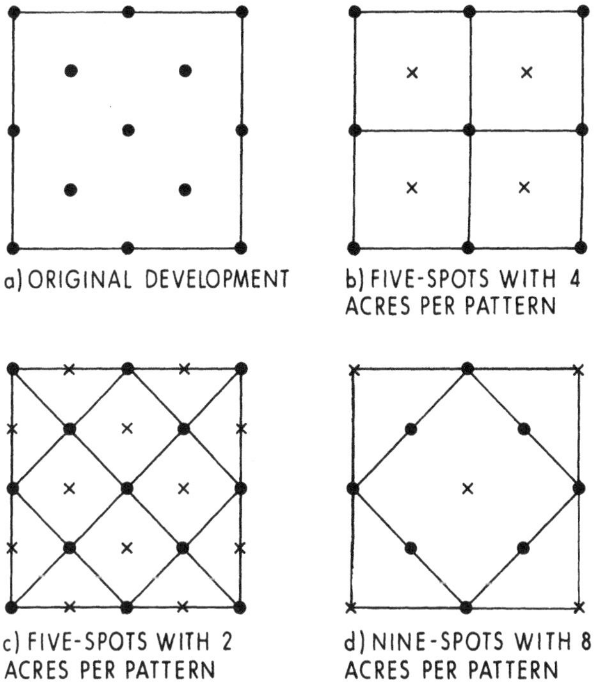

Fig. 7.12 – Well patterns considered in Example 7.2.

TABLE 7.7 – DESIGN DATA – EXAMPLE 7.2

Duration of Steam Injection (years)	i/A_p at Various Steam Temperatures (B/D-acre) 400°F	450°F	500°F
0.88	153	176	197
1.33	112	130	149
1.78	90.9	106	122
2.67	67.6	78.9	91.7
4.44	49.6	57.5	67.8
	Oil/Steam Ratio at Various Steam Temperatures		
	400°F	450°F	500°F
0.888	0.232	0.201	0.175
1.33	0.209	0.181	0.158
1.78	0.192	0.164	0.143
2.67	0.168	0.143	0.123
4.44	0.141	0.119	0.102
	Steam Temperature		
Quantity	400°F	450°F	500°F
C_w, Btu/lbm-°F	1.02	1.03	1.05
f_{hv}	0.667	0.619	0.564
p_s, psig	233	408	666
$L_v/C_w\Delta T$	2.53	2.03	1.62
t_{cD}	2.05	1.41	0.940

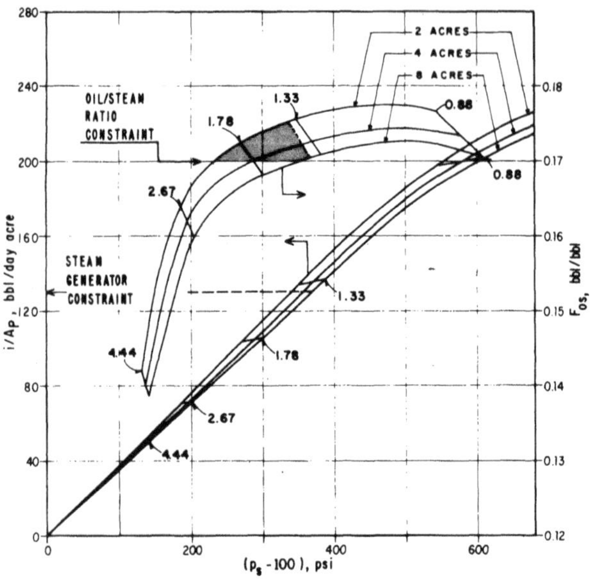

Fig. 7.13 – Conditions considered in designing a steamflood (Example 7.2).

The length of the side of a five-spot pattern, defined in Table 4, is related to the spacing in acres per pattern by $L' = 208.71\sqrt{A}$.

For Option 1, $A=4$ acres, and the injection rate is calculated as

$$i=\frac{(7.082\times 10^{-3})(28.2\text{ ft})(2{,}920\text{ md})\ \Delta p}{(2)(185\text{ cp})\ln\left(\dfrac{208.7\sqrt{4}}{1/4}-0.964\right)}$$

$$=0.244\ \Delta p\,,$$

where i is in barrels per day.

Since the maximum allowed injection pressure is 780 psi and the operator does not think the producers can be operated at bottomhole pressures of less than 100 psi, the maximum pressure drop between wells is about 680 psi. Thus, it appears that the maximum attainable injection or production rate is about 166 B/D. Consideration of the reduction in pattern flow resistance obtainable by steamsoaking all wells indicates production rates of 500 B/D can be achieved – provided the flow resistance of the pattern can be reduced by two-thirds. This appears to require small cyclic steam injection volumes.

The 500-B/D rate considered achievable with the help of cyclic steam injection still assumes that 185-cp oil is the only flowing phase except in the neighborhood of the wells. In the neighborhood of the wells, the flow resistance is considered to be negligible after cyclic steam injection. It is realized that a pattern flow resistance essentially based on the premise that 185-cp oil is the only flowing phase in the bulk of the reservoir is pessimistic. Note also that in this reservoir the thickness of the larger sand interval is only 17 ft, which is unlikely to cause any pronounced bypassing. If bypassing were significant, injectivity would be governed by the transmissivity of the water and steam instead of that of the oil (see Chap. 4).

The operator considers that with cyclic steam injection the effective viscosity of the oil would be reduced slightly because of mild increases in temperature, and that bypassing may be slightly more pronounced. These combined effects lead the operator to estimate that 1,000 B/D is achievable under these conditions, with somewhat lower values prevailing at the higher injection rates. The final estimates of i/A_P vs. Δp for the 4-acre five-spot is plotted in Fig. 7.13. Estimates, obtained in a similar manner, for the other patterns under consideration also are plotted in Fig. 7.13. Such estimates are consistent with regional experience.

At the shallow depth and good injectivity expected in this reservoir, it is reasonable to assume that steam could be delivered downhole at 80% quality.

As a working guide, it is considered that steam injection will continue until the steam zone occupies 50% of the pattern volume. From Eqs. 7.10 and 7.11, it follows that

$$V_s=0.5\ (35\text{ ft})\ A_P=\left(\frac{\text{acre-ft/cu ft}}{43{,}560}\right)$$

$$\cdot\frac{(i\text{ B/D})(62.4\text{ lbm/cu ft})(5.615\text{ cu ft/bbl})}{(38.8\text{ Btu/cu ft-°F})}$$

$$\times C_w\left[1+\frac{(0.8)L_v}{C_w\Delta T}\right]t\ E_{h,s}\,,$$

where A_P is the pattern area in acres and t is time in days. This can be rearranged to give the injection rate per pattern area as a function of steam temperature and of the duration of the steam injection period:

$$i/A_P=\frac{8.44\times 10^4}{\left[C_w\left(1+\dfrac{(0.8)\,L_v}{C_w\Delta T}\right)t\ E_{h,s}\right]}\,.$$

Values of i/A_P were calculated at 400, 450, and 500°F steam temperatures and for steam injection periods ranging from 0.88 to 4.4 years. Values of C_w, $L_v/C_w\Delta T$, and E_h were evaluated as in Example 7.1. Key results, including the steam pressure, are listed in Table 7.7. This information is plotted in Fig. 7.13 as curves of i/A_P vs. Δp, but only within the range of similar values given by the injectivity estimates. The intersections of these two sets of i/A_P vs. Δp curves give the time required to attain a steam zone volume of 50% of the pattern volume, when the indicated steam pressure and injection rates are used.

A shorter injection period requires a higher steam injection rate and a larger steam injection capacity. If the reservoir is very large, development can proceed in phases, with the equipment (including the steam generators) being moved or used as the phases are developed. This allows the use of high rates for a short time over a small part of the field. When the equipment is used as the phases develop, its amortization is spread out over long times and large oil recoveries.

Other considerations may be used in delimiting the range of operating conditions. Oil/steam ratio vs. Δp curves calculated from Eqs. 7.9, 7.10, 7.15, and 7.16

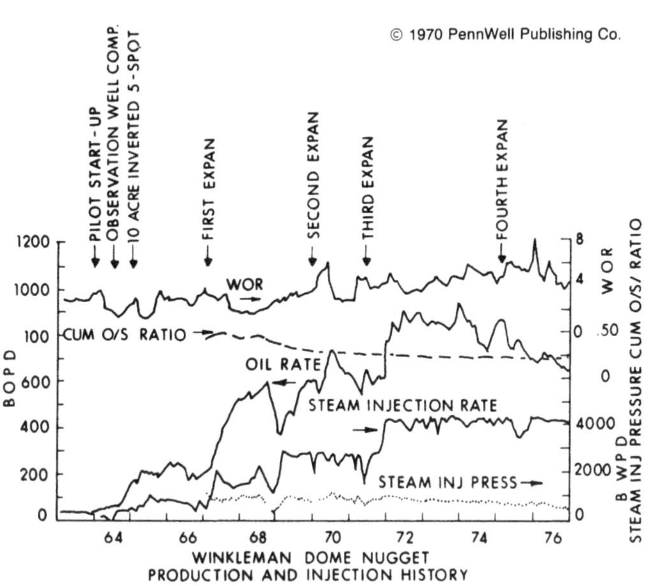

Fig. 7.14–Winkleman Dome Nugget production and injection history.[31]

TABLE 7.8 – SELECTED MAJOR STEAM DRIVE PROJECTS

Project and Location	Date Started	Reference (Chap. 7)
Schoonebeek, The Netherlands	1960	19
Tia Juana, Venezuela	1961	23
East Coalinga, CA	1963	15
Winkleman Dome, WY	1964	31,32
Kern River, CA	1964	18
Inglewood, CA	1965	17
Slocum, TX	1967	20,33
Kern River, CA	1968	4,32
Mount Poso, CA	1970	32,34
Smackover, AR	1971	21,32
Kern River, CA	1972	29
Tia Juana Este Venezuela	1975	32

also are listed in Table 7.7. To calculate from Eq. 7.16 the equivalent steam injected, it was assumed that the heat content of the steam at the boiler outlet was the same as that of the steam entering the reservoir. These calculated oil/steam ratios are considered to be insensitive to the well pattern. These oil/steam ratio data were plotted vs. Δp, but only the intervals corresponding to the injectivity results are shown in Fig. 7.13. These curve segments, which have a negative slope, also are labeled according to the time it takes to attain a steam zone that is 50% of the pattern volume. As shown in Fig. 7.13, the oil/steam ratio vs. Δp curve exhibits a maximum for each well pattern. If the minimum acceptable oil/steam ratio is 0.14 bbl oil per barrel of equivalent steam injected, the average steam pressure during the life of the project would be at least 330 psig for the 2-acre five-spot well pattern. Higher average steam injection pressures would be required for the longer well spacings.

Other factors also affect the operation. In this project the operator is willing to drill new wells as required but wants to make use of existing steam generators that will be available from another of his projects. These will have enough capacity to provide an injection rate of 130 B/D-acre. The field is so small that all of it must be developed at the same time, and this imposes an upper limit of 130 B/D-acre. This steam-capacity constraint corresponds to an average steam pressure during the life of the project of no more than 445 psig for the 2-acre five-spot well pattern and slightly more (up to 480 psig) for the larger well patterns. The hatched area in Fig. 7.13 indicates the range of calculated acceptable oil/steam ratios for the patterns under consideration, and the steam injection pressures required to obtain them. And for each injection pressure and well pattern there is a corresponding injection rate obtained from the i/A_p vs. Δp curves. This approach then places limits on the injection pressure (and temperature) of the steam, on the injection rate, on the expected cumulative oil/steam ratio, on the duration of the steam injection period, and on the number of injection wells required. The project should be a technical success if it is operated within the prescribed set of operating conditions. The final choice of operating conditions will be governed by economic and legal factors and will likely be modified by the project response.

The procedure discussed in Example 7.2 can be used either to help design an actual project or to screen candidates for steam drives and obtain a likely range of operating conditions. In the latter case, the range of operating requirements, and their impact on economics, may be refined with tools capable of considering more detailed information on reservoir and crude properties (i.e., thermal numerical simulators). But if the operator is experienced enough with steam drives and with the suggested design procedures, the procedure may serve as adequate guidelines to design a project, given attractive economics.

This example (1) identifies the principal factors affecting the technical design of a steam drive, (2) discusses the interaction between these factors, (3) provides examples of constraints that limit the practical range of the process variables the operator can affect, and (4) stresses the importance of economics in the final selection of the operating conditions.

7.5 Examples of Field Applications

Matheny[30] gives a survey of the 1979 steam-drive projects. The largest commercial steam drive is Getty Oil Co.'s project at Kern River, with oil production rates of about 52,650 bbl reported for 1979.[30] The steam drives with the largest production rates in 1979 are identified in Table 15.2. There have been many steam drive pilots, and far fewer large-scale projects.

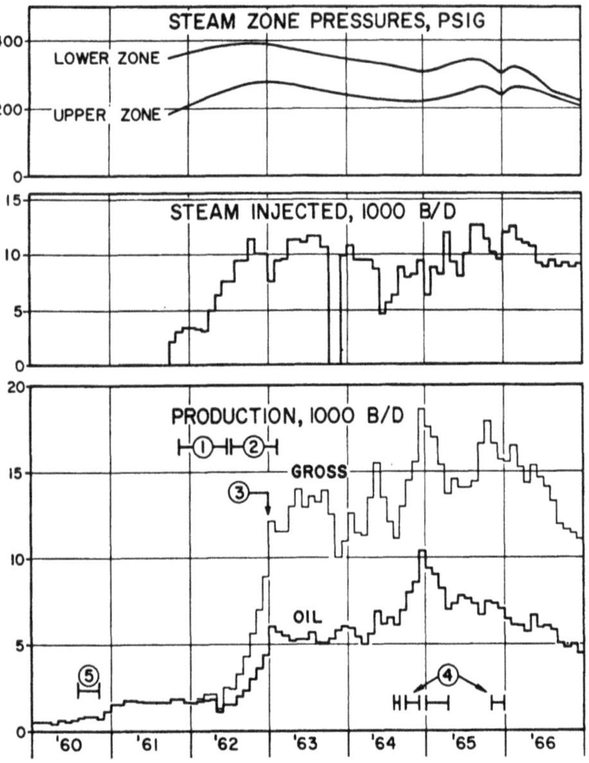

① INCREASED WATER CUT IN MOST TEST WELLS
② INCREASED TEMPERATURE IN MOST TEST WELLS
③ FIRST WATER CUT INCREASE IN FIRST-ROW SURROUNDING WELLS
④ STEAM SOAKING OF SOME PRODUCTION WELLS
⑤ COMPLETION OF 12 NEW PRODUCTION WELLS

Fig. 7.15 – Overall steam drive project performance, Tia Juana field, Venezuela.[24]

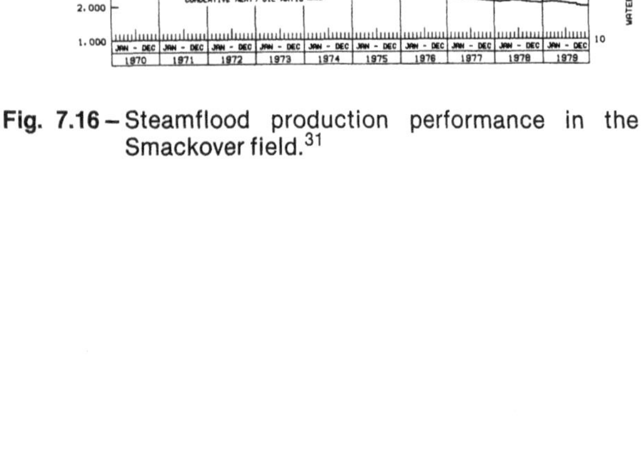

Fig. 7.16 – Steamflood production performance in the Smackover field.[31]

Table 7.8 lists some of the major projects. Spacing per injection well ranges from about 2.5 acres at Kern River[18] to more than 15 acres at Schoonebeek.[19] In most projects, the steam pressures range from 400 to 600 psi, with a high of 1,350 psi at Winkleman Dome.[30,31] (Of the projects reported in Table 7.1, Brea had a steam pressure of 2,000 psi.) The permeability at Winkleman Dome is the lower, about 650 md, whereas all the other projects are from 2 to 40 times as permeable. With the exception of the Winkleman Dome project, which has a porosity of 0.25, all have porosities of 0.30 or greater. Oil viscosities at reservoir temperature range from 280 cp at Mount Poso to about 4,000 cp at Kern River.[18] On the whole these projects represent the larger of the early steamflood operations – those that generally have achieved oil production responses of about 1,000 B/D and higher. Of these, the Slocum, Kern River (1968), and East Coalinga projects have been converted to cold-water injection. The purpose of the water injection is to reduce operating costs while maintaining (and even increasing) pressure gradients in the reservoir and increasing the heat utilization. Injected water will be heated by the heat stored in the reservoir and will recover some of the heat lost to the adjacent formations during steam injection. Alternative ways to terminate a steamflood include blowing the pressure down and injecting foam.

Examples of production performance of steam drives are given in Figs. 7.14, 7.15, and 7.16, corresponding to the projects at Winkleman Dome, Tia Juana, and Smackover fields. The production performance at Winkleman Dome reflects the four expansions that followed an initial evaluation pilot. A gradual growth is characteristic of most large steam drive projects, not only to test the technology but also to generate income for further expansion and reduce peak personnel requirements. Note that at Winkleman Dome, the oil/steam ratio generally decreased with time. This is the opposite of the behavior observed at the Smackover project (Fig. 7.16). The performance graph of the latter project distinguishes between the total oil recovery and that resulting from the steam injection. Also, total rate of heat injection is given in millions of Btu's per month rather than in barrels of steam (as condensed water) per day. Both ways of expressing the steam injection rate require additional information to find the rate of injection of steam *vapor*. The performance at Tia Juana (Fig. 7.15) is shown to indicate that steam drives often are combined with cyclic steam injection to improve production rates and reduce the flow resistance between wells. This cyclic steam injection usually is carried out before the steam drive is started, but as the figure shows, it also can be used during the steam drive itself. This treatment more likely would be used in those reservoirs containing high-viscosity oils, where the flow resistance is relatively high.

References

1. Willman, B.T., Valleroy, V.V., Runberg, G.W., Cornelius, A.J., and Powers, L.W.: "Laboratory Studies of Oil Recovery by Steam Injection," *J. Pet. Tech.* (July 1961) 681-

690; *Trans.*, AIME, **222**.

2. Volek, C.W. and Pryor, J.A.: "Steam Distillation Drive – Brea Field, California," *J. Pet. Tech.* (Aug. 1972) 899-906.
3. Wu, C.H. and Brown, A.: "A Laboratory Study on Steam Distillation in Porous Media," paper SPE 5569 presented at the 1975 SPE Annual Technical Conference and Exhibition, Dallas, Sept. 28-Oct. 1.
4. Blevins, T.R. and Billingsley, R.H.: "The Ten-Pattern Steamflood, Kern River Field, California," *J. Pet. Tech.* (Dec. 1975) 1505-1514; *Trans.*. AIME, **259**.
5. Harmsen, G.J.: "Oil Recovery by Hot-Water and Steam Injection," paper PD9 presented at Eighth World Pet. Cong., Moscow, June 13-16, 1971.
6. Miller, C.A.: "Stability of Moving Surfaces in Fluid Systems With Heat and Mass Transport; III. Stability of Displacement Fronts in Porous Media," *AIChE J.* (May 1975) 474-479.
7. Muskat, M.: *The Flow of Homogeneous Fluids Through Porous Media*, J.W. Edwards Inc., Ann Arbor, MI (1946) 458-465.
8. Hagoort, J., Leijnse, A. and van Poelgeest, F.: "Steam-Strip Drive: A Potential Tertiary Recovery Process," *J. Pet. Tech.* (Dec. 1976) 1409-1419.
9. Combarnous, M. and Sourieau, P.: "Les Méthodes Thermiques de Production des Hydrocarbures," *Revue*, Inst. Français du Pétrole (July-Aug. 1976) Chap. 3, 543-577.
10. Mandl, G. and Volek, C.W.: "Heat and Mass Transport in Steam-Drive Processes," *Soc. Pet. Eng. J.* (March 1969) 59–79; *Trans.*, AIME, **246**.
11. Farouk Ali, S.M.: *Oil Recovery by Steam Injection*, Producers Publishing Co. Inc., Bradford, PA (1970).
12. Neuman, C.H.: "A Mathematical Model of the Steam Drive Process—Applications," paper SPE 4757 presented at the 1975 SPE California Regional Meeting, Ventura, April 2–4.
13. Marx, J.W. and Langenheim, R.H.: "Reservoir Heating by Hot Fluid Injection," *Trans.*, AIME (1959) **216**, 312-315.
14. Myhill, N.A. and Stegemeier, G.L.: "Steam Drive Correlation and Prediction," *J. Pet. Tech.* (Feb. 1978) 173-182.
15. van Lookeren, J.: "Calculation Methods for Linear and Radial Steam Flow in Oil Reservoirs," *Soc. Pet. Eng. J.* (June 1983) 427–439.
16. Afoeju, B.I.: "Conversion of Steam Injection to Waterflood, East Coalinga Field," *J. Pet. Tech.* (Nov. 1974) 1227-1232.
17. Hearn, C.L.: "The El Dorado Steam Drive – A Pilot Tertiary Recovery Test," *J. Pet. Tech.* (Nov. 1972) 1377-1384.
18. Blevins, T.R., Aseltine, R.J., and Kirk, R.S.: "Analysis of a Steam Drive Project, Inglewood Field, California," *J. Pet. Tech.* (Sept. 1969) 1141-1150.
19. Bursell, C.G.: "Steam Displacement – Kern River Field," *J. Pet. Tech.* (Oct. 1970) 1225-1231.
20. van Dijk, C.: "Steam-Drive Project in the Schoonebeek Field, The Netherlands," *J. Pet. Tech.* (March 1968) 295-302; *Trans.*, AIME, **243**.
21. Hall, A.L. and Bowman, R.W.: "Operation and Performance of the Slocum Thermal Recovery Project," *J. Pet. Tech.* (April 1973) 402-408.
22. Smith, R.V., Bertuzzi, A.F., Templeton, E.E., and Clampitt, R.L.: "Recovery of Oil by Steam Injection in the Smackover Field, Arkansas," *J. Pet. Tech.* (Aug. 1973) 883-889.
23. French, M.S. and Howard, R.L.: "The Steamflood Job, Hefner Sho-Vel-Tum," *Oil and Gas J.* (July 17, 1967) 64-66.
24. de Haan, M.J. and Schenk, L.: "Performance Analysis of a Major Steam Drive Project in the Tia Juana Field, Western Venezuela," *J. Pet. Tech.* (Jan. 1969) 111-119; *Trans.*, AIME, **246**.
25. Jones, J.: "Steam Drive Model for Hand-Held Programmable Calculators," *J. Pet. Tech.* (Sept. 1981) 1583-1598.
26. Baker, P.E.: "Effect of Pressure and Rate on Steam Zone Development in Steamflooding," *Soc. Pet. Eng. J.* (Oct. 1973) 274-284; *Trans.*, AIME, **255**.
27. Davies, L.G., Silberberg, I.H., and Caudle, B.H.: "A Method of Predicting Oil Recovery in a Five-Spot Steamflood," *J. Pet. Tech.* (Sept. 1968) 1050-1058; *Trans.*, AIME, **243**.
28. Kendall, G.H.: "Importance of Reservoir Description in Evaluating In-Situ Recovery Methods for Cold Lake Heavy Oil – Part 1, Reservoir Description," paper 7620 presented at the CIM 27th Annual Technical Meeting, Calgary, Canada, June 7-11, 1976.
29. Duerksen, J.H. and Gomaa, E.E.: "Status of the Section 26C Steamflood, Midway-Sunset Field, California," paper SPE 6748 presented at the 1977 SPE Annual Technical Conference and Exhibition, Denver, Oct. 9–12.
30. Matheny, S.L. Jr.: "EOR Methods Help Ultimate Recovery," *Oil and Gas J.* (March 31, 1980) 74-124.
31. Pollock, C.B. and Buxton, T.S.: "Winkleman Dome Steam Drive a Success," *Oil and Gas J.* (Aug. 10, 1970) 151-154.
32. *Enhanced Oil Recovery Field Reports*, Society of Petroleum Engineers, Dallas (Sept. 1980) **6**, 1.
33. *Enhanced Oil Recovery Field Reports*, Society of Petroleum Engineers, Dallas (June 1979) **5**, 1.
34. Stokes, D.D., Brew, J.R, Whitten, D.G., and Wooden, L.G.: "Steam Drive as a Supplemental Recovery Process in an Intermediate-Viscosity Reservoir, Mount Poso Field, California," *J. Pet. Tech.* (Jan. 1978) 125-131.

Chapter 8
In-Situ Combustion

In situ is Latin for "in place." Thus, in-situ combustion is simply the burning of fuel where it exists in a reservoir. The term is applied to recovery processes in which air, or more generally an oxygen-contaning gas, is injected into a reservoir, where it reacts with organic fuels. The heat generated then is used to help recover unburned crude.

The first four sections of this chapter cover fuel, oxygen/fuel reactions, kinetics, and production models for the dry combustion process. Section 8.5 expands on these subjects for wet combustion. Section 8.6 discusses reverse combustion.

8.1 The Fuel

The fuel actually burned in forward in-situ combustion is not the crude oil in the reservoir. Rather the fuel is primarily the carbon-rich residue resulting from thermal cracking and distillation of the residual crude near the combustion front. Naturally occurring coal, if present in the rock, also can contribute to the fuel available for combustion. The amount of fuel present per unit bulk volume of reservoir is an extremely important parameter in combustion operations, for it generally determines the air required to burn a unit bulk volume of reservoir. In wet combustion operations, however, the addition of sufficient water may reduce the temperature to the point that the fuel will not burn completely. Where the combustion of all the fuel is unnecessary, incomplete combustion may be an advantage since it reduces the amount of air required to burn through a unit bulk volume of reservoir. The process also can be carried out under partially quenched conditions (see Section 8.5) so that a midfraction of the crude is burned or oxidized at low temperatures. The reverse combustion process (see Section 8.6) also consumes a midfraction of the crude oil in place.

The usual way to determine the fuel that would be burned in a given reservoir is by experimentation. This is done by conducting a combustion test in laboratory tubes using a sandpack or core under conditions simulating those in the reservoir. The sand or core and the crude oil should be from the reservoir being considered. Ideally, the contents of the tube are heated to simulate the advance of the temperature front, and air is injected when the combustion temperature is reached. At that point, combustion occurs and the combustion front is advanced through the tube by continued air injection. The fuel burned then is determined from the amount of oxygen consumed and the combustion front velocity.

Air consumption is affected not only by reactions with organic fuels but also by reactions between oxygen and some minerals in the reservoir, such as pyrite. Where minerals react, special care is required to determine the amount of fuel burned. Experimentally determined variations in the fuel burned – variations that can be correlated with reservoir samples from zones of different mineralogy or lithology – deserve special interpretation. These variations should be compared with those due to the reproducibility of the experimental process and analyzed for possible statistical significance. That is seldom done.

The fuel burned in a reservoir also is known as the fuel content, the fuel consumption, or the fuel availability. For combustion at high temperatures, the terms are interchangeable. Not all companies have the equipment required for determining the air requirements and the fuel burned. Some service companies and consulting firms have those facilities. The field user of the combustion process generally is interested in (1) the amount of air required to burn a unit bulk volume of reservoir rock and (2) the amount of crude available for displacement from the burned zone. Although it is assumed that any unburned crude is displaced, careful analyses of the burned matrix usually show some organic residue. In such cases, the estimate of fuel available – the organic residue plus the fuel burned – cannot, in principle, be used interchangeably with the estimate of fuel burned. However, this organic residue is generally negligible where combustion occurs at high temperatures. Organic residues also are found after conventional core extraction.

In the absence of experimental values for the fuel burned in the field under study, correlations can be used. These, however, exhibit large deviations due to variations in crude and rock properties. The effect of the rock matrix can be seen in Fig. 8.1, and the effect of crude gravity is shown in Figs. 8.2 and 8.3. Note that fuel consumption data from Alexander *et al.*[1]

(shown in Fig. 8.2) exhibit much more scatter than those of Showalter[2] shown in Fig. 8.3. Also note that those authors express the fuel burned in different units. The trend, however, is the same; there is less fuel burned as the API gravity increases. In 1965, Wohlbier[3] published a thorough analysis of the available results on the amount of fuel burned. Fig. 8.4, taken from his work, shows the large variation in fuel burned reported by various investigators. As these results show, API gravity and crude density are not satisfactory parameters for making good estimates of the fuel burned.

The correlation between fuel burned and air consumed is generally excellent for combustion at high temperatures, as shown in Fig. 8.5. At lower temperatures some fraction of the consumed oxygen generally is reacted with the crude without generating carbon oxide gases or water. This type of oxidation reaction would give rise to air requirements greater than the 189 scf of air per pound of fuel burned obtained from the slope of Fig. 8.5. There is no specific temperature separating the high temperature behavior exemplified by Fig. 8.5 from the lower-temperature oxidation reaction. But a temperature of approximately 650°F is consistent with observations. This is referred to as the minimum active combustion temperature, even though it has no exact definition and may be off a few degrees from 650°F.

Operational variables also affect the amount of fuel available for combustion. At low air fluxes, the active combustion temperature may not be reached. Then some oxygen passes through the combustion zone and is taken up by the crude. It is known (see Fig. 8.6) that low-temperature oxidation of the crude increases the fuel available for high-temperature combustion. At very high air fluxes, the residence time of the oxygen in the high-temperature reaction zone is so reduced that not all the oxygen can react. Although the residence time in the low-temperature oxidation region would be relatively long, the reaction rates are low enough to allow unreacted oxygen to pass through in the effluent gas, especially in short laboratory tube runs. Fig. 8.7 from Martin *et al.*[4] shows the injected air requirement to have a minimum of about 240 scf per cubic foot of sand at an air flux of about 480 scf/D-sq ft. The effect of increased pressure on fuel content, shown in Fig. 8.8, is to increase the fuel content slightly. Adding water to the injection stream cools the fuel and reduces the amount of fuel burned; it also generates additional steam and, thus, improves sweep and displacement efficiencies (see Fig. 8.9 taken from Garon and Wygal[5]).

Nelson and McNiel[6] proposed a correction to the amount of fuel burned in the laboratory when the porosity of the material in the tube run, ϕ_E, is different from the porosity of the reservoir being considered, ϕ. This correction is given by

$$m_R = \frac{1-\phi}{1-\phi_E} m_E, \quad \ldots\ldots\ldots\ldots\ldots\ldots\ldots (8.1)$$

where m_R is the mass of fuel burned per unit bulk reservoir volume, and m_E is the mass of fuel burned per unit bulk volume in the laboratory experiment.

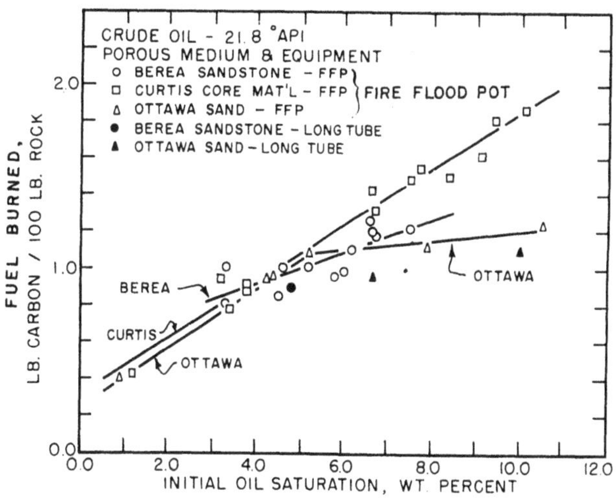

Fig. 8.1 – Effect of initial oil saturation on fuel burned.[1]

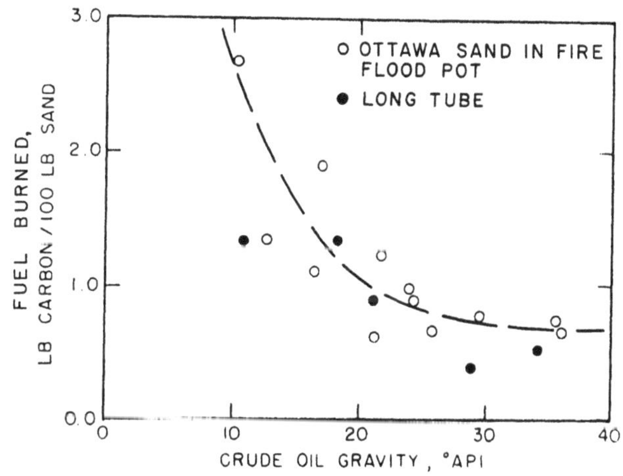

Fig. 8.2 – Correlation of fuel burned with crude-oil gravity.[1]

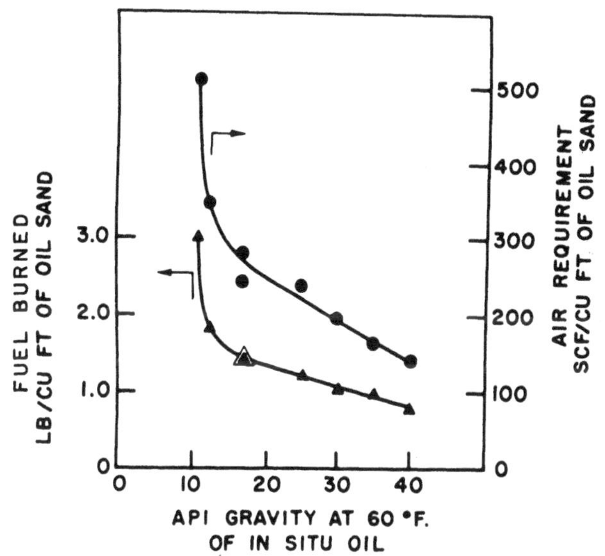

Fig. 8.3 – Combustion-drive fuel burned and air requirements vs. oil gravity.[2]

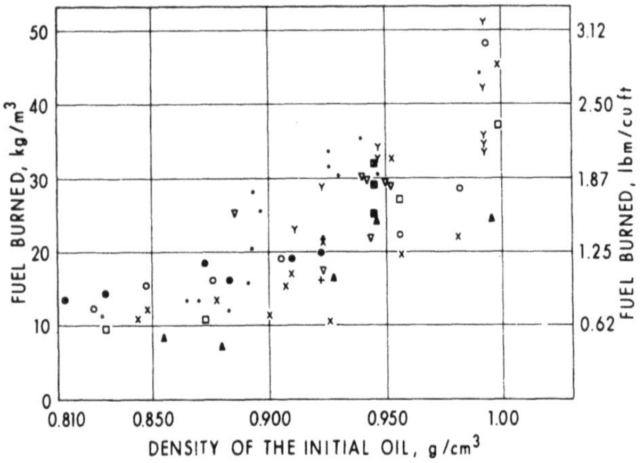

Fig. 8.4 – Fuel burned as a function of the density of the initial oil.[3]

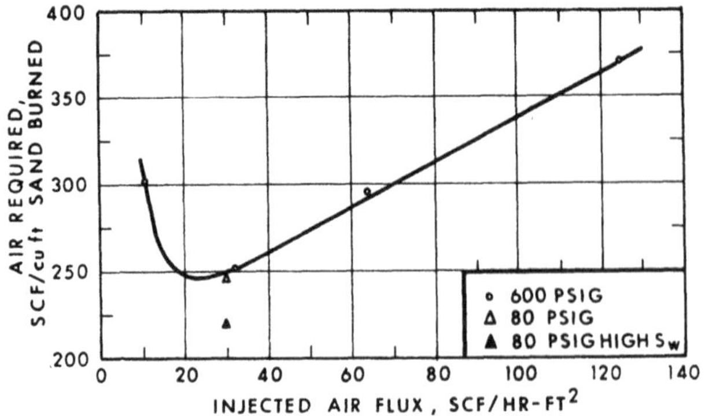

Fig. 8.7 – Effect of injected air flux on the air required per cubic foot of sand burned.[5]

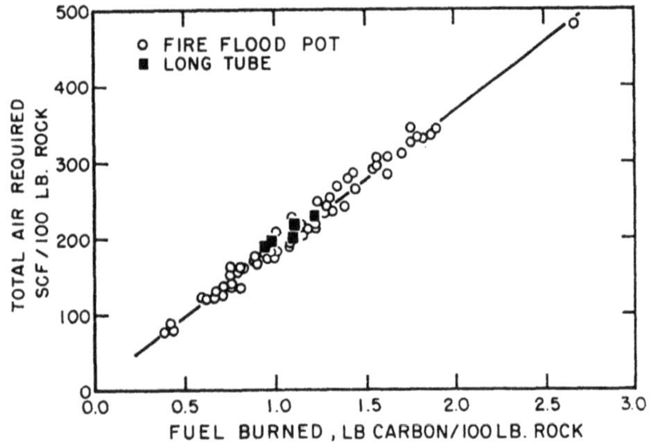

Fig. 8.5 – Effect of fuel burned on total air requirement.[1]

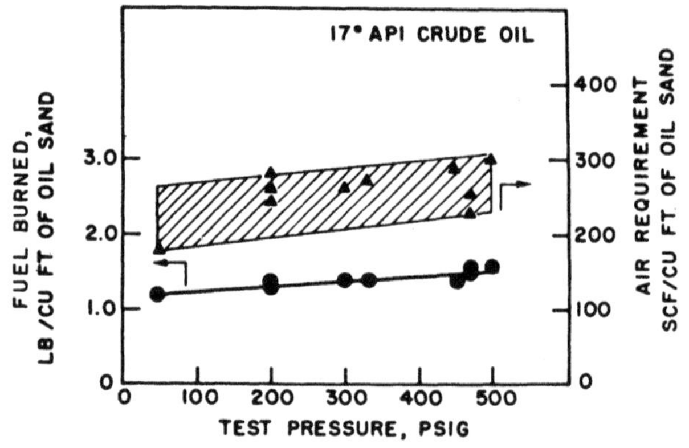

Fig. 8.8 – Combustion-drive fuel burned and air requirements vs. test pressure.[2]

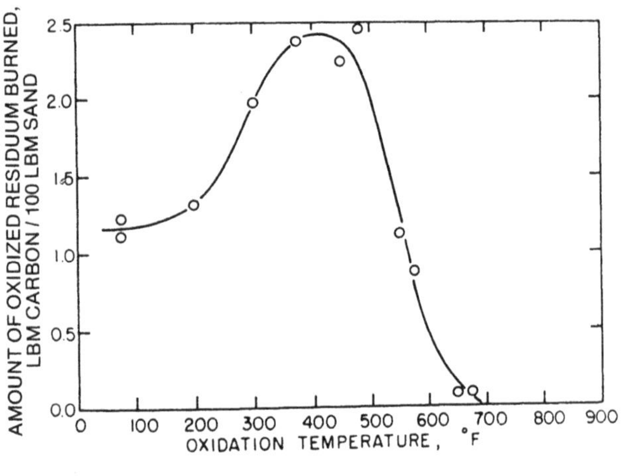

Fig. 8.6 – Effect of low-temperature oxidation on fuel burned at 800°F.[1]

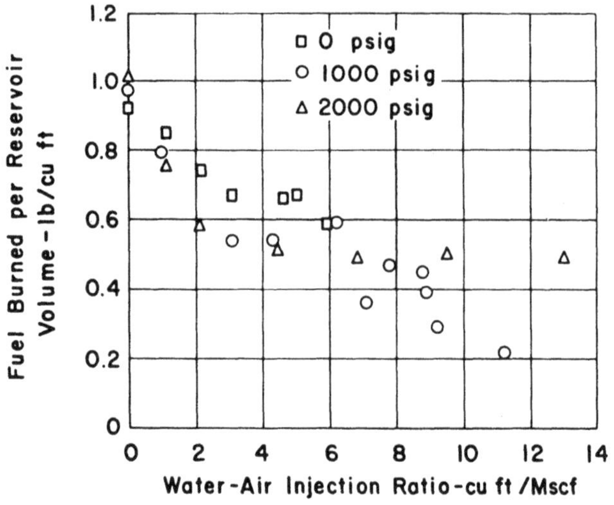

Fig. 8.9 – Fuel burned as a function of water/air ratio for Crude H.[5]

This correction is based on the observation that the amount of fuel required to heat a unit volume of the matrix to a given temperature increases as the amount of inorganic matter (which is proportional to $1-\phi$) increases.

The equivalent oil saturation burned can be expressed in terms of the fuel burned (given in mass per unit bulk volume) by

$$S_{oF}=\frac{m_R}{\phi\rho_o}, \quad \text{(8.2)}$$

where ρ_o is the oil density.

8.2 Oxygen/Fuel Reactions

At temperatures more than about 650°F, reactions between oxygen and crude-derived organic fuels result in the formation of CO_2, CO, and H_2O as the principal reaction products. These reactions remove carbon from the fuel, thus breaking the fuel carbon chain. At less than about 650°F, water and oxygenated organic compounds are the principal reaction products. These partial oxidation products generally consist of carboxylic acids, aldehydes, ketones, alcohols, and hydroperoxides and are characterized (by Burger and Sahuquet[7]) as being the result of low-temperature oxidation (LTO).

The stoichiometry of the high-temperature oxidation (HTO) process is given by

$$CH_x+(1-0.5m'+0.25x)O_2\rightarrow(1-m')CO_2+m'CO+\frac{x}{2}H_2O, \quad \text{(8.3)}$$

when the sulfur, nitrogen, and oxygen contents of fuel are ignored. Here, x is the average number of hydrogen atoms per carbon atom, known as the atomic H/C ratio, and m' is the mole ratio of CO to $(CO+CO_2)$.

Analysis of gas produced in field operations has shown, in addition to hydrocarbon gases, the presence of N_2, CO_2, CO, O_2, H_2, A, and H_2S. As related to the combustion process, nitrogen, oxygen, and argon are considered to come from the injected air; the carbon oxides are considered to be the products of high-temperature oxidation; hydrogen is considered to be released by thermal cracking of the crude in an oxygen-deficient zone just downstream of the combustion region; and any sulfur in excess of that associated with naturally-occurring hydrogen sulfide is considered to arise from desulfurization reactions in oxygen-deficient zones. Analyses of produced carbon oxides and free oxygen often are used to determine the type and amount of fuel being burned. Often it is assumed that the minerals and the water in the reservoir are inert. They are not. Poettmann *et al.*[8] showed that the presence of pyrite increases air requirements. Naturally occuring carbonates and sulfates do decompose at high temperatures, as well as react with acid products formed by low-temperature oxidation. Analyses of gases produced in steam soaks and drives show the presence of carbon oxides (sometimes in concentrations too large to be accounted for by the oxygen content of the crude), hydrogen sulfide, and hydrogen; this suggests reactions between the steam and the crude. There are very few published interpretations of the composition of gases produced from *field operations*, especially as they may be affected by minerals or by water/crude reactions. But unless the most careful interpretation is desired (which seldom is warranted in field operations), it is reasonable to assume that the minerals and water are inert where there are insignificant amounts of carbonates and pyrites. The solubility of reaction products in reservoir liquids also would affect the composition of the effluent gas, especially at early times.

Even though the stoichiometry given by Eq. 8.3 only approximates the reactions involving oxygen, carbon, and hydrogen because of (1) LTO reactions, (2) reactions involving minerals, and (3) water/organic-fuel reactions, it generally is used as a basis for estimating an equivalent H/C ratio of the fuel burned. The hydrogen-to-carbon ratio of the burned fuel (the H/C ratio) may be expressed both as an atomic ratio and as a mole ratio. When the stochiometry given by Eq. 8.3 is valid, the value of x is determined by measuring the mole ratio of CO to $CO+CO_2$, denoted by m', and the mole ratio of total oxygen reacted to CO_2 generated, denoted by $\hat{n}$. Then, when the reaction follows Eq. 8.3, x is given by

$$x=4(1-m')\hat{n}+2m'-4. \quad \text{(8.4)}$$

But it is difficult, even under laboratory conditions, to determine the amount of oxygen reacted and the amount of carbon and hydrogen oxides arising solely from the oxygen/fuel reaction. This, in principle, requires careful material balances on the O_2, H_2, and C present in the reactive system, including the often-neglected oxygen content of the crude and the contributions resulting from mineral alterations. In practice, these contributions usually are ignored and the mole ratio of total oxygen reacted to CO_2 generated, $\hat{n}$, is determined from a simple oxygen balance and the assumption that there is no accumulation of these gases in the system. For air,

$$\hat{n}=\frac{0.21i_a-qc_{O_2}}{qc_{CO_2}}, \quad \text{(8.5)}$$

where 0.21 is the fractional mole (or volume) content of oxygen in air, i_a is the air injection rate, q is the gas production rate (both on a dry basis), and c is the fractional molar (or volume) concentration in the effluent. The subscripts O_2 and CO_2 refer to oxygen and carbon dioxide. Since nitrogen and argon, which essentially compose the other 0.79 of the air, may be considered inert, it follows that

$$0.79i_a=qc_{N_2}. \quad \text{(8.6)}$$

With these assumptions, then, the equivalent atomic H/C ratio of the fuel burned may be calculated to be

$$x=4(1-m')\left(\frac{0.27c_{N_2}-c_{O_2}}{c_{CO_2}}\right)+2m'-4. \quad \text{(8.7)}$$

This equation is essentially the same as Eq. 8 of Dew and Martin,[9] except here it is not assumed that the dry effluent gas contains only nitrogen, oxygen, and the carbon oxides.

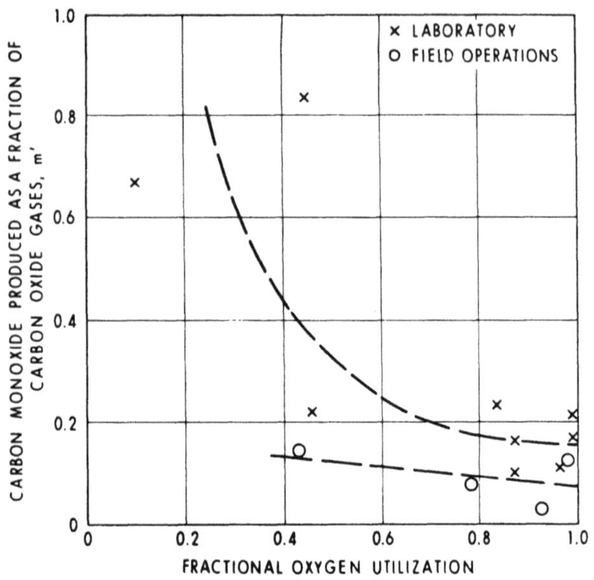

Fig. 8.10 – Variation of m' with oxygen utilization (based on data of Wöhlbier[3]).

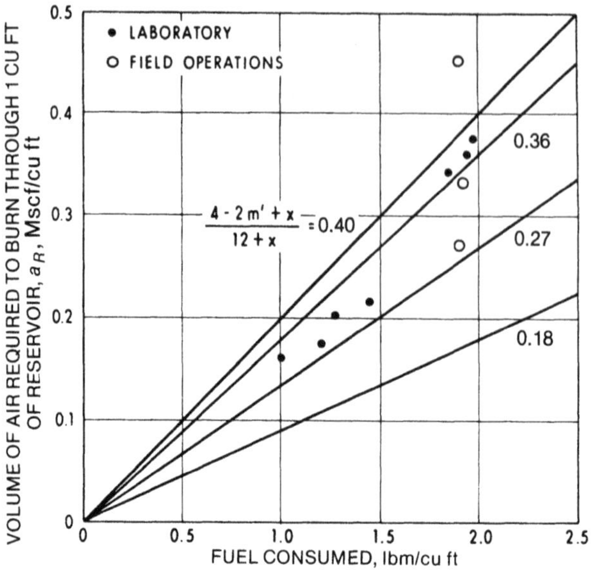

Fig. 8.11 – Effect of fuel burned and its composition on air requirements (based on data of Wöhlbier[3]).

Moss *et al.*[10] have pointed out that the atomic H/C ratio of the fuel burned is a function of how the process is operated. Where little or no carbonaceous residue remains in or near the burned zone, which is indicative of high-temperature oxidation reactions, the value of x is relatively low. As the contribution of low-temperature reactions increases – reactions that may be caused by low oxygen fluxes or by increased rates of heat loss – the value of x increases.

Both in the laboratory and in the field, the range of x is typically from about one to two, although a few values above two have been reported.[11,12] Values of m' appear to be related to the type of combustion taking place (LTO or HTO), which may be indicated by the oxygen concentration in the effluent stream. Values of m' vs. the fractional oxygen use are plotted in Fig. 8.10; the data are taken from Wohlbier's[3] summary of reported results. Note that field projects typically exhibit lower carbon monoxide concentrations in effluent gas than do laboratory studies.

Nelson and McNiel[6] have explained how the composition of the effluent gas can be used to determine the amount of fuel burned and the amount of water generated in the combustion process. As discussed in Section 8.1, there are many reports on the amount of fuel burned. The amount of water generated in the combustion process, which generally is not reported, can be determined from the stoichiometry of the HTO reaction. Expressed as an equivalent water saturation in the burned zone, it is given by

$$S_{wF}=\frac{0.319xa_R}{(4-2m'+x)\phi}, \qquad (8.8)$$

where a_R is the volume of air at standard conditions required to burn through a unit volume of reservoir in thousand standard cubic feet per cubic foot. The mass of water produced per unit bulk volume is directly proportional to the mass of fuel burned. A correction for differences between laboratory and reservoir porosities can be applied as indicated in Eq. 8.1. As a rule of thumb, the volume of water generated during combustion is equal to the equivalent volume of crude burned.

Air Requirements

Once the equivalent atomic H/C ratio of the burned fuel is available, the volume of air required to burn 1 lbm of fuel is found from Eq. 8.3, according to which $(1-0.5m'+0.25x)$ moles of oxygen react with 1 mole of fuel. A mole of gas occupies 0.379 Mscf, and a mole of fuel weighs $12+x$ lbm. Therefore, the volume of air required to burn through a cubic foot of reservoir, a_R, is calculated from

$$a_R=\langle 10^{-3}\text{ Mscf/scf}\rangle\frac{451m_R}{(12+x)}\cdot(4-2m'+x)\text{ scf/cu ft}, \qquad (8.9)$$

where m_R is the fuel burned per unit bulk reservoir volume in lbm/cu ft.

As pointed out by a number of investigators,[6,9,10] the value of the air requirement a_R is one of the most important parameters affecting the design and economics of a combustion project. Eq. 8.9 is presented in graphic form in Fig. 8.11. Air requirements summarized by Wohlbier[3] from published laboratory and field investigations also are included in this graph.

Although reported as the amount of air required to burn a unit volume of formation, the air requirement is, in fact, a measure of the air injected to burn a unit volume of formation. Some air is stored in the burned volume. The stored air is proportional to the average air density in the burned zone. The pressure there often is approximated by the injection pressure, p_{inj}, but the average temperature generally is not well known. The injected air required to burn through a unit bulk is given by

$$a_R^* = \left[a_R + (10^{-3}\ \text{Mscf/cu ft})\left(\frac{T_{sc,ab}}{\bar{T}_{ab}}\right)\right.$$
$$\left.\cdot\left(\frac{p_{\text{inj},ab}}{p_{sc,ab}}\right)\phi\right] \div E_{O_2}, \quad \ldots\ldots\ldots\ldots\ldots\ldots (8.10)$$

where p_{sc} and T_{sc} are the pressure and temperature at standard conditions, the subscript ab denotes absolute values, and E_{O_2} is the oxygen consumption or utilization efficiency. In the absence of better information, the temperature ratio may be approximated by 0.5.

Heat of Combustion

The heat released by the reaction indicated in Eq. 8.3 is given, after Burger and Sahuquet,[7] by

$$\Delta h_a = \frac{94.0 - 67.9m' + 31.2x}{1 - 0.5m' + 0.25x} \text{Btu/scf air}, \quad \ldots (8.11)$$

when the water of combustion condenses. Results are plotted in Fig. 8.12 for several values of the parameter β, which is the carbon monoxide/carbon dioxide mole ratio of the effluent gas. (The parameter β is the reciprocal of the parameter m introduced by Dew and Martin[9] to describe the combustion reaction). Over the range of typical values ($1<x<2$, $0<\beta<1$), the average value of Δh_a is close to 100 Btu per standard cubic foot of air reacted, which is a good rule of thumb for estimating the amount of heat generated. A value of 100 Btu per standard cubic foot of air reacted is also a good approximation ($\pm 10\%$) for the heat generated under LTO conditions, except when significant quantities of alcohols, and especially hydroperoxides, are formed. The amount of heat liberated by burning a pound of fuel is obtained by multiplying Eq. 8.11 by a_R/m_R (the air required to burn a pound of fuel), which gives

$$\Delta h_F = \frac{1{,}800}{12+x}(94.0 - 67.9m' + 31.2x) \text{ Btu/lbm}, \quad \ldots\ldots\ldots\ldots\ldots\ldots (8.12)$$

which is graphed in Fig. 8.13. In Figs. 8.12 and 8.13 and Eqs. 8.11 and 8.12, it has been assumed that water formed during combustion has condensed.

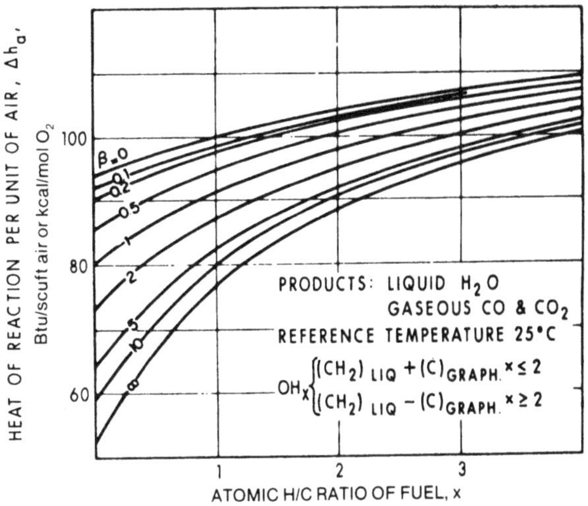

Fig. 8.12—Heat released (kcal/mol O_2 or Btu/scf air) as a function of the H/C ratio of the fuel and the CO/CO_2 ratio in the produced gases.[7]

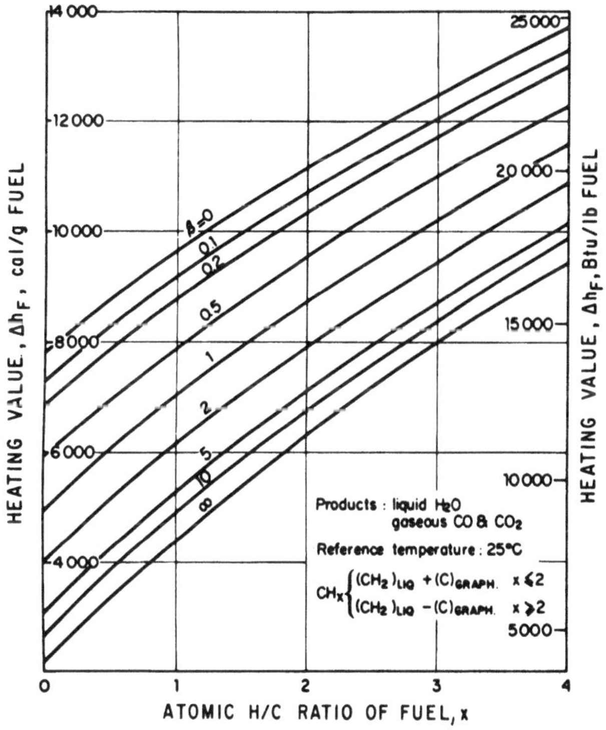

Fig. 8.13—Heat of combustion (cal/gm and Btu/lbm CH_x) as a function of the H/C ratio of the fuel and the CO/CO_2 ratio in the produced gases.[7]

8.3 Kinetics

Kinetics deals with the dynamics of reaction processes—i.e., the order, rate, and mechanisms of the reactions.

Fig. 8.14, taken from Burger and Sahuquet,[7] shows the concentration of carbon dioxide and carbon monoxide in the effluent gas and the concentration of oxygen consumed while air is passed through a mixture of crude and sand being subjected to a linearly increasing temperature. Note the two

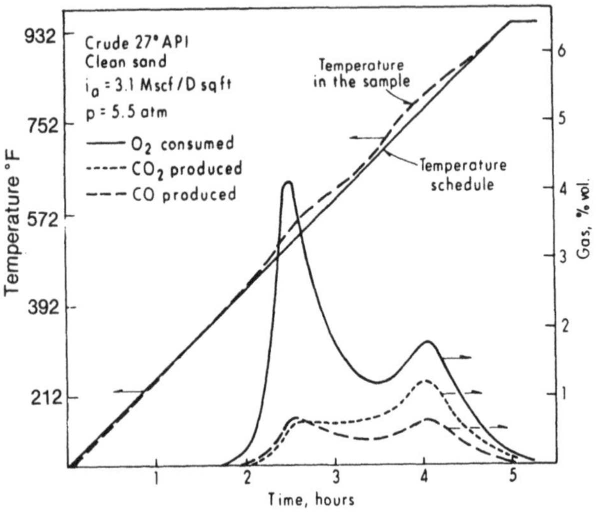

Fig. 8.14 – Oxidation of a crude oil in a clean sand.[7]

sets of concentration peaks. The first (at about 485°F) corresponds to a low-temperature oxidation of the crude, one in which relatively little CO_2 is formed even though O_2 consumption is high. Note that in the neighborhood of this concentration peak, the amount of oxygen consumed exceeds that recovered as carbon oxide gases. The second peak (at about 750°F) corresponds to the combustion of a fuel with a low H/C ratio. Almost all the oxygen consumed at high temperatures can be accounted for by the produced carbon oxide gases. Since the production of carbon oxide gases represents the removal of carbon, the reactions associated with the second peak are controlled by the simultaneous availability of fuel and oxygen at high temperatures.

The fuel is said to be burning when conditions associated with the second peak prevail – i.e., the amount of oxygen consumed is essentially balanced by the amount in the produced carbon oxide gases. In the low-temperature region, the fuel is being oxygenated rather than burned; a smoldering rather than a burning takes place.

HTO reactions usually are found during the frontal advance (see Chap. 4) of a dry combustion front. Smoldering reactions are likely to be found (1) under bypass conditions, (2) in wet combustion processes, and (3) in dry combustion operations conducted in thin sands or at low air injection rates.

The main interest in studying the kinetics of in-situ combustion is to determine the conditions required to achieve ignition and maintain combustion. The rate of oxygen reacted per unit mass of fuel, K, can be written as

$$-\frac{1}{m_o}\frac{dm_{O_2}}{dt}=K=p_{O_2}^{n}A_c e^{-E/RT_{ab}}, \quad \ldots . (8.13)$$

where m_{O_2} is the mass of oxygen consumed per unit bulk reservoir volume, m_o is the mass of oil per unit bulk reservoir volume, p_{O_2} is the partial pressure of oxygen (atm), E is the activation energy (Btu/lbm mol), R is the universal gas constant (=1.986 Btu/°R·lbm mol), T_{ab} is the absolute temperature (°R), A_c is the pre-exponential constant (sec^{-1} atm^{-n}), and n is the order of the reaction with respect to oxygen. During low-temperature oxidation, the fuel is probably the crude itself.

Reported values of the constants A_c, n, and E obtained experimentally for temperatures up to 475°F have been summarized by Burger and Sahuquet.[11] Table 8.1 lists these and other reported values for low and intermediate temperatures. The activation energy E ranged from 30,300 to 36,400 Btu/lbm mol, with an average value of 32,200 Btu/lbm mol. Values of the pre-exponential constant A_c vary widely, and the order of the reaction n ranges from 0.31 to 0.75. Surface area and type of mineral in the rock matrix influence oxidatiom and combustion kinetics and appears to have a strong effect on the amount of fuel burned.[13,14]

Kinetic reactions also can be written for the fuel and for any reactant (e.g., carbon, hydrogen, and sulfur). Fassihi *et al.*[13] present a thorough survey of published kinetic data, and emphasize the importance of clay content and previous history on the kinetics of oxygen/fuel reactions. Additional kinetic data are reported by Thomas *et al.*[15]

Spontaneous Ignition

Before a field combustion project is started, air generally is injected either to determine or to increase the injectivity. It is undesirable for the crude to ignite spontaneously before one is ready, as discussed in

TABLE 8.1 – KINETIC PARAMETERS FOR OXIDATION AND COMBUSTION OF CRUDES

Crude (°API)	A_c (sec^{-1}atm^{-n})	E (Btu/lbm mol)	n	Temperature Range (°F)	Reference
N.R.	3,080	31,600	0.46*	140 to 250	16
N.R.	925	30,800	0.57	140 to 250	16
N.R.	498	31,600	0.79	140 to 300	16
N.R.	84,800	36,600	0.48	140 to 250	16
N.R.	1,210	30,900	0.45	140 to 250	16
N.R.	7,380	33,800	0.31	140 to 250	16
27	310	30,800	0.65	194 to 300	17
18	16,580	36,400	0.6	174 to 284	11
22	1,200	30,400	0.6	122 to 284	11
19.9	**	30,300	0.5	270 to 475	63
27.1	**	31,350	0.75	250 to 450	63
27.1	**	31,360	0.75	250 to 450	63

*For South Belridge crude.[63]
**Not available in these units.

Sections 8.6 and 11.5. The earliest attempt at determining the time at which spontaneous ignition occurs after air injection is started was made by Tadema and Weijdema.[16] Their model of the process is rather idealized. They neglected all bulk heat transfer so that the energy balance obtained by neglecting spatial derivatives in Eq. 3.27 reduces to

$$\frac{dT}{dt}=\langle 86{,}400\ \mathrm{s/D}\rangle \cdot \frac{\phi S_o \rho_o \Delta h_{O_2} A_c\, p_{O_2}^n}{M_R} e^{-E/RT_{ab}}, \quad \ldots\ldots (8.14)$$

where Δh_{O_2} is the heat released per unit mass of O_2 reacted, S_o and ρ_o are the saturation and density of the oil in the reservoir, and M_R is the volumetric heat capacity of the reservoir. As reported by Tadema and Weijdema,[16] the ignition delay time (measured from the start of air injection) can be estimated from the approximate analytical expression:

$$t_{ig}=\left[2.04\times10^{-7} M_R T_{abi}^2 (1+2RT_{abi}/E)\cdot Re^{E/RT_{abi}}\right] \Big/ \left[\Delta h_a \phi S_o \rho_o A_c E\, p_{O_2}^n\right]. \quad \ldots\ldots (8.15)$$

In this equation, T_{abi} is the initial absolute temperature of the reservoir, air injection is considered, and the fact that Δh_{O_2} is proportional to Δh_a has been used.

Ignition times are very sensitive to the value of E/RT_{abi}. As indicated in Example 8.1, a 1°F error in the reservoir temperature would result in an error of several percent in the estimated ignition time.

Eq. 8.14, it must be remembered, was derived on the assumption of no bulk heat transfer. Burger,[17] using a numerical simulator, has investigated the effect of the flow geometry (both radial and linear) and of heat transfer in the direction of flow by convection and conduction. As indicated in Fig. 8.15, Burger's calculated ignition times are only a few percent longer than those given by Eq. 8.15, except for reservoirs at low initial temperature, where the differences may be as much as 100%. Heat losses to caprock and base rock also would tend to increase ignition times beyond those calculated by Burger. This factor has not been investigated yet, but it is doubtful it would have much of an effect except in very thin or low-temperature reservoirs containing slowly reacting crudes.

Burger[17] also has investigated the distance from the injection well at which spontaneous ignition occurs. The relatively cool injected air transfers heat away from the immediate vicinity of the wellbore, so that ignition always occurs a few feet away (see Fig. 8.16). His model, however, does not take into account how the distance at which ignition occurs is affected by the cooling associated with the stripping of water by the injected gas. As reported by Howard *et al.*,[18] the cooling by the injected air caused spontaneous ignition to occur at distances between 50 and 115 ft in the Charco Redondo field.

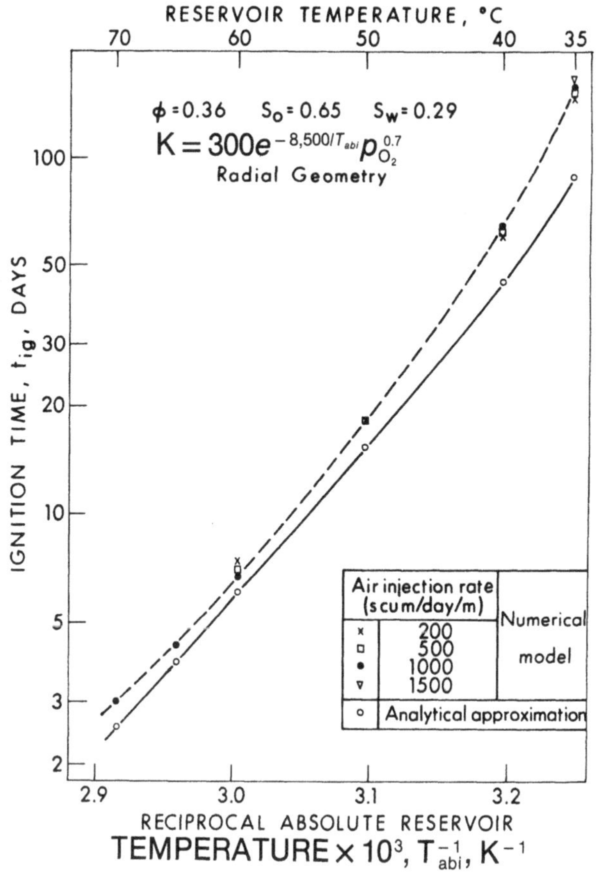

Fig. 8.15 – Effect of reservoir temperature and air injection rate on time to obtain spontaneous ignition.[17]

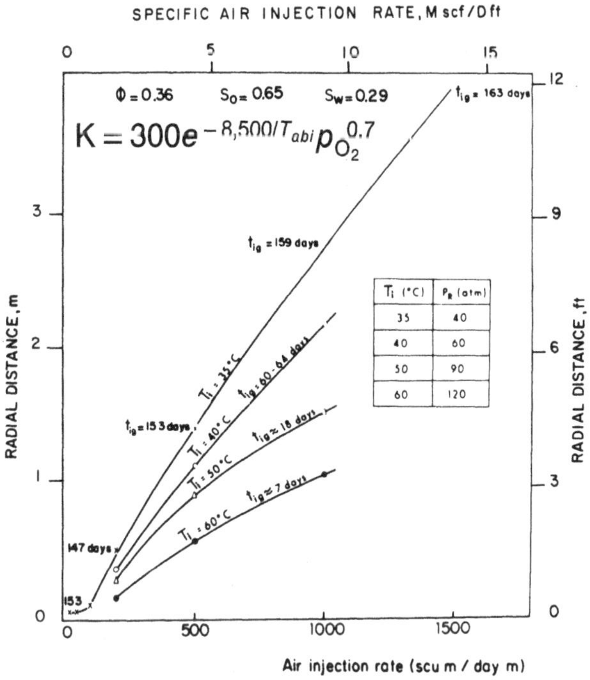

Fig. 8.16 – Effect of injection rate and reservoir temperature on distance at which spontaneous ignition occurs.[17]

Example 8.1 – Estimation of Ignition Time

Estimate the time required to obtain spontaneous ignition at South Belridge field (Gates and Ramey[19]) given the following.

Formation temperature T_i, °F	85
Air injection pressure p_{inj}, psig	210
Heat of oxidation Δh_a, Btu/scf	100
Porosity ϕ	0.37
Oil saturation S_o	0.60
Oil density ρ_o, lbm/cu ft	60.7
Observed ignition time t_{ig}, days	106
Volumetric reservoir heat capacity M_R, Btu/cu ft-°F	40

From these data, we obtain

$$T_{abi} = 460 + 85 = 545\text{°R},$$

and

$$p_{O_2} = (0.209)(210 + 14.7)/(14.7) = 3.20 \text{ atm}.$$

From Table 8.1, we obtain

$$n = 0.46,$$

$$A_c = 3{,}080 \text{ s}^{-1} \text{ atm}^{-0.46},$$

and

$$E/R = 15{,}900\text{°R}.$$

From Eq. 8.15,

$$t_{ig} = \{(2.04 \times 10^{-7})(40)(545)^2 \cdot [1 + 2(545)/15{,}900](4.68 \times 10^{12})\} \div [(0.37)(0.6)(60.5)(100) \cdot (3{,}080)(3.20)^{0.46}(15{,}900)]$$

$$= 108 \text{ days}.$$

The calculated value of 108 days is in good agreement with the observed value of 106 days. Note that a 1°F (0.6°C) difference in the temperature used for the reservoir would have resulted in a 5.5-day change in the calculated ignition time.

Relationship Between Air Flux and Combustion Front Velocity

It follows from the definition of the air requirement a_R^* that, on a local basis, the relationship between the rate of advance of a combustion front v_b and the air flux u_a (in the same direction) is

$$v_b = E_{O_2} u_a / a_R^*, \quad \text{.....................} \quad (8.16)$$

where E_{O_2} is the oxygen consumption efficiency. The air requirement a_R^* is proportional to the amount of fuel burned per unit volume of reservoir (see Eq. 8.9 or Fig. 8.11), whose rate of burning is controlled by the ambient temperature (Eq. 8.13).

For sufficiently large fluxes, the rate at which energy is liberated at a combustion front is large compared with the rate at which heat is lost to the nearby reservoir and adjacent formations. The temperatures, thus, will remain high at the advancing combustion front, ensuring a high rate of reaction between fuel and oxygen and a continued vigorous process. At lower air fluxes, however, the rate at which heat is generated is correspondingly lower. And for very low rates, it is possible that the rate of heat loss to the surrounding mass exceeds the rate at which heat is being generated. This would lead to a reduction in temperature and a further drop in the rate of reaction and heat release. Thus, it is argued that there is a minimum air flux below which combustion cannot be maintained. This air flux is very difficult to determine, since it is highly sensitive to the geometry of flow, the previous thermal history of the project, and the low-temperature oxidation characteristics of the fuel contacted by the oxygen in the reservoir. Nevertheless, there seems to be a consensus that active combustion (HTO) cannot be supported at average air fluxes less than about 20 scf/D-sq ft, when oxygen utilization is complete.[3,6]

8.4 Dry Forward Combustion

The most commonly used form of the combustion process is simple air injection. It is called dry combustion to distinguish it from wet combustion, in which water and air are injected into the reservoir.

Model of the Process

The conceptual model that is used most commonly to describe, understand, and calculate the recovery by the dry forward combustion process is the frontal advance (FA) model. Within a horizontal reservoir, all phenomena and displacements are independent of vertical position. In the FA model, the combustion front fills the entire vertical extent of the reservoir, and all fluid transport is horizontal. Since temperatures in the oil bank are not too high (the downstream surface of the oil bank moves much faster than the temperature fronts), the growth of the oil bank increases the resistance to flow. In the case of high initial oil saturations of viscous crudes, the resistance to flow can be great enough to reduce the air injection rate to the point that the rate of heat losses exceeds the rate at which heat is being generated by combustion. The resultant cooling only aggravates the situation. The system becomes plugged. Plugging, it must be remembered, does not always occur in the FA model. But the flow resistance in the FA model is higher than in any other model. The FA model is easy to conceptualize and understand; it was proposed in a manner and at a time when other displacement processes (e.g., waterflooding) were being investigated similarly; it is very well represented by simple laboratory combustion tube experiments; and it lends itself to the separation of chemical and displacement mechanisms. The FA model is representative of field behavior where there is no tendency for fluids to bypass through high-permeability zones or where pressure gradients everywhere are large compared with the gravitational potential gradient ($g\Delta\rho/g_c$) – conditions that often may not prevail. Despite these limitations, the FA model has served as a useful springboard from which much significant information and insight on combustion processes have been obtained and developed.

Bypass models, though more realistic, are inherently more complicated than FA models. Not only do they increase the number (by at least one) of space variables that need to be considered, but they allow different mechanisms and displacement processes to occur within the vertical extent of the reservoir. By and large, simple bypass models capable of yielding quantitative results have not been developed adequately. Numerical reservoir simulators, which are far from simple, can include all pertinent phenomena. This section will touch on selected phenomena associated with bypassing in combustion processes, but the presentation will emphasize FA models. A general discussion on the effect of the severity of bypassing on reservoir mechanisms (where no bypassing leads to FA models) is given in Sections 4.1 and 4.2.

The locations of the various zones in a dry combustion process relative to each other and to the injector can be depicted as in Fig. 8.17, which represents an FA model. Seven zones are recognized in this figure, which is taken from Tadema.[20] Starting from the inlet end, these are (1) the burned zone, (2) the combustion zone, (3) the evaporation (and cracking) zone, (4) the condensation zone, (5) the water bank, (6) the oil bank, and (7) the virgin zone. These zones are moving in the direction of the air flow so that they reach any point in the reservoir in reverse order. The saturations and temperature profiles for these zones also are given in Fig. 8.17.

The burned zone characteristically is filled with air and may contain as much as 2% of unburned solid organic fuel (as equivalent crude saturation). Generally, the higher the combustion temperature, the lower the amount of unburned fuel present. The presence of this organic residue is not necessarily obvious from visual examination. The color of the burned zone is typically off-white, with light hues in grays, browns, yellows, and reds. This is the zone that has been subjected to the highest temperatures for the longest period and usually exhibits mineral alteration upon detailed analysis.[21] Because of heat losses to adjacent formations and the influx of relatively cold air, the temperature in the burned zone increases with distance from the injector.

It is in the combustion zone that temperatures are highest. Here the combination of fuel, oxygen, and high temperature leads to rapid and exothermic reactions. It is generally idealized that in high-temperature reactions all the oxygen is reacted within the combustion zone. Combustion gases (carbon oxides) are considered to be generated within the combustion zone, starting at its upstream boundary. As has been stated (Section 8.2), the fuel burned is fairly rich in hydrogen. Solid at room temperature, the fuel is presumed to be solid also while burning. Water generated by the combustion is generally present as superheated steam and flows along with the other vapors into and through the evaporation zone.

In the evaporation zone, the organic matter is considered to be fluid (although different from the original crude). Several things happen to the crude in this zone. It is thermally cracked (pyrolyzed), distilled by the gas stream, and decarboxylated (-CO groups break off on mild heating, giving off CO_2).

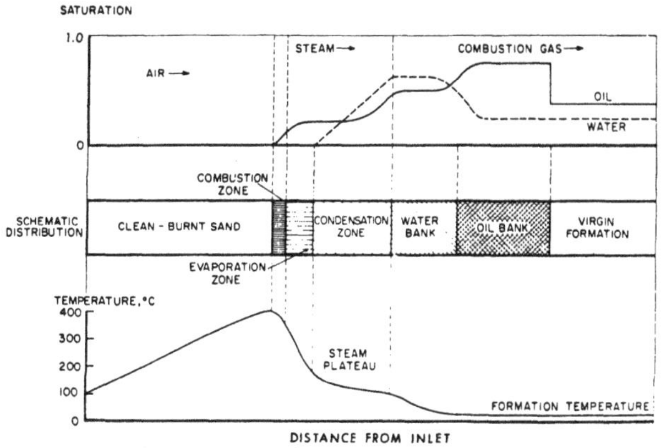

Fig. 8.17 – Schematic saturation and temperature distribution, dry combustion process.[20]

Hydrocarbon and organic gases are given off, the atomic hydrogen/carbon ratio of the remaining organic matter is reduced, and the remaining organic matter shrinks in volume. Although the temperatures are lower than in the combustion zone, they are enough to keep water in the state of superheated steam.

The extent of the condensation zone coincides with that of the (saturated) steam zone. Because the pressure drop and changes in gas phase composition are not large within this zone, the temperature is fairly uniform and leads to what commonly is called the steam plateau. Some of the hydrocarbon vapors entering this zone condense and dissolve in the crude, thus reducing its viscosity. Decarboxylation and visbreaking also may occur in this zone, depending on its temperature. The oil in place is displaced by the advancing steam zone, so that the evaporation zone only "sees" the oil remaining in the steam zone.

Ahead of the steam condensation front (the downstream boundary of the condensation zone), there is a water bank which is at a temperature intermediate between those of the initial reservoir and the steam zone. The water bank is characterized by water saturations somewhat higher than those in the oil bank. The higher water saturations are thought to result from the fact that the mobility of the oil is higher in this zone than in the virgin zone, in accordance with the fractional-flow relationship. Some oil is displaced from the water bank, and all the oil that has been displaced from the upstream zones (except that fraction present as hydrocarbon gases) accumulates in the water and oil banks, with the oil saturations reaching the highest values in the oil bank, where the temperatures are lowest.

Downstream of the oil bank lies the part of the reservoir as yet unaffected – except by the solution of CO_2 and the flow of combustion gases – and labeled in Fig. 8.17 the virgin formation. If there is mobile water saturation initially in the system, the oil in the oil bank will displace water and form a second water bank (not identified in Fig. 8.17) between the oil bank and the virgin reservoir.

This description would suggest that the zones are sharply divided and easily identifiable, which is not

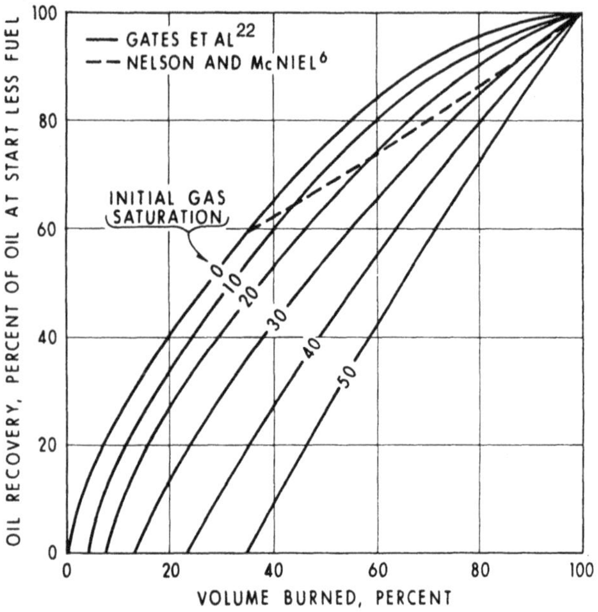

Fig. 8.18 – Estimated oil recovery vs. volume burned (based on data of Refs. 6 and 22).

at all so in all situations. There are transitions between all the zones, and although it is usually possible to identify the predominant phenomena occurring within a zone, the mechanisms grade across the transition zones between adjacent zones.

Where bypassing occurs, the sequence of zones is generally different from that described above. Differences arise mainly from the fact that the convective oxygen, fluid, and heat fluxes are different at various points of the combustion surface. No general description is possible, but the relative importance of the various zones shown in Fig. 8.17 may be very different at various locations of the same project. For example, even within a homogeneous sand the denser fluids would tend to flow and collect near the bottom, and the lighter fluids near the top. The condensation zone, then, would tend to be larger near the top than near the bottom of the sand. In extreme cases, there would be essentially no water or oil banks near the top and no steam zone near the bottom of the sand. But every case would be different.

Air Requirements

In a continuing field operation, the air requirement per unit volume of reservoir burned can be determined from information about the cumulative injected air required to burn a known reservoir volume. Although the cumulative air injected is known, the corresponding volume burned must be estimated from temperature responses at nearby wells or from well tests designed for that purpose (see Section 11.4). Cores obtained from the project area sometimes are used to help determine the extent of the burned zone. The time of arrival of the combustion front at a well also can be used to estimate the velocity of the combustion front, from which the air requirement may be estimated from Eq. 8.16. Once the air requirement a_R^* is estimated by these means, the corresponding amount of fuel burned, m_R, is estimated from Eqs. 8.9 and 8.10 and the composition of the effluent gas. Fuel consumption also may be estimated from analyses of temperature responses in observation wells (see Section 13.4) and from analyses of the water cut of produced fluids.[22] The fuel consumption and specific air requirement also may be estimated from laboratory experiments, as discussed in Sections 8.1 and 8.2. Often, both field and laboratory approaches are used to aid the analyses and to complement and support the interpretations.

The experimental approach also can be used in projects that are in the planning stage. Or correlations such as those given by Figs. 8.1 through 8.5 may be used.

Once the specific air requirement for a project has been accepted, the total air injection required to burn a bulk reservoir volume V_{Rb} is given by

$$G_a = \langle 43.56 \text{ Mscf/acre-ft}\rangle a_R^* V_{Rb} / \bar{E}_{O_2} \text{ MMscf}, \qquad (8.17)$$

where $\bar{E}_{O_2}$ is the average oxygen consumption efficiency. Generally, before the combustion front arrives at the producers, the values of $\bar{E}_{O_2}$ are in the range of 0.95 to 1.0.

The value of the cumulative air injected per volume of oil produced, denoted by F_{ao}, then is given by

$$F_{ao} = \langle 10^3 \text{ Mscf/MMscf}\rangle \cdot G_a(t)/N_p(t) \text{ Mscf/bbl}, \qquad (8.18)$$

where N_p is the cumulative oil production. A simple model for estimating the cumulative oil production required in this equation is given in the next section.

Production Response

Empirical,[6] analytical,[23] numerical,[3,24-26] experimental,[27-30] correlative,[31] and hybrid[22,32] methods are available to estimate expected recoveries from combustion operations.

An example of an analytical method is that of Wilson *et al.*,[23] which is applicable to frontal displacement models in one dimension. It gives estimates of the fluid saturations downstream of the combustion front and explains how these affect the rates of growth of the oil and water banks and the rates of oil and water production during the operation. Though idealized, it provides insight into the three-phase flow phenomena that occur downstream of a combustion front, and is especially suitable for studying plugging due to the buildup of a cold oil bank.

The numerical methods range in applicability from tank to three-dimensional models. Some are process simulators[33]; others appear to be able to simulate the reservoir response from several wells. Three-dimensional reservoir simulators capable of describing the production response of combustion or other thermal processes have been developed.[24,25]

A hybrid method based on field data and analysis of experimental results from tube runs is that first mentioned by Gates *et al.*[22] and subsequently discussed in more detail by Fassihi *et al.*[32] The heart of the method is represented by a plot of oil recovery, expressed as a percent of the oil in place at the start of the fireflood less the cumulative fuel burned vs. the volume burned, expressed as a percent of the total volume. Fig. 8.18 shows the effect of initial gas saturation on the oil recovery. As discussed in Section 4.3, free gas or highly mobile water saturations serve as storage space for the displaced oil, which delays the oil production. Fassihi *et al.*[32] provide a program for a hand-held calculator to compute the performance of the in-situ combustion process. Oil production rates and instantaneous air/oil ratios can be estimated from the slope of the curves, taking into consideration the air required to burn through a unit reservoir volume and the fraction of the injected air actually being used. The oxygen consumption efficiency is not predicted by this method. Once oil production starts, rates are higher the longer the delay. Notice that for low initial gas saturations, the amount of oil produced initially is greater than that displaced from the burned volume. Some of the oil downstream of the combustion front is being produced as a result of the imposed pressure gradients. Although there are fundamental differences between the displacements from tube experiments (which are presumably frontal advance in character) and displacements in the field (where bypassing often is observed), the method has given good results. The results presented in Fig. 8.18 were obtained for conditions in the South Belridge field and are thought applicable to similar reservoirs.

Here, an empirical method is discussed for estimating the ultimate production response. The method, patterned after that of Nelson and McNiel,[6] is essentially a tank mass balance and is easy to apply. It provides only the cumulative production of oil and water at the end of the project and the average rates and life of the operation.

Nelson and McNiel Method.[6] The total oil and water production is given by

$$N_p = \langle 7{,}758 \text{ bbl/acre-ft}\rangle \phi [V_{Rb}(S_{oi}-S_{oF}) + 0.4(V_P - V_{Rb})S_{oi}], \quad \ldots\ldots\ldots (8.19)$$

and

$$W_p = \langle 7{,}758 \text{ bbl/acre-ft}\rangle V_{Rb}\phi(S_{wi}+S_{wF}), \quad \ldots\ldots\ldots (8.20)$$

where S_{oi} is the initial oil saturation, V_{Rb} is the volume of the burned zone, V_P is the volume of the well pattern, S_{oF} is the equivalent oil saturation burned, and S_{wF} is the equivalent water saturation (based on the volume of the burned zone) resulting from the combustion process. Values of S_{oF} are related to the reported values of fuel burned per unit volume by Eq. 8.2, and values of S_{wF} can be found from the composition of the effluent gas in a combustion process using Eq. 8.8. Nelson and McNiel also have shown how to obtain the fuel burned per unit volume from effluent gas composition.

In Eq. 8.19, the quantity $\phi V_{Rb}(S_{oi}-S_{oF})$ represents the oil displaced from the burned zone. The last term in that equation represents the contribution to the oil production resulting from those parts of the well pattern that lie outside the burned zone. The assumption is that 40% of the original oil in place will be produced from the unburned pattern volume. "Core analysis from field experiments shows that, in a heavy oil reservoir, more than half of the oil in those portions of the pattern either bypassed vertically or not reached by the front may be produced by a combination of hot gas drive and gravity drainage. For the purpose of this design procedure, an average recovery efficiency of 40 percent is assumed for the unburned portion of the reservoir."[6] Clearly this could occur only after substantial heating of the reservoir, so that Eq. 8.19 would apply only after a substantial fraction of the well pattern had been burned.

The water production is composed of the water initially in place in the burned zone and that formed as a result of combustion.

The cumulative oil recovery can be used together with the air requirement given by Eq. 8.17 to obtain the estimated ratio of injected air to produced oil, as indicated in Eq. 8.18. Oil/water production ratios are obtained directly from Eqs. 8.19 and 8.20. Rates are constant over the project life and are obtained by dividing the cumulative recovery by the project life. The project life is determined by dividing the total air requirement by the air injection rate. The factor that must be estimated to obtain all these results is the size of the burned volume at the end of the project.

The Nelson and McNiel method can be presented as in Fig. 8.18. The dashed line shows the results for an initial oil saturation of 0.55 and a burned oil saturation of 0.08. The dashed line is shown only for large values of the burned volume, the applicable range. As one might expect, recoveries obtained by Nelson and McNiel's empirical method agree with those obtained by the method of Gates *et al.* only for low initial gas saturations.

Nelson and McNiel[6] are silent on general methods for estimating the volume burned at the end of a project – a value that can vary widely. Where gravity override is significant, the thickness burned often averages 2 to 10 ft over the burned volume. Where bypassing is not along the top of the sand, the thickness burned may be slightly higher, but the amount of fuel burned may be relatively great because of oil draining into the combustion zone. Combustion generally proceeds until about breakthrough (the arrival) of the combustion front at nearby producers. For frontal advance in a five-spot pattern, the combustion front arrives at a producer when about 50% of the pattern has burned; this is inferred from the correlations provided by Craig[34] for the effect of breakthrough sweep efficiency at large unfavorable mobility ratios. Similar results have been reported for very unfavorable mobility ratios by other investigators whose studies also have

TABLE 8.2 – EFFECT OF RATE ON VOLUME BURNED AT BREAKTHROUGH[6]

i_D	Five Spot Areal Sweep Efficiency at Breakthrough (%)
3.39	50.0
4.77	55.0
6.06	57.5
∞	62.6

included other well patterns.[35] Where bypassing is significant, operations often are continued carefully (see Section 11.5) beyond breakthrough. Thus, the volume burned at the end of a project can vary from a few percent to about 50% of the process volume.

Design

Dry combustion displacement processes are alternatives to steam drives. Conditions that would suggest using combustion rather than steam include: (1) high reservoir pressure – e.g., greater than 2,000 psi, (2) excessive heat losses from steam injection wells – e.g., in reservoirs deeper than 4,000 ft, (3) lack of a fresh water supply or water treatment costs that make the use of steam too expensive, (4) serious clay swelling problems due to fresh condensate, and (5) unattractive or prohibited use of any fuel to fire steam generators. As in any displacement process, sand continuity between injectors and producers is an absolute requirement.

The design of a combustion project, like that of any injection operation, must consider how the reservoir flow resistance and the injection pressure limitations imposed by prudent operation practices affect the injection rate. Since the rate of growth of the burned zone is essentially proportional to the air injection rate, the maximum air injection rate determines the *minimum life* of the project. If the minimum life is unacceptably long from an investment (or any other) point of view, ways must be considered either to prudently increase the levels of the injection pressure or to reduce the flow resistance to air, or both. Flow resistance to air may be reduced by increasing the gas saturation before ignition, usually by injecting air for weeks or months before ignition, by reducing the well spacing, or through conventional thermal stimulation treatments. If air is injected for a prolonged period before ignition, there is a risk that ignition will occur before an adequately low resistance to air flow is established. The increased oil mobility resulting from the combustion then may move oil into the gas flow channels and reduce the air injection rate. Some increase in flow resistance may occur anyway once the project is properly started, especially where gravity override is not extensive. It may be detrimental to the project if partial plugging occurs before the operator has been able to establish an adequately low flow resistance to air before ignition.

An additional factor that must be considered in combustion operations is the concept of a minimum air flux. According to this concept, vigorous burning cannot occur for long where the local rate of heat losses exceeds the rate of heat generation. Here, vigorous burning is associated with high-temperature reactions described by Eq. 8.3. Since in any well pattern there are regions of essentially no flow, it follows from the minimum air flux concept that it is *never* possible to completely burn a reservoir. Nelson and McNiel[6] have provided some information on the areal sweep efficiency in a repeated five-spot well pattern at the time the combustion front reaches the producers (i.e., at breakthrough), and in the absence of gravity override. Their results, which are reported to represent an infinite mobility ratio, are reproduced in Table 8.2. As expected, the fraction of the pattern burned at breakthrough increases as the dimensionless air injection rate i_D increases. This dimensionless injection rate is defined as follows.

$$i_D = i_a / Lhu_{min}, \quad \text{(8.21)}$$

where i_a is the air injection rate, L is the distance between injector and producer of a five-spot pattern, h is the formation thickness, and u_{min} is the minimum air flux. It should be pointed out, however, that although the trend of higher breakthrough sweep efficiencies with increasing i_D is to be expected, the maximum value of the sweep efficiency reported by Nelson and McNeil is approximately 20% higher than the average values reported by others.[34] The trend, however, is informative.

On the basis of field and laboratory studies, Nelson and McNiel concluded that the minimum air flux is the one that results in a rate of advance of a combustion front of 0.125 ft/D.[6] Thus, from Eq. 8.16,

$$u_{min} = 0.125\, a_R^* / E_{O_2}, \quad \text{(8.22)}$$

where a_R^* is the air required to burn a unit volume of formation, and E_{O_2} is the oxygen consumption efficiency.

Where the injection pressure is limiting, the maximum air injection rates for the well patterns under consideration would be estimated as indicated by Nelson and McNiel[6] for the five-spot or as indicated in Chap. 4. This determines the value of i_D, from which the breakthrough sweep efficiency is determined. Ultimate oil recoveries are estimated on the basis that all the oil displaced from the burned zone at breakthrough plus a fraction of the oil outside the burned zone are produced. The fraction of the oil recovered from the unburned zone obviously depends on the reservoir, the crude, the well locations, and the operation practices. In those operations where gravity override has prevailed and the reservoir is not too thick (e.g., no more than about 50 ft), substantial reductions in oil saturations below the combustion zone have been measured after a prolonged operation. Nelson and McNiel[6] have proposed using a recovery fraction of 40% for the oil initially in the reservoir outside the unburned zone.

On the basis of a 40% factor, the oil recovered from the unburned zone can approximate that recovered from the burned zone. Obviously, the use of estimates of this sort can lead to large variations in the calcuated design performance. The resultant uncertainties must be taken into account in such preliminary design calculations. Where bypassing is

TABLE 8.3 – SELECTED MAJOR IN-SITU COMBUSTION PROJECTS

Project	Operator	Date Started	References
West Newport, CA	Kadane, General Crude	1958	64
Midway-Sunset, CA	Mobil	1960	38
Miga, Venezuela	Gulf	1964	53
South Belridge, CA	Mobil	1964	22
Suplacu, Rumania	Ministry of Mines, of Petroleum, and Rumanian Geology	1967	40,52
Bellevue, LA	Getty	1967	54,65,66
Trix-Liz, TX	Sun	1968	67
Midway-Sunset, CA	Chanslor-Western	1972	39
Brea Olinda, CA	Union	1972	66,68

not pronounced and the oil is very viscous, recovery from outside the burned zone could be appreciably smaller than the 40% cited above. Furthermore, the resistance to the injected air in such a case would be appreciably higher than where there is override, which would have other adverse effects on performance.

Examples of Field Applications

The in-situ combustion process has been applied in a wide variety of fields. Most of the projects have been small, aimed at assessing the commercial merits of the process. There are numerous large-scale examples,[36,37] but they are only a small fraction of all the field projects that have been carried out.

Table 8.3 lists selected large in-situ combustion projects. All appear to be technical successes. The depth of these projects below surface ranges from 250 ft at Suplacu to more than 4,000 ft at Miga. Reservoir thickness ranges from a net 9.1 ft at Trix-Liz to a gross of 500 ft at Midway-Sunset.[38] Porosities range from 0.23 at Miga to 0.38 at Bellevue. Average permeabilities range from 300 md at Brea Olinda to 5,000 md at Miga. Crude viscosities at ambient reservoir temperatures range from 26 cp at Trix-Liz to 2,000 cp at Suplacu. Oil gravities range from 11.5°API at Midway-Sunset[39] to 24°API at Trix-Liz, and the dip in the formations ranges from less than 4° at Miga, South Belridge, Suplacu,[40] and Bellevue to as much as 45° at Brea Olinda and Midway-Sunset.[38]

A number of these projects (West Newport, South Belridge, Suplacu, and Bellevue) were expanded from pilot projects. In the other cases, the operators were confident enough of the process in their own circumstances to commit a large effort from the start. It is significant that with the exception of the Bellevue project (and perhaps the one at Suplacu), the operations listed in Table 8.3 do *not* use a regular system of developed well patterns such as the five-, seven-, or nine-spots. In general, the ratio of producers to injection well ranges from approximately 2:1 at Bellevue to about 10:1 in several other projects.

The contribution from outlying wells (those beyond the confines of an isolated pattern) is shown in Fig. 8.19 for the West Newport project from 1958 through 1964. The performance of this project, also shown in Fig. 8.19, is characteristic of the process.

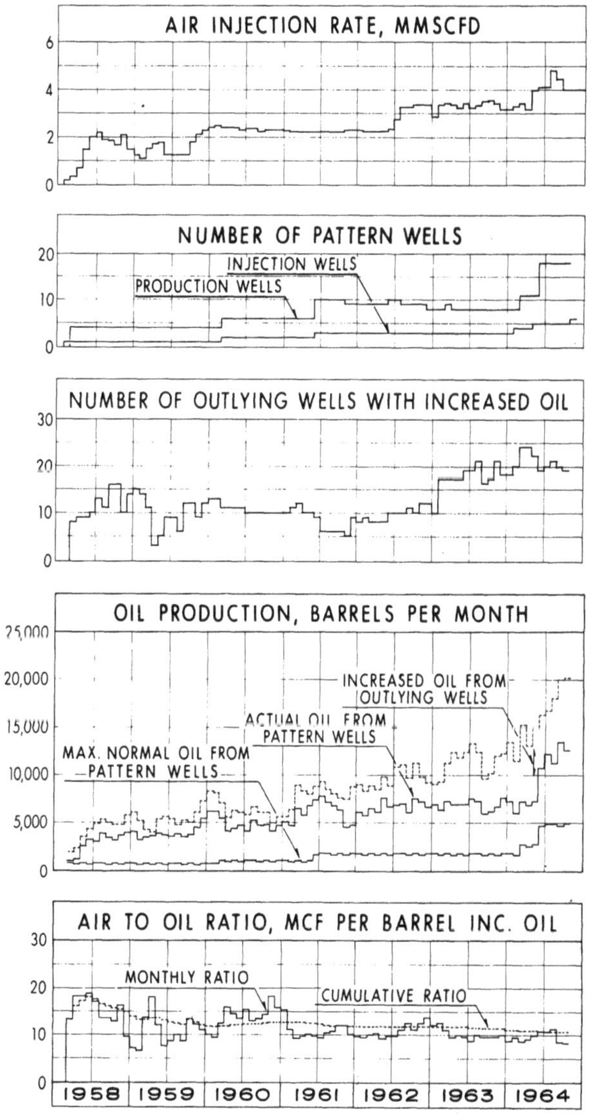

Fig. 8.19 – Operational and production history of West Newport combustion project (1958-1964).[64]

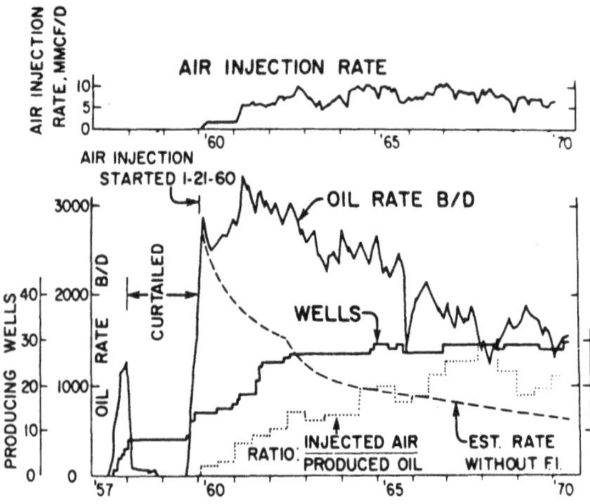

Fig. 8.20 – Moco zone performance.[38]

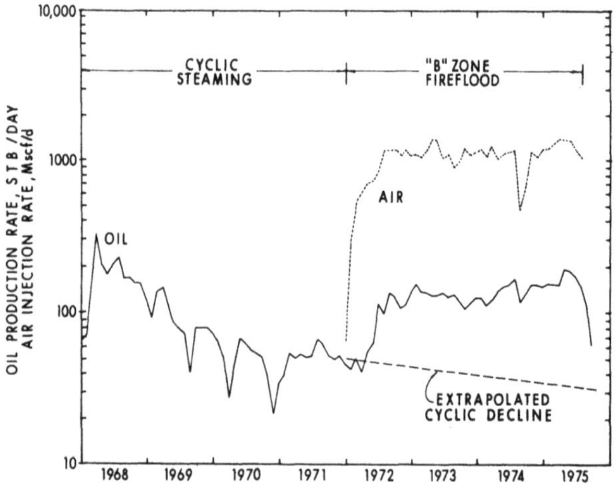

Fig. 8.21 – 154 pattern injection and production performance.[39]

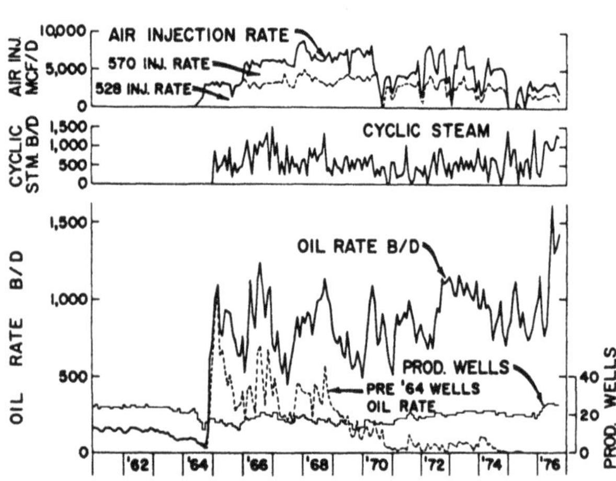

Fig. 8.22 – Section 12 in-situ combustion project oil production, air injection and cyclic steam rates.[22]

Other projects, such as those at Midway-Sunset,[38,39] have been carried out in conjunction with extensive use of steam or other production mechanisms. Fig. 8.20 (taken from Gates and Sklar[38]) shows that production rates during primary were substantial and, thus, primary production mechanisms probably contributed significantly to the early thermal oil production response at Midway-Sunset. Fig. 8.21 (taken from Counihan[39]) shows that before combustion was begun, there was extensive reservoir heating resulting from several years of cyclic steam injection. Fig. 8.22 (taken from Gates *et al.*[22]) gives data for South Belridge and shows that the results of the combustion project at South Belridge are combined with those of a continued program of cyclic steam injection undertaken to maintain "the producibility of wells before temperature arrival from the in-situ combustion process."

At Bellevue, Miga, and Suplacu, the dry combustion operations first were carried out for substantial periods of time, and then water was injected into the burned zone (1) through a well bypassed by the combustion front (Miga), (2) after individual patterns had developed considerable burned volumes (Bellevue), and (3) as part of a wet combustion process (Suplacu). Wet combustion is discussed in the following section. It is obvious that successful dry combustion projects have been carried out under a wide range of conditions.

8.5 Wet Combustion

Wet combustion is the general name given to the process in which water passes through the combustion front along with the air (or other reaction supporting gas). The process always has been applied to forward combustion. The water entering the combustion zone may be in either the liquid or the vapor phase, or both. Ideally, the water is injected along with the air but is injected intermittently with the air when the flow resistance to two-phase flow near the injection well is too high to achieve the desired injection rates.

Model of the Wet Combustion Process

The most comprehensive description of various situations that may develop in the wet combustion process (and one that is mathematical) is that of Beckers and Harmsen.[41] The description offered here is highly idealized and emphasizes the recovery aspects of the wet combustion process. For a theoretical description of the heat transfer mechanisms prevailing under a wide range of water/air ratios, refer to Beckers and Harmsen[41]; for a descriptive and graphic presentation of the wet combustion process, refer to Dietz.[27]

Although the addition of water to a combustion process alters its performance in significant and practical ways, no new mechanisms affecting displacement are brought into play. Control of the wet combustion process (everything else being the same) is attained through the injected water/air ratio, F_{wa}. In this section, we first describe the wet combustion process in terms of a frontal advance model (see Sections 4.1, 4.2, and 8.4); then we discuss some consequences of bypassing.

Fig. 2.6 illustrates schematically the temperature, steam zone, and oil saturations resulting from a range of injected water/air ratios. Fig. 2.6a represents the base case (dry combustion) where there is no injection of water. At low injected water/air ratios, the injected water is converted to steam as it moves toward the combustion front. The part of the injected water that is not stored in the growing burned zone enters the combustion front as superheated steam. In such cases the injected water fails to recuperate all the heat from the burned zone, as shown in Fig. 2.6b. At the optimal water/air ratio, essentially all the heat is transferred from the burned zone and across the advancing combustion front (Fig. 2.6c).

At even higher water/air ratios, represented in Fig. 2.6d, the water entering the reaction zone is in the liquid state, and its cooling effect reduces the peak temperature to the point that only low-temperature reactions occur. At sufficiently high water/air ratios, only low-temperature reactions will occur, and these will occur over a wide reaction zone. Fig. 2.6d shows the reaction zone to be almost as extensive as the steam zone; however, the size of the reaction zone is governed by the availability of oxygen. At the upstream side, the cooling by the water prevents the complete consumption of the fuel, and for this reason it is called "partially quenched" combustion. Dietz[27] has been one of the foremost proponents of partially quenched combustion.

As can be seen from Figs. 2.6a through 2.6d, the size of the steam zone increases as the injected water/air ratio increases, as does the amount of heat recovered from the burned zone. The increasing size of the steam zone results in a more rapid displacement of the oil, so that less air injection is required to obtain a given amount of oil than when no water is injected. Any reduction in the fuel burned resulting from the injection of water also would reduce the air requirements, provided that the rate of heat generation would be high enough to maintain the necessary reactions. In these figures, the downstream slope of the temperature peak resulting from the high-temperature reactions is shown to become less steep as the increases in injected water/air ratio result in higher rates of convective heat transfer. It is customary, as shown in the upper part of Fig. 8.23, to represent the temperature profile as a square wave, with the high-temperature zone corresponding to the steam zone.

Any temperature rise due to burning of fuel is added to the base temperature downstream of the combustion front – i.e., the temperature of the steam zone. The effects of high pressure on the wet combustion process include (1) increased reaction rates (see Eq. 8.13) and (2) higher steam temperatures[42] and smaller volume of the steam zone, which for fixed air- and water-injection rates lead to higher peak temperatures than can be obtained at low pressures. This may be the reason for the relatively high peak temperatures observed by Ejiogu *et al.*[43] in laboratory experiments conducted at about 1,000 psi.

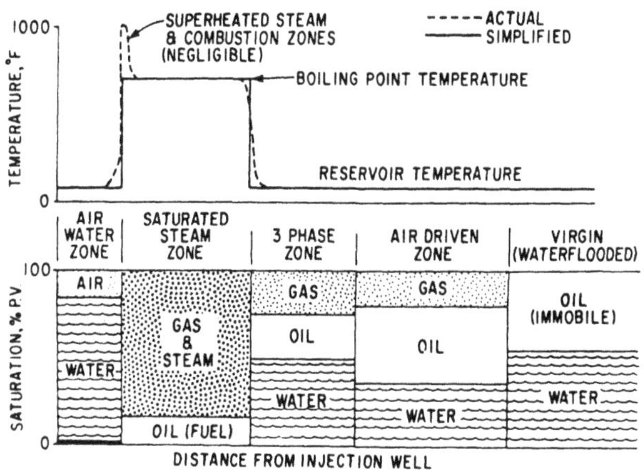

Fig. 8.23 – Schematic drawing of COFCAW saturation and temperature profiles.[44]

Fig. 8.23 (taken from Parrish and Craig[44]) shows the zones associated with a near-optimal wet combustion process and shows temperature and saturation distributions. The main difference in the saturation distributions between the dry combustion process (shown in Fig. 8.17) and the wet combustion process is the identification of a significant steam zone between the combustion front and the three-phase zone, one large enough that its contribution no longer can be ignored.

Before heat breakthrough, the recuperation and redistribution of heat resulting from the water injection does not change the total rate of heat losses to the caprock and base rock in FA systems. This is indicated by the work of Prats[45] (see Section 5.2), who shows that the reduced heat losses to (and even heat recuperation from) the caprock and base rock adjacent to the burned zone are exactly offset by increased heat losses from the downstream zones. If the heat in the reservoir were entirely within the steam zone downstream of the combustion front, then the *size* of the steam zone would be *exactly* that given by Marx and Langenheim.[46] As discussed by Holst and Karra,[47] the size of the steam zone is smaller than that given by Marx and Langenheim if a significant amount of heat remains upstream of the reaction front.

Air and Water Requirements

The amount of air required to advance the combustion front through 1 cu ft of reservoir is affected by the injected water/air ratio. Fig. 8.24 gives experimental laboratory values of the air requirement a_R^* vs. the injected water/air ratio F_{wa} reported in four studies. Values reported by Garon and Wygal[5] and by Burger and Sahuquet[48] are plotted as given. Values ascribed to Dietz and Weijdema[49] are obtained by assuming equivalence between the air requirement a_R^* and the ratio of air flux to combustion front velocity u_a/v_b. Values ascribed to Parrish and Craig[44] are obtained by assuming that the air requirement is equal to the product of the injected-air/produced-oil ratio F_{ao} times the oil

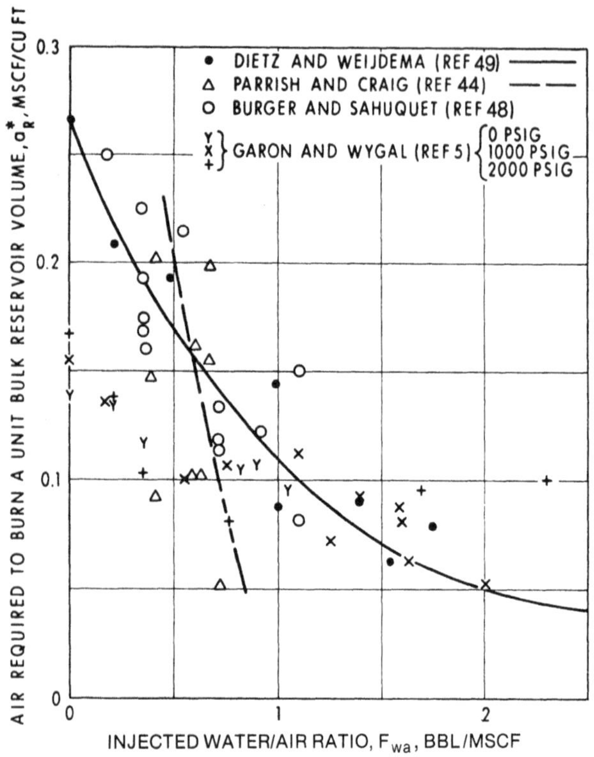

Fig. 8.24 – Effect of injected water/air ratio on air required to burn unit reservoir volume (based on data from Refs. 5, 44, 48, and 49).

produced per unit bulk rock volume. There are differences in the reported air requirements for wet combustion, just as there are for dry combustion. The data of Burger and Sahuquet and of Dietz and Weijdema, which are fairly close together, have a curve drawn through them. A dashed curve, steeper than the first, has been drawn through the data of Parrish and Craig. The air required to burn a unit reservoir volume is proportional to the amount of fuel burned, as seen in Fig. 8.5 for dry combustion. The reasons less fuel is burned as the water/air ratio increases are that (1) less oil is encountered by the advancing combustion front as a result of the improved displacement and stripping efficiency of the larger steam zones and (2) the water quenches the reactions at the higher water/air ratios.

There are two useful relations involving the air requirement a_R^*. One is Eq. 8.16, which indicates that the combustion front moves faster (at a constant air injection rate) the lower the value of a_R^*. This means that less air is required to produce a given reservoir as the water/air ratio increases, provided of course that air still reacts with fuel to generate heat at practical rates. The second relation, obtained from Eq. 8.18 and a generalization of Eq. 8.19, is

$$F_{ao}E_{cb} = \langle 5.615 \text{ cu ft/bbl}\rangle \cdot \frac{a_R^*}{\phi(S_{oi}-S_{oF})\bar{E}_{O_2}}, \quad \ldots\ldots (8.23)$$

where E_{cb} is the oil recovery expressed as a fraction of the oil displaced from the burnt zone. E_{cb} is the slope of the line from the origin to a point on any curve in Fig. 8.18, and can be larger than one.

This second relation indicates that the product of the air/oil ratio and oil recovery efficiency in a given reservoir is directly proportional to the air requirement. This shows that for the same oil recovery efficiency the air/oil ratio is decreased by the use of wet combustion. In the range of $F_{wa}=0.5$ to 1.0 bbl/Mscf, the reduction in the air/oil ratio is approximately 50% over that obtained from dry combustion.

In addition to reducing the amount of air required to move a combustion front through a reservoir unit volume, the wet combustion process also generates a significant steam zone ahead of the combustion front. This steam zone usually leaves only residual oil in its swept path, so that it is generally uneconomical to continue a wet combustion project once the response to the steam zone becomes unattractive. This means that air injection can be stopped early.

The size of the steam zone in laboratory experiments generally is attained under conditions such that heat losses from the experimental equipment are low. Kuo[50] and Holst and Karra[47] provide information from which to estimate the size of the steam zone in a wet combustion operation. Both take into account heat losses from the steam zone to the adjacent formations, using models in which there is no temperature variation within the reservoir normal to the plane of fluid flow. Their results are applicable to both linear and radial flows. Kuo's results are analytical. Those of Holst and Karra (shown in Fig. 8.25) were obtained numerically. This figure shows that the dimensionless steam volume given by the vertical ordinate reaches a fixed value at large times, at least for values of $f_{\dot{Q}}<0.8$, where $f_{\dot{Q}}$ is the fraction of the total generated heat that has passed downstream through the advancing combustion front. It is a quantity assumed to be constant during the entire operation.

Kuo gives the heat downstream of an advancing combustion front resulting from convective heat transfer. If this heat were solely in a steam zone, its equivalent volume would be given by

$$\langle 43{,}560 \text{ cu ft/acre-ft}\rangle \frac{M_R V_s \Delta T t_D}{Q_i}$$

$$= e^{t_D}\cdot \operatorname{erfc}\left[(2F_v-1)\sqrt{\frac{t_D}{4F_v(F_v-1)}}\right]-1$$

$$+(2F_v-1)\operatorname{erf}\sqrt{\frac{t_D}{4F_v(F_v-1)}}. \quad \ldots\ldots (8.24)$$

The dimensionless time t_D is given by

$$t_D = 4\frac{M_S^2}{M_R^2}\frac{\alpha_S t}{h_t^2}, \quad \ldots\ldots (5.6)$$

and the dimensionless parameter

$$F_v = v_T/v_b \quad \ldots\ldots (8.25)$$

is the ratio of the velocity of the convective heat front to that of the combustion front. Eq. 8.24 also leads to fixed values of the dimensionless steam zone volumes at large times for all cases except $F_v = \infty$.

With some analysis, it is possible to show that when F_v and $f_{\dot{Q}}$ are related by

$$8(F_v - 1) = \pi f_{\dot{Q}}^2/(1 - f_{\dot{Q}}), \quad \ldots\ldots\ldots\ldots\ldots (8.26)$$

the results of Kuo[50] and Holst and Karra[47] coincide at large times.* Evaluation of Eq. 8.24 for values of F_v corresponding to values of $f_{\dot{Q}}$ used in Fig. 8.25 shows that they are indistinguishable from each other. Thus, in the range of parameters shown in Fig. 8.25, the results of Kuo and of Holst and Karra are essentially the same. For values of the dimensionless time less than about 1.0 and values of $f_{\dot{Q}}$ less than 0.9, however, there are discernible differences. These are not shown but may be assessed from early-time results given by Holst and Karra vs. those given by Eq. 8.24. The quantity $f_{\dot{Q}}$ used here represents the quantity $1 - \tilde{B}$ used by Holst and Karra.

The agreement between the results of the two models lends support to the assumption that at long times a constant fraction of the generated heat passes through the combustion zone in a wet combustion process. At short times, the approximation does not appear to be as good. Thus, Eq. 8.26 should prove a relatively good relationship between F_v and $f_{\dot{Q}}$ at large times ($t_D > 1$).

From the definitions of v_T (see Eq. 3.15) and the assumption that the injected and flowing water/air ratios were equal, it follows that

$$F_v = \langle 10^3 \text{ scf/Mscf} \rangle \frac{a_R^*}{E_{O_2}} \frac{M_w}{M_R} \cdot \left[\langle 5.615 \times 10^{-3} \text{ Mscf/bbl} \rangle F_{wa} + M_a / M_w \right], \quad \ldots\ldots\ldots\ldots (8.27)$$

where M_a and M_w are the volumetric heat capacities of air and water, respectively. Thus, the dimensionless parameters F_v and $f_{\dot{Q}}$ required to determine the size of the steam zone from Eq. 8.24 and Fig. 8.25 can be calculated from parameters related to the combustion process. Note that the effect of F_{wa} on a_R^* is such (Fig. 8.24) that v_b increases faster than v_T as the water/air ratio increases – i.e., F_v decreases as F_{wa} increases.

In cases of practical interest, where bypassing of the injected gas usually occurs because of gravity override or through zones of relatively high effective permeability, the injected water may not transfer the heat from the burned zone so efficiently as described for the FA model. As described in Sections 2.3 and 4.1, gravity separation between the injected water and air may cause the combustion to be dry near the top of the sand and completely quenched near the bottom. Although these extremes may occur, the experimental evidence[28] is that wet combustion in three-dimensional scaled physical models with some gravity overlay still has all the advantages of the process described by the FA model. Field comparisons between dry and wet combustion also indicate the wet combustion process yields better results.[74] The question is the extent to which bypassing affects the correlations and theoretical results available from the FA model. Unless numerical simulation or the results of scaled physical models can be employed, the only current alternative is to use the results discussed for frontal drive models; these have been used successfully in planning and designing field projects.

*This result was obtain by D.G. Whitten, Shell Develoment Co., Houston (1970).

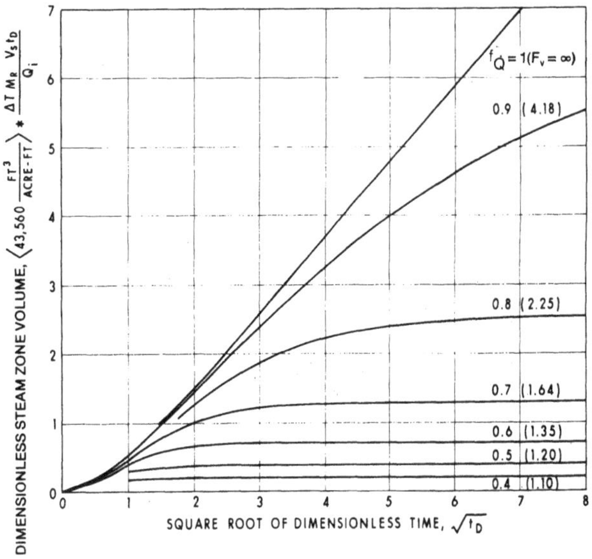

Fig. 8.25 – Dimensionless volume of steam zone downstream of wet combustion front.[47]

Performance Prediction

From the saturation in the burned and steam zones indicated in Fig. 8.23, the oil produced in a wet combustion project N_p can be expressed as

$$N_p = \langle 7{,}758 \text{ bbl/acre-ft} \rangle \left[V_{Rb} (S_{oi} - S_{oF}) \phi + V_s (S_{oi} - S_{or}) \phi \right] \frac{h_n}{h_t} E_c, \quad \ldots\ldots\ldots (8.28)$$

where the first term in the brackets represents the oil displaced from the burned volume, and the second term represents that from the steam zone. From the comparison between calculated and observed recoveries for steam drives shown in Fig. 7.7, it is estimated that the capture efficiency for the wet combustion process is, on the average, less than 70%. Fig. 8.18 suggests E_c is lower the higher the initial gas saturation or where a pronounced oil bank develops.

The volume burned can be related to the cumulative air injected by

$$V_{Rb} = \langle 0.0230 \text{ acre-ft/Mscf} \rangle G_a \bar{E}_{O_2} / a_R^* \quad \ldots (8.29)$$

and the volume of the steam zone can be found from Fig. 8.25 or Eq. 8.24. It is realistic to consider that air injection does not continue beyond the time at which the volume of the burned and steam zones equal that of the pattern:

$$V_{Rb} + V_s = V_P. \quad \ldots\ldots\ldots\ldots\ldots\ldots\ldots (8.30)$$

Additional elements of wet combustion performance prediction are illustrated by means of an example.

TABLE 8.4 – DATA FOR EXAMPLE 8.2

Gross sand thickness, ft	20
Net/gross sand ratio	0.9
Porosity	0.33
Permeability, md	600
Initial oil saturation	0.65
Oil viscosity at 75°F,cp	2,000
Depth, ft	650
Reservoir pressure, psig	125
Reservoir temperature, °F	75

TABLE 8.5 – WET COMBUSTION DESIGN DATA FOR EXAMPLE 8.2

Maximum air injection pressure, psig	465
Compressor capacity per injector, Mscf/D	385
Average steam zone pressure, psia	200
Average steam zone temperature, °F	382
Air required to burn a volume of reservoir, Mscf/cu ft	0.127
Oil saturation burned	0.05
Steam flood residual oil saturation	0.10
Relative permeability to air	0.10
Relative permeability to water	0.15
Thermal diffusivity of caprock, sq ft/D	0.8
Thermal conductivity of caprock, Btu/ft-°F-D	24
Volumetric heat capacity of reservoir, of steam zone, and of adjacent formations, Btu/cu ft-°F	39
Heat generated per scf of air reacted, Btu/scf	10^2
Pattern area, acres	5.5
Water/air ratio, bbl/Mscf	0.7
Well radius, ft	⅓
Oxygen consumption efficiency, %	100

TABLE 8.6 – ESTIMATED PERFORMANCE FOR WET COMBUSTION PROJECT – EXAMPLE 8.2

Air injection rate, Mscf/D	385
Water injection rate, bbl/D	270
Maximum steam zone volume, acre-ft	24.7
Volume burned, gross acre-ft	85.3
net acre-ft	76.8
Project air/oil ratio, Mscf/bbl	4.1
Pattern life, years	3.02
Cumulative air injection, MMscf	425
Cumulative water injection, bbl	268,000
Average oil production rate, B/D	95
Cumulative oil produced, bbl	104,000
Average water production, B/D	221
Cumulative water production, B/D	243,000
Average oil cut, bbl of oil per gross bbl	0.30
Oil recovery efficiency	0.63

Example 8.2 – Estimation of Ultimate Performance From a Wet Combustion Project

Reservoir data are given in Table 8.4 and combustion operational data applicable to the reservoir are in Table 8.5. The estimated ultimate performance is summarized in Table 8.6. Details of selected calculations are given in the following paragraphs for a frontal displacement (no bypassing) model in a five-spot well pattern.

Steam Zone Volume at Steady State. From Eq. 8.27, data given in Tables 8.4 and 8.5, the definition of volumetric heat capacity, and data provided in Appendix B, we find that

$$F_v = (10^3 \text{ scf/Mscf})\Big\{(0.127 \text{ Mscf/cu ft}) \cdot (55.4 \text{ Btu/cu ft-°F})\Big[(0.005615 \text{ Mscf/bbl}) \cdot (0.7 \text{ bbl/Mscf}) + \frac{(0.5 \text{ Btu/scf-°F})}{(55.4 \text{ Btu/cu ft-°F})}\Big]\Big\} \div 39 \text{ Btu/cu ft-°F} = 2.34.$$

For the water/air ratio of 0.7 bbl/Mscf, which is in the optimal wet combustion range (see Fig. 2.6c), we find (from Eq. 8.26) that about 81% of all the generated heat is transported ahead of the combustion front. The steady-state (maximum) size of the steam zone is found by evaluating Eq. 8.24 as $t_D \to \infty$ and gives

$$(43{,}560 \text{ cu ft/acre-ft})\left(\frac{M_R V_s \Delta T_s t_D}{Q_i}\right)_{ss} = 2(F_v - 1) = 2.68.$$

Using $Q_i = \dot{Q}_i t$, the definition of t_D given in Eq. 5.6, and data given in Tables 8.4 and 8.5, the steady-state steam zone volume is

$$V_{s,ss} = \Big[2.68(385 \times 10^3 \text{ scf/D})(100 \text{ Btu/scf})\Big] \div \Big[(39 \text{ Btu/cu ft-°F})(307°\text{F}) \cdot (43{,}560 \text{ cu ft/acre-ft})(8 \times 10^{-3} D^{-1})\Big] = 24.7 \text{ acre-ft}.$$

The constant $8 \times 10^{-3} D^{-1}$ appearing in the numerator is the conversion factor between dimensionless and real time. Note that in this calculation it has been assumed the air injection rate is limited only by compressor capacity. Estimates of the steady-state steam zone volume would need to be redetermined if the injection rates were found to be injectivity limited.

Time Required to Achieve 90% of Steady-State Steam Zone Volume. In principle, it takes an infinite amount of time to attain the maximum steam zone volume. In practice, a large fraction of the maximum steam zone is attained in a reasonable time. Since the dimensionless value of the maximum steam zone volume is 2.68, it follows that the dimensionless steam zone volume corresponding to 90% of its maximum values is 2.41. From Eq. 8.24, it can be found by trial and error that this occurs at a dimensionless time of 4.4, which corresponds to a real time given by

$$t=\frac{(20\text{ ft})^2\,4.4}{4(0.8\text{ sq ft/D})}=550\text{ days or }1.5\text{ years}.$$

Thus, the time required to develop a full steam zone volume is rather short and will not affect the project life.

Volume Burned. The operation ends when the volumes of the burned and steam zones equal that of the well pattern. From this, the gross volume of the burned zone at abandonment can be found to be

$$(V_{Rb})_{max}=(5.5\text{ acres})(20\text{ ft})-24.7\text{ acre-ft}$$
$$=85.3\text{ acre-ft}.$$

Air and Water Requirements. Since it takes 0.127 Mscf of air to burn a cubic foot of reservoir and the volume burned is 85.3 acre-ft, the total air requirement is

$$G_a=(0.127\text{ Mscf/cu ft})(43{,}560\text{ cu ft/acre-ft})$$
$$\cdot(85.3\text{ acre-ft})\times(18\text{ ft}/20\text{ ft})$$
$$=425{,}000\text{ Mscf},$$

where only the net acre-feet of the burned zone actually is burned. The water requirements will be

$$W_i=(425{,}000\text{ Mscf})(0.7\text{ bbl/Mscf})$$
$$=298{,}000\text{ bbl}.$$

Because water is not added until combustion is well developed, the actual amount of water may be some 10% less, or about 268,000 bbl. This is equivalent to 1.06 net pattern PV.

Project Life. From the equation for flow of gas in a five spot (Eqs. 4.3 and 4.8), it can be found that the steady-state air injection rate, attainable where the relative permeability to gas is 0.1, is

$$i_a=(1.352\times10^{-6})\frac{(600\text{ md})(0.1)(18\text{ ft})(520°\text{R})}{(2)(0.02\text{ cp})(800°\text{R})}$$
$$\cdot\left[\frac{(479.7\text{ psia})^2-(64.7\text{ psia})^2}{\ln\dfrac{\sqrt{(5.5\text{ acres})(43{,}560\text{ sq ft/acre})}}{(1/3\text{ ft})}-0.964}\right]$$
$$=845\text{ Mscf/D}.$$

This far exceeds the available compressor capacity of 385 Mscf/D per well which the operator plans to use. Accordingly, the estimated project life is

$$t=\frac{425{,}000\text{ Mscf}}{385\text{ Mscf/D}}=1{,}100\text{ days}$$
$$=3.02\text{ years}.$$

The corresponding water injection rate is

$$i_w=(385\text{ Mscf/D})(0.7\text{ bbl/Mscf})$$
$$=270\text{ B/D},$$

or 15 B/D per net foot of injection interval. Using Eqs. 4.2 and 4.3 and a relative permeability to water of 0.15, we find that the steady-state water injection rate is

$$i_w=(7.08\times10^{-3})\frac{(600\text{ md})}{(2)(1\text{ cp})}$$
$$\cdot(0.15)(18\text{ ft})\frac{(465-50)\text{psi}}{6.33}$$
$$=376\text{ B/D},$$

so that no injectivity problems are expected.

Oil Recovery. The oil recovery is estimated to be

$$N_p=(7{,}758\text{ bbl/acre-ft})[(85.3\text{ acre-ft})(0.65-0.05)$$
$$\cdot(0.33)+(24.7\text{ acre-ft})(0.65-0.10)(0.33)]$$
$$\cdot(18\text{ ft}/20\text{ft})(0.7)=104{,}000\text{ bbl}$$

at an average rate of

$$q_o=\frac{104{,}000\text{ bbl}}{1100\text{ days}}=94.6\text{ B/D}.$$

The oil recovery efficiency is

$$E_R=(104{,}000)\div[(7{,}758\text{ bbl/acre-ft})$$
$$\cdot(99\text{ net acre-ft})(0.33)(0.65)]=0.63$$

Air/Oil Ratio. The air/oil ratio is estimated to be

$$F_{ao}=\frac{425{,}000\text{ Mscf}}{104{,}000\text{ bbl}}=4.1\text{ Mscf/bbl}.$$

Water Production. Water production can be estimated by assuming an increase in the final water saturation in the reservoir of 0.10, up to 0.40. The result is

$$W_p=268{,}000-(7{,}758\text{ bbl/acre-ft})(110\text{ acre-ft})$$
$$\cdot(0.33)(0.10)(18\text{ ft}/20\text{ ft})$$
$$=243{,}000\text{ bbl}$$

or an average oil cut of

$$f_o=\frac{104{,}000\text{ bbl}}{104{,}000\text{ bbl}+243{,}000\text{ bbl}}=0.30.$$

The average water production rate is 221 B/D. If the water retained in the reservoir results in a saturation increase of more than 0.1, then the water production would be lower and the oil cut higher.

The air and water rate capacities for the well pattern have been calculated on the assumption that the same fluids are being produced – a requirement imposed by the steady-state condition considered in the calculations. Since (1) oil is being produced along with the combustion gas and water, (2) the oil is significantly more viscous than the cold water even at steam temperatures, and (3) the *total* average liquid production rate is 311 B/D, then it is likely that some fluids may not be produced at a fast enough rate to maintain the calculated injection schedules unless some remedial work, such as cyclic steam injection, is applied at the producers.

This example illustrates (1) how to estimate the steady-state size of the steam zone and the time required to attain it, (2) that the size of the steam

TABLE 8.7 – SELECTED MAJOR WET COMBUSTION PROJECTS

Project	Operator	Date Started	References
Schoonebeek, Holland	Shell	1962	27
Iola, KS	Layton	1964	70
E. Tia Juana, Venezuela	Shell	1965	27, 69
Sloss, NE	Amoco	1967	71, 72, 73
Bellevue, LA	Getty	1969	54, 65, 66
Suplacu, Rumania	Ministry of Mines, of Petroleum, and Rumanian Geology	1973	52

zone may be a significant fraction of the well pattern volume, (3) how to estimate the project recovery and performance, and (4) that productivity can be a problem that may need attention during the project life.

Although the average rates may be quite representative of the actual performance, the values of the maximum rates and their duration are not given by these simple calculations.

Design

Wet combustion should be considered an alternative to dry combustion in all cases, since it could reduce the air requirements and accelerate the production response. However, it should not be used in formations where flow resistance is marginally acceptable for dry combustion, because the addition of water will increase the flow resistance further. Neither should it be used where the injected water would interact adversely with formation clays or other minerals to reduce the reservoir injectivity. The effectiveness of the wet combustion process decreases where gravity override is expected to be important, especially in thick, massive intervals having good vertical continuity and high permeability.

In thin sands, where there is significant heat loss to adjacent formations, wet combustion processes should be considered. The effect of the injected water in recuperating heat from the burned zone and the zones adjacent to it is particularly important where sands are thin. In such cases a high-temperature combustion process may not be sustainable.

Besides considering the items discussed in the section on design for dry combustion, special attention should be given to the flow capacity of the wells used in the operation. Adding water to the air reduces the relative permeability to the two fluids. Accordingly, the flow resistance is increased in the burned zone, especially near the injectors. Similarly, the production will include a larger proportion of water (i.e., the water cut will increase). Since adding water can reduce the project life, the average oil production rate tends to increase. The relatively high oil production rates and water cuts also tend to increase the pressure gradients near producers. Furthermore, the addition of water lowers temperatures at the producers – one of the advantages of wet combustion. However, this is accompanied by a slight increase in the crude viscosity and, thus, in the flow resistance. This increase in flow resistance is one reason for preferring a higher ratio of producers to injectors, even higher than in dry combustion operations.

The flow resistance may be high enough in some instances to require stimulation of both injectors and producers and even to require the injection of alternate slugs of water and air to reduce two-phase flow effects near injection wells.[52]

The hardware in the injectors may be arranged especially to reduce downhole separation of the air and water. Because of lower temperatures at the wells and less acidic waters owing to lower fuel consumption and increased dilution by the injected water, corrosion of producing wells is not so severe in wet combustion projects. This reduced corrosion may allow the use of more existing wells than would be possible if no water were added.

On the other hand, water disposal problems will increase in wet combustion projects.

In reservoirs where clay/water problems may arise, the injection of water must be considered carefully. If the burned zone is subjected to sufficiently high temperatures (e.g., 1,200°F), the clays often are fired to the point that they no longer will swell when contacted with water. But in a wet combustion process, the resultant temperatures may be so low that clay swelling cannot be avoided unless a compatible formation water is injected. Although little is known about the effect of the water on clay swelling in the steam zone downstream of the combustion zone, no adverse consequences have been reported.

Examples of Field Applications

It was pointed out earlier that the dry in-situ combustion project at Suplacu had been converted to a wet combustion project[52] and that water had been injected into a well bypassed by the combustion front at Miga[53] and into individual well patterns after considerable burned volumes had developed at Bellevue.[54] In these last two projects, the water appears to have been injected primarily to minimize or prevent the burned zones from being resaturated by operations continuing in nearby areas. Although some of the elements of wet combustion are present (water injection, heat recuperation, and transfer), they are not considered wet combustion projects, because the water injection required to transfer the heat through an *advancing* combustion front does not appear to have been part of the operation.

Surveys limited to wet combustion operations have been made by Craig and colleagues,[29,55] but wet combustion projects normally are included in general surveys of combustion operations.

Dry combustion has been practiced for about 10 years longer than wet combustion, which may account for the fact that large wet combustion projects are not so plentiful as dry combustion projects. Table 8.7 lists selected wet combustion projects, all of which appear to be technical successes.

Sloss and East Tia Juana represent extremes in reservoir and crude oil conditions: the reservoir thickness, porosity, average permeability, crude viscosity, and API gravity of the crude at Sloss are 14.3 ft, 0.19, 191 md, 0.8 cp, and 38.8°API, and at East Tia Juana they are 128 ft, 0.40, 5000 md, 2500 cp, and 12.5 °API. The depth below surface of the projects in Table 8.7 ranges from 250 ft at Suplacu to 6,200 ft at Sloss. The injected water/air ratio ranged from 0.13 bbl/Mscf at Iola to 1.34 bbl/Mscf at Sloss. The ratio of injected air to produced oil ranged from 2 Mscf/bbl at East Tia Juana to 30 Mscf/bbl at Iola.

The Sloss project, conducted in a previously waterflooded reservoir, had an estimated remaining oil saturation at the start of the project of no more than 0.40. In addition, the porosity is less than 0.20, the crude viscosity is 0.8 cp, and the depth is 6,200 ft. Fig. 8.26 shows the response from this project. Fig. 8.27 shows the performance at Schoonebeek.

In summary, wet combustion projects have been carried out successfully under widely different reservoir and operating conditions and geographic and geologic settings.

8.6 Reverse Combustion

In reverse combustion, the combustion front moves opposite to the direction of air flow. Combustion is initiated at the production well, and the combustion front moves against the air flow. In a frontal displacement model, the crude that is displaced must pass through the burning combustion zone and through the hot burned zone. A schematic diagram of the process is given in Fig. 8.28, where four zones are identified. The burned zone in this figure (taken from Burger and Sahuquet[11]) shows a hot zone immediately downstream of the combustion front. The extent of the high-temperature region in the burned zone depends (among other things) on the rate of heat losses to the adjacent formations. As the crude is displaced through the combustion front, it is cracked; the light ends vaporize and the heavy ends contribute a residue. As the vapors approach cooler sections of the burned zone, some condensation occurs, and liquid oil and water may exist near the outlet. The region upstream of the combustion zone is heated by heat conduction, which leads to low-temperature oxidation reactions and the generation of considerable heat at significant rates. Upstream of this zone is the initial zone, unaffected by the process except by the flow of air.

Because no oil bank builds up, the total flow resistance of this process can only decrease with time. For this reason, the process is particularly well suited for very viscous crudes, where the buildup of an oil bank would likely increase the flow resistance significantly. We should point out two phenomena – glossed over in the foregoing description – that significantly alter the applicability of reverse combustion. One of these is the possibility of spontaneous ignition. Dietz and Weijdema[56] have pointed out that it is seldom feasible to avoid

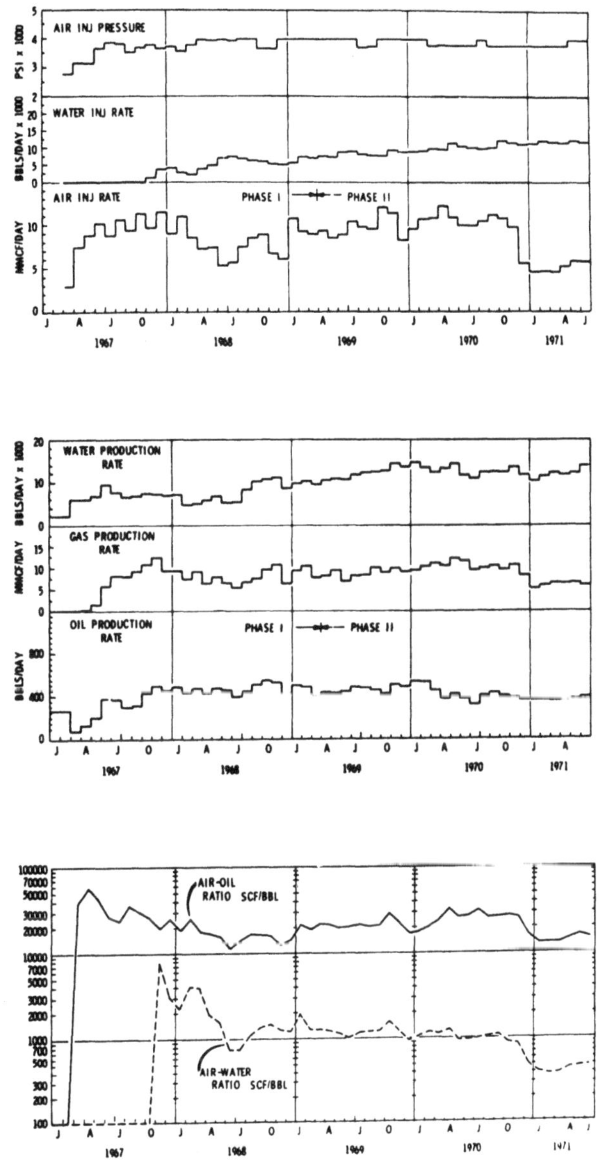

Fig. 8.26 – Operational and production history at Sloss field (1967-1971).[71]

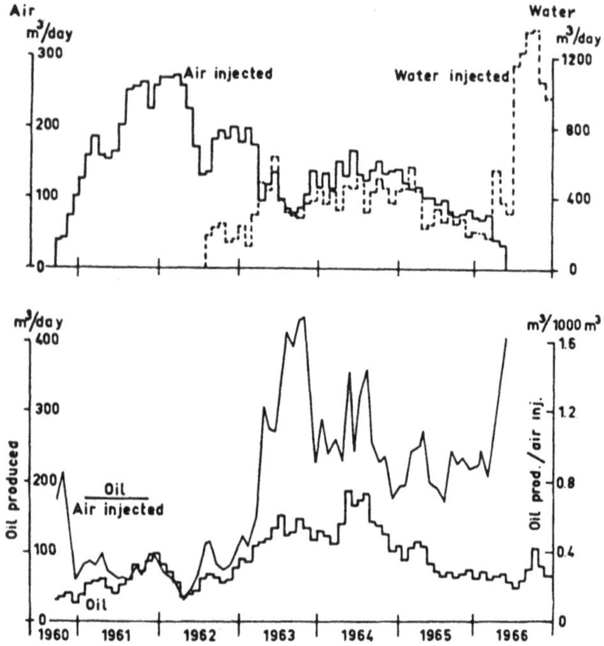

Fig. 8.27 – Schoonebeek underground combustion reservoir performance.[27]

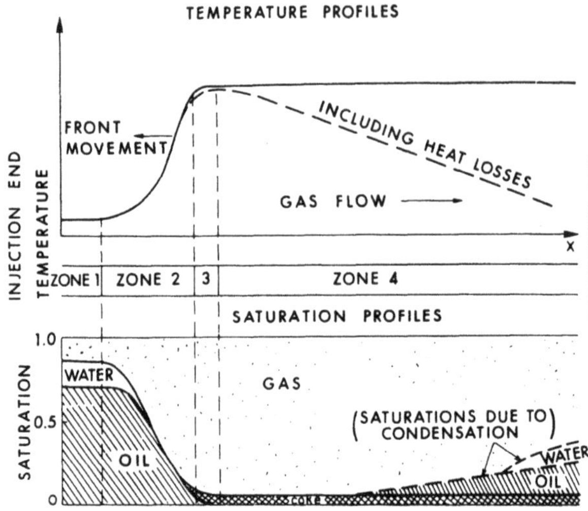

Fig. 8.28 – Schematic representation of temperature and saturation profiles during reverse combustion.[11]

spontaneous ignition for very long unless the reservoir temperature is unusually low. The natural reactivity of crudes, coupled with the surrounding reservoir temperatures found in most situations would, in most cases, combine to provide spontaneous ignition in a matter of months. And, of course, once this happens the oxygen would be consumed close to the injection point and no longer would be available to maintain reverse combustion. There are exceptions to this – i.e., where the project life is shorter than the ignition time. An example is the field test at Bellamy, where Trantham and Marx[57] used a very short well spacing (about 15 ft) and obtained a short operating life (a few weeks). Another example is at Athabasca,[58] where reverse combustion has been used to decrease the flow resistance between wells 100 ft apart. This apparently succeeds because of the low initial temperature of the formation (probably about 60°F) which results in relatively long ignition times (see Eq. 8.15).

The second phenomenon is the inherent instability of reverse combustion. A recent study by Gunn and Krantz[59] concludes that the diameter of the combustion channel would tend to be limited to a few feet, the preferred dimension depending on the reservoir and crude properties and on the operating conditions. It is likely that the preferred dimension of the burned zone is larger than that of the combustion tubes in which the process has been studied experimentally[11,60,61] and may be about the same as the thickness of the sand intervals (6 ft) in the Bellamy field test. It is not surprising that the effects of these instabilities have not been reported in the literature covering experimental laboratory results, although they have been demonstrated amply in in-situ coal gasification field projects.

The formation of narrow channels in reverse combustion operations does not mean reverse combustion cannot be used successfully. But it means that (1) the frontal displacement model is not applicable, (2) steps may have to be taken to delay ignition (e.g., by inhibiting the slow oxidation reaction), and (3) additional work is required to determine how to use these burned channels in a full-scale operation, such as has been investigated in field experiments in the Asphalt Ridge tar sands.[62]

References

1. Alexander, J.D., Martin, W.L., and Dew, J.N.: "Factors Affecting Fuel Availability and Composition During In-Situ Combustion," *J. Pet. Tech.* (Oct. 1962) 1154-1164; *Trans.*, AIME, **225**.
2. Showalter, W.E.: " Combustion Drive Tests," *Soc. Pet. Eng. J.* (March 1963), 53-58; *Trans.*, AIME, **228**.
3. Wohlbier, R.: "Theoretical Studies of Petroleum Recovery Operations with Forward Combustion, Based on a Computer Model for a Linear System," *Erdoel-Erdgas-Z.* (1965) No. 11, 435-463.
4. Martin, W.L., Alexander, J.D., and Dew, J.N.: "Process Variables of In-Situ Combustion," *Trans.*, AIME (1958) **213** 28-35.
5. Garon, A.M. and Wygal, R.J. Jr.: "A Laboratory Investigation of Wet Combustion Parameters," *Soc. Pet. Eng. J.* (Dec. 1974) 537-544.
6. Nelson, T.W. and McNiel, J.S. Jr.: "How to Engineer an In-Situ Combustion Project," *Oil and Gas J.* (June 5, 1961) 58-65.
7. Burger, J.G. and Sahuquet, B.C.: "Chemical Aspects of In-Situ Combustion – Heat of Combustion and Kinetics," *Soc. Pet. Eng. J.* (Oct. 1972) 410-422; *Trans.*, AIME, **253**.
8. Poettmann, F.H., Schilson, R.E., and Surkalo, H.: "Philosophy and Technology of In-Situ Combustion in Light Oil Reservoirs," *Proc.*, Seventh World Pet. Cong., Mexico City (1967) **3**, 487.
9. Dew, J.N. and Martin, W.L.: "Air Requirements for Forward Combustion," *Pet. Eng.* (Dec. 1964) 82-86.

10. Moss, J.R., White, P.D., and McNiel, J.S. Jr.: "In-Situ Combustion Process – Results of a Five-Well Field Experiment in Southern Oklahoma," *J. Pet. Tech.* (April 1959) 55-64; *Trans.*, AIME, **216**.
11. Burger, J. and Sahuquet, B.: *Les Méthodes Thermiques de Production des Hydrocarbures*, Revue IFP (March-April 1977) Chap. 5, 141-188.
12. Buesse, H.: "Experimental Investigations of Fuel Recovery by Underground Partial Combustion of Petroleum Reservoirs," *Erdoel, Erdgas A.* (1971) **87**, No. 12, 414-427 (English translation).
13. Fassihi, M.R., Brigham, W.E., and Ramey, H.J. Jr.: "Reaction of Kinetics of In-Situ Combustion: Part 1—Observations, Part 2—Modeling," *Soc. Pet. Eng. J.* (Aug. 1984) 399–416.
14. Vossoughi, S. *et al.*: "Automation of an In-Situ Combustion Tube and Study of the Effect of Clay on the In-Situ Combustion Process," *Soc. Pet. Eng. J.* (Aug. 1982) 493–502.
15. Thomas, G.W., Buthod, A.P., and Allag, O.: "An Experimental Study of the Kinetics of Dry, Forward Combustion," Fossil Energy Rep. No. BETC-1820-1, U.S. DOE (Feb. 1979).
16. Tadema, H.J. and Weijdema, J.: "Spontaneous Ignition of Oil Sands," *Oil and Gas J.* (Dec. 14, 1970) 77-80.
17. Burger, J.: "Spontaneous Ignition in an Oil Reservoir," *Soc. Pet. Eng. J.* (April 1976) 73-81.
18. Howard, C.E., Widmeyer, R.H., and Haynes, S. Jr.: "The Charco Redondo Thermal Recovery Pilot," *J. Pet. Tech.* (Dec. 1977) 1522-1532.
19. Gates, C.F. and Ramey, H.J. Jr.: "Field Results of South Belridge Thermal Recovery Experiment," *Trans.*, AIME (1958) **213**, 236-244.
20. Tadema, H.J.: "Oil Production by Underground Combustion," *Proc.*, Fifth World Pet. Cong., New York (1959) Sec. II.
21. Hardy, W.C. and Raiford, J.D.: "In-Situ Combustion in a Bartlesville Sand – Allen County, Kansas," *Proc.*, Tertiary Oil Recovery Conf., Wichita, KS, Oct. 23-24, 1975; Inst. of Mineral Resources Research, U. of Kanasas, Lawrence.
22. Gates, C.F., Jung, K.D., and Surface, R.A.: "In-Situ Combustion in the Tulare Formation, South Belridge Field, Kern County, California," *J. Pet. Tech.* (May 1978) 798-806; *Trans.*, AIME, **265**.
23. Wilson, L.A., Wygal, R.J., Reed, D.W., Gergins, R.L., and Henderson, J.H.: "Fluid Dynamics During an Underground Combustion Process," *Trans.*, AIME (1958) **213**, 146-154.
24. Crookston, R.B., Culham, W.E., and Chen, W.H.: "A Numerical Simulation Model for Thermal Recovery Processes," *Soc. Pet. Eng. J.* (Feb. 1979) 37-58, *Trans.*, AIME, **267**.
25. Coats, K.H.: "In-Situ Combustion Model," *Soc. Pet. Eng. J.* (Dec. 1980) 533-554.
26. Acharya, U.K. and Somerton, W.H.: "Theoretical Study of In-Situ Combustion in Thick Inclined Oil Reservoirs," paper SPE 7967 presented at the 1979 SPE California Regional Meeting, Ventura, April 18–20.
27. Dietz, D.N.: "Wet Underground Combustion, State of the Art," *J. Pet. Tech.* (May 1970) 605-617; *Trans.*, AIME, **249**.
28. Binder, G.G., Jr., Elzinga, E.R., Tarmy, B.L., and Willman, B.T.: "Scaled Model Tests of In-Situ Combustion in Massive Unconsolidated Sands," *Proc.*, Seventh World Pet. Cong., Mexico City (1977) **3**, Paper PD-12(4), 477-485.
29. Buxton, T.S. and Craig, F.F. Jr.: "Effect of Injected Air-Water Ratio and Reservoir Oil Saturation on the Performance of a Combination of Forward Combustion and Waterflooding," *AIChE Symposium Series* (1973) **69**, No. 127, 27-30.
30. Garon, A.M., Geisbrecht, R.A., and Lowry, W.E. Jr.: "Scaled Model Experiments of Fireflooding in Tar Sands," paper SPE 9949 presented at the 1980 SPE Annual Technical Conference and Exhibition, Dallas, Sept. 21–24.
31. Brigham, W.E., Satman, A., and Soliman, M.Y.: "A Recovery Correlation for Dry In-Situ Combustion Processes," *J. Pet. Tech.* (Dec. 1980) 2131-2138.
32. Fassihi, M.R., Gobran, B.D., and Ramey, H.J. Jr.: "Algorithm Calculates Performance of In-Situ Combustion," *Oil and Gas J.* (Nov. 16, 1981) 90–98.
33. Hwang, M.K., Jines, W.R., and Odeh, A.S.: "An In-Situ Combustion Process Simulator With a Moving-Front Representation," *Soc. Pet. Eng. J.* (April 1982) 271–79.
34. Craig, F.F. Jr.: *The Reservoir Engineering Aspects of Waterflooding*, Monograph Series, Society of Petroleum Engineers, Dallas (1971) **3**.
35. Ramey, H.J. Jr. and Nabor, G.W.: "A Blotter-Type Electrolytic Model Determination of Areal Sweeps in Oil Recovery by In-Situ Combustion," *Trans.*, AIME (1954) **201**, 119-123.
36. Chu, C.: "State-of-the-Art Review of Fireflood Field Projects," *J. Pet. Tech.* (Jan. 1982) 19-36.
37. Farouq Ali, S.M.: "A Current Appraisal of In-Situ Combustion Field Tests," *J. Pet. Tech.* (April 1972) 477-486.
38. Gates, C.F. and Sklar, I.: "Combustion as a Primary Recovery Process – Midway Sunset Field," *J. Pet. Tech.* (Aug. 1971) 981-986; *Trans.*, AIME, **251**.
39. Counihan, T.M.: "A Successful In-Situ Combustion Pilot in the Midway-Sunset Field, California," paper SPE 6525 presented at the 1977 SPE California Regional Meeting, Bakersfield, April 13–15.
40. Burger, J., Aldea, Gh., Carcoama, A., Petcovici, V., Sahuquet, B., and Delye, M.: "Recherches de Base Sur La Combustion In-Situ et Résultats Récent sur Champ," *Proc.*, Ninth World Pet. Cong., Tokyo (1975) **3**, 279-289.
41. Beckers, H.L. and Harmsen, G.J.: "The Effect of Water Injection on Sustained Combustion in a Porous Medium," *Soc. Pet. Eng. J.* (June 1970) 145-163; *Trans.*, AIME, **249**.
42. Satman, A., Brigham, W.E., and Ramey, H.J. Jr.: "An Investigation of Steam Plateau Phenomena," paper SPE 7965 presented at the 1979 SPE California Regional Meeting, Ventura, April 18–20.
43. Ejiogu, G.J., Bennion, D.W., Moore, R.G. and Donnelly, J.K.: "Wet Combustion – A Tertiary Recovery Process for the Pembina Cardium Reservoir," *J. Cdn. Pet. Tech.* (July-Sept. 1979) 58-65.
44. Parrish, D.R. and Craig, F.F. Jr.: "Laboratory Study of a Combination of Forward Combustion and Waterflooding – The COFCAW Process," *J. Pet. Tech.* (June 1969) 753 761; *Trans.*, AIME, **246**.
45. Prats, M.: "The Heat Efficiency of Thermal Recovery Processes," *J. Pet. Tech.* (March 1969) 323-332; *Trans.*, AIME, **246**.
46. Marx, J.W. and Langenheim, R.H.: "Reservoir Heating by Hot Fluid Injection," *Trans.*, AIME (1957) **216**, 312-315.
47. Holst, P.H. and Karra, P.S.: "The Size of the Steam Zone in Wet Combustion," *Soc. Pet. Eng. J.* (Feb. 1975) 13-18.
48. Burger, J.G. and Sahuquet, B.C.: "Laboratory Research on Wet Combustion," *J. Pet. Tech.* (Oct. 1973) 1137-1146.
49. Dietz, D.N. and Weijdema, J.: "Wet and Partially Quenched Combustion," *J. Pet. Tech.* (April 1963) 411-415; *Trans.*, AIME, **243**.
50. Kuo, C.H.: "A Heat Transfer Study for the In-Situ Combustion Process," paper SPE 2651 presented at SPE 44th Annual Meeting, Denver, Sept. 28-Oct. 1, 1969.
51. Myhill, N.A. and Stegemeier, G.L.: "Steam Drive Correlation and Prediction," *J. Pet. Tech.* (Feb. 1978) 173-182.
52. Gadelle, C.P. *et al.*: "Heavy-Oil Recovery by In-Situ Combustion—Two Field Cases in Rumania," *J. Pet. Tech.* (Nov. 1981) 2057–66.
53. Terwilliger, P.L., Clay, R.R., Wilson, L.A. Jr., and Gonzalez-Gerth, E.: "Fireflood of the $P_{2\text{-}3}$ Sand Reservoir in the Miga Field of Eastern Venezuela," *J. Pet. Tech.* (Jan. 1975) 9-14.
54. "Getty Expands Bellevue Fire Flood," *Oil and Gas J.* (Jan. 13, 1975) 45-49.
55. Craig, F.F. Jr. and Parrish, D.R.: "A Multipilot Evaluation of the COFCAW Process," *J. Pet. Tech.* (June 1974) 659-666.
56. Dietz, D.N. and Weijdema, J.: "Reverse Combustion Seldom Feasible," *Prod. Monthly* (May 1968) 10.
57. Trantham, J.C. and Marx, J.W.: "Bellamy Field Tests: Oil from Tar by Counterflow Underground Burning," *J. Pet. Tech.* (Jan. 1966) 109-115; *Trans.*, AIME, **237**.
58. Giguere, R.J.: "An In-Situ Recovery Process for the Oil Sands of Alberta," *Energy Process Can.* (Sept.-Oct. 1977) **70**, No. 1, 36-40.

59. Gunn, R.D. and Krantz, W.B.: "Reverse Combustion Instabilities in Tar Sands and Coal," *Soc. Pet. Eng. J.* (Aug. 1980) 267-277.
60. Reed, R.L., Reed, D.W., and Tracht, J.H.: "Experimental Aspects of Reverse Combustion in Tar Sands," *Trans.*, AIME (1960) **219**, 99-108.
61. Wilson, L.A., Reed, R.L., Reed, D.W., Clay, R.R., and Harrison, N.H.: "Some Effects of Pressure on Forward and Reverse Combustion," *Soc. Pet. Eng. J.* (Sept. 1963) 127-137.
62. Johnson, L.A., Fahy, L.J., Romanowski, L.J., Barbour, R.V., and Thomas, K.P.: "An Echoing In-Situ Combustion Oil Recovery Project in a Utah Tar Sand," *J. Pet. Tech.* (Feb. 1980) 295-305.
63. Dabbous, M.K. and Fulton, P.F.: "Low-Temperature Oxidation Reaction Kinetics and Effects on the In-Situ Combustion Process," *Soc. Pet. Eng. J.* (June 1974) 253-262.
64. Koch, R.L.: "Practical Use of Combustion Drive at West Newport Field," *Pet. Eng.* (Jan. 1965) 72-81.
65. Cato, R.W. and Frnka, W.A.: "Getty Oil Reports Fireflood Pilot is Successful Project," *Oil and Gas J.* (Feb. 12, 1968) 93-97.
66. *Enhanced Oil Recovery Field Reports*, Society of Petroleum Engineers, Dallas, TX (issued biannually).
67. Buchwald, R.W. Jr., Hardy, W.C., and Neinast, G.S.: "Case Histories of Three In-Situ Combustion Projects," *J. Pet. Tech.* (July 1973) 784-792.
68. Showalter, W.E. and MacLean, M.A.: "Fireflood at Brea-Olinda Field, Orange County, California," paper SPE 4763 presented at the 1974 SPE Improved Oil Recovery Symposium, Tulsa, April 22–24.
69. Rincón, A.: "Proyectos Pilotos de Recuperación Térmica en la Costa Bolívar, Estado Zulia," *Proc.*, Simposio Sobre Crudos Pesados, U. del Zulia, Inst. de Investigaciones Petroleras, Maracaibo, Venezuela, July 1-3, 1974.
70. Smith, M.W.: "Simultaneous Underground Combustion and Water Injection in the Carlyle Pool, Iola Field, Kansas," *J. Pet. Tech.* (Jan. 1966) 11-18.
71. Parrish, D.R., Pollock, C.B., and Craig, F.F. Jr.: "Evaluation of COFCAW as a Tertiary Recovery Method, Sloss Field, Nebraska," *J. Pet. Tech.* (June 1974) 676-686.
72. Buxton, T.S. and Pollock, C.B.: "The Sloss COFCAW Project – Further Evaluation of Performance During and After Air Injection," *J. Pet. Tech.* (Dec. 1974) 1439-1448; *Trans.*, AIME, **257**.
73. Parrish, D.R., Pollock, C.B., Ness, N.L., and Craig, F.F. Jr.: "A Tertiary COFCAW Pilot Test in the Sloss Field, Nebraska," *J. Pet. Tech.* (June 1974) 667-675; *Trans.*, AIME, **257**.
74. Joseph, C. and Pusch, W.H.: "A Field Comparison of Wet and Dry Combustion," *J. Pet. Tech.* (Sept. 1980) 1523-1528.

Chapter 9

Cyclic Steam Injection and Other Thermal Stimulation Methods

This chapter deals with methods for stimulating the productivity of wells by thermal means. Thus, it has some basic elements in common with the other SPE monographs on well stimulation. At a production well, pure stimulation treatment may be defined as any operation (not involving perforating or recompleting) carried out with the intent of increasing the posttreatment production rate without changing the driving forces in the reservoir. The pure stimulation treatment may eliminate damage present in the neighborhood of the wellbore, for example, by circulating hot solvent to dissolve asphaltene and other organic precipitates from perforations. Or it may reduce the flow resistance into an undamaged wellbore, for example, by reducing the crude viscosity by means of a heated wellbore. Several of the more common thermal stimulation techniques *do* change the driving forces in the reservoir, sometimes intentionally. Therefore, they should be considered combination stimulation and displacement processes. For example, cyclic steam injection clearly reduces the flow resistance near the wellbore, but it also (1) enhances the depletion mechanism by causing gas dissolved in the crude to become less soluble as the temperature increases and (2) increases the amount of water retained in the reservoir through a combination of temperature-induced wettability changes and hysteresis in the relative permeability curves in a manner analogous to a countercurrent imbibition displacement process. Recognition of some of the mechanisms that occur naturally during cyclic steam injection have led to modifications to enhance them – e.g., the addition of a noncondensable gas to the injected steam.[1]

9.1 Cyclic Steam Injection

Mechanisms

The mechanisms involved in the production of oil during cyclic steam injection are diverse and complex. There is no doubt that a reduction in the viscosity of the crude in the heated zone near the well greatly affects the production response. Changes in surface forces with increasing temperature probably play a role in the preferential retention of the condensed water in the reservoir rather than the more viscous oil, a fact that has been noted in several studies.[2,3] Other factors, such as different phase compressibilities and flashing of water to steam (thus reducing the relative permeability to water) may also favor water retention. The retention of the condensed water may be estimated from production performance data such as those given in Fig. 9.1. (See also Table 9.1.) Another class of mechanisms is associated with the generation of a noncondensable gas phase. This may result from the fact that the solubility of gases in the formation liquids decreases with increasing temperature (see Fig. 9.2 for solubility of hydrocarbon gases in water; see Figs. B.4 through B.6 for solubility of hydrocarbon gases in crudes), but also may result from chemical reactions. Chemical reactions that generate gases include (1) decarboxylation of the crude (decomposition of $-CO$ radicals to form CO_2), (2) formation of H_2S from sulfur-containing radicals in the crude, (3) formation of H_2, CO, CH_4, and CO_2 from reactions between water and crude, and (4) formation of CO_2 by decomposition and reactions of carbonate minerals and dissolved bicarbonates. Published gas analyses before and after cyclic steam injection are essentially nonexistent. But changes in gas composition have been reported in steam drives, and it is inferred that the same mechanisms occur at least during the steam injection phase of a cyclic steam injection project. This additional gas provides an extra driving force to produce oil during the production phase of the project. Even if the reduced solubility of gases at the higher temperatures does not release free gas, the solution-gas-expansion effect will be enhanced, and all liquids will expand on heating. Furthermore, some of the water between the mineral grains and the oil phase may flash to steam as the pressure is reduced during the production cycle, thus adding to the driving force.

By and large, there must be a driving force present

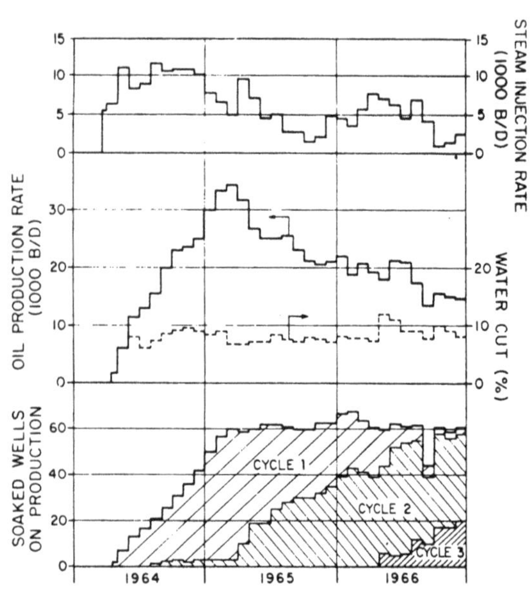

Fig. 9.1 – Overall project performance – total steam injection and oil production rates for all soaked wells vs. time – Tia Juana field, Venezuela.[2]

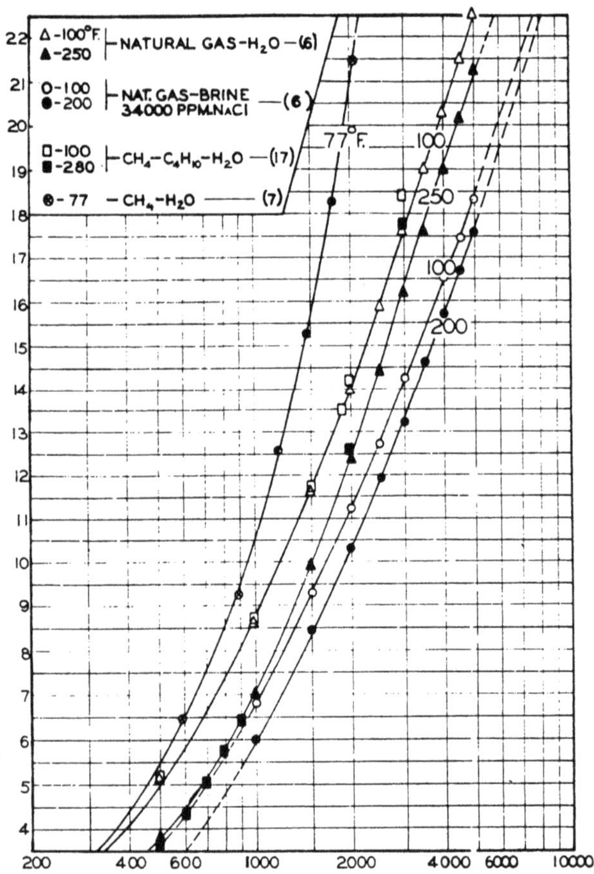

Fig. 9.2 – Solubility of natural gases in water and brine.[4]

in the reservoir initially if cyclic steam injection is to succeed. In other words, it is not sufficient merely to reduce the flow resistance in the reservoir. Gravity drainage and solution-gas drive are often highly important in providing driving forces during the production phase. And during soaking and production, condensation of steam tends to reduce the pressure at and near the well, thus promoting flow. Along the eastern shore of Lake Maracaibo, an important driving force is compaction, which is particularly effective when an extensive area is subjected to thermal operations.[2,5]

Performance Prediction

The performance of a cyclic steam injection operation is sensitive to the acting production mechanisms, to the reservoir and fluid properties near the well, and to the operating variables. The applicability of predictive methods depends on the proper representation of the reservoir and crude properties, even assuming that the total interaction between the operating variables and the reservoir are properly taken into account. One reservoir property that is not known precisely but that significantly affects the production response of cyclic steam injection operations is the relative permeability to the flowing fluids. Coats *et al.*[6] reported that relative permeabilities require hysteresis modifications (i.e., they are different during injection and backflow) in order to match the performance of a multicycle operation. Relative permeabilities, especially with imbibition/drainage/temperature effects, are not available normally. So it is highly questionable if a prediction of performance in an area where there has been no cyclic steam injection is likely to be even moderately accurate. Where cyclic steam injection production is available, likely values of the reservoir and crude properties can be determined through history-matching procedures.[6] These then can be used to predict the behavior of subsequent steam injection cycles in the same well, or to predict the performance of nearby wells under different operating conditions.

These comments are even more applicable to the simplified or desk-top predictive methods, which of necessity consider only the most pertinent but easily tractable factors. The simpler predictive methods are based on specific models of how the cyclic steam injection works (in contrast with the thermal reservoir numerical simulators, which in principle provide solutions to the differential equations describing conservation of mass and energy in three dimensions). Owens and Suter,[3] for example, consider the production rate ratio before and after steam injection to be

$$\frac{q_{oh}}{q_o}=\frac{\mu_o(T_i)}{\mu_o(T)}, \quad \text{(9.1)}$$

where the viscosity of the oil changes with time as heat is lost. They did not predict how the viscosity would increase with time during production; they assumed that the temperature affecting the viscosity would decline in the same manner as the wellhead temperature so that a few measurements would

TABLE 9.1 – STEAM-SOAK INJECTION-PRODUCTION DATA, TYPICAL WELL – BUENA FE FEE LEASE, MIDWAY SUSNET FIELD[12]

	Cycle						
	First	Second	Third	Fourth	Fifth	Sixth	Seventh
Steam injection, bbl	6,332	6,179	5,170	4,592	5,618	5,420	5,664
Injection time, days	11	15	12	8	9	12	7
Injection rate, B/D	453 to 666	118 to 536	275 to 497	514 to 600	522 to 729	200 to 663	709 to 791
Injection temperature, °F	350	298 to 334	307 to 338	298 to 327	307 to 324	298 to 307	378 to 415
Injection pressure, psig	115 to 125	50 to 95	60 to 100	50 to 90	60 to 80	50 to 60	175 to 275
Soak time, days	8	3	3	3	6	3	2
Oil production during cycle, bbl	5,585	8,784	5,702	5,898	7,218	5,137	7,393
Water production during cycle, bbl	1,854	3,074	2,800	3,185	4,576	4,382	3,940
Total time, days	120	216	162	192	298	262	340
Cumulative oil produced, STB	5,585	14,369	20,071	25,969	33,187	38,324	45,717
Cumulative water injection, bbl	6,332	12,511	17,681	22,273	27,891	33,311	38,975
Cumulative water production, bbl	1,854	4,928	7,728	10,913	15,489	19,871	23,811
Water production/water injection	0.293	0.394	0.437	0.490	0.555	0.597	0.611
Cumulative time, days	120	336	498	694	992	1,254	1,594
Average rate per cycle, B/D	55.3	44.3	38.8	31.9	25.5	20.8	22.3
Oil/steam ratio, vol/vol	0.88	1.42	1.10	1.28	1.28	0.95	1.31
Steam/oil ratio, vol/vol	1.13	0.70	0.91	0.78	0.78	1.06	0.77

provide a basis for extrapolating and predicting the subsequent temperature and production response.

Boberg and Lantz's model[7] encompasses additional key elements. They considered the temperature distribution in the reservoir after steam injection, which they represented by a step function. Their method can be used to estimate the oil production response when the injected steam enters several sand intervals. But only the case where steam enters a single interval is discussed here.

The area of the heated volume is given by Eq. 5.8, and the area beyond it is considered to remain at the initial reservoir temperature T_i. The temperature within the heated zone is allowed to decrease by both vertical and horizontal conduction and by removal of energy by means of the produced fluids. Fig. 9.3 shows the geometry for the case of a single production interval. The average temperature within the heated zone, which is not considered to change in size, is given by

$$\bar{T}=T_i+(T_s-T_i)\,[f_{VD}f_{HD}(1-f_{pD})-f_{pD}], \quad (9.2)$$

where the injected steam temperature is measured at downhole conditions, f_{HD} and f_{VD} are time-dependent dimensionless quantities that account for conduction losses in and normal to the reservoir plane, respectively, and f_{pD} is a time-dependent dimensionless quantity that accounts for the energy removed by means of produced fluids. The quantities f_{VD} and f_{HD} for one heated interval are given in graphic form in Fig. 9.4. Note that Boberg and Lantz consider the volumetric heat capacities to be the same in the oil sands and surrounding shales, and that the dimensionless time is defined differently for each of these two quantities. For vertical heat losses, the dimensionless time t_D is given by Eq. 5.6 with $M_R=M_S$. Vertical heat losses depend on thermal properties of both the reservoir and its adjacent formation. On the other hand, radial heat losses depend only on the thermal properties of the reservoir.

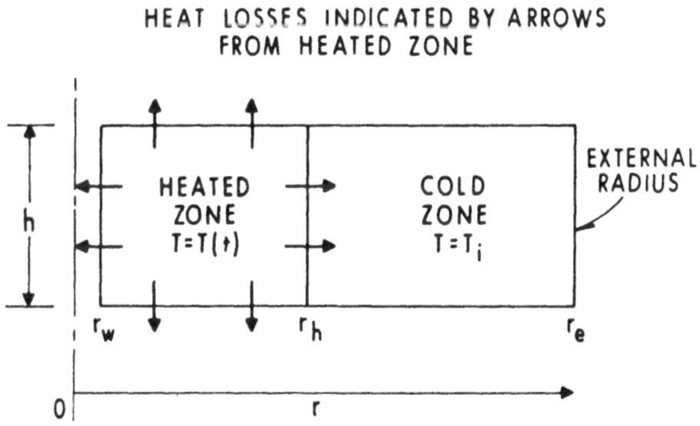

Fig. 9.3 – Schematic representation of Boberg and Lantz's thermal model of the cyclic steam injection process for a single interval.

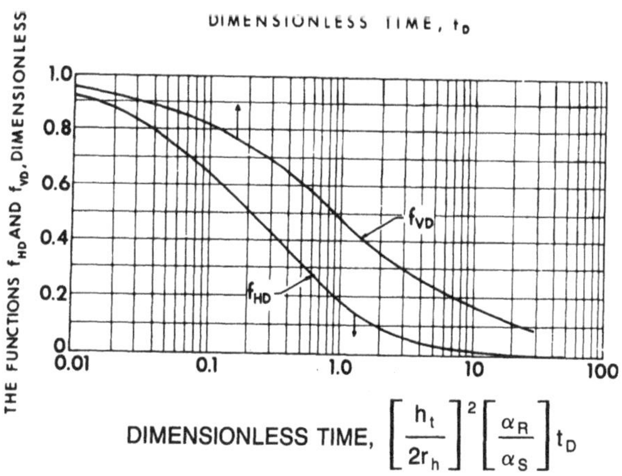

Fig. 9.4 – Horizontal and vertical heat loss functions for the case of a single sand and equal volumetric heat capacities in the sand and adjacent formations.[7]

TABLE 9.2 – EXPRESSIONS FOR F_1 AND F_2 IN EQ. 9.6 – RADIAL FLOW

Flow Conditions	F_1	F_2
Steady State	$\dfrac{\ln\left(\frac{r_h}{r_w}\right)+s_h}{\ln\left(\frac{r_e}{r_w}\right)+s}$	$\dfrac{\ln\left(\frac{r_e}{r_h}\right)}{\ln\left(\frac{r_e}{r_w}\right)+s}$
Semisteady State	$\dfrac{\ln\left(\frac{r_h}{r_w}\right)-\frac{r_h^2}{2r_e^2}+s_h}{\ln\left(\frac{r_e}{r_w}\right)-½+s}$	$\dfrac{\ln\left(\frac{r_e}{r_h}\right)-½+\frac{r_h^2}{2r_e^2}}{\ln\left(\frac{r_e}{r_w}\right)-½+s}$

The relationship between the function f_{pD} and the energy removed by means of produced fluids is

$$f_{pD}=\frac{1}{2Q_{\max}}\int_0^{t_p}\dot{Q}_{p,dh}dt, \quad \ldots\ldots\ldots\ldots (9.3)$$

where the rate of heat removal from the reservoir, $\dot{Q}_{p,dh}$, must be measured or estimated at downhole conditions and the heat content of the reservoir at the end of steam injection, $Q_{\max}$, is determined from Eq. 5.10.

Producing time t_p is measured from the termination of steam injection. $\dot{Q}_{p,dh}$ is related to the production rates and average temperature of the heated zone by

$$\dot{Q}_{p,dh}=\left[<5.615 \text{ cu ft/bbl}>\left(q_{oh}M_o+q_{wh}M_w+q_sM_w+\frac{q_s\rho_wL_v}{\bar{T}-T_i}\right)+<10^3\ scf/Mscf>\cdot q_{gh}M_g\right](\bar{T}-T_i). \quad \ldots\ldots (9.4)$$

The subscript h on the production rates is used to distinguish the heated production from the unheated. All rates are at subsurface conditions. Bottomhole oil and gas rates are not difficult to estimate from surface ones. But the water and steam production rate must be corrected from that measured at the surface to account for heat losses. The water, gas, and steam contents of the production are not predicted by this model. Values for these quantities must be available from other wells, or trends must be established from the early response of the well. The ratio of oil production rate after steam injection to the rate before injection is given by

$$\frac{q_{oh}}{q_o}=\frac{J_h}{J}, \quad \ldots\ldots\ldots\ldots (9.5)$$

where the stimulated to unstimulated productivity ratio (or reciprocal flow resistance ratio) is given by

$$\frac{J_h}{J}=\frac{1}{(\mu_{oh}/\mu_o)F_1+F_2}, \quad \ldots\ldots\ldots\ldots (9.6)$$

where F_1 and F_2 are constants that arise from radial flow geometry and type of flow considered and the

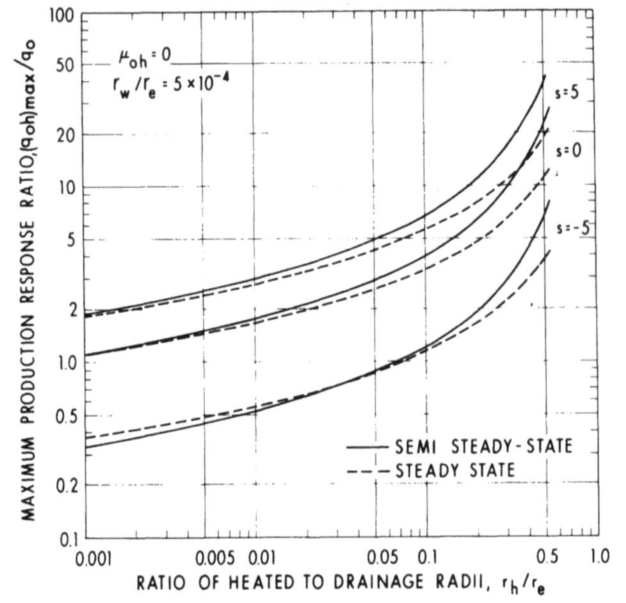

Fig. 9.5 – Influence of heated zone on maximum production response from cyclic steam injection.

oil viscosity is evaluated at the average temperature of the heated zone. Boberg and Lantz give expressions for F_1 and F_2 for steady-state and semisteady-state conditions; they are listed in Table 9.2.

In these expressions, the skin factor after steam stimulation, s_h, is distinguished from that prior to steam stimulation, s.

Because the average temperature in the heated zone decreases during production, the oil rate ratio also decreases with time. Predictions of the performance of a cyclic steam injection operation can be calculated by hand if a few time steps are sufficient to describe the problem. First, the factors f_{VD} and f_{HD} are determined from Fig. 9.4 at the end of the soak period. Also, since there is no production, $f_{pD}=0$. Eq. 9.2 is used to calculate the average temperature in the heated zone, from which the oil production rate is calculated with Eqs. 9.5 and 9.6. This oil production rate then is used during the next time period to calculate the amount of energy removed through production. At the end of a selected time interval, values of f_{VD} and f_{HD} are calculated again and the process is repeated as required. The user may reduce the size of the time intervals until the results are relatively insensitive to these changes.

The model of Boberg and Lantz assumes that before stimulation there is reservoir energy capable of producing some oil at the initial reservoir temperature T_i. It assumes that the reservoir energy causes oil to be moved into the heated zone, where resistance to its flow is reduced. This model is basically a stimulation model, the response of which is controlled by the productivity ratio given by Eqs. 9.5 and 9.6, and Table 9.2. Fig. 9.5 shows the ratio of oil production rate after stimulation to the rate before stimulation for the limiting case where the viscosity of the hot oil approaches zero, which is given by the case where the productivity ratio is equal

TABLE 9.3 – MODELS OF THE CYCLIC STEAM INJECTION PROCESS

Model	Direction of Fluid Flow*	Flow Regime**	Additional Calculations Needed?	Vertical Extent of Hot Zone	Radial Temperature Profile
Owens and Suter[3]	r	i	No	Total	Uniform
Boberg and Lantz[7]	r	i, sss	No	Total	Step
Martin[9]	r	i, sss	Yes	Total	Complex
de Haan and van Lookeren[2]	r	Transient	No	Total†	Step
Seba and Perry[10]	r,z	i + gravity drainage	No	Total	Constant $<r_h$ Variable $>r_h$
Closmann *et al.*[11]	r,z	i	Yes	Any	Complex

*r = radial; z = vertical.
**i = incompressible flow; sss = semisteady state (rate of pressure decline is everywhere the same).
†Method represents steam overlay in terms of equivalent radial flow.

to $1/F_2$. Obviously, this yields the maximum response obtainable.

Because there is no flow resistance within the heated zone in this case, the stimulation ratios shown in Fig. 9.5 do not depend on s_h. Results are presented for these three skin values: $s = -5$, 0, and 5. The results for $s = -5$ require some discussion, because the steady-state stimulation ratio for this case is less than one for heated radii less than $0.074r_e$. For low values of μ_{oh}/μ_o, Eq. 9.6 can give stimulation ratios less than one when the radius of the heated zone r_h is less than the effective wellbore radius, r_{wa}. The latter is related to the actual wellbore radius and skin by

$$r_{wa}=r_we^{-s}. \qquad (9.7)$$

Thus, Eq. 9.6 can be used to investigate the extent of the heated zone that would yield adequate stimulation ratios. This is especially important in wells initially having negative skins, where significant stimulation ratios may be difficult to obtain. Similar comments apply to semisteady-state conditions. Note that the stimulation ratio under semisteady-state conditions is somewhat higher than for steady state for cases of practical interest – i.e., for stimulation ratios higher than one. Also note that the stimulation ratio is higher the more positive the skin. Naturally, if realistic hot-oil viscosities (i.e., viscosities greater than zero) are considered, the stimulated production rate ratio given in Fig. 9.5 would be lower.

Heat remaining in the reservoir at the start of a new steam injection cycle is considered as equivalent heat injected during that cycle. Eq. 9.2 is used to calculate average temperatures remaining in the formation.

As noted in Burns' early review[8] of the cyclic steam injection process, the model of Owens and Suter is consistent with that of Boberg and Lantz only when the radius of the heated zone coincides with the external boundary of the drainage volume of the well.

Martin,[9] de Haan and van Lookeren,[2] Seba and Perry,[10] and Closmann *et al.*[11] have developed different models to account for the performance of the cyclic steam injection process. To simplify the calculations, all of these and the two discussed earlier make some approximation about the nature of the heat and fluid flows. Note that these studies preceded the availability of numerical thermal reservoir simulators. Even today, however, these models offer indispensable insight into the process and can be used either to make simple sensitivity studies or to screen prospective projects. Table 9.3 summarizes the main characteristic features of the models. Those of Martin[9] and Closmann *et al.*[11] may require substantial additional computations. Surprisingly, only de Haan and van Lookeren attempt to account for transient effects. Most of the models consider radial fluid flow only, but those of Seba and Perry[10] and of Closmann *et al.*[11] consider both radial and vertical flows. In all cases, only horizontal reservoirs are considered. Only Seba and Perry[10] treat gravity drainage as a producing mechanism.

Design

Because cyclic steam injection is relatively easy and inexpensive to implement, it customarily is tested directly in the field, even where there has been no previous experience of the process. Where an operator has access to the results of previous experience, or where it is known that others have had success, the object is not to prove the process but to optimize it. Optimizing also generally is done in the field, sometimes aided by models such as the ones just discussed or, more rarely, by numerical thermal reservoir simulators.

Optimizing requires some data or basis for estimating the interrelation of the amount of steam and heat injected, the duration of the soak phase, and the production response for a sequence of steam injection and production cycles. Most of this information is obtained in the field. Optimizing is most difficult where the number of wells is small. A single well cannot provide information for its optimum operation, although knowledge gained from the analysis of a first-cycle performance may indicate changes in the operations of subsequent cycles likely to result in improved production. Where the response from a large number of wells is available from similar formations over a range of operational variables, it is customary to sort the production response by categories. These categories may include (1) reservoir characteristics (e.g., permeability, porosity, gross sand thickness, net sand thickness, and oil viscosity at initial reservoir conditions), (2) cumulative steam injection, (3) average steam quality, (4) duration of soak phase, and (5) number of the cycle.

Typical responses then are assigned for each of the categories, and an optimized cyclic steam injection operation is calculated using the operator's criteria, and applied to each well taking into consideration its location in the reservoir. The more sophisticated optimizing procedures would use updates on the

TABLE 9.4 – SUMMARY OF TYPICAL WELL PERFORMANCES – FIRST-CYCLE STEAM SOAK OPERATIONS IN VARIOUS FIELDS OF CALIFORNIA[8]

		Production Rate (B/D)				Net	Steam Injected			Oil Recovered (bbl)		Additional Oil Recovered
		Before Steam,	After Steam,*	q_h		Sand Open	Total	Per Foot of Sand	Cycle Period		Per Barrel	Per Barrel of Steam
Field	Zone	q	q_h	q	End	(ft)	(bbl)	(bbl)	(months)	Total	of Steam	(bbl)
Huntington Beach	TM	15	160	11	25	40	4,500	112	15	29,000	6.5	5.0
San Ardo	Lombardy	25	360	14	35	220	14,000	64	18	50,000	3.6	2.8
Kern River	China	3	140	47	15	22	4,400	200	6	11,600	2.62	2.5
Midway-Sunset	Potter (A)	10	110	11	25	250	6,000	24	5	9,240	1.54	1.29
Kern River	Kern River	14	65	4.6	20	220	6,500	30	5	4,730	0.73	0.43
Coalinga	Temblor	3	52	17.3	15	107	9,000	84	5	4,300	0.48	0.40
Midway-Sunset	Tulare	5	56	11	10	240	12,000	50	6	4,640	0.38	0.31
Midway-Sunset	Potter (B)	5	35	7	10	250	7,700	31	4	3,000	0.39	0.29
White Wolf	Reef Ridge	30	85	2.4	30	75	14,000	187	4	6,750	0.48	0.23
Poso Creek	Etchegoin	7	20	3	10	80	6,700	84	6	2,660	0.40	0.21

*Average first 30 days.

performance of each cycle to refine procedures. Table 9.1 shows typical well responses by cycle obtained in the Midway Sunset field and discussed by Rivero and Heintz.[12]

Optimizing is subject to many constraints and can be narrow or general. A simple but typical decision resulting from an optimizing program is this: A steam generator of sufficient capacity to supply steam to two wells will become available in two weeks, and during that time seven wells will be candidates for steam injection cycles. Which wells should be selected for steam injection? A separate question would be whether additional steam generation capacity is justified. Criteria for optimizing are invariably economic ones.[13] Typically, the intent is to maximize the economic benefit from each cycle, but the economic benefit may be defined in several ways. Rivero and Heintz[12] compare two economic yardsticks on cyclic steam injection operations: the cumulative average daily profit and present worth. But any yardstick suitable to the operator can be used. The optimizing procedures resulting from the various economic criteria are sensitive to the reservoir and crude properties. It is doubtful that these could be generalized to all reservoirs and for all operators.[14,15]

Difficulties in implementing and evaluating cyclic steam injection may arise – e.g., where the injectivity is poor, the crude is very viscous, the reservoir is deep, the reservoir pressure is high, and where large gas caps or bottomwater zones are present. Considerable effort may then be required to optimize the process.[16]

Examples of Field Applications

Cyclic steam injection is the most common of the thermal recovery processes. It has been used extensively in the heavy oil fields in California[8,17] and in the Bolivar coast of Venezuela.[18,19] There are indications it may become increasingly popular in the heavy oils of Alberta and Saskatchewan.[16,19]

Table 9.4 (taken from Burns[8]) summarizes applications and results from first-cycle response in several fields in California. Reported additional oil recovery per barrel of steam injected ranges from 0.21 bbl at Poso Creek to 5.0 bbl at Huntington Beach. Reported total oil recovery per barrel of steam injected, which includes contributions due to natural production mechanisms, is slightly larger – from 0.38 to 6.5 bbl. The ratio of oil production rate after stimulation to that before ranges from 2.4 to 47 but centers on a factor of 12. Net sand intervals range from 22 to 250 ft, and the steam injection range from 24 to 260 bbl per foot of net sand. Total steam injection per well ranges from 4,400 to 14,000 bbl. No correlation is available between the average response and operating variables.

In Venezuela, the amount of steam injected per cycle is generally larger – in the neighborhood of 50,000 bbl per cycle.[2,7,18] The oil recovery per cycle also is larger than that in California. An example of a cyclic steam injection project performance from the Tia Juana field in Venezuela is given in Fig. 9.1. Table 9.5 shows cyclic steam injection results from several projects in the Bolivar Coast of Venezuela. The oil produced per barrel of steam injected is, on the average, higher than that for California fields shown in Table 9.4. The higher oil/steam ratios common in the Bolivar Coast are considered to be influenced significantly by compaction of the oil-bearing sands during production, the importance of which cannot be overemphasized in that area.

Oil production response can vary widely over a series of cycles. Table 9.1 shows no major trend in the average oil production from seven cycles in the Midway Sunset field. Table 9.6 shows that the average oil recovery per unit of steam injected, expressed in barrels of oil per 10^6 Btu injected, declined slightly over four cycles. Table 9.7 shows the effect of structural position and well completion (indicated in parentheses) on the oil response by cycle in Yorba Linda.[21] The amount of oil produced per barrel of steam injected tends to be smaller for the upstructure wells, when calculated at the time the cycles were terminated. The downstructure wells benefit from gravity drainage, and their rate of decline of this ratio is not discernible.

Surveys of the various thermal recovery processes, including cyclic steam injection, appear in the literature regularly. These surveys are a rich source of information regarding the extent of the application of thermal methods, especially in the U.S. Examples of such surveys are those prepared by *Oil and Gas J.*[22-24] and *Pacific Oil World.*[25] Surveys that are more in-depth but not so extensive have been prepared by others, notably Farouq Ali.[26,27] The latter provide some analyses and interpretation of the

TABLE 9.5 – CYCLIC STEAM INJECTION EXPERIENCE IN THE BOLIVAR COAST OF VENEZUELA[20]

Project	Cumulative Steam Injection (ton × 10^6)	Number of Wells	Cumulative Oil Productivity (bbl × 10^6)	Cumulative Incremental Oil Production (bbl × 10^6)	Incremental Oil/Steam Ratio (bbl/bbl)	Production Resulting From Compaction (% of stock-tank OOIP)
Tia Juana						
B/C-3	0.576	63	25.3	18.6	5.14	19.4
C-2/3/4	0.268	13	7.02	5.12	3.04	
D-2/E-2	1.96	69	49.8	35.5	2.88	18.9
D/E-3	0.687	69	28.2	16.7	3.86	19.3
G-2/3	0.887	123	50.7	37.7	6.77	22.2
H-6	0.422	29	16.5	12.5	4.71	29.6
H-7	0.227	13	9.23	7.13	5.01	16.1
J-6	0.075	9	4.87	4.33	9.16	14.6
J-7	1.21	125	44.6	36.8	4.85	
F-7	1.28	135	46.0	36.2	4.52	10.7
M-6	0.923	95	27.1	20.4	3.51	18.0
D-6	0.982	103	46.7	44.3	7.19	14.1
Lagunillas						
T-6	0.615	31	28.4	18.8	4.86	23.4
V-7	0.685	116	24.4	17.0	3.96	20.4
W-6	1.01	37	22.1	11.2	1.76	21.2
W-6E	1.08	47	24.3	17.1	2.51	

processes, always provide references, and often include results of efforts outside the U.S.

9.2 Other Cyclic Injection Processes

Steam and hot water are the only heated fluids reported to have been injected for several cycles to stimulate wells, although other heated fluids (including mixtures of lease crude and refinery-gas oil, and noncondensable gases) are reported to have been used in at least one cycle.[28,29] Forward combustion, which requires injection of air, also has been used as a stimulation procedure.[28,30,31] It appears that these processes have been used only once in any one well. Farouq Ali[28] reports hot-water stimulation to "have met with varied success," but the best results he gives (25 B/D of oil per 10^9 Btu injected) is equivalent to about 0.01 B/D oil per barrel of steam injected, which is only one-twentieth of the lowest value reported in Table 9.4 for cyclic steam injection. Still, there may be reasons why steam cannot be injected into a well – e.g., the mechanical condition of the well,[32] the lack of a steam generator or the permits to operate one, or high injection pressures. But in the Morichal field, Venezuela, Araujo[33] reports a response averaged over 31 stimulations in 21 wells equivalent to an oil/steam ratio of 3 bbl/bbl, apparently the result of removal of asphaltic and scale deposits downhole, as well as reductions in oil viscosity.

In the Northeast Butterly Oil Creek Unit, OK, Holke and Huebner[32] report an incremental oil production of 8,200 bbl resulting from the injection of 28,348 bbl of hot water at 450°F and 1,600 psi. Total Btu injection was 4.3 billion, which is comparable to an oil/steam ratio of 0.8 bbl/bbl, and is typical of responses for cyclic steam injection given in Table 9.4. Another hot-water stimulation of 300°F maximum injection temperature gave an even higher oil production response per unit of energy injected. But a third well was unsuccessful, apparently because the injected water went below the water/oil contact.

The sensitivity of conditions affecting hot-water stimulation response has been studied by means of numerical simulation by Diaz-Muñoz and Farouq Ali.[34]

In combustion stimulation, the fire front moves from the wellbore a distance of 10 to 20 ft. The high temperatures resulting from the combustion process, usually in excess of 1,000°F in the vicinity of an injection-ignition well, help remove some types of plugging agents from the vicinity of the wellbore. Not only would organic solids and precipitates be removed by burning, but the high temperatures would dehydrate and perhaps stabilize clays, cause small fractures in consolidated reservoir rock, and thus restore and even increase its absolute permeability.[31] And of course the oil viscosity is reduced by the increased temperature. White and Moss[31] report one barrel of oil produced per 10^6 Btu "generated in the formation." At 3.5×10^5 Btu per barrel of steam, this is equivalent to 0.35 bbl of oil produced per barrel of steam, or a steam/oil ratio of 2.9 vol/vol. Although the data are limited, they are comparable with the lower range of oil/steam ratios reported in Table 9.4 for the cyclic steam injection process.

9.3 Wellbore Heating

The use of downhole heaters, a practice dating from the last century, is the oldest method of increasing production by thermal means. The most commonly used tools are electric heaters and gas burners. They have been used successfully to increase oil production rates from reservoirs containing viscous or paraffinic crudes.

Mechanisms

The principal mechanisms affecting oil production while using downhole heaters are the reduction of the crude viscosity and the resolubilizing (or prevention of precipitation) of asphaltenes and other organic solids in the crude. The mechanisms that apply in

TABLE 9.6 – SUMMARY OF PERFORMANCE THROUGH FOUR HUFF 'N' PUFF CYCLES AS OF OCT. 1, 1970, TM SAND, HUNTINGTON BEACH OFFSHORE FIELD[15]

	Cycle 1	Cycle 2	Cycle 3	Cycle 4
Number of wells	24	18	11	4
Total oil recovery, STB	694,000	556,250	271,150	116,900
Average cycle length, months	14	18	15.3	14.5
Average oil recovery per well, STB	28,900	30,900	24,650	29,225
Average quality of steam injected, %	71.4	69.3	75.1	78.5
Average volume of steam injected, bbl	9,590	8,130	10,190	11,760
Steam injected, bbl/ft	213.9	191.3	232.1	267.3
Heat input, MMBtu/ft	64	63.8	89.4	98.4
Ratio of oil recovered to steam injected, STB/bbl	3	3.8	2.4	2.5
Oil recovery per foot of sand, STB/ft	645	737	560	665
Oil recovery, STB/MMBtu	10.1	11.2	6.3	6.8

cyclic steam injection would apply here to a lesser degree.

Heat is transferred through the reservoir and away from the well primarily by conduction. Production of reservoir fluids brings heat back from the reservoir, decreasing the net rate at which heat is transferred from the wellbore into the reservoir. This can be a particularly serious problem in wells producing at high water cuts. But increasing the temperature around the well, or rather reducing the viscosity of the crude near the wellbore, will reduce the water cut, as indicated by the fractional flow relation given by Eqs. 6.4 and 6.5. Combustion products entering the reservoir from gas burners will increase the rate of heat injection into the reservoir, but only slightly, since their mass rate of flow is relatively small. Downhole heaters are usually left "on," except where there are light paraffinic crudes, in which case intermittent operation is more economical.

The placement of downhole heaters is extremely important. It is essential to heat the interval having the highest potential for increased production. Considerations should include the degree of productivity damage and the available driving force. Also, heaters must not be placed above zones producing fluids at high rates. Production from lower intervals inevitably will carry heat up the producing string and reduce the amount of heat available for transfer into the shallower interval of interest.

The rate of heat generated by borehole heaters is limited by the maximum temperature at which the heaters can be operated safely. Gas burners usually have higher temperature ratings than electric heaters. Both are used to improve the production rate of low-rate producers. Typical improvements in oil production rate range from a few percent to as much as a factor of three but are generally closer to a factor of about two. Even with a total production response of 100 B/D at a 25% water cut, which is abnormally high, and a desired maximum crude temperature rise of 400°F, which is also unnecessarily high, the rate of heat removed by the produced fluids is found from

$$\dot{Q}_{p,dh} = \langle 5.615 \text{ cu ft/bbl} \rangle \left(q_o M_o + q_w M_w \right) \Delta T_{pdh} . \quad \ldots\ldots (9.8)$$

TABLE 9.7 – EFFECT OF COMPLETION TYPE AND STRUCTURAL POSITION ON WELL PERFORMANCE, YORBA LINDA FIELD[21]

Type of Well and Completion	Cycle No.	Steam Injected (bbl)	Oil Produced (bbl)	Extrapolated Oil Produced (bbl)	Extrapolated Oil/Steam Ratio
Upstructure (full)	1	24,500	13,800	19,400	0.79
	2	27,300	10,700	14,100	0.52
	3	19,200	9,100	11,200	0.58
	4	17,200	10,100	12,900	0.75
	5	15,800	1,900	*	–
Upstructure (short)	1	16,500	20,100	21,400	1.30
	2	15,700	10,000	20,600	1.31
	3	15,900	11,800	15,200	0.96
Upstructure (scab)	1	–	–	–	–
	2	–	–	–	–
	3	15,100	11,300	14,200	0.94
	4	14,700	9,900	13,200	0.90
	5	16,400	6,800	9,300	0.57
Downstructure (full)	1	22,300	23,200	13,800	1.42
	2	20,400	18,900	23,300	1.14
	3	15,000	17,300	20,700	1.38
	4	13,500	13,800	18,100	1.34
	5	16,400	17,800	17,100	1.09
Downstructure (short)	1	13,900	18,900	22,200	1.60
	2	15,300	15,100	25,600	1.67
	3	15,100	11,100	23,800	1.48

*Insufficient data to extrapolate the curve.

to be (5.615 cu ft/bbl) (100 B/D)[(0.75) 24 Btu/cu ft °F + (0.25) 63 Btu/cu ft °F] (400°F), or about 7.6 × 10^6 Btu/D. If only 10% of the heater output is used to heat the produced fluid, the rate of heat generation would need to be 10 times as large:

$$\dot{Q}=7.6\times10^7 \text{ Btu/D}.$$

This is equivalent to the net heat content of about 220 B/D of steam. We see, then, that downhole heaters inherently are limited to rates of energy output that are low compared with those usually attained by cyclic steam injection. For this example, the results are equivalent to an oil/steam ratio of 0.34 vol/vol or a steam/oil ratio of 2.9 vol/vol.

Maximum temperatures must be limited for a number of reasons: (1) the equipment must operate for practical periods of time without requiring repairs to insulation and heating elements, (2) hot spots that could lead to metal failure in gas burners must be prevented, (3) high temperatures can increase corrosion rates in certain downhole environments, and (4) excessive heat can cause coking or precipitating of organic solids which may impair productivity and inhibit the proper transfer of heat from heater to fluids, thus leading to burnouts.

Design and Prediction

Schild[35] presented calculated temperature distributions and production responses "for the simple case of a well producing oil by a radial drive and in the steady state." Steady state, it should be noted, results from the model used. Its existence stems directly from the fact that heat losses to the adjacent formations are considered to be nil. No one appears to have reported the effect of heat losses on results. The dimensionless steady-state temperature distribution for radial heat and fluid flow can be represented by[35]

$$\Delta T_D=\frac{T-T_i}{\Delta T_{pdh}}=\frac{T-T_i}{T_{pdh}\quad T_i}=\left(\frac{\Delta r}{r_w}\right)^{-b}, \quad \ldots (9.9)$$

where

$$b=\dot{Q}_{p,dh}/2\pi h_t \lambda_R \Delta T_{pdh}, \quad \ldots (9.10)$$

T_i = temperature of the undisturbed reservoir, °F,
T_{pdh} = resultant temperature of the produced fluids at the heated interval, °F,
T = temperature anywhere in the reservoir, °F,
$\Delta r/r_w$ = distance from the wellbore expressed as a multiple of the wellbore radius,
λ_R = thermal conductivity of the reservoir, Btu/ft-D-°F, and
$\dot{Q}_{p,dh}/\Delta T_{pdh}$ = rate per unit temperature at the heater at which heat is removed by the produced fluids (can be calculated using Eq. 9.4 or 9.8), Btu/D°-F.

The producing downhole fluid temperature T_{dh} can be measured from wellhead fluid temperatures using procedures outlined in Section 10.3 or can be estimated from Eq. 9.9 after making the substitution

$$\dot{Q}_{p,dh}=(1-\eta_h)\dot{Q}, \quad \ldots (9.11)$$

where the heater efficiency η_h accounts for the fraction of the heater output $\dot{Q}$ entering the formation.

Radial temperature profiles in the reservoir near the heated well, given by Eq. 9.9, are plotted in Fig. 9.6. As the value of the parameter b increases – i.e., as the mass production rate at steady state increases – the heated region around the wellbore decreases in radial extent. This is because the produced fluids convect heat toward the well, countering the effect of conduction. At low values of the parameter b, the amount of heat in the reservoir at steady state is substantial and can amount to more than 30% of the heat required to bring the entire drainage volume within a radius of 500 ft to the well temperature T_{pdh}. The amount of heat stored in the reservoir is so large that it would take more than 150 years to store the heat in the reservoir for the case where the parameter b is 0.2, without taking into consideration heat losses outside the heated reservoir volume. Obviously, steady state would not be reached for low values of the parameter. For the general case, the minimum time to approximate steady state (on the basis of the steady-state heat content in the reservoir and assuming no heat losses to the adjacent formations) is given by[35]

$$t_{\min}=\frac{2\pi h_t r_w^2 M_R \Delta T_{pdh}(r_{eD}^{2-b}-1)}{\eta_h \dot{Q}(2-b)}, \quad \ldots (9.12)$$

unless $b=2$, in which case it is

$$t_{\min}=\frac{2\pi h_t r_w^2 M_R \Delta T_{pdh}}{\eta_h \dot{Q}} \ln r_{eD}, \quad \ldots (9.13)$$

where

$$r_{eD}=r_e/r_w, \quad \ldots (9.14)$$

and r_e is the drainage radius of the flow system.

Sheinman *et al.*[36] also report expressions for the unsteady-state temperature distribution during both continuous and intermittent wellbore heating. Their results, which have been discussed by Farouq Ali,[28] consider a *constant* production rate while the temperature approaches steady-state conditions. But for a fixed maximum bottomhole producing temperature, the unsteady-state production response would be expected to be lower than the steady-state response given by Schild.

In Schild's[35] analysis of the steady-state oil production rate, the crude viscosity was considered to be of the form

$$\ln\mu_o=\ln\mu_o(T_i) - \Delta T_D\left[\ln\mu_o(T_i)-\ln\mu_o(T_{pdh})\right]. \quad \ldots (9.15)$$

With this expression for the crude viscosity, he was able to eliminate the temperature variable from Eqs. 9.9 and 9.15 to express the crude viscosity in terms of radial distance Δr, and to integrate the flow resistance as a function of that distance to obtain the improvement in crude production rate by assuming

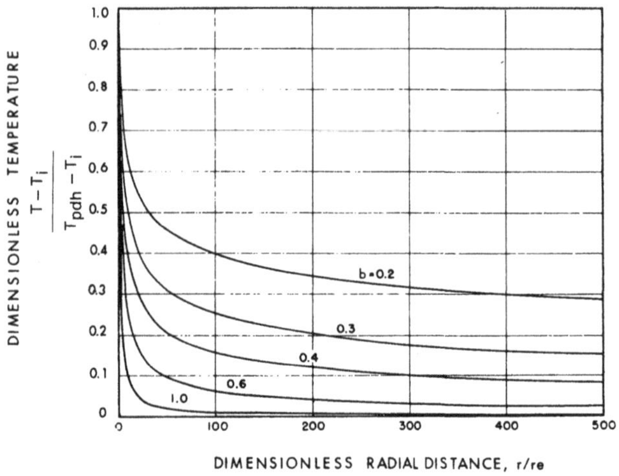

Fig. 9.6 – Idealized steady-state temperature distribution resulting from bottomhole heaters.[35]

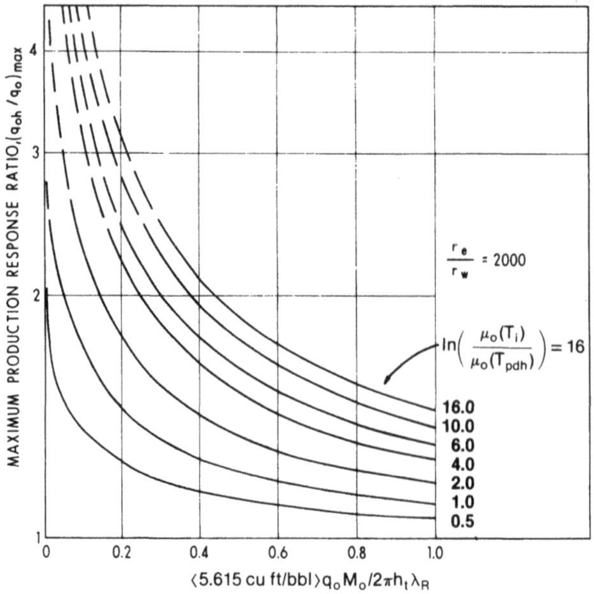

Fig. 9.7 – Steady-state oil production rate ratio vs. dimensionless unstimulated oil production rate.

that oil is the *only* fluid produced. The results are presented in Fig. 9.7. The dashed "portions of the curves in this ... figure correspond to theoretical steady-state situations which are of little practical interest because it takes too long a time period to approach the steady state and to get the benefit of the full production increase due to borehole heating."[35] This figure shows that when the production increase is due only to viscosity reduction, the steady-state ratio of the stimulated to unstimulated production rates is higher the lower the unstimulated production rate and is higher the higher the unheated/heated viscosity ratio. The latter means that a relatively higher response ratio would be obtained the more viscous the crude.

As discussed in connection with the steady-state temperature distributions shown in Fig. 9.6, the reservoir can be heated more rapidly as the production rate decreases. It can be heated very rapidly if heated fluid is injected, which is the basis for the cyclic injection stimulation processes. Obviously, the sequence of heating and producing can be alternated to optimize results. Simultaneously heating and producing – sometimes with intermittent heating – appears to be the common practice when there is no impairment. Where asphaltenes, paraffins, and other organic precipitates have or may come out of solution within the reservoir, some operators favor shutting in production while maintaining the heater in operation. This procedure allows slightly higher temperatures to be reached within the reservoir. The oil production response is generally insensitive to the sequence of producing and heating and depends primarily on the amount of thermal energy in the reservoir, except where there is reservoir impairment by temperature-sensitive organic solids.

Some alternative methods are to inject small volumes of heated crude or to circulate heated water and oil in the well. In the latter method, the fluids must be circulated at high rates to reduce heat losses to a tolerable level. Otherwise, the bottomhole temperature would increase only slightly, especially in deep reservoirs. Sometimes, small volumes of heated crude are injected in batches to dissolve organic precipitates. In this case it is customary to let the crude "soak in" for a few days before putting the well back on production.

The processes for borehole heating discussed in this section are characterized by low rates of heat injection into the reservoir and by relatively small improvements in crude production response. The choice of which approach to use is based on the nature of the problem, cost effectiveness of the available energy, the condition of the well, and (where available) local experience.

Published procedures for predicting stimulated performance exhibit large differences from actual performance.[37,38] Farouq Ali[28,30] and McNiel and Nelson[38] discuss the use of electrical heaters, gas burners, and hot-fluid circulation systems.

Examples of Field Applications

Well stimulation by wellbore heating appears to have been on the wane since the discovery of the cyclic steam injection process. But the method may be applicable in areas where burning fuel is either restricted or prohibited, where water is scarce or prohibitively expensive to treat, or in reservoirs that are too deep or that contain clays or minerals sensitive to steam condensate. Survey articles by Farouq Ali[28,30] and McNiel and Nelson[38] as late as 1973 discuss a number of field applications.

Fig. 9.8 shows the fieldwide increase in production at the Cut Bank field, MT, along with the number of heater installations.[39] Electric heaters, each with a 5-kW rating, were used in this operation. Treated wells about doubled in oil production. Paraffin plugging is

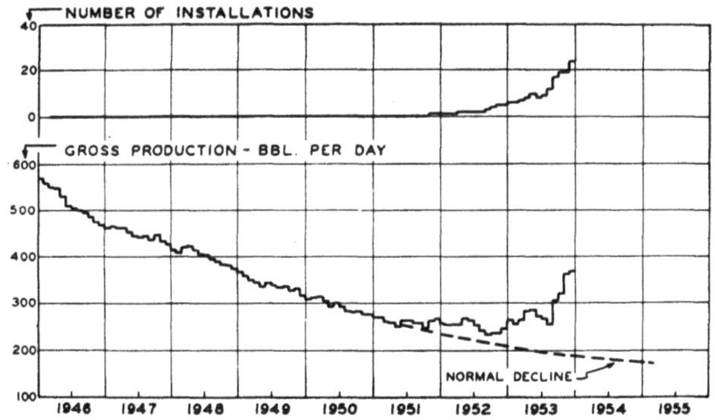

Fig. 9.8 – Production performance of formation-heater installations.[39]

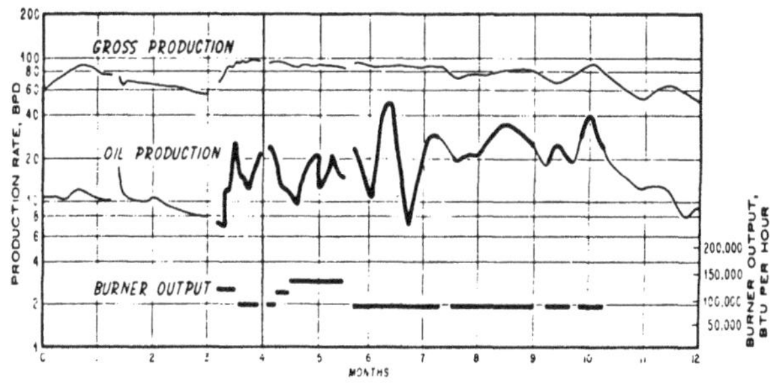

Fig. 9.9 – Production performance of commercial burner installation at 2,870 ft. Average production increase due to heat stimulation was 12 BOPD.[37]

suspected at Cut Bank. Although the response shown in Fig. 9.8 is due to continuous heating, it later was found that the production "could be sustained with intermittent heating."[38] For the average increase of 8 B/D in oil production per well, the equivalent oil/steam ratio amounts to 188 bbl of oil per barrel of steam, based on the 5-kW rate of energy consumption. The high oil/steam ratio suggests that removal of impairment was an important factor in the success of the project.

The use of gas burners has been reported by Brandt *et al.*[37] for three fields in California at depths ranging from 1,378 to 3,177 ft. Gas burners were considered applicable to depths of 5,000 ft. Fig. 9.9 gives the production history of an application at a depth of 2,870 ft. The oil production rate about doubled during the heating period and generally increased with time even though the rate of heat generaton was, on the average, highest during the first 9 weeks. This suggests that the reservoir response had not yet reached a near steady-state condition. The gross production rate decreased during the heating period, so that the water cut decreased during the stimulation period. "Except for occasional power failures in the field, and one well pulling operation, the burner operated continuously for about seven months."[37] Oil production declined to its prestimulation level about a month after the burner was shut off. The rate of burner energy output was about 90,000 Btu/hr, which is equivalent to the injection of 5.4 B/D of steam. This well produced an 11°API crude containing 18 wt% asphalt before stimulation.

Wellbore heating does not appear to have been practiced widely outside of the U.S., except in the USSR. Table 9.8 summarizes results in six fields containing paraffinic crudes of medium to high API gravity. Production increases between 36 and 170% are reported. Equivalent oil/steam ratios range from 5.4 to 276 bbl of incremental oil produced per barrel of steam.

Factors leading to effective stimulation by wellbore heating appear to be (1) high viscosity of the hydrocarbon phase entering the wellbore and (2) high effective pour point of the crude and its components.

References

1. Pursley, S.A.: "Experimental Studies of Thermal Recovery Processes," *Proc.*, Simposio Sobre Crudos Pesados, U. del Zulia, Inst. de Investigaciones Petroleras, Maracaibo, Venezuela (1974).
2. de Haan, H.J. and van Lookeren, J.: "Early Results of the First Large-Scale Steam Soak Project in the Tia Juana Field, Western Venezuela," *J. Pet. Tech.* (Jan. 1969) 101-110; *Trans.*, AIME, **246**.
3. Owens, W.D. and Suter, V.E.: "Steam Stimulation – Newest Form of Secondary Petroleum Recovery," *Oil and Gas J.* (April 26, 1965) 82-87, 90.
4. McKetta, J.J. Jr. and Katz, D.L.: "Phase Relationships of Hydrocarbon-Water Systems," *Trans.*, AIME (1947) **170**, 34-43.
5. Giusti, L.E.: "Experiencias de la C.S.V. con la Inyección Alternada de Vapor en la Costa Bolívar, Estado Zulia," *Proc.*, Simposio Sobre Crudos Pesados, U. del Zulia, Inst. de Investigaciones Petroleras, Maracaibo, Venezuela (1974).
6. Coats, K.H., Ramesh, A.B., Todd, M.R., and Winestock, A.G.: "Numerical Modeling of Thermal Reservoir Behavior," *Proc.*, CIM Canada/Venezuela Oil Sands Symposium, Edmonton, Alta. (1977) Special Vol. **17**, 399–410.
7. Boberg, T.C. and Lantz, R.B.: "Calculation of the Production Rate of a Thermally Stimulated Well," *J. Pet. Tech.* (Dec. 1966) 1613-1623; *Trans.*, AIME, **237**.
8. Burns, J.: "A Review of Steam Soak Operation in California," *J. Pet. Tech.* (Jan. 1969) 25-34.
9. Martin, J.C.: "A Theoretical Analysis of Steam Stimulation," *J. Pet. Tech.* (March 1967) 411-418; *Trans.*, AIME, **240**.
10. Seba, R.D. and Perry, G.E.: "A Mathematical Model of Repeated Steam Soaks of Thick Gravity Drainage Reservoirs," *J. Pet. Tech.* (Jan. 1969) 87-94; *Trans.*, AIME, **246**.
11. Closmann, P.J., Ratliff, N.W., and Truitt, N.E.: "A Steam-Soak Model for Depletion-Type Reservoirs," *J. Pet. Tech.* (June 1970) 757-770; *Trans.*, AIME, **249**.
12. Rivero, R.T. and Heintz, R.C.: "Resteaming Time Determination – Case History of a Steam Soak Well in Midway Sunset," *J. Pet. Tech.* (June 1975) 665-671.
13. Bentsen, R.G. and Donohue, D.A.T.: "A Dynamic Programming Model of the Cyclic Steam Injection Process," *J. Pet. Tech.* (Nov. 1969) 1582-1596; *Trans.*, AIME, **246**.
14. Adams, R.H. and Khan, A.M.: "Cyclic Steam Injection Project Performance Analysis and Some Results of a Continuous Steam Displacement Pilot," *J. Pet. Tech.* (Jan. 1969) 95-100; *Trans.*, AIME, **246**.
15. Yoelin, S.D.: "The TM Sand and Steam Stimulation Project," *J. Pet. Tech.* (Aug. 1971) 987-994; *Trans.*, AIME, **251**.
16. Buckles, R.S.: "Steam Stimulation Heavy Oil Recovery at Cold Lake, Alberta," paper SPE 7994 presented at the 1979 SPE California Regional Meeting, Ventura, April 18–20.

TABLE 9.8 – FIELD TEST RESULTS FOR WELL STIMULATION BY ELECTRIC HEATERS IN THE USSR[28]

	Field Location					
	Knibyshav	Ishimbsl	Arianab	Uzitkistan	Azarbaldjan	Turkistan
Number of wells	70	–	–	–	–	–
Producing interval, ft	1,706 to 2,132	2,953 to 3,937	4,183 to 4,327	1,247 to 2,461	991 to 2,543	1,804 to 4,921
Permeability, md	10 to 150	–	50	–	–	–
Completion	Openhole	Openhole	Openhole	Cased	Cased	Cased
Heater location, ft	1,713 to 2,126	3,599 to 3,645	4,052 to 4,167	1,266 to 2,132	984 to 2,362	2,238
Heating time, days	2 to 10	5 to 12	3 to 8	3 to 4	8 to 14	3 to 11
Heat input, MMBtu	2.3 to 7.3	4.0 to 12.0	2.8 to 6.7	4.3 to 8.2	6.5 to 11.5	3.2 to 9.3
Maximum temperature, °F	280	176	338	248	190	316
Oil gravity, °API	35.2	34.0	27.5	42.1	25.7	~35
Oil rate (before heating), B/D	12.6	22.1	29.5	14.0	3.8	
Oil rate (after heating), B/D	19.4	30.1	77.0	29.4	6.9	53.8
Percent increase	54	36.4	170	111	80	–
Total production per well, bbl	3,109	1,878	–	2,604	273	–
Additional production per well, bbl	1,090	128	3,274	1,402	122	652
Duration of stimulated production, days	16 to 511	40 to 190	35 to 153	60 to 200	30 to 114	–

17. *Annual Review of California Oil and Gas Production*, Conservation Committee of California Oil Producers, Los Angeles (1970-1978).
18. *Proc.*, Simposio Sobre Crudos Pesados, U. del Zulia, Inst. de Investigaciones Petroleras, Maracaibo, Venezuela (1974).
19. *Proc.*, CIM Canada/Venezuela Oil Sands Symposium, Edmonton, Alta. (1977) Special Vol. **17**.
20. Borregales, C.: "Inyeccion Alternada de Vapor en la Costa Bolívar," I Simposio de Crudos Extra-Pesados, Petroleos de Venezuela, Maracay, Oct. 13-15, 1976.
21. Stokes, D.D. and Doscher, T.M.: "Shell Makes a Success of Steam Flood at Yorba Linda," *Oil and Gas J.* (Sept. 2, 1974) 71-76.
22. Bleakley, W.B.: "Journal Survey Shows Recovery Projects Up," *Oil and Gas J.* (March 25, 1974) 69-74.
23. Noran, D. and Franco, A.: "Four Production Reports," *Oil and Gas J.* (April 5, 1978) 107-138.
24. Noran, D.: "Growth Marks Enhanced Oil Recovery," *Oil and Gas J.* (March 27, 1978) 113-140.
25. *Annual Review, Pacific Oil World*, January issue of each year, 1970-1978.
26. Farouq Ali, S.M.: "Current Status of Steam Injection as a Heavy Oil Recovery Method," *J. Cdn. Pet. Tech.* (Jan.-March 1974) 54-68.
27. Farouq Ali, S.M. and Meldau, R.F.: "Current Steamflood Technology," *J. Pet. Tech.* (Oct. 1979) 1332-1342.
28. Farouq Ali, S.M.: "Well Stimulation by Downhole Thermal Methods," *Pet. Eng.* (Oct. 1973) 25-35.
29. Nelson, T.W. and McNiel, J.S. Jr.: "Oil Recovery by Thermal Methods," Part 1, *Pet. Eng.* (Feb. 1959) B27-32.
30. Farouq Ali, S.M.: "Well Stimulation by Thermal Methods," *Prod. Monthly* (April 1968) 23-27.
31. White, P.D. and Moss, J.T.: "High-Temperature Thermal Techniques for Stimulating Oil Recovery," *J. Pet. Tech.* (Sept. 1965) 1007-1011.
32. Holke, D.C. and Huebner, W.B.: "Thermal Stimulation and Mechanical Techniques Permit Increased Recovery from Unconsolidated Viscous Oil Reservoir," paper SPE 3671 presented at the 1971 SPE California Regional Meeting, Los Angeles, Nov. 4–5.
33. Araujo, J.: "Estimulación Cíclica con Agua Caliente en las Arenas del Grupo I – Campo Morichal," I Simposio de Crudos Extra-Pesados, Petroleos de Venezuela, Maracay, Oct. 13-15, 1976.
34. Diaz-Muñoz, J. and Farouq Ali, S.M.: "Effectiveness of Hot-Water Stimulation of Heavy-Oil Formations," *J. Cdn. Pet. Tech.* (July-Sept. 1975) 66-76.
35. Schild, A.: "A Theory for the Effect of Heating Oil-Producing Wells," *Trans.*, AIME (1957) **210**, 1-10.
36. Sheinman, A.B., Malofeev, G.E., and Sergeev, A.I.: "The Effect of Heat on Underground Formations for the Recovery of Crude Oil," Nedra Publishing House, Moscow (1969) 252-262; translated by Marathon Oil Co. (1973).
37. Brandt, H., Poynter, W.G., Hummell, J.D.: "Stimulating Heavy Oil Reservoirs With Downhole Air Gas Burners," *World Oil* (Sept. 1965) 91-95.
38. McNiel, J.S. Jr. and Nelson, T.W.: "Thermal Methods Provide Three Ways to Improve Oil Recovery," *Oil and Gas J.* (Jan. 19, 1959) 86-98.
39. Allen, H.E. and Davis, R.K.: "Electric Formation Heaters Boost Cut Bank Production," *Oil and Gas J.* (June 14, 1954) 125-127.

Chapter 10

Heat Losses From Surface and Subsurface Lines

This chapter considers heat losses from surface pipes used to bring heated fluid to or from the wellheads, and heat losses to the earth formations penetrated by wells carrying hot fluids. The consequence of wellbore heat losses on wellbore fluid temperature is analyzed, and methods for estimating downhole temperatures and steam qualities are presented. Heat losses from the reservoir to adjacent formations, and by production of hot fluids, are discussed in Chap. 5. Heat losses from heaters and steam generators and the interaction between casing temperature and well completion are presented in Chap. 11.

Heat losses through pipes, whether surface lines or wells, usually are estimated at steady-state conditions in oilfield operations. The procedures for estimating heat losses may appear laborious. But they can be made easily with the aid of a simple engineering pocket calculator by following the procedures and referring to the examples given here. In some cases, the procedures must be applied iteratively until there is consistency between the assumed and the calculated values. Repeating the procedures does indeed make the calculation process somewhat longer but certainly no more complicated than a single-pass calculation. And if the pocket calculator is programmable, the iterative calculation procedure can be automated.

The steady-state rate of heat loss per unit length of pipe is directly proportional to the temperature difference and is inversely proportional to the overall specific thermal resistance of the system. The temperature difference normally is known, at least in surface pipes. The high temperature is usually that of the injected (or produced) fluid. The low temperature is usually the ambient one at the surface of the earth. Generally, the mean ambient temperature over the period of interest is used to account for diurnal and even seasonal variations.

Estimating the overall specific thermal resistance of the system is the cumbersome part. The overall specific thermal resistance is the sum of all specific thermal resistances in the system. The insulation, and the pipe itself, have thermal resistance, which is inversely proportional to the thermal conductivity of the material and is a function of the system geometry. Also, at the boundary betweeen any two contiguous media, there is a thermal resistance related to a quantity known as the film coefficient of heat transfer. Graphs and equations for estimating the necessary film coefficients of heat transfer are provided in Section B.8 of Appendix B.

Often, in practice, the resistance of one component dominates the total thermal resistance of the system. When such is the case, and the dominant contributor is identified, the calculations can be simplified greatly. A general approach has been followed in this chapter, showing in some detail how to estimate the various contributions to the overall specific thermal resistance. Examples illustrate the relative importance of some of the elements in specific applications. For the general case, and when there is doubt as to the contribution of terms likely to be ignored, the full treatment is recommended.

10.1 Heat Losses From Surface Lines

Even though heat losses from surface lines in hot-fluid injection operations may be a small fraction of the total heat injected, it is generally worthwhile to use insulation to reduce heat losses to save both fuel and money. That savings can be significant will be demonstrated by means of a steam injection example.

The basic equation used to calculate heat losses per unit length of pipe, $\dot{Q}_{ls}$, is

$$\dot{Q}_{ls} = (T_b - T_A)/R_h, \quad \ldots\ldots\ldots\ldots\ldots (10.1)$$

where R_h usually is represented as $1/(2\pi r U)$, U being the overall coefficient of heat transfer and r being an arbitrary radius that usually coincides with the radius of one of the surfaces across which the heat loss is being determined. Here, R_h is the specific thermal resistance (thermal resistance per unit length of pipe) and is given in units of (Btu/ft-D-°F)$^{-1}$. T_b is the bulk temperature of the fluid in the pipe in degrees Fahrenheit, T_A is the ambient temperature of the atmosphere in degrees Fahrenheit and $\dot{Q}_{ls}$ is the rate of heat loss per unit length of pipe in Btu/ft-D. Rates of heat loss during transient periods can be several times larger than at steady state. Transient effects generally are neglected in calculations of heat losses from thermal lines, since the transient phase is often of short duration (of the order of less than a day). Transient effects would be most important in steam-soak operations having short steam injection cycles. For the purpose of estimating heat losses, the use of steady-state conditions in both the injection

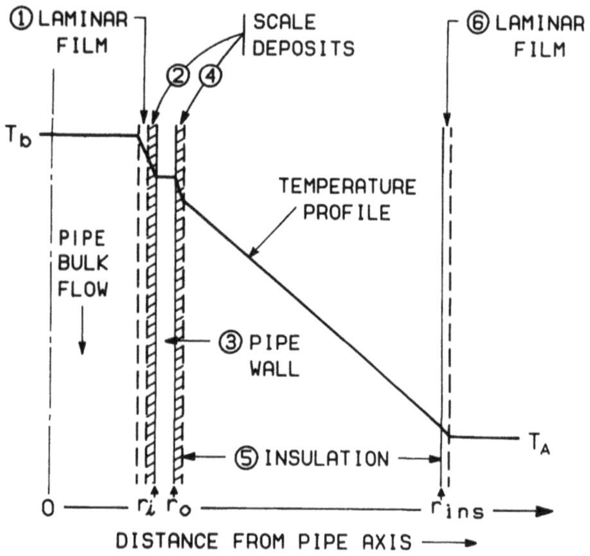

Fig. 10.1 – Schematic representation of resistance to heat transfer and of temperature profile in a suspended surface pipe (not to scale).

and production phases of steam soaks is recommended. For improved estimates of heat losses under transient conditions, refer to McAdams.[1]

For a pipe covered with insulation, the specific thermal resistance of heat loss is given as follows.

$$R_h = \frac{1}{2\pi}\left[\frac{1}{h_f r_i} + \frac{1}{h_{Pi} r_i} + \frac{1}{\lambda_P}\ln\frac{r_o}{r_i} + \frac{1}{h_{Po} r_o} + \frac{1}{\lambda_{ins}}\ln\frac{r_{ins}}{r_o} + \frac{1}{h_{fc} r_{ins}}\right]. \quad \ldots (10.2)$$

Here, h_f is the film coefficient of heat transfer between the fluid inside the pipe and the pipe wall, h_{Pi} is the coeffient of heat transfer across any deposits of scale or dirt at the inside wall of the pipe, h_{Po} is the coefficient of heat transfer across the contact between pipe and insulation, h_{fc} is the coefficient of heat transfer due to forced convection (air currents) at the outer surface of the insulation, r_i is the inner radius of the pipe, r_o is the outer radius of the pipe and (essentially) the inner radius of the insulation, r_{ins} is the external radius of the insulation, and λ_P and λ_{ins} are the thermal conductivities of the pipe and insulation. Coefficients of heat transfer are expressed in Btu/sq ft-D-°F, radii in feet, and thermal conductivities in Btu/ft-D-°F. Since the temperature on the surface of most insulated lines is low, radiation is usually insignificant and is not included in Eq. 10.2. Insulation is always protected. The thermal resistances introduced by a protective sheath around the insulation also are not included in Eq. 10.2.

The physical significance of each of the six terms in the right side of Eq. 10.2 is illustrated in Fig. 10.1. Each of the six terms is proportional to a thermal resistance in the system affecting heat losses. Fig. 10.1 is a schematic representation of an insulated pipe carrying a hot fluid and shows six resistances to heat flow, one for each of the terms on the right side of Eq. 10.2. Adjacent to the inner surface of the pipe is a low-velocity fluid film (1). Because of its low velocity this film has heat transfer characteristics different from those of the flowing bulk fluid and accounts for the introduction of the film coefficient of heat transfer h_f. Note that the resistance to heat flow across this film decreases as the value of the coefficient of heat transfer increases. Scale or dirt deposits at the inside (2) and outside (4) pipe walls lead to coefficients of heat transfer h_{Pi} and h_{Po}, respectively. Heat transfer through the pipe wall (3) and the insulation (5) is by conduction. A low-velocity fluid film at the exterior surface of the insulation (6), which affects heat losses to the atmosphere by forced convection, leads to the coefficient of heat transfer h_{fc}.

The representation of the total resistance to heat transfer as the sum of resistances in series follows directly from the assumption that steady state prevails. This means that the rate at which heat is being transferred across each resistance is the same at any period of time. Also, the total temperature drop is the sum of the temperature drops across each resistance. Accordingly, the fractional temperature drop is equal to the fraction of the total thermal resistance. For example, the temperature at the inside surface of the pipe T_i can be found from

$$\frac{T_b - T_i}{T_b - T_A} = \frac{1}{2\pi r_i}\left(\frac{1}{h_f} + \frac{1}{h_{Pi}}\right)/R_h. \quad \ldots (10.3)$$

Here, the temperature drop between the bulk fluid and the inside wall of the pipe is controlled by two thermal resistances: a laminar film and a scale deposit. The resistances due to these effects are inversely proportional to the film coefficients of heat transfer h_f and h_{pi}, respectively.

Similar use of this concept for all the thermal resistances allows the determination of the shape of the temperature profile when the total temperature drop is known.

Values of the thermal conductivity of steel and insulation necessary to determine the value of the specific thermal resistance R_h are given in Section B.6 of Appendix B. The determination of the coefficients of heat transfer is discussed at some length in a number of textbooks.[1-3] Section B.8 of Appendix B provides the means to determine values of the pertinent coefficients of heat transfer. Film coefficients inside pipes are provided for turbulent flow conditions, which typically prevail in recovery projects based on hot-fluid injection. Film coefficients of heat transfer discussed in Section B.8 are for (1) condensing steam, (2) hot water, and (3) hot gases – all flowing inside pipes – and for cooling of surface lines by (4) forced convection (as would occur in lines exposed to winds), and (5) natural convection and radiation (as would occur inside buildings). Heat transfer coefficients due to deposits of scale and dirt also are discussed.

It should be pointed out that adding more insulation does not necessarily reduce the rate of heat losses further. This can be illustrated best by using in Eq. 10.2 the following expression for the forced convective heat transfer coefficient discussed in Section B.8.

$$h_{fc} r_{ins} = 18 v_W^{0.6} r_{ins}^{0.6} \text{ . . . . (10.4)}$$

Here, v_W is the wind speed in miles per hour. Differentiating the resultant expression for R_h with repect to r_{ins} and equating to zero gives

$$r_{ins} = 3.5 \times 10^{-3} \lambda_{ins}^{1.67} / v_W \text{ . . . . (10.5)}$$

as the outer radius of the insulation in feet, for which the thermal resistance is a minimum. Thus, r_{ins} should exceed the value calculated from Eq. 10.5. Before discussing why there is an optimum insulation thickness, it is necessary to point out that Eq. 10.4 is not applicable where the wind speed drops to zero. Also, if another expression for calculating the forced convection coefficient of heat transfer is used, a different result would be obtained.

Now, the reason for the optimum thickness of insulation is that the rate of heat losses to the atmosphere increases as the radius of the exposed surface increases, whereas the rate of heat losses through insulation decreases with increasing thickness. Since these two effects counteract each other, it is necessary to check that the planned insulation actually will reduce the rate of heat loss below that which would be obtained with less insulation.

Another method sometimes used for insulating surface pipe is to bury it in the ground. This is illustrated schematically in Fig. 10.2. In these cases the calculation of heat transfer can become complicated because of the absence of radial heat flow from the pipe.

Then, an approximate thermal resistance for buried pipes is given by Rohsenow and Hartnett[4] by using $\ln(r_{ins}/r_o) = \cosh^{-1}(D/r_o)$ and $h_{fc} = \infty$ in Eq. 10.2. Rates of heat loss then are based on this approximate thermal resistance.

The rate of heat losses on start-up is appreciably greater for buried than for suspended pipes. Thus, as an insulation method, pipe burial is not recommended when hot-fluid injection is of relatively short duration (as in some cyclic steam projects), where steady-state heat losses calculated by Eq. 10.2 indicate a near marginal operation, or where there are frequent and significant changes in the moisture content of the soil.

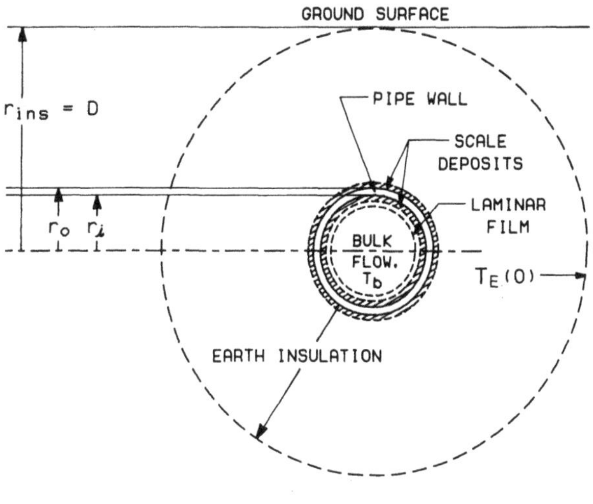

Fig. 10.2—Schematic representation of idealized thermal resistances in buried surface lines (not to scale).

Example 10.1—Calculations of Heat Losses From Surface Lines

Steam at 550°F is to be injected through 4-in. N-80 pipe at a rate of 229 B/D. Find the steady-state heat loss per year per 100 ft of pipe when the pipe is (1) not insulated—i.e., bare—and (2) insulated with 3 in. of calcium silicate. The average yearly temperature is 60°F, and the prevailing winds have an average velocity of 20 mph normal to the injection line.

The following data apply.

r_i = 1.774 in. = 0.1478 ft (from Table B.15).
r_o = 2.000 in. = 0.1667 ft.
r_{ins} = 5.000 in. = 0.4167 ft.
λ_p = 600 Btu/ft-D-°F.
λ_{ins} = 0.96 Btu/ft-D-°F.
h_f = 48,000 Btu/sq ft-D-°F.
h_{fc} = 154 Btu/sq ft-D-°F, from Eq. 10.4.
$h_{Pi} = \infty$.
h_{Po} = 48,000 Btu/sq ft-D-°F.

With insulation, the overall specific thermal resistance is calculated from Eq. 10.2:

$$R_h = \frac{1}{2\pi}\left[\frac{1}{(48{,}000)(0.1478)} + 0 + \frac{\ln(2.000/1.774)}{600} + \frac{1}{(48{,}000)(0.1667)} + \frac{\ln(5.00/2.00)}{0.96} + \frac{1}{154(0.4167)}\right]$$

$$= \frac{1}{2\pi}(0.000141 + 0 + 0.000201 + 0.000125 + 0.954 + 0.0156)$$

$$= 0.154(\text{Btu/ft-D-°F})^{-1}.$$

$$\dot{Q}_{ls} = (T_b - T_A)/R_h = (550-60)/(0.154) = 3{,}180 \text{ Btu/ft-D}.$$

Accordingly, the amount of heat lost from a 100-ft length of pipe over a period of 1 year would be

$$Q_l = (3{,}180)(365)(100) = 1.16 \times 10^8 \text{ Btu}.$$

Without insulation and at a surface temperature near 550°F, radiation heat losses would be important. The sum of the coefficients of heat transfer

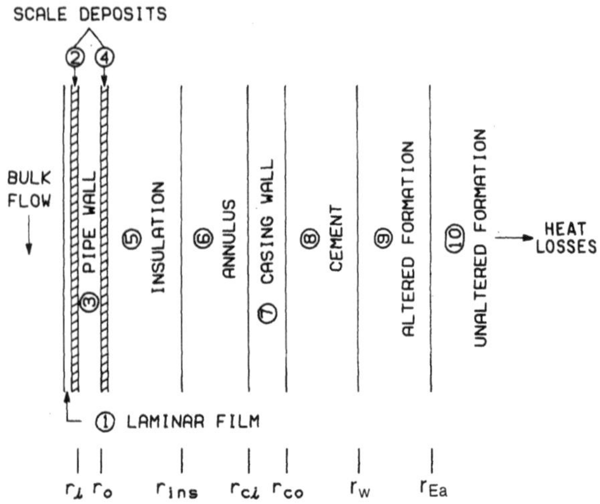

Fig. 10.3 – Schematic representation of resistances to heat transfer in wells (not to scale).

due to radiation and free (or natural) convection for a horizontal pipe is given in Table B.14 for several pipe sizes and temperatures. For a 2-in. radius and 550°F surface temperature, this combined coefficient of heat transfer is about 110 Btu/sq ft-D-°F. The contribution of the natural convection is negligible under these conditions and can be ascertained from Eq. B.62. Thus, the effective coefficient of heat transfer at the bare pipe surface in the open is estimated as the sum of the 110 Btu/sq ft-D-°F found from Table B.14 and the coefficient of heat transfer due to forced convection calculated from Eq. B.60. For this purpose, it is considered that h_{fc} can be estimated adequately from Eq. 10.4, which gives $h_{fc}=220$ Btu/sq ft-D-°F. Thus, the coefficient of heat transfer due to radiation and forced convection is estimated to be 330 Btu/sq ft-D-°F. Accordingly,

$$R_h=\frac{1}{2\pi}\left[\frac{1}{(48{,}000)(0.1478)}+0+\frac{\ln(2.000/1.774)}{600}+\frac{1}{110(0.1667)}+0\right]$$

$$=\frac{1}{2\pi}(0.000141+0+0.000200+0.0545)$$

$$=2.94\times10^{-3}(\text{Btu/ft-D-°F})^{-1}.$$

$$\dot{Q}_{ls}=(T_b-T_A)/R_h$$

$$=(550-60)/(2.94\times10^{-3})$$

$$=1.67\times10^{5}\text{ Btu/ft-D}.$$

Thus, the amount of heat lost from a 100-ft length of pipe over a period of 1 year would be

$$Q_l=(1.67\times10^{5})(365)(100)$$

$$=6.21\times10^{9}\text{ Btu}$$

when the pipe is bare.

Thus, the insulation would reduce heat losses by a factor of about 50. At 6.0 × 10^6 Btu per barrel of fuel, the reduction in yearly heat losses resulting from insulating the pipe would amount to more than 1,000 bbl of fuel for a 100-ft length of pipe.

If the surface lines are to be buried 42 in., and the thermal conductivity of the dry soil, λ_E, is 8.9 Btu/ft-D-°F, the value of R_h would be found from

$$R_h=\frac{1}{2\pi}\left[\frac{1}{h_f r_i}+\frac{1}{h_{Pi} r_i}+\frac{\ln(r_o/r_i)}{\lambda_P}+\frac{1}{h_{Po} r_o}+\frac{\cosh^{-1}(D/r_o)}{\lambda_E}\right]$$

$$=\frac{1}{2\pi}\left(0.000141+0+0.000201+0.000125+0.42\right)$$

$$=6.7\times10^{-2}(\text{Btu/ft-D-°F})^{-1}.$$

$$\dot{Q}_{ls}=(T_b-T_A)/R_h$$

$$=(490)/(6.7\times10^{-2})=7.3\times10^{3}\text{ Btu/ft-D}$$

and the yearly heat losses per 100 ft of buried line would be

$$Q_l=(7.3\times10^{3})(365)(100)$$

$$=2.7\times10^{8}\text{ Btu}.$$

This is about 2.5 times as great as the loss from the insulated pipe and is significantly lower than that from the bare pipe.

The example shows two points. One is that, in the absence of insulation, radiation heat transfer to the atmosphere contributes significantly to the effective thermal resistance R_h. The other is that when insulation is used, it essentially accounts for all the resistance to heat losses from steam lines. If we had calculated the heat losses from the resistance to heat flow provided by the insulation only, the yearly heat loss per 100 ft of line would be the 1.18×10^{8} Btu, which is less than 2% greater than that obtained by considering all contributions to the thermal resistance. For steam injection, then, it is usually adequate to consider the insulation as the only thermal resistance affecting the rate of heat loss. Such may not be the case for hot water and gas.

The film coefficient of heat transfer may affect the value of R_h to a greater extent when hot water is injected than when steam is injected and to an even greater extent when hot gases are injected. It is recommended in these cases that all terms in Eq. 10.2 be used until the relative importance of the terms for the particular injection system being considered has been established.

10.2 Heat Losses From Wells

Heat losses from wells never reach a steady state. They attain, as pointed out by Ramey[5] and Willhite,[6] a quasisteady state in which the rate of heat loss is a monotonically decreasing function of time. This function of time, which will be discussed later in more detail, is a measure of how fast the earth can conduct heat away from the well. Heat losses from the well to the earth still are characterized by Eq. 10.1, where in this case the ambient temperature is the geothermal temperature and, thus, a

TABLE 10.1 – TIME FUNCTION $f(t_D)$ FOR THE RADIATION BOUNDARY CONDITION MODEL[6]

	$2\pi R'_h \lambda_E$													
t_D	100	50	20	10	5.0	2.0	1.0	0.5	0.2	0.1	0.05	0.02	0.01	0
0.1	0.313	0.313	0.314	0.316	0.318	0.323	0.330	0.345	0.373	0.396	0.417	0.433	0.438	0.445
0.2	0.423	0.423	0.424	0.427	0.430	0.439	0.452	0.473	0.511	0.538	0.568	0.572	0.578	0.588
0.5	0.616	0.617	0.619	0.623	0.629	0.644	0.666	0.698	0.745	0.772	0.790	0.802	0.806	0.811
1.0	0.802	0.803	0.806	0.811	0.820	0.842	0.872	0.910	0.958	0.984	1.00	1.01	1.01	1.02
2.0	1.02	1.02	1.03	1.04	1.05	1.08	1.11	1.15	1.20	1.22	1.24	1.24	1.25	1.25
5.0	1.36	1.37	1.37	1.38	1.40	1.44	1.48	1.52	1.56	1.57	1.58	1.59	1.59	1.59
10.0	1.65	1.66	1.66	1.67	1.69	1.73	1.77	1.81	1.84	1.86	1.86	1.87	1.87	1.88
20.0	1.96	1.97	1.97	1.99	2.00	2.05	2.09	2.12	2.15	2.16	2.16	2.17	2.17	2.17
50.0	2.39	2.39	2.40	2.42	2.44	2.48	2.51	2.54	2.56	2.57	2.57	2.57	2.58	2.58
100.0	2.73	2.73	2.74	2.75	2.77	2.81	2.84	2.86	2.88	2.89	2.89	2.89	2.89	2.90

function of depth. And in this case, of course, the specific thermal resistance is time dependent, reflecting the variable effective thermal resistance of the earth.

A representation of the typical elements offering resistance to heat losses from the wellbore is given in Fig. 10.3. For the insulated tubing held concentrically within the casing shown in this figure, the heat resistance elements are combined to obtain the overall coefficient of heat loss:

$$R_h = \frac{1}{2\pi}\left[\frac{1}{h_f r_i} + \frac{1}{h_{Pi} r_i} + \frac{\ln(r_o/r_i)}{\lambda_P} + \frac{1}{h_{Po} r_o} + \frac{\ln(r_{ins}/r_o)}{\lambda_{ins}} + \frac{1}{h_{rc,an} r_{ins}} + \frac{\ln(r_{co}/r_{ci})}{\lambda_p} + \frac{\ln(r_w/r_{co})}{\lambda_{cem}} + \frac{\ln(r_{Ea}/r_w)}{\lambda_{Ea}} + \frac{f(t_D)}{\lambda_E}\right]. \quad (10.6)$$

The first five terms have been discussed in the preceding section in connection with surface lines. The last five terms represent, in order of appearance, the resistance to radiation and convection in the annulus, the resistance of the casing, the resistance of the cement, the resistance of an altered zone (resulting from drying due to high temperatures) in the earth, and the variable resistance of the earth. Additional resistances that should be included for consistency and that are omitted for brevity include those of scale deposits on each surface of the casing and of any sheath (usually metallic) used to support the insulation or to keep it dry. Different well designs will, of course, lead to a different expression for determining the overall thermal resistance R_h. The intent here is to show how R_h can be determined from a consideration of the resistances to heat transfer present in a well under consideration.

In Eq. 10.6, $h_{rc,an}$ is the radiation and convection coefficient of heat transfer for the annulus, r_{ci} and r_{co} are the inner and outer casing radii, r_w is the wellbore radius, r_{Ea} is the radius of the altered zone in the earth near the wellbore, λ_{cem} is the thermal conductivity of the cement, λ_{Ea} and λ_E are the thermal conductivities of the altered and unaltered earth, and $f(t_D)$ is the time function that reflects the thermal resistance of the earth. Coefficients of heat transfer are expressed in Btu/sq ft-D-°F, radii in feet, and thermal conductivities in Btu/ft-D-°F. The function $f(t_D)$ is dimensionless, and the dimensionless time is discussed later. The function $f(t_D)$ and the radiation-convection coefficient of heat transfer in the annulus, $h_{rc,an}$, are the only additional quantities requiring discussion. All other quantities appearing in Eq. 10.6 are similar to those discussed in connection with heat losses from surface lines. Like all other coefficients of heat transfer, $h_{rc,an}$ is discussed in detail in Section B.8 of Appendix B.

The function $f(t_D)$ has been discussed by a number of authors (Ramey,[5] Willhite,[6] Moss and White,[7] Carslaw and Jaeger,[8] Huygen and Huitt,[9] and López and Rivera[10]) and has been evaluated by solving the heat conduction equation in the earth in a plane normal to the wellbore. Implicit in the solution is the assumption that the thermal conductivity of the earth can be considered to be isotropic in the plane normal to the wellbore and that radial conductive heat transfer is adequate to desribe heat losses into the earth. If there is an altered earth zone beyond the wellbore to a radius r_{Ea}, the $f(t_D)$ function considers heat conduction into the earth from r_{Ea}, otherwise from the wellbore at r_w. The temperature at the wall of the wellbore (say there is no altered earth zone) is not constant in time. Accordingly, the boundary condition used to obtain $f(t_D)$ is governed by the thermal resistance between the injection string and the borehole wall, R'_h, which is related to R_h and $f(t_D)$ by

$$R'_h = R_h - \frac{f(t_D)}{2\pi\lambda_E}, \quad (10.7)$$

where $f(t_D)$ is represented in terms of the dimensionless time:

$$t_D = \alpha_E t / r_w^2, \quad (10.8)$$

or if there is an altered zone,

$$t_D = \alpha_E t / r_{Ea}^2. \quad (10.9)$$

Here α_E is the thermal diffusivity of the earth in square feet per day, and t is the time from start of heating in days. The value of $f(t_D)$ is not sensitive to values of $R'_h \lambda_E$ for values of $t_D \geq 100$ and is then given (Ramey[5]) by

$$f(t_D) \simeq \tfrac{1}{2}\ln(t_D) + 0.403 \quad (10.10)$$

for $t_D \geq 100$. For values of $t_D \leq 100$, Willhite[6] has published a table of the f function for a range of values of both t_D and $R'_h \lambda_E$. Those values are given here in Table 10.1.

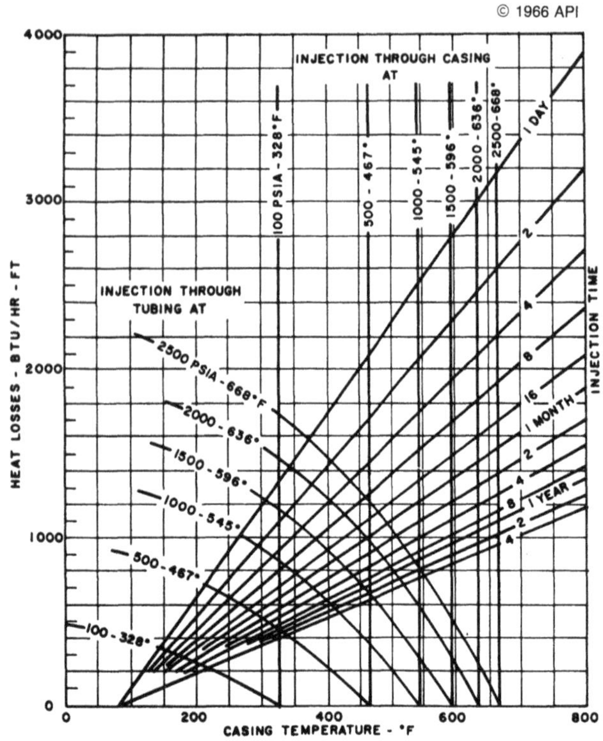

Fig. 10.4 – Rate of heat loss and casing temperature for injection conditions in 2½-in. tubing and 7-in. casing.[9]

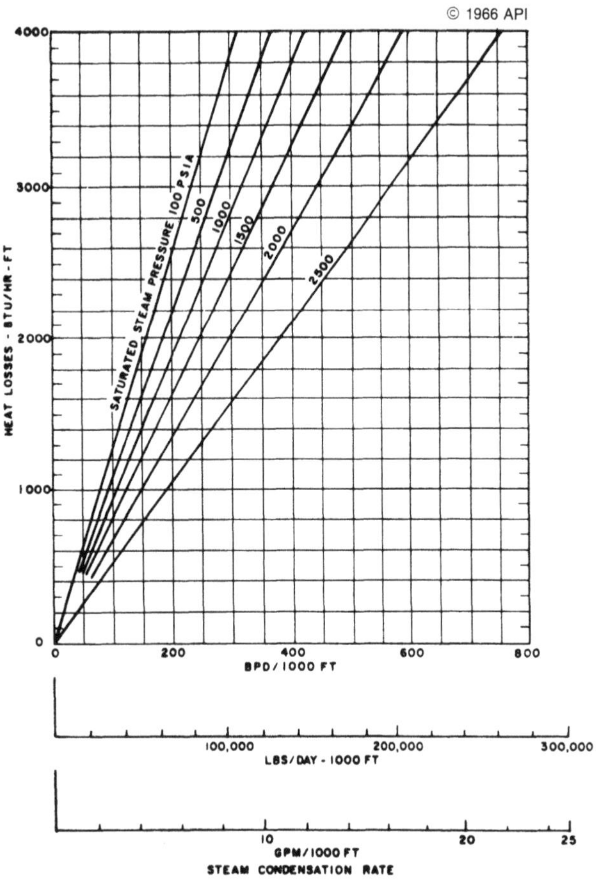

Fig. 10.5 – Rate of heat loss vs. steam condensation rate.[9]

Huygen and Huitt[9] were among the first to consider radiation effects on heat losses. They present the variation, with time, of specific heat loss rate and casing temperature at a given injection temperature for typical combinations of tubing and casing sizes. Fig. 10.4 is a typical result, applicable to a reservoir having an initial temperature of 80°F. For example, steam injection for 4 days through 7-in. casing (or its annulus) at 1,000 psia (or 545°F) gives a specific-heat-loss rate of 1750 Btu/hr-ft (and a casing temperature of 545°F, of course). Injection through 2½-in. tubing gives a specific-heat-loss rate of 900 Btu/hr-ft and a casing temperature of 320°F. These specific-heat-loss rates may be converted to steam condensation rates by means of Fig. 10.5. For the specific-heat-loss rate of 900 Btu/hr-ft, the condensation rate of steam is 100 B/D per 1,000 ft of casing.

Example 10.2 – Calculations for Heat Losses From an Injection Well

Steam at 600°F is injected down 3.5-in. tubing set on a packer in 9⅝-in., 53.5-lbm/ft N-80 casing. The annulus contains a stagnant gas at zero gauge pressure at the wellhead, and the casing is cemented to surface in a 12-in. hole. The tubing is insulated with 1 in. of calcium silicate, the insulation being held in place and sealed from accidental entry of liquids in the annulus by a very thin sheath of aluminum. A temperature survey in the well indicates a mean subsurface temperature of 100°F over the 1,000-ft depth. Estimate the rate of heat loss 21 days after steam injection is started, as well as the casing temperature. There is no altered zone near the borehole.

From the results obtained in Example 10.1 for the effect of insulation on heat losses from steam injection in surface pipes, thermal resistances across films, deposits, and metals are neglected so that the working equation to determine R_h becomes

$$R_h = \frac{1}{2\pi}\left[\frac{\ln(r_{ins}/r_o)}{\lambda_{ins}} + \frac{1}{h_{c,an} r_{ins}} + \frac{\ln(r_w/r_{co})}{\lambda_{cem}} + \frac{f(t_D)}{\lambda_E}\right].$$

The following data apply.

$$\begin{aligned}
r_o &= 1.75\text{ in.} = 0.1458\text{ ft.}\\
r_{ins} &= 2.75\text{ in.} = 0.2292\text{ ft.}\\
r_{ci} &= 4.27\text{ in.} = 0.3556\text{ ft.}\\
r_{co} &= 4.81\text{ in.} = 0.4010\text{ ft.}\\
r_w &= 6.00\text{ in.} = 0.5000\text{ ft.}\\
\alpha_E &= 0.96\text{ sq ft/D.}\\
\epsilon_{ins} &= \epsilon_{ci} = 0.9.\\
\lambda_E &= 24\text{ Btu/ft-D-°F.}\\
\lambda_{cem} &= 12\text{ Btu/ft-D-°F.}\\
\lambda_{ins} &= 0.96\text{ Btu/ft-D-°F.}
\end{aligned}$$

Heat transfer across the gas-filled annulus is by radiation and natural convection. Radiation is sensitive to the temperature levels and emissivities (ϵ) of the surfaces. The temperature at the surface of the insulation (T_{ins}) and that at the inner radius of the casing (T_{ci}), together with the emissivities at these surfaces (ϵ_{ins} and ϵ_{ci}), affect the radiation heat losses across the annular space between the insulated tubing and casing.

The calculation of the thermal resistance R_h depends on the coefficient of annular heat transfer ($h_{rc,an}$), which is estimated from Eq. B.63 using parameters given by Eqs. B.64 through B.70. These parameters, in turn, depend on the thermal resistance R_h. Thus, an iterative procedure to be outlined here is required to find both R_h and $h_{rc,an}$.

Step 1. To start, make an initial assumption, such as that the sum of all the thermal resistances is twice that due to the insulation:

$$R_h=\frac{2}{2\pi}\left[\frac{\ln(r_{ins}/r_o)}{\lambda_{ins}}\right]=\frac{1}{\pi}\left[\frac{\ln(2.75/1.75)}{0.96}\right]$$
$$=0.150\ (\text{Btu/ft-D-°F})^{-1}.$$

Step 2. Calculate $f(t_D)$ at 21 days.

$$t_D=\alpha_E t/r_w^2=(0.96)(21)/(0.5)^2$$
$$=80.6.$$

Since this is less than 100, Table 10.1 is used to estimate the value of $f(80.6)$. For the initial evaluation of the f function we take $R_h'=R_h$, so that

$$2\pi R_h'\lambda_E=2\pi(0.150)24=23.$$

From Table 10.1, a value of $f(80.6)=2.60$ is estimated for $2\pi R_h'\lambda_e=23$, and $t_D=80.6$.

Step 3. Calculate T_{ci} from Eq. B.68.

$$T_{ci}=100+(600-100)(1.061)$$
$$\cdot\left[0+\frac{\ln(6.0/4.81)}{12}+0+\frac{2.60}{24}\right]$$
$$=100+500(1.061)(0.127)=167°\text{F}.$$

Step 4. Calculate T_{ins} from Eq. B.70.

$$T_{ins}=600-(600-100)(1.061)$$
$$\cdot\left[0+0+0+0+\frac{\ln(2.75/1.75)}{0.96}\right]$$
$$=600-500(1.061)(0.471)=350°\text{F}.$$

Step 5. Calculate $h_{rc,an}$ from Eqs. B.63 through B.66. The average temperature in the annulus is needed to estimate air properties and is

$$\bar{T}_{an}=\tfrac{1}{2}(T_{ins}+T_{ci})=258°\text{F}.$$

For air in the annulus,

$$\rho_a=0.076\left(\frac{460+60}{460+258}\right)=0.055\ \text{lbm/cu ft}$$

(from Table B.1),

$$\mu_a=0.023\ \text{cp (from Fig. B.41)},$$

and

$$\lambda_a=0.45\ \text{Btu/ft-D-°F (from Fig. B.72)}.$$

Considering air as an ideal gas,

$$\beta_a=\frac{1}{460+\bar{T}_{an}}=1.39\times10^{-3}\,°\text{F}^{-1}$$

(from Eq. B.28).

The Grashof number N_{Gr} needed to evaluate the effective thermal conductivity of the air in the annulus is evaluated using Eq. B.66:

$$N_{Gr}=\left[7.12\times10^7(0.3556-0.2292)^3(0.055)^2\right.$$
$$\left.\cdot(1.38\times10^{-3})(350-167)\right]\div(0.023)^2$$
$$=2.08\times10^5.$$

From Eq. B.65, the apparent thermal conductivity of air in the annulus is

$$\lambda_{a,an}=0.049(0.45)(2.08\times10^5)^{0.333}(0.92)$$
$$=1.20\ \text{Btu/ft-D-°F}.$$

The radiation temperature function is obtained from Eq. B.64:

$$F(350,182)=1.51\times10^9\,°\text{F}^3.$$

Then, $h_{rc,an}$ is estimated from Eq. B.63:

$$h_{rc,an}=4.11\times10^{-8}\left[\frac{1}{0.9}+\frac{2.75}{4.27}\right.$$
$$\left.\cdot\left(\frac{1}{0.9}-1\right)\right]^{-1}(1.51\times10^9)$$
$$+\frac{1}{0.2292}\frac{1.20}{\ln(4.27/2.75)}$$
$$=3.48\times10^{-8}(1.51\times10^9)+9.92(1.20)$$
$$=64.5\ \text{Btu/sq ft-D-°F}.$$

Step 6. Calculate R_h using Eq. 10.6.

$$R_h=\frac{1}{2\pi}\left[0+0+0+0+0.471\right.$$
$$\left.+\frac{1}{64.5(0.2292)}+0+0.0460+0+\frac{2.60}{24}\right]$$
$$=0.110\ (\text{Btu/ft-D-°F})^{-1}.$$

Since the initially assumed and calculated values of R_h (obtained in Steps 1 and 6) do not agree, Steps 2 through 6 are repeated until the difference in successive approximations of R_h (and of other intermediate results of interest) is small.

In redoing Step 2, the values of $f(80.6)=2.60$ and $R_h=0.110$ are used in calculating R_h' from Eq. 10.7. The new value of R_h' then is used to obtain a new value of $f(80.6)$. The results of the various steps through three iterations are shown in Table 10.2. The iterations were carried out until two consecutive iterations gave the same value of the thermal resistance R_h. Accordingly, the rate of heat loss from the well at 21 days is

$$\dot{Q}_l=(1{,}000\ \text{ft})\dot{Q}_{ls}=10^3(T_b-T_A)/R_h$$
$$=(600-100)10^3/(0.112)$$
$$=4.46\times10^6\ \text{Btu/D}.$$

At an equivalent energy content of 6.0×10^6 Btu/bbl of fuel, the daily heat loss from such an insulated well corresponds to less than 1 bbl of fuel.

In this example, the thermal resistance R_h is essentially the same for all four iterations. This

TABLE 10.2—SUMMARY OF RESULTS FOR EXAMPLE 10.2

Step	Quantity	Iteration 1	2	3	4
1	R_h(Btu/ft-D-°F)$^{-1}$	0.150	0.110	0.113	0.112
2	t_D	80.6			
	R_h'(Btu/ft-D-°F)$^{-1}$	0.150	0.0928	0.0956	0.0946
	f(80.6)	2.60	2.62		
3	T_{ci}, °F	167	192	190	190
4	T_{ins}, °F	350	259	268	265
5	$\bar{T}_{an}$, °F	258	226	229	228
	ρ_a, lbm/cu ft	0.055	0.058	0.057	
	μ_a, cp	0.023	0.022		
	λ_a, Btu/ft-D-°F	0.45	0.44		
	β_a, °F^{-1}	1.39×10^{-3}	1.46×10^{-3}	1.45×10^{-3}	
	N_{Gr}	2.08×10^5	4.82×10^4	1.09×10^5	1.05×10^5
	$\lambda_{a,an}$, Btu/ft-D-°F	1.20	0.719	1.03	1.01
	F, °R^3	1.51×10^9	1.29×10^9	1.31×10^9	1.30×10^5
	$h_{rc,an}$, Btu/D-sq ft-°F	64.5	52.0	55.9	55.3
6	R_h(Btu/D-ft-°F)$^{-1}$	0.110	0.113	0.112	0.112

means that a single pass would have been sufficient to estimate the thermal resistance R_h. On the other hand, the temperatures at the inner and outer surfaces of the annulus require two iterations before reasonably good answers are obtained. This suggests that one iteration is sufficient for estimating R_h, but two are required to determine the temperature profile. If there is doubt, however, the procedure shoud be continued until convergence is attained.

As an alternative to using a packer to keep steam from entering the annulus, natural gas sometimes is injected slowly down the annulus. Since the heat transfer coefficient $h_{c,an}$ given by Eq. B.63 was developed for natural convection in the annulus, a different heat transfer coefficient should be used for the injection of gas (forced convection) down the annulus.

In the heat loss calculations discussed thus far, the entire well is considered as a unit. That is, it is assumed that the temperature in the injection string does not vary with depth; the same holds for the temperature in the earth. We know that the earth's temperature varies little over a vertical distance of 1,000 ft. Where steam is present, the temperature in a shallow steam injection well is essentially constant. Then, the use of a single value of R_h may provide a close enough estimate of heat losses and wellbore temperatures.

Where the value of R_h varies significantly with depth, a different approach should be taken. One approach is to calculate the temperature distribution with depth and use that information to determine heat losses from a number of appropriate intervals in the well. Methods for estimating temperature profiles in wells are discussed in the following section.

10.3 Temperature Profiles in Wells

Procedures for estimating well temperature profiles in thermal[5,11,12] and conventional[13-15] operations are numerous. For the flow of noncondensable fluids in vertical injection and production wells, the basis for obtaining approximate temperature profiles is to assume that (1) the system is at steady state, except for conduction heat losses into the earth, (2) variations in thermal properties are negligible, (3) friction pressure drop is negligible, (4) vertical heat transfer is by convection only, and (5) heat losses to the formation are radial.

On the basis of these assumptions, the temperature distribution in an injection well as a function of both depth and time is given by

$$T(D,t)=\left[T_E(D)-g_G w_i C R_h-\frac{w_i g R_h \delta}{g_c J}\right]+\left[T_{inj}-T_E(0)+g_G w_i C R_h+\frac{w_i g R_h \delta}{g_c J}\right]\exp\left(-\frac{D}{w_i C R_h}\right). \quad (10.11)$$

Here, $T_E(0)$ is the mean surface temperature of the earth, and g_G is the geothermal gradient, so that the undisturbed temperature of the earth as a function of depth D is given by

$$T_E(D)=T_E(0)+g_G D. \quad (10.12)$$

C is the heat capacity of the flowing fluid, w_i is the mass rate of injection in lbm/D, and δ is a constant that takes on the value of zero for liquids and a value of one for gases. Although the development considers liquids to be incompressible and gases to be ideal, Eq. 10.11 is used to estimate temperature profiles for any fluid encountered in thermal operations. The quantity $g/g_c J$ is a constant, having the value 1/778 Btu/ft-lbm. T_{inj} is the injection temperature at the surface, which should be essentially constant.

Ramey,[5] from whom these results are adapted, has provided comparisons between measured and calculated temperature profiles. Fig. 10.6, reprinted from his publication, is an example of such comparisons. According to Ramey[5]: "During the year and a half this test was operated, the temperature of the injected gas was increased to almost 500°F, and the gas-injection rate varied from 10 to 215 Mcf/D. Gas was injected down 3-in. tubing. The annulus between the tubing and the 7-in. casing was filled with Perlite."

For production, the mass production rate w_p is used instead of w_i. Whenever a single value of R_h can be used to represent the rate of heat loss at a given time, the temperature at depth D resulting from the production of liquid at downhole temperature $T_p(t)$ from a reservoir at depth D_R is given by

$$T(D,t)=T_E(D)+g_G w_p C R_h+\left[T_p(t)-T_E(D_R)-g_G w_p C R_h\right]\exp\left[\frac{-(D_R-D)}{w_p C R_h}\right]. \quad (10.13)$$

This expression was used by Romero,[13] whose principal objective was to use the wellhead producing temperature $T(0,t)$ to monitor changes in the production rates in conventional (nonthermal) production operations. For these conditions, the wellhead production temperature is calculated by setting $D=0$ and $T_p(t)=T_E(D_R)$ in Eq. 10.13.

Although the stimulus for developing Eqs. 10.11 and 10.13 to calculate temperature profiles in wells was provided by thermal recovery processes, the equations are applicable for all processes involving fluid injection and production. Eq. 10.11, for example, has been used to calculate the bottomhole temperature in water injection wells[5] and for estimating the bottomhole temperature before well stimulation treatments.[13]

Example 10.3 – Temperature Profile During Hot-Water Injection

Hot water at 400°F and 4,800 B/D is injected down 7-in., 23-lbm casing initially designed for the production of unheated fluids from a reservoir at 3,000 ft. The bottom 500 ft of casing is cemented in a 9-in. borehole traversing soft formations. Surface pipe and cement are considered too short to be significant. The geothermal temperature is $T_E(D)=70°\text{F}+(0.0083\ °\text{F/ft})D$. Find the expression for the temperature distribution at 75 days after hot-water injection started, and use it to determine the temperature of the fluid entering the formation.

The following data apply.

$$r_{ci} = 3.183 \text{ in.} = 0.2653 \text{ ft.}$$
$$r_{co} = 3.500 \text{ in.} = 0.2917 \text{ ft.}$$
$$r_w = 4.500 \text{ in.} = 0.3750 \text{ ft.}$$

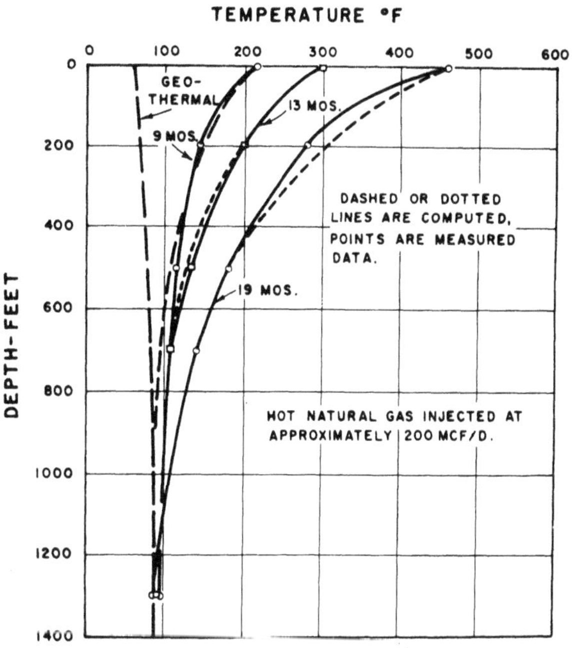

Fig. 10.6 – Comparison of measured and calculated temperature profiles.[5]

$$\lambda_P = 600 \text{ Btu/ft-D-°F.}$$
$$\lambda_{cem} = 12 \text{ Btu/ft-D-°F.}$$
$$\lambda_E = 34 \text{ Btu/ft-D-°F.}$$
$$\alpha_E = 0.8 \text{ sq ft/D.}$$
$$T_E(3{,}000) = 94.9°\text{F.}$$
$$g_G = 0.0083°\text{F/ft.}$$
$$T_E(0) = 70°\text{F.}$$

It is assumed (1) that the soft formations above the cement have packed around the casing so that the effective radius controlling conductive heat losses into the earth is $r_w=r_{co}$, (2) that the effect of the surface pipe and cement can be ignored on the heat losses from the upper 2,500 ft, (3) that the bottom 500 ft deserves a separate calculation, and (4) that there is no altered zone in the earth as a results of the operations.

In Example 10.1, the film coefficient of heat transfer for steam was given. In this example, the film coefficient of heat transfer for water, which is required to determine the specific thermal resistance R_h, is calculated from information given in Section B.8. Its determination depends on the Reynolds number values, which are given by Eq. B.54.

$$N_{Re}=0.0616\frac{(62.4 \text{ lbm/cu ft})(4{,}800 \text{ B/D})}{(0.13 \text{ cp})(0.2653 \text{ ft})}$$
$$=5.3\times10^5.$$

The Reynolds number exceeds 2,100, so that (1) turbulent flow prevails in the injection well and (2) Eq. B.57 can be used to calculate the film coefficient of heat transfer.

$$h_f=1.6\,(4{,}800)^{0.8}(0.2653)^{-1.8}$$
$$=1.5\times10^4 \text{ Btu/sq ft-D-°F.}$$

This value of h_f is about the same as the one given in Example 10.1 which was found to have a very negligible effect. Accordingly, the contribution of the film coefficient of heat transfer for water to the overall specific thermal resistance R_h is neglected in this example compared with that of the earth. Likewise, the contributions due to film deposits and metals, which also have been shown to be small, are neglected. Thus, the overall specific thermal resistance in the *upper* interval is given by

$$R_h = \frac{1}{2\pi} \frac{f(t_D)}{\lambda_E}.$$

In the *upper* interval the value of t_D is found from

$$t_D = \frac{\alpha_E t}{r_{co}^2} = \frac{(0.8)(75)}{(0.2917)^2} = 705,$$

and since $t_D > 100$, Eq. 10.10 can be used to evaluate $f(852)$:

$$f(852) = ½\ln 852 + 0.403 = 3.68.$$

The overall specific thermal resistance then is

$$R_h = \frac{1}{2\pi} \frac{3.68}{34}$$

$$= 0.017\,(\text{Btu/ft-D-°F})^{-1}.$$

In the *lower* 500 ft, a similar approach gives

$$t_D = \frac{\alpha_E t}{r_w^2} = \frac{(0.8)(75)}{(0.3750)^2} = 427,$$

$$f(427) = 3.43.$$

Here the overall specific thermal resistance R_h, which now includes the contribution due to the cement sheath, is found from

$$R_h = \frac{1}{2\pi}\left[\frac{\ln(r_w/r_{co})}{\lambda_{cem}} + \frac{f(427)}{\lambda_E}\right]$$

$$= \frac{1}{2\pi}(0.0209 + 0.101)$$

$$= 0.019\,(\text{Btu/ft-D-°F})^{-1}.$$

The dominant contribution to the overall specific thermal resistance is that of the earth. These small differences in the values of R_h would have essentially no effect on the calculated bottomhole injection pressure. But the example is continued to illustrate *how* to handle a change in R_h.

In the *upper* interval, then, the expression for the temperature distribution in the wellbore is obtained from Eq. 10.11 using $\delta = 0$ to represent hot-water injection. The calculation is as follows.

$$w_i CR_h = (4800\text{ B/D})(62.4\text{ lbm/cu ft})$$
$$\cdot(5.615\text{ cu ft/bbl})(1.08\text{ Btu/lbm-°F})$$
$$\cdot(0.017\text{ ft-D-°F/Btu})$$
$$= 31{,}000\text{ ft}.$$

The expression for the temperature (in °F) then is found to be

$$T(D) = 70 + (0.0083)(D) - (0.0083)$$
$$\cdot(31{,}000) + [400 - 70 + (0.0083)$$
$$\cdot(31{,}000)]\exp(-D/31{,}000)$$
$$= -190 + 0.0083D$$
$$+ 590\exp(-D/31{,}000),$$

and at $D = 2{,}500$ ft,

$$T(2{,}500) = -190 + 21 + 544 = 375\text{°F}.$$

In the *lower* interval, $w_i CR_h = 36{,}600$ ft. The temperature of the fluid entering the lower interval is 375°F, and the temperature distribution is found by moving the origin of the coordinate system down to a depth of 2,500 ft. Then,

$$T(D) = T_E(2{,}500) + g_G(D - 2{,}500)$$
$$- g_G w_{inj} CR_h + \left[T_{inj} - T_E(2{,}500)\right.$$
$$\left. + g_G w_{inj} CR_h\right]\exp\left(-\frac{(D-2{,}500)}{w_{inj} CR_h}\right)$$

for $D \geq 2{,}500$ ft. Substitution of the values of the parameters gives

$$T(D) = [70 + 0.0083D - (0.0083)(36{,}600)]$$
$$+ [375 - 70 - 0.0083(2{,}500)$$
$$+ 0.0083(36{,}600)]\exp\left(-\frac{D-2{,}500}{36{,}600}\right)$$
$$= -234 + 0.0083D + 630\exp(-D/36{,}600).$$

At the reservoir $D = 3{,}000$ ft, and $T(3{,}000) =$ 371°F, which compares with a temperature of 370°F that would have been obtained if the parameters applicable to the upper interval were used to calculate the bottomhole injection temperature. Using the overall specific thermal resistance obtained for the lower interval over the entire well depth to obtain the bottomhole temperature gives a reservoir injection temperature of 375°F.

This example illustrates (1) that the overall specific thermal resistance R_h controlling heat losses from wells is controlled primarily by the function $f(t_D)$, which represents the thermal resistance of the earth, (2) how changes in the overall specific thermal resistance with depth are handled to determine downhole temperatures, and (3) that even at the relatively great injection rate of 4,800 B/D and rather shallow depth of 3,000 ft, injection of 400°F water at the wellhead results in a bottomhole temperature of about 371°F. Similar stepwise procedures can be applied when the geothermal or fluid temperature varies significantly with depth. Alternative methods for treating temperature changes along the flow conduit from which heat losses occur are discussed by McAdams[1] and Ramey.[5]

10.4 Quality of Steam as a Function of Depth

During the injection of steam, it is important to know the quality of the steam delivered to the formation – i.e., the mass fraction of steam vapor in a steam/water system. Satter[11] has shown that when the overall specific thermal resistance R_h and the steam temperature do not vary appreciably with depth, the steam quality at any depth $f_s(D)$ can be related to that at the surface, $f_s(0)$, by

$$f_s(D)=f_s(0)\left(\dot{Q}_v-\frac{D_1 g_G D}{R_h}+\frac{g_G D^2}{R_h}\right)\Big/\dot{Q}_v, \quad \text{(10.14)}$$

where

$$\dot{Q}_v=w_i L_v f_s(0). \quad \text{(10.15)}$$

is the rate of injection of latent heat. The parameter D_1, which has units of length, is defined by

$$D_1=\left[T_{\text{inj}}-T_E(0)-\frac{w_i R_h g}{g_c J}\right]\Big/g_G. \quad \text{(10.16)}$$

This is the primary equation for calculating the quality of steam as a function of depth in an injection well.

The depth at which all the injected steam has condensed (the hot-water point of Satter[11]) is found by setting $f_s(D)=0$ in Eq. 10.14 and solving the quadratic expression in D. The result is

$$D_c=D_1\left(1-\sqrt{1-\frac{2\dot{Q}_v R_h}{g_G D_1^2}}\right). \quad \text{(10.17)}$$

For the assumed conditions, the temperature of the fluid in the well is equal to the steam injection temperature down to the depth D_c given by Eq. 10.17. For depths greater than D_c, where only hot water flows, Eq. 10.11 must be used to obtain the temperature.

In many cases it is found that

$$2\dot{Q}_v R_h \ll g_G D_1^2. \quad \text{(10.18)}$$

Whenever this is the case, the hot-water point may be approximated by

$$D_c=\frac{w_i L_v f_s(0) R_h}{T_{\text{inj}}-T_E(0)-\dfrac{R_h w_i g}{g_c J}}. \quad \text{(10.19)}$$

A more general approach to determining temperature and steam quality profiles (and one that can reduce the number of assumptions required to develop results similar to those given by Eqs. 10.11, 10.13, and 10.19) is to solve numerically the pertinent differential equations. This normally is done by dividing the length of the wellbore into N intervals and using the proper value of thermal properties and R_h within each interval. The heat losses from and temperature drop across each interval then are summed to obtain the heat losses and temperatures to any depth as a function of time. Such numerical procedures have been used by Satter[11] to study the temperature distribution resulting from both superheated and saturated steam injection using the process conditions given in Table 10.3. Figs. 10.7 and 10.8, reproduced from Satter's publication, show temperature and steam quality distributions for superheated steam injection and the location of the hot-water point (zero quality steam) for injection of

TABLE 10.3 – PROCESS PARAMETERS RELATED TO FIGS. 10.6 and 10.7

Surface temperature $T_E(0)$, °F	75
Geothermal temperature gradient g_G, °F/ft	0.011
Thermal conductivity of the earth λ_E, Btu/D-ft-°F	24
Thermal diffusivity of the earth α_E, sq ft/D	1.1
Inside diameter of casing r_{ci}, in.	5.989
Outside diameter of casing r_{co}, in.	6.625
Inside diameter of tubing r_i, in.	2.441
Outside diameter of tubing r_o, in.	2.875
Annulus is filled with air	

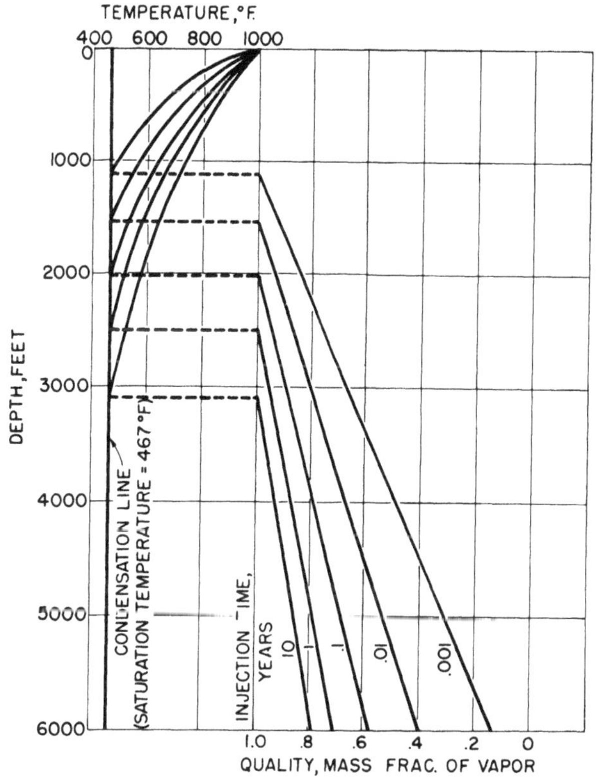

Fig. 10.7 – Temperature and steam quality distributions resulting from injection of superheated steam at 120,000 B/D, 1,000°F, and 500 psia.[11]

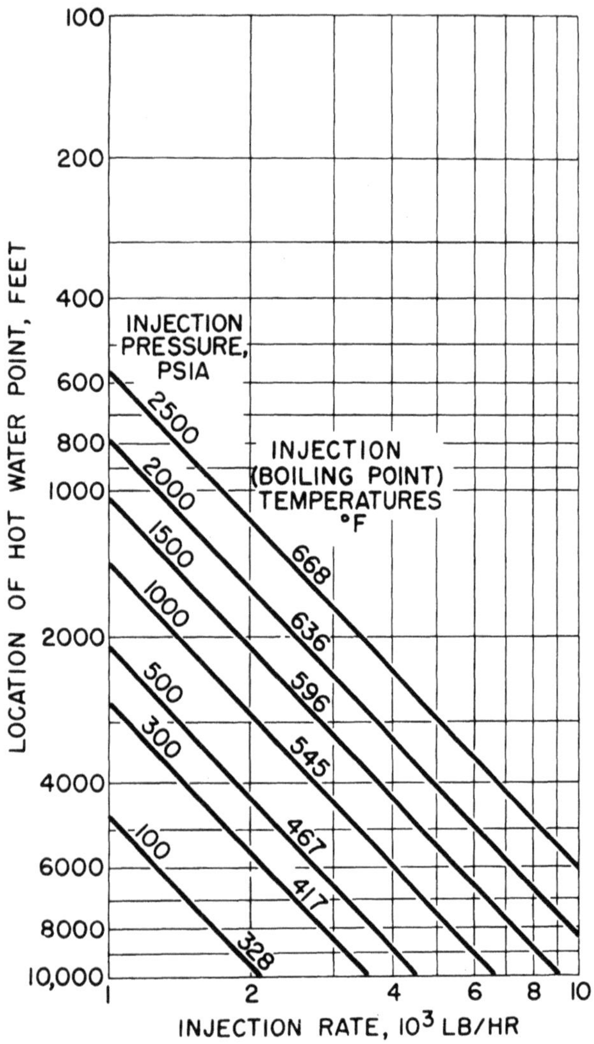

Fig. 10.8 – Hot-water points as a function of injection rate and pressure after injecting saturated steam for 1 year.[11]

saturated (100% quality) steam. Fig. 10.7 shows that the minimum depth at which superheated steam can exist above its saturation temperature of 467°F increases rapidly for early times and very slowly for large times. Once there is saturated steam, the temperature remains constant while the quality decreases with depth. Fig. 10.8 shows that the depth of the hot-water point after 1 year of saturated steam injection is essentially directly proportional to the injection rate. Other numerical programs, such as the one described by Earlougher,[12] also include pressure friction losses and the weight of the fluid column in calculating the temperature and quality of the steam as a function of depth and time.

Smith and Weinbrandt[16] programmed the heat loss equations in steam injection wells for a hand-held calculator to give the downhole steam quality, steam enthalpy, and heat injection rate. The calculations, which are simplified somewhat, are applicable to the case where there is no insulation and low-pressure air is used to insulate the tubing. They consider both concentric and eccentric tubings.

References

1. McAdams, W.H.: *Heat Transmission*, third edition, McGraw-Hill Book Co. Inc., New York City (1954).
2. Nelson, W.L.: *Petroleum Refinery Engineering*, fourth edition, McGraw-Hill Book Co. Inc., New York City (1958).
3. Jakob, M.: *Heat Transfer*, John Wiley and Sons Inc., New York City (1950) **1**.
4. Rohsenow, W.M. and Hartnett, J.P.: *Handbook of Heat Transfer*, McGraw-Hill Publishing Co. Inc., New York City (1973) 3-121.
5. Ramey, H.J. Jr.: "Wellbore Heat Transmission," *J. Pet. Tech.* (April 1962) 427-440; *Trans.*, AIME, **225**.
6. Willhite, G.P.: "Over-all Heat Transfer Coefficients in Steam and Hot Water Injection Wells," *J. Pet. Tech.* (May 1967) 607-615.
7. Moss, J.T. and White, P.D.: "How to Calculate Temperature Profiles in a Water-Injection Well," *Oil and Gas J.* (March 9, 1959) 174-178.
8. Carslaw, H.S. and Jaeger, J.C.: *Conduction of Heat in Solids*, Oxford U. Press, Amen House, London (1950).
9. Huygen, H.H.A. and Huitt, J.L.: "Wellbore Heat Losses and Casing Temperatures During Steam Injection," *Drill. and Prod. Prac.*, API (1967) 25-32; also in *Prod. Monthly* (Aug. 1966) 2-8.
10. López C., F.F. and Rivera R., J.: "Simulación Matemática de los Mecanismos de Transferencia de Calor hacia las Formaciones que atraviezan los Pozos Injectores de Vapor," Publication No. 72BH/089, Inst. Mexicano del Petroleo (Nov. 1971).
11. Satter, A.: "Heat Losses During Flow of Steam Down a Wellbore," *J. Pet. Tech.* (July 1965) 845-851; *Trans.*, AIME, **234**.
12. Earlougher, R.C. Jr.: "Some Practical Considerations in the Design of Steam Injection Wells," *J. Pet. Tech.* (Jan. 1969) 79-86; *Trans.*, AIME, **246**.
13. Romero-Juárez, A.: "A Simplified Method for Calculating Temperature Changes in Deep Oil Well Stimulations," paper SPE 5888, Society of Petroleum Engineers, Richardson, TX (1976).
14. Eickmeier, J.R, Ersoy, D., and Ramey, H.J. Jr.: "Wellbore Temperatures and Heat Losses During Production or Injection Operations," *J. Cdn. Pet. Tech.* (April-June 1970) 115-121.
15. Wooley, G.R.: "Computing Downhole Temperatures in Circulation, Injection, and Production Wells," *J. Pet. Tech.* (Sept. 1980) 1509-1522.
16. Smith, D.D. and Weinbrandt, R.M.: "Calculation of Unsteady State Heat Loss for Steam Injection Wells Using a TI-59 Programmable Calculator," paper SPE 8914 presented at the 1980 SPE California Regional Meeting, Pasadena, April 9–11.

Chapter 11

Facilities, Operational Problems, and Surveillance

This chapter discusses operational requirements and typical problems associated with thermal oil recovery projects. It points to some field experience and discusses diagnostic techniques used to monitor and interpret the progress of a field operation. The wide diversity of crudes, reservoirs, geological settings, thermal methods, and initial and operating conditions does not allow a full and complete development of this important subject. Rather, selected items and references typifying thermal operations are discussed.

To set the stage, the injection of steam in a steam drive and the injection of air in a dry forward combustion project are considered. The injection fluid – steam or air – will be followed from its source through its disposal or recycling. This is illustrated in Figs. 11.1 and 11.2 for steam and air, respectively. In these figures, the path of the water is treated in more detail than that of the air. Surface components associated with the injection well make up the surface injection system, which is discussed in Section 11.1. Subsurface components are discussed in Section 11.2, and the surface production system is discussed in Section 11.3.

Rich sources of information on subsurface and well completion techniques are the reports of Gates and Holmes,[1] and those of Giusti,[2] Burkhill and Leal,[3] and Rincón[4] that appear in the *Proceedings* of the Heavy Oil Symposium held in Maracaibo in July 1974. Noran[5] gives a brief summary of the special equipment and procedures used in thermal operations.

11.1 Surface Injection System

Water Treatment

Water treatment[6-12] is affected by the source and chemical composition of the water. Waters from rivers and lakes tend to have high concentrations of dissolved oxygen and suspended solids and low concentrations of dissolved solids. Subsurface waters tend to have the opposite characteristics.

Although not all these procedures necessarily are applied in every water-treating case, it is generally appropriate to do the following.

1. Use a bactericide—usually liquid or gaseous chlorine. Less frequently, methylene bisthiocyanate (MBT) and barium peroxide are used. Quarternary amines and diamines should not be used because they tend to plate out on the generator tubes.
2. Filter out the suspended solids and bacterial debris.
3. Remove dissolved oxygen by adding catalyzed sulfite or by stripping it with sweet natural gas or steam.
4. Prevent iron pickup by coating metal components, controlling pH (usually above 9), or using nonmetal piping. Iron pickup increases with water velocity and is likely to be most severe in the tubing string of a water-supply well. If the casing should prove a source of iron, cathodic protection may help. Nonmetallic flowlines may be used where their strength is adequate at the maximum operating pressure and temperature.
5. Remove silica and scale-forming ions such as calcium and magnesium by ion-exchange methods. Typically, a scale inhibitor is used as an additional precaution.

Bacteria can cause well-plugging problems, and the sulfate-reducing bacteria sometimes found in subsurface waters can generate hydrogen sulfide. In the presence of iron, iron sulfide is formed, and it tends, along with bacterial bodies, to plate out on the ion-exchange resin in the water softeners. This reduces the effectiveness of the ion-exchange resin. Suspended solids and oil in the water can deposit in the ion-exchange resin beds and in the steam generators and reduce their effectiveness. Insoluble ferric iron compounds can affect the ion-exchange resins adversely, and soluble iron compounds can collect in the generator pipes and impair their heat transfer. Removal of oxygen (and acid gases like carbon dioxide and hydrogen sulfide) reduces corrosion rates. Silica deposits and scale can reduce heat transfer and cause hot spots in steam generators.

Not all these problems necessarily will exist in every operation; thus, not all the five steps listed may be required to maintain a high degree of efficiency. However, in certain circumstances, water-treatment facilities may be more complex than indicated.

Produced water also can be handled and treated for reuse in steam generators.[13,14] Although produced water often is more difficult and expensive to treat than source water because of increased

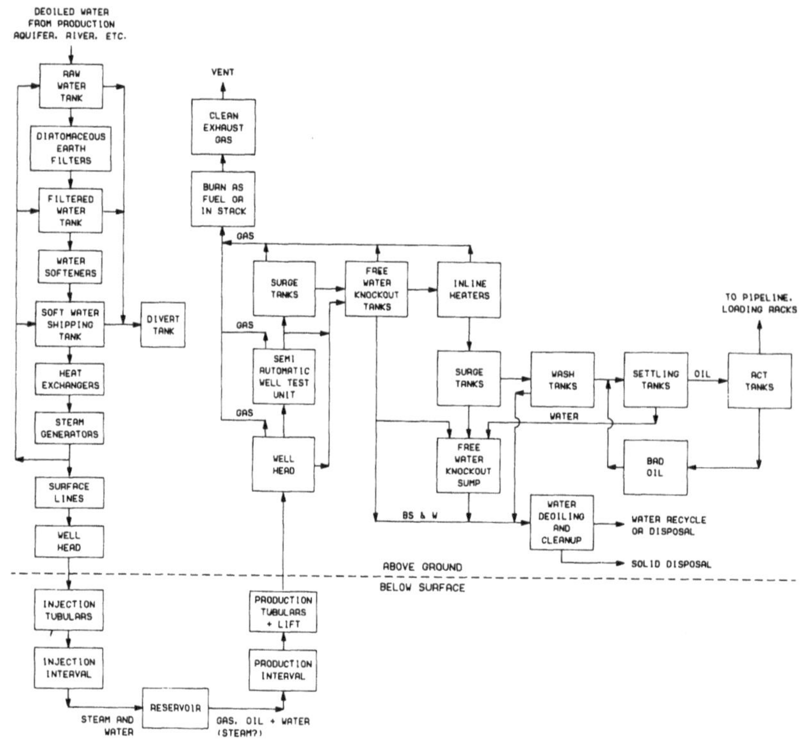

Fig. 11.1 – Schematic flow diagram for water.

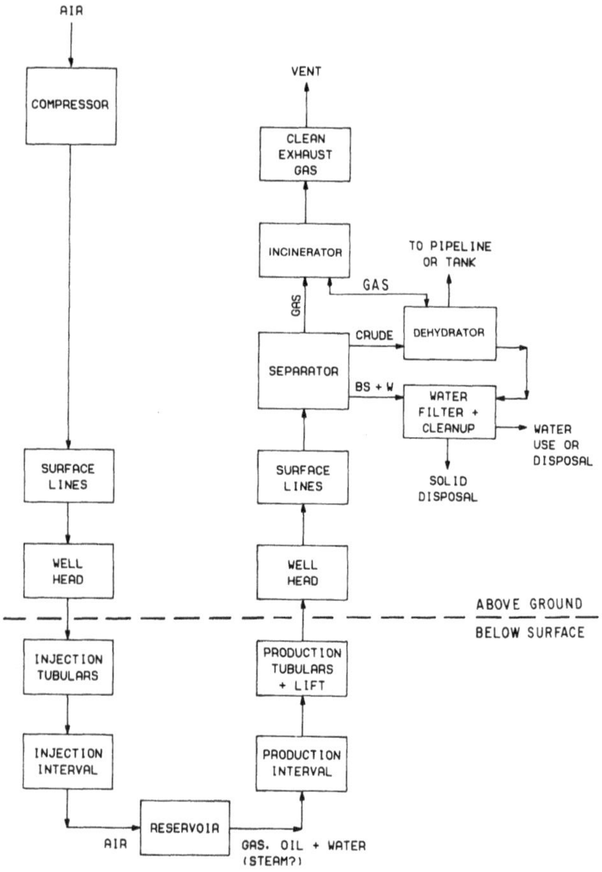

Fig. 11.2 – Schematic flow diagram for air.

hardness, dissolved solids, salinity, and oil content, it may be the only water available in the required amounts.

Steam Generation

The softened water usually is gravity-fed from a storage tank to a positive-displacement pump that forces the water through the steam generator. The storage tank preferably should be galvanized and have a nitrogen blanket to reduce iron pickup and dissolved oxygen.

Steam generators of the once-through type are used almost exclusively for thermal oil recovery.[15] They were developed for such use and differ from a true boiler in that they have no steam-separating drum requiring recirculation and blowdown. Because the generators do not have a separating drum, the maximum steam quality must be limited to about 80% to prevent the deposition of dissolved solids or scale on the coils and to reduce the possibility of localized film boiling and subsequent tube failure. There are some generators, however, that use unsoftened or produced water.[16]

Fig. 11.3 is a schematic representation of a gas-fired once-through or single-pass steam generator. Typical thermal efficiencies of such units are about 80 to 85%, with most of the lost heat going through the stack. Stack heat losses can be reduced by using heat exchangers in the convective section (economizers) to preheat the feedwater and by using special burners that allow reducing the excess air to less than 5%. The amount of heat recoverable often is limited by the corrosive effects of SO_3 contained in the exhaust fuel oil, which condenses at lower temperatures. Thus, it is desirable to operate stacks at temperatures higher than the acid dewpoint, even though this limits the efficiency of the generators when sulfur-containing fuels are used. Usually this temperature is about 350°F. Stack gas normally is scrubbed through either a caustic system or an ammonium hydroxide system. Most generator manufactures can provide stack-gas scrubbers[17] for their units and nearly all have an energy and mass balance flow chart for their generators (see, for example, Fig. 11.4).

Efficient flame patterns in the radiant section of the generator are obtained easily with gaseous and liquid fuels. But it is important that liquid fuels have

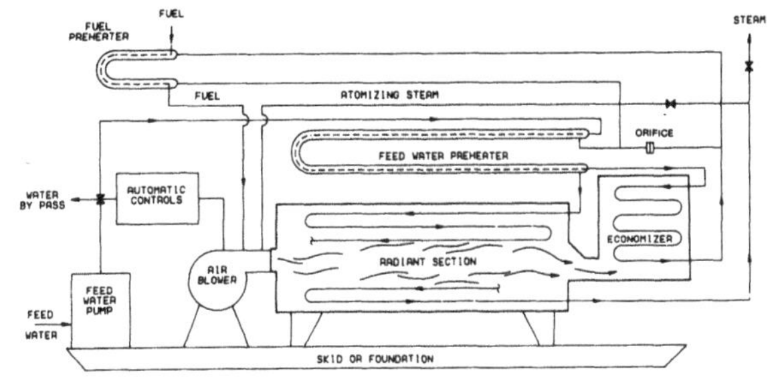

Fig. 11.3 – Flow scheme through a single-pass helical coil steam generator.

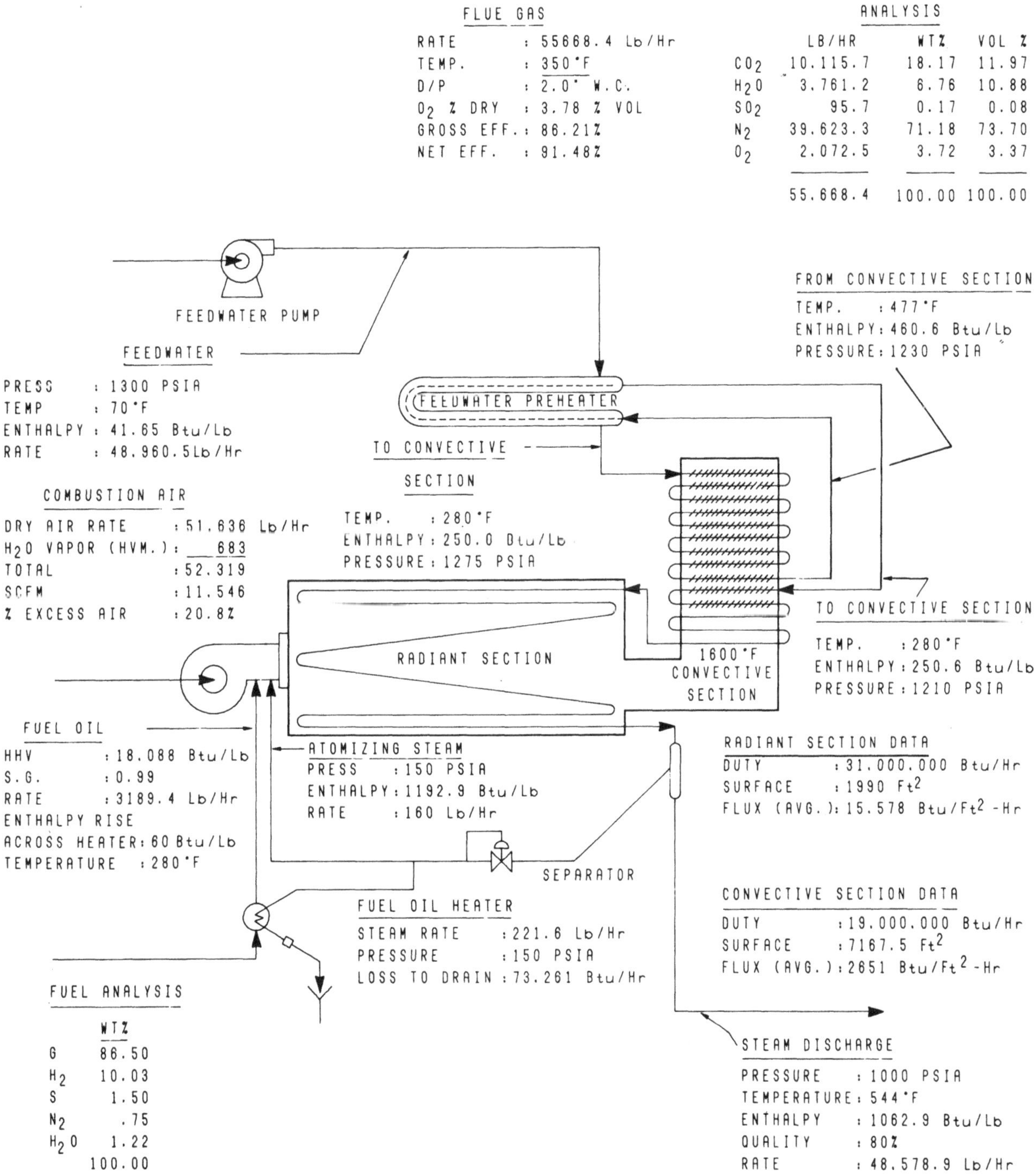

Fig. 11.4 – Process flow diagram and thermodynamic heat balance for a 50×10^6 Btu/hr steam generator (data from Thermotics Inc.).

constant properties – such as viscosity, volatility, heating value, and dissolved gas and water content. Coal-fired oilfield generators are not common because of intrinsic drawbacks related to burner design and to reduced heat transfer to the coils resulting from soot coatings. However, efforts are being made to develop and improve the operational and thermal efficiencies of coal-fired generators for oilfield use.

Loss in the thermal efficiency of a unit is usually an indication that heat transfer to the coils is impaired. In gas-fired heaters, where soot should not be a problem, this loss in thermal efficiency generally indicates that scale has formed inside the coils. Scale usually is removed by circulating dilute acid solutions through the unit until the effluent acid concentration indicates the reaction is complete.[18] The unit is, of course, "cold" during such a treatment. When the tubes are coiled vertically, as opposed to horizontally as shown in Fig. 11.3, high circulation rates are preferred to minimize accumulation of hydrogen against the top portions of the coils and assure even acidizing.

The efficiency of a generator may be determined from the enthalpy of the produced steam relative to the total energy used to generate it. The enthalpy of the produced stream is affected by its pressure and quality and is proportional to the feedwater rate. As long as there are no branches in the effluent line, the steam quality at any point downstream of a generator operating at stable conditions commonly is calculated as the increase in the chloride concentration in the liquid phase relative to that in the feedwater, divided by the chloride concentration in the liquid phase at the point of interest. The method is based on the conservation of the chloride ion. Any other ion that remains in solution and does not react or decompose also could be used. Concentrations are best determined chemically but can be correlated with electrical conductivities and dissolved solids.

Orifice meters are used by some operators. Then the steam quality is given by the square of the ratio of the apparent mass flow rate through the orifice meter (assuming the stream is 100% steam vapor) to the mass feedwater rate. Moreover, the steam qualities obtained with orifice meters can be correlated with those obtained chemically.

Steam generators can be instrumented in varying degrees to ensure the safe and efficient delivery of steam of predetermined pressure and temperature.[18] Important elements of steam generators are automatic safety shutdown features – which include instruments to control steam quality, excessive tube temperatures, hard water, power failure, blower failure, low feedwater rates, and high and low steam pressure or temperature. They protect the equipment by shutting the fuel if the feedwater stops flowing or if there is some other malfunction or abnormal condition.

Steam generators and heaters are generally available at three nominal coil pressure ratings: 1,000, 1,500, and 2,500 psig. For pressure ratings of 2,700 and 2,800 psig, molybdenum steels are used, and stainless steel is required for higher pressures. More commonly, the ratings of positive-displacement feedwater pumps limit the pressure at which steam and hot water can be delivered.

Typical capacities of steam generators and heaters range from 10 million to 50 million Btu/hr, corresponding to throughput rates of about 600 and 3,200 BWPD, respectively. A generator is not designed to be operated at less than half its peak flow rate. Small units are available on trailer mounts, and all field units are normally available on skid mounts. Some manufacturers also provide complete portable water-treating facilities; others can arrange for them. As a rule, manufacturers can meet any unusual operating and pollution-control requirements.

Steam-generator and heater failure can be kept to a minimum with the proper use of fail-safe instruments, especially at the start of a project. As with any piece of equipment, the design affects the operating difficulty and the downtime required for repairs, as well as the frequency of repairs. Where possible, it is advisable to compare the refractory and tube replacement records when considering the purchase of steam generators and heaters.

Surface lines from steam generators and heaters normally are insulated to reduce heat losses.[19] Lines may be buried, laid on the ground, or supported off the ground. The last approach is preferred for long-term installations since allowance must be made for expansion due to changes in temperature. This usually is done by providing various configurations of loops and anchors and installing sliding shoes at intermediate points.[18] When uninsulated lines are buried, it is often advisable to lay them in sand-filled large-diameter conduit pipe to reduce excessive heat losses through wet soil. Calculation of heat losses from surface lines is discussed in Chap. 10.

For hot-water flooding, large-capacity field heaters normally are used, and these are located close to the injection wells to minimize heat losses and temperature reductions.[20] Typical heat injection rates per heater[21] range from 5 million to 50 million Btu/hr. Steam generators have been modified successfully to heat water,[22] but their thermal efficiency decreases unless they are sized and operated properly.

Compressors

Reciprocating compressors commonly are used in combustion projects. Except in the smallest pilots, the compressors generally are mounted in a centrally located permanent compressor station, and for maximum flexibility, several of them are operated in parallel.[23] Routine maintenance and repair and major overhauls then can be carried out with minimum disturbance to the project. Spare compressor capacity sometimes is made available.

Properly selecting air compressors – considering type, number of stages, and type of engine – requires a thorough understanding of the project's operating requirements and the environmental constraints.[24,25] Electric-motor-driven compressors[26] are fairly quiet and emit no exhaust gases but may have high energy costs. These advantages and disadvantages must be weighed against those of gas or diesel engines. The choice of engine is affected by local conditions, including environmental limitations and the availability and dependability of a fuel source. Selecting an air compressor requires the skill of a compressor engineer and the competent advice of the manufacturer.

Some guidelines for estimating compressor requirements are considered here. Compression generally is done in stages. Each stage is a compression element consisting of a single cylinder compressing on only one side of a piston (single-acting) or on both sides (double-acting). Spring-loaded intake and discharge check valves open only when the proper pressure differential exists across the valve. Discharge valves open when the pressure in the cylinder is slightly above the discharge pressure; conversely, the intake valves open when it is slightly below the intake pressure. When more than one stage is used, the discharged air is cooled; the volume of air then is scrubbed to remove liquids before it is fed into the next.

In determining the number of stages to be used, some rules of thumb as well as cost assessments are applied. The stage compression ratio F_{SC} represents the ratio of the discharge to intake absolute pressure per stage:

$$F_{SC}=\frac{p_{ab,\,\mathrm{dis}}}{p_{ab,\,\mathrm{int}}}. \qquad (11.1)$$

One of the factors inhibiting the use of high compression ratios is the discharge temperature $T_{ab,\,\mathrm{dis}}$. For air, the discharge temperature in any stage is given approximately by

$$T_{ab,\,\mathrm{dis}}=T_{ab,\,\mathrm{int}}F_{SC}^{\;0.29}. \qquad (11.2)$$

The value of the discharge temperature given by this equation is the one that would be obtained if no cooling were used. Even though the system is always cooled, the temperature of the air inside the cylinder during the compression stroke approaches that given by Eq. 11.2. To minimize the possibility that the piston lubricant will lose its properties upon reaction with the oxygen and even catch fire, it is important to limit the maximum air temperature to 250 to 300°F, depending on the compression ratio and pressure level. Special, relatively nonoxidizing lubricants often are used in air compressors to maintain both safety and lubricant integrity.[27,28] They must be used with care, however, as lubricant vapors in the compressed air can condense and accumulate in portions of surface lines and may lead to fires and even explosions.

The temperatures of the compressed air entering a subsequent stage are usually 15 to 20°F higher than that of the cooling water entering the interstage cooler. Values of the discharge temperature vs. stage compression ratio are given in Fig. 11.5 for several values of the intake temperature. As can be seen in this figure, the prudent stage compression ratio is affected greatly by the value of the intake temperature, so that both the climatic conditions in an area and the practical degree of cooling become important factors. Over the range of intake temperatures from 0 to 120°F, the prudent stage compression ratio ranges from 4.5:1 to 2.5:1.

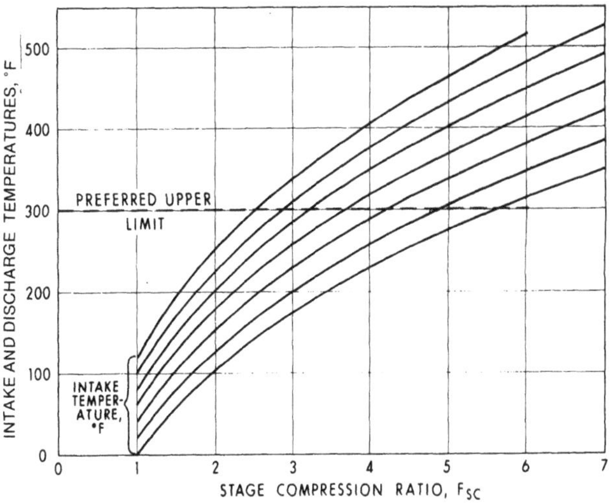

Fig. 11.5—Effect of compression on discharge temperature.

From the standpoint of ideal power requirements, it is desirable to have the same compression ratio in all stages. When the maximum injection pressure is given, the overall (total) compression ratio of the compressor unit is

$$F_{\mathrm{comp}}=\frac{p_{ab,\,\mathrm{fin}}}{p_{ab,\,\mathrm{suc}}}, \qquad (11.3)$$

and the compression ratio per stage is

$$F_{SC}=F_{\mathrm{comp}}^{\;1/n_{CS}}, \qquad (11.4)$$

where n_{CS} is the number of stages, $p_{ab,\mathrm{suc}}$ is the absolute intake pressure at the first stage, and $p_{ab,\,\mathrm{fin}}$ is the absolute discharge pressure from the last stage. If, for example, the overall compression ratio is 9:1, then choosing $n_{CS}=2$ (two stages) gives a compression ratio per stage of 3:1, which is reasonable. An injection pressure of 1,000 psig is equivalent to an overall compression ratio of approximately 69:1 (depending on the mean intake pressure). Two stages would give a compression ratio per stage of 8.3:1, which would be very high. Three stages would result in a value of 4.1:1, which is within the preferred range. The suitablity of using three stages in this case would be influenced greatly by the intake temperatures, as Fig. 11.5 shows.

Once the number of stages is determined, the compressor flow capacity in 1,000 Mscf/D at 60°F and 14.7 psi is found approximately from

$$q=2.95\times10^{-5}V_{\mathrm{str}}\left(\frac{p_{ab}}{z\,T_{ab}}\right)_{\mathrm{suc}}n_{str}E_{\mathrm{comp}}, \qquad (11.5)$$

where V_{str} is the volume displaced by the piston over its maximum extension (adjustments must be made for double-acting pistons),[24] z is the deviation or supercompressibility factor of air, n_{str} is the number of strokes per minute, and E_{comp} is the volumetric compressor efficiency. V_{str} represents the first-stage piston displacement volume and is expressed in cubic inches.

The volumetric efficiency of a compressor is af-

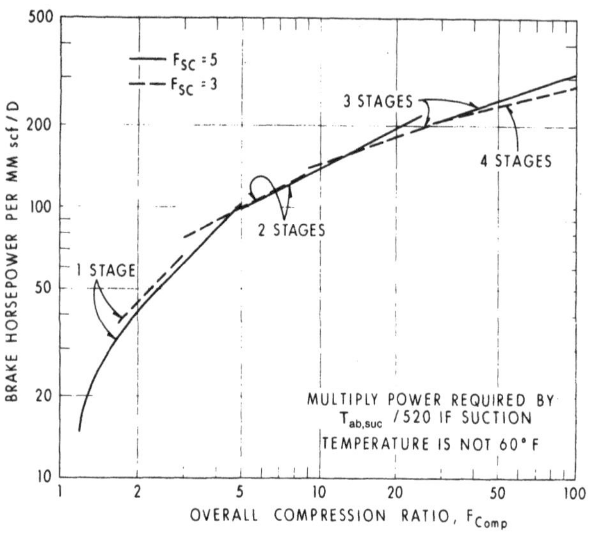

Fig. 11.6 – Power requirements.

fected primarily by the cylinder clearance, which is the fraction of the cylinder volume not traversed by the piston. Some clearance between the piston and the end of the cylinder is necessary for a number of reasons, including ease of valve operation; also, it gives rise to a pocket of compressed air left in the cylinder as the piston starts the intake stroke. The volumetric efficiency is estimated from

$$E_{comp} = 0.96 - 0.01F_{comp} - \left(\frac{z_{suc}}{z_{fin}} F_{comp}^{0.71} - 1 \right) f_{cy}, \quad \ldots\ldots (11.6)$$

where f_{cy} is the fractional cylinder clearance, normally in the range of 0.04 to 0.15.

From the compressor capacity q_{SC}, the brake horsepower of the compressor is approximated from

$$P = 0.043\, n_{CS} F_{SC}^{1/n_{CS}} F(n_{CS}) T_{ab,\,suc}\, q \quad \ldots\ldots (11.7)$$

where the factor $F(n_{CS})$ allows for the interstage pressure drop resulting from n_{CS} interstages of cooling, and the power P is in horsepower. The horsepower requirements per 1,000 Mscf of compressed air calculated from Eq. 11.7 are plotted in Fig. 11.6 vs. the overall compression ratio for stage compression ratios of 3:1 and 5:1, assuming a suction temperature of 60°F. Ref. 24 gives the following values for $F(n_{CS})$: $F(1) = 1.0$. $F(2) = 1.08$, and $F(3) = 1.10$.

Aftercoolers (on the last stage) sometimes are omitted or used at reduced efficiency. The temperature of the surface lines then can be appreciable,[29] requiring the use of expansion loops.

Moisture condensed in the injected air sometimes can corrode the well tubing and impair injectivity. Lubricating oil carried in the air from the compressor cylinders also can cause impairment. To remove these vapors requires condensation, which is one of the purposes of the interstage and aftercoolers. Where the moisture of the intake air is high, glycol has been used to dry it. However, using glycol has some undesirable side effects that lead to the formation of corrosive organic acids. This has prompted the installation of "gravity-operated centrifugal scrubbers equipped with dehumidifiers at the top" in at least one tropical combustion project.[30]

11.2 Wells

Well Design Considerations

Discussion of well designs will be limited to the effect of temperatures on thermal stresses in the casing and, briefly, to criteria for designing failure-free casing systems.

Leutwyler and Bigelow[31] and Willhite and Dietrich[32] have published excellent discussions on the design of steam-injection wells. Although they directed their studies toward wells used in cyclic steam-injection operations, where the temperature cycles are most pronounced, the same considerations apply to wells in all thermal recovery operations.

The injection or production of hot fluid can be interrupted by mechanical failures – such as heater coil burnouts or disruptions in fuel or feedstream supply – and by workovers to restore injectivity or productivity or to correct an unexpected situation – such as an undesirable injection or production profile.

If the shutdown of hot fluid injection period is long, the wellbore temperatures will fall, leading to tensile casing stresses (to be discussed). In the case of workovers, where injection or circulation of extraneous fluids may be required to control the well, cooling may be fast and severe. Since workovers almost certainly will be needed to correct impairment or mechanical failures, it is well to consider the effect of heating and cooling in designing wells for any thermal operation.

Casing tends to elongate when heated. For casing cemented on bottom, buckling is almost inevitable in unsupported sections of any significant length unless the casing is free to expand through a packing gland at the surface pipe. In all but very shallow wells, hole deviations and formation sloughing will cause enough loading at intermittent points to build up compressive stresses and possibly buckle the heated casing.

The stresses that result when heated casing cools can be understood with the aid of Fig. 11.7. As long as the compressive stress on the casing has not exceeded the yield strength of the steel, or the casing does not get stuck in an expanded position, cooling will not cause difficulties; the stress is reversible and follows the Line A-B. If the yield strength of the steel is exceeded, however, the reduction in compression will follow a line such as C-D when the casing cools. The decompression cycle parallels A-B but is displaced by the difference between the maximum temperature increase (ΔT_{max}) and the temperature increase at which the steel yields (ΔT_{yp}). If the temperature drops below this value (which is denoted by ΔT_{te}), the casing and joints actually will be in tension. Casing joints then will separate if the tensile load exceeds the joint-pullout strength.

An obvious way to prevent joint pullout is to ensure that the compressive casing stress does not exceed its yield strength. In new completions this can be done by selecting a suitable combination of materials, by controlling casing temperature, and by cementing the casing in tension. In existing wells, one can control only the casing temperature.

Table 11.1 compares calculated minimum tensions resulting from heat/cool cycles in carbon steel with field observations regarding failure of wells completed with 7-in., 20-lbm/ft, J-55 short-thread casing. The joint strength in tension of this casing[32] is 254,000 lbf, which is equivalent to 44,300 psi (based on the cross-sectional area of the casing wall). The field data are consistent with the calculations, thus supporting this design theory.

If casing lacks lateral support, it can buckle. Buckling, in turn, can reduce the temperature increase at which the casing reaches its yield strength and, thus, increase the likelihood of joint pullout on subsequent cooling. For a ½-in. radial clearance in a rigid borehole, Table 11.2 gives calculated allowable temperature increases for the buckling of a 7-in., 23-lbm/ft, 30-ft length of casing whose yield strength would not exceed that of carbon steel. The allowed temperature increase for a perfectly supported casing also is given. As can be seen, buckling significantly reduces the temperature increase at which yield strength is reached.

The length and clearance of unsupported casing in old wells depend on the cementing practice, the number and distribution of hole washouts, the tendency of the formations to slough, and the hole deviation. In new wells, lateral deformation (and buckling) can be reduced significantly by drilling a relatively straight hole and by cementing to surface with a temperature-stable cement.

Another way to prevent casing buckling in new wells is to use material with a higher yield strength than carbon steel. This, of course, is more expensive. And one must weigh the possibility that existing wells will fail against the cost of properly designing and completing new wells. If the choice is to drill new

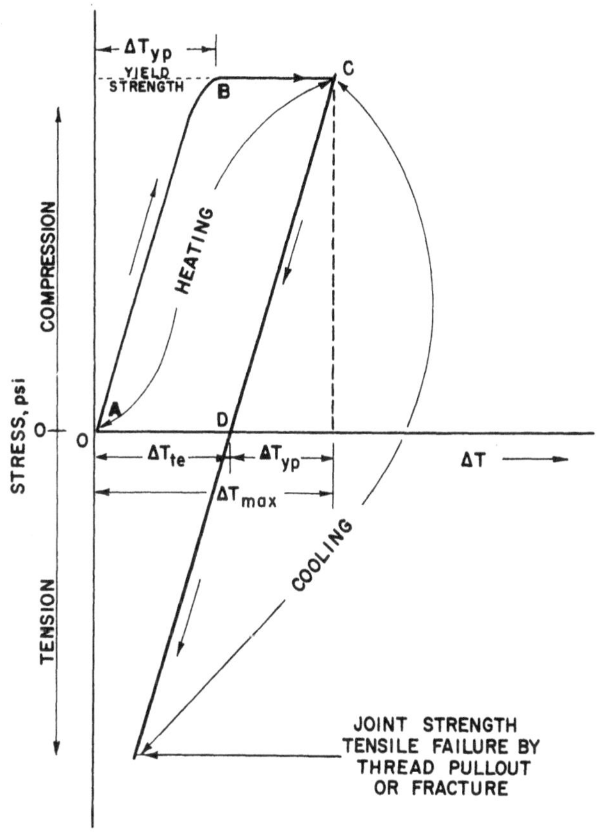

Fig. 11.7 – Stress history of casing overstressed during steam injection.[32]

TABLE 11.1 – TEMPERATURE CHANGES/TENSILE FORCES RESULTING FROM COOLED PIPE[32]

Casing Temperature (°F)	ΔT_{max} (°F)	ΔT_{yp} (°F)	ΔT_{te} (°F)	Tension When Cool (psi)	Remarks
500	400	275	125	25,000	No failure
550	450	275	175	35,000	No failure
600	500	275	225	45,000	Failure
650	550	275	275	55,000	Failure

TABLE 11.2 – ALLOWABLE TEMPERATURE CHANGES FOR BUCKLING AND COMPRESSIVE LOADING[32]

Material Grade (spec)	Minimum Yield Stress (psi)	Allowable Temperature Change (Δ°)	
		Buckling*	Uniform Loading
J	55,000	237	275
N	80,000	346	400
S	95,000	410	475

*7-in. 23-lbm/ft, 30 ft long with ½-in. radial clearance.

wells, decisions must be made about insulation, the quality of materials, and cementing techniques.

Bissoondatt[33] discusses the effect of temperature cycling resulting from cyclic steam injection operations and how to modify well design to avoid failures, using examples from a commercial operation.

There is one type of concentric tubing whose annulus is filled with insulation and whose inner string is tensioned as the outer string is welded[34]; it is designed for equal stress at a particular steam injection temperature. In some installations, the outer conduit is not welded to the inner tubing but is free to expand to mitigate thermal stresses.[35] Insulation must be protected from moisture over a substantial fraction of tubing if it is to remain effective. Sodium and calcium silicate often are used as the annular insulation.

Examples of injection well designs without insulation are shown in Fig. 11.8. These wells are only 500 ft deep. Production wells were designed in similar fashion using either 7-in., 22-lbm/ft, K-55 casing or 5-in., 14 lbm/ft, K-55 casing and using Class H cement modified with Perlite, silica flour, and calcium chloride. Silica flour and calcium chloride are used to reduce the loss in cement strength at high temperature. Deep wells in steam projects almost always are insulated, not only to reduce the likelihood of casing failure but also to reduce heat losses. A well design using a small gas flow in the tubing-casing annulus to insulate the casing is shown in Fig. 11.9.

The successful development of a downhole steam generator[36] would avoid the need for insulation in steam injection wells. But insulation still could be needed during the production phase of cyclic steam operations unless it is safe to expose the well to the producing temperatures, which are generally

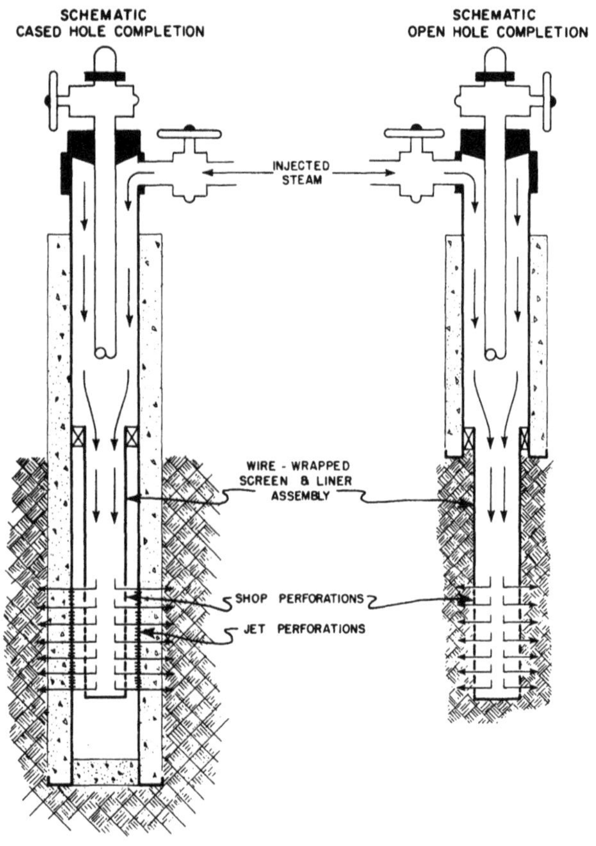

Fig. 11.8 – Schematics of well completions.[37]

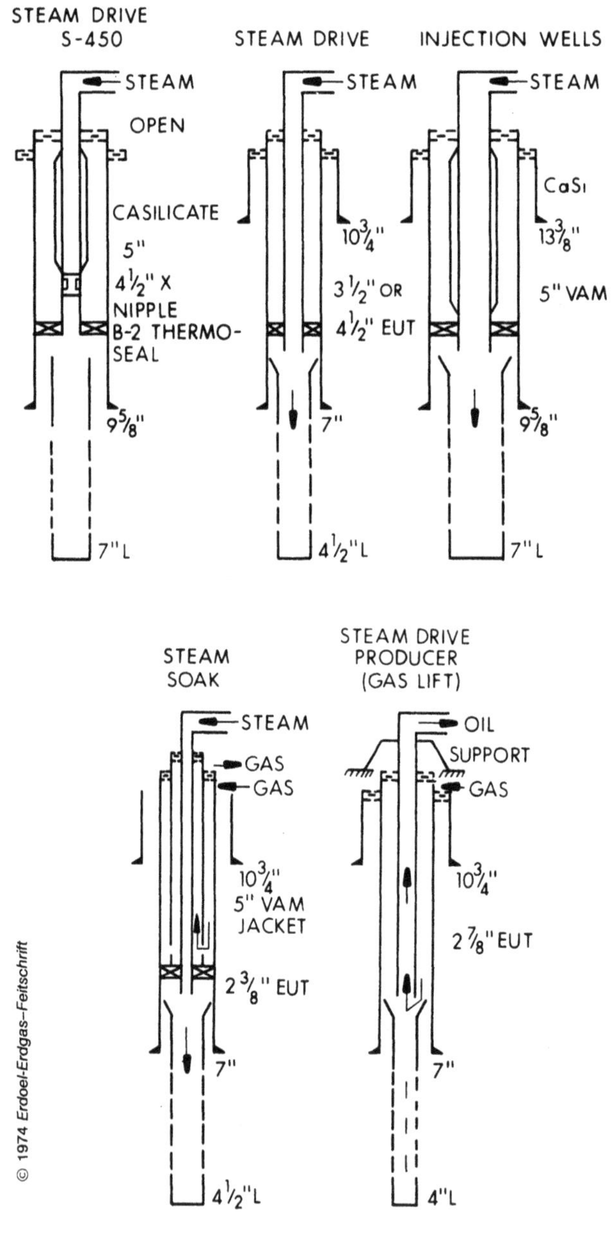

Fig. 11.9 – Schematics of well completions.[38]

significantly lower than the injection temperatures.

Well designs in which the casing is uncemented near the surface – and, thus, is free to expand – are shown in Fig. 11.10. In this "thermal sandwich" test, in which the depths are shallow, no insulation was used. Other casing completions allowing expansion have been discussed by Bleakley.[39]

One other general aspect of well design that must be considered is the need for materials and equipment that can survive and operate at high temperatures – materials such as cement,[1] thread dope (welding is an alternative), expansion sleeves or joints, elastomers and other sealants in packers, and any equipment that must engage or disengage under varying amounts of thermally induced loads. In steam injection processes, such equipment may be subjected to pressures of 2,000 psi and temperatures up to 550°F.

Well Completions

A large proportion of the reservoirs to which thermal recovery methods are applied have unconsolidated or friable formations. Well completions are designed to provide low resistance to fluids and at the same time keep sand from entering the wellbore. Although production wells receive the most attention, it is also advisable to control sand in injection wells. When, because of equipment malfunction or any other reason, injection pressure in injection wells decreases, the well may backflow. Where there is no space below the injection interval to accomodate the sandfill, the injectivity will decrease when operations resume. Well completions also are designed to promote or ensure the desired distribution of the injection fluid among multiple sands or sand

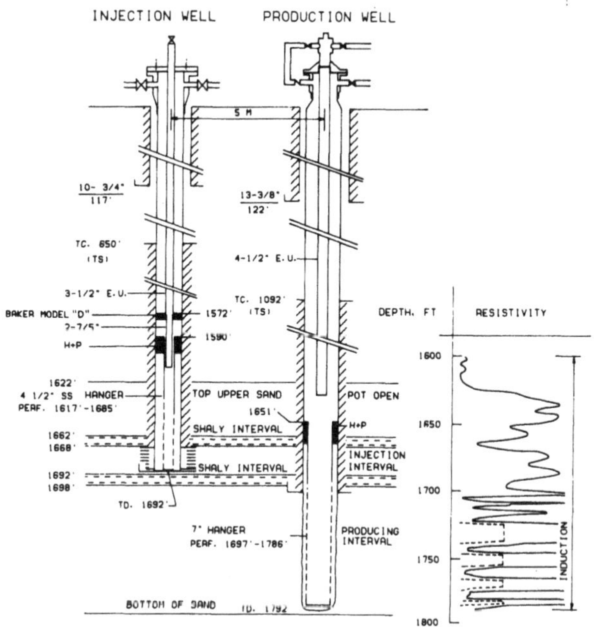

Fig. 11.10 – Schematic of well completion in thermal "sandwich" project.[4]

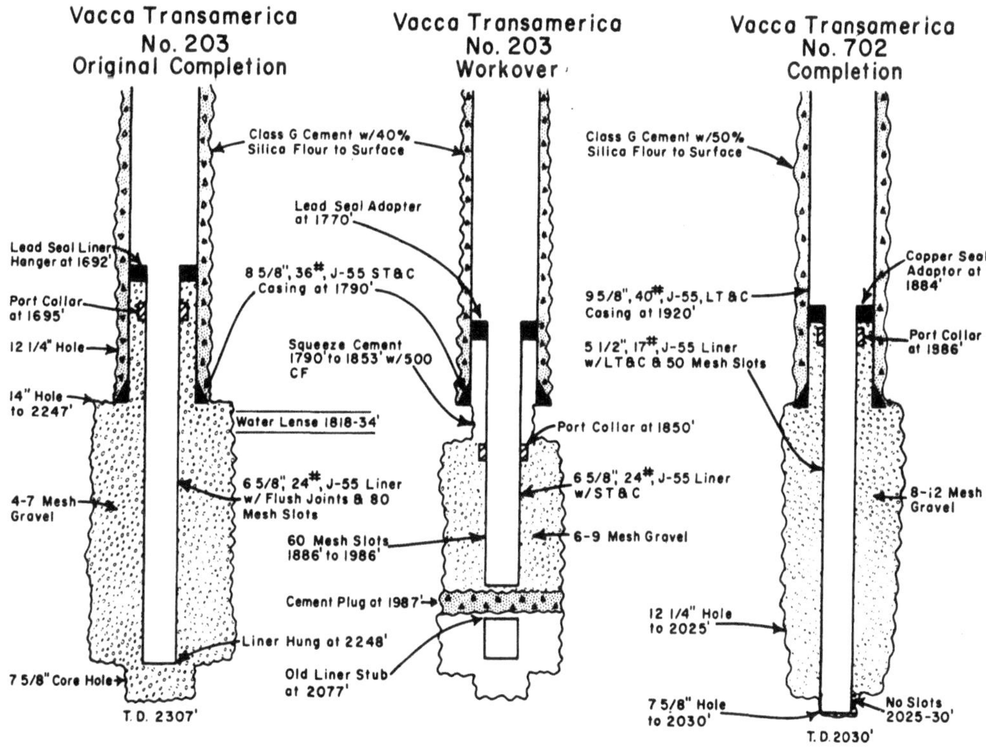

Fig. 11.11 – Vacca Transamerica well completions.[40]

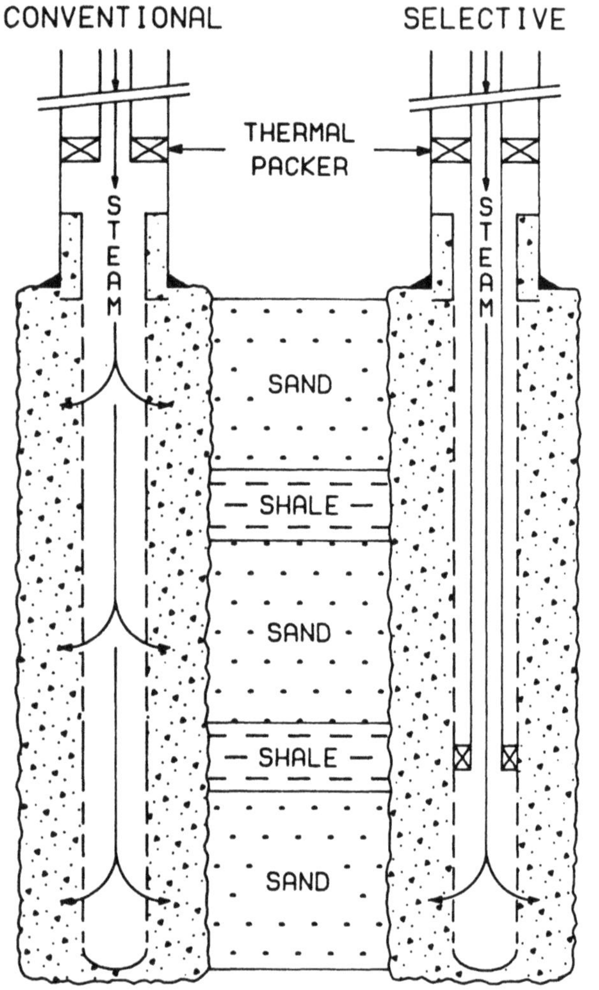

Fig. 11.12–Schematic diagram of conventional and selective completions in gravel-packed wells.[3]

stringers. Sometimes it also is desirable to separate producing intervals, as, for example, when two sands are at different pressures (referred to a common datum).

Fig. 11.8 shows a sand exclusion technique using a wire-wrapped liner assembly, hung either inside perforated casing or from casing inside open hole. The same technique was used in the production wells. Another practice is to gravel-pack a preslotted liner in an underreamed hole (see Fig. 11.11). Slot-opening and gravel sizes are important in effective sand control.[41-43] Saucier[41] recommends that "the ratio of pack median grain size to formation median grain size should be between five and six where there is severe flow disturbance." The slots are then sized to retain the gravel. Because some gravel is relatively soluble at the steam-condensate conditions prevailing downhole in injection wells, voids can develop in the gravel pack.[44] *Local experience is a strong factor in proper gravel-pack design.* To reduce the likelihood of parting due to stresses, liner joints sometimes are welded,[45] especially in cyclic steam injection operations.

In combustion projects, and particularly where there are multiple zones and spontaneous ignition is either planned or likely to occur, the casing often is cemented through the entire injection interval to provide mechanical strength at high temperatures. After perforation, a slotted liner may be used to minimize sand entry in the event of casing failure.[46]

Other mechanical and chemical types of sand control have been discussed by Suman.[47] In wells where steam is injected, the hot, high-pH liquid water from the generator and that condensed as a result of heat losses will tend to dissolve the sand grains in the reservoir, the gravel in the gravel pack,[37] and even the consolidating agent. Perhaps because of dissolution, methods for consolidating sand grains have not been very successful in hot-fluid injection wells where temperatures are high.[43]

Profile control is a much more difficult problem, especially for unconsolidated sands separated by thin shale stringers. As the "conventional" completion shown in Fig. 11.12 shows, the injected steam enters each of the sands according to its relative flow resistance. There is no control of the relative amount of steam entering each sand. In fact, with the completion as shown, the lower sand can be expected to receive relatively wet steam compared with the upper sand even when the flow resistances are equal, since the liquid water will tend to move to the lower part of the wellbore. Depending on the quality of the steam entering the injection interval (and other factors), the effective liquid/vapor level may be above the first shale stringer, effectively cutting off the lower sand from steam injection. One way to remedy this downhole steam separator effect is to extend the tubing below the lowermost injection interval (or as close to it as feasible) so that the steam coming up the annulus can distribute the liquid water and vapor more uniformly.[1] The "selective" steam injection completion shown in Fig. 11.12 aims at injecting steam into the lowest sand interval, which is usually the least depleted and the one having the highest flow resistance (everything else being the same). Although the figure shows steam entering the formation, it must be realized that part of the steam can move up through the gravel pack behind the packers and enter the upper sands. The effectiveness of such a selective completion obviously depends on the amount of steam leaking behind the packer through the gravel pack.

Profile correction falls in three different categories, and in each the condition and completion of the well must be considered.

In the case of set-through completions where the well is perforated, selective injection is a straightforward matter of using straddle packers (and perhaps bridge plugs). Essentially any mechanical means that delivers the fluid in the desired interval will be successful if there are a competent cement job and adequate separation of the sands within the reservoir.

Gates and colleagues[45,48] report the successful use of limited-entry perforation techniques in a properly cemented casing to control the injection profile in steam and air injection wells in the South Belridge field. Showalter and Maclean[49] also report using limited-entry techniques in the combustion project at Brea-Olinda field. With these methods, it is necessary to test and ensure that each perforation is open; the technique works best in low-pressure reservoirs.

In underreamed and gravel-packed wells, profile correction is much more difficult. Two choices are available. One is to place an impermeable plug over a short interval of the gravel-packed region between the liner and the sandface. A packer, bridge plug, or other mechanical device then can be set inside the liner to divert flow. Attempts have been made to seal the gravel pack across the shale intervals by injecting hardening sealants such as polymers, cement, mud, and other chemicals. Burkhill and Leal[3] report success using a mixture of bentonite, silica flour, barites, cement, and other chemicals in the Bolivar Coast of Venezuela. Furfuryl alcohol is an example of a low-viscosity chemical that can be catalyzed to rock hardness and that has been moderately successful in field applications.[50,51] Chemicals also have been used successfully to shut off thief zones in cyclic steam injection wells.[48]

Instead of forming an impermeable plug within the gravel pack opposite a shale section, it is somctimes desirable to seal the matrix in the thief zone. Many chemicals are available to seal thief zones – either temporarily or permanently – to divert the injected fluid into desired intervals. But few are economically attractive and easy to place. And if a permanent seal is desired, few chemicals can withstand the effects of high temperatures over prolonged times. Lignosulfonate forms a gel that can plug thief zones,[52] but it loses its effectiveness with prolonged exposure to high temperatures.

The addition of surfactants to injected steam forms steam foams that have been used to preferentially plug or increase the flow resistance of thief zones.[53] Because the foam degrades rapidly at temperatures of 500 to 550°F, the treatments are repeated periodically to provide new sealant. Wells should be completed initially to facilitate profile control. Possible steps that would aid later operations include the use of external casing packers and cementing zones between gravel-packed regions.[54] The latter technique requires relatively large nonpay zones within the total production interval.

Fluid Lifting

Produced fluids almost always can be lifted to the surface. The trick is to do so efficiently and in such a way as to promote the desired reservoir response. This usually entails maintaining a low wellbore pressure to make efficient use of the presssure difference available from the reservoir.

Production of unheated fluid is conventional in the oil industry.[55] Special problems arise where the viscosity or temperature of the crude is high at downhole conditions. Where crude viscosity is high, diluent sometimes is added through hollow sucker rods or side strings to improve rod fall and pump performance. (However, hollow sucker rods are not practical at depths greater than about 4,000 ft.) Other uses of diluent for improving lifting performance are discussed by Franco.[30] The diluent generally is obtained from the light ends of the

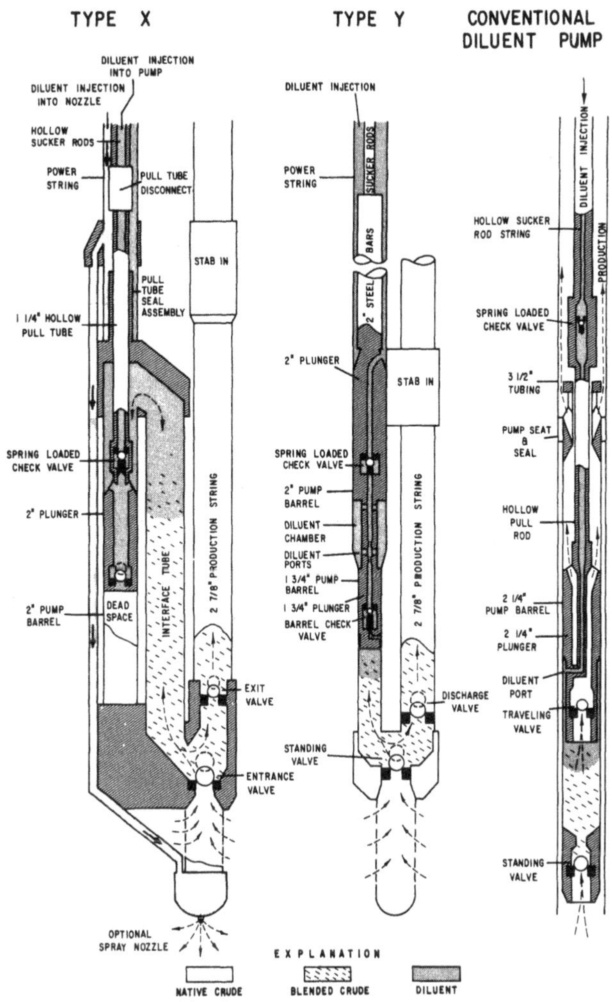

Fig. 11.13 – Diagrams of pumps capable of lifting sand-laden heavy crude.[56]

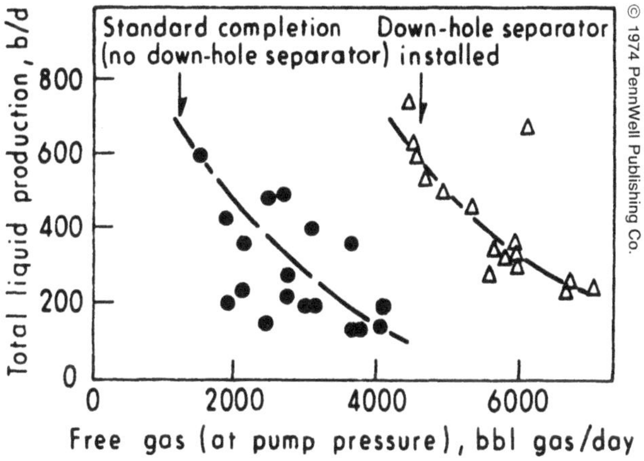

Fig. 11.14 – Effect of downhole gas separation.[57]

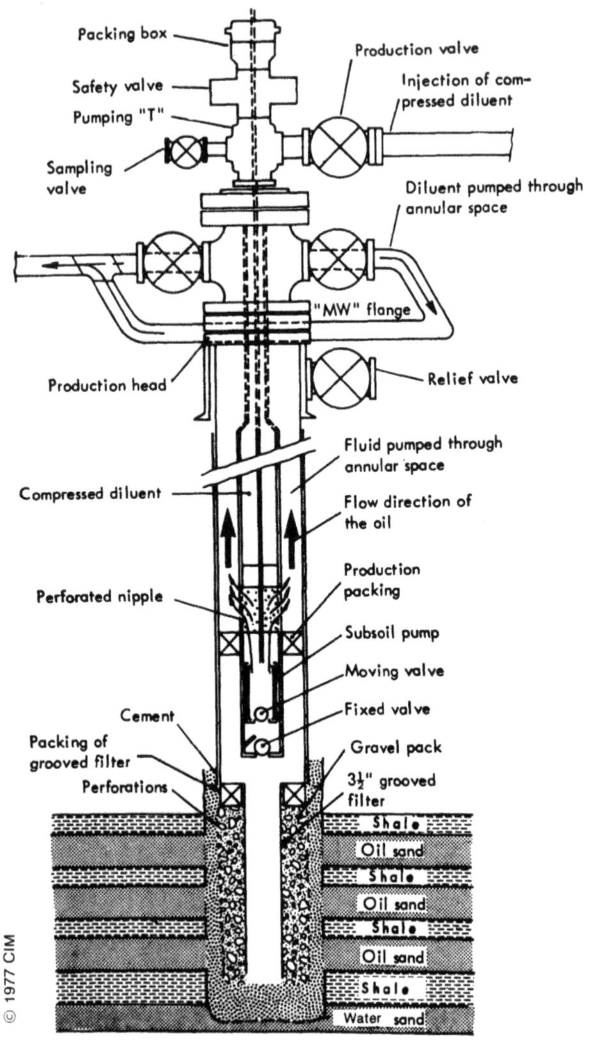

Fig. 11.15 – Jobo Field – typical well completion.[59]

produced crude by means of a topping unit, but lighter crudes also have been used.

Fig. 11.13 shows cross sections of a conventional diluent pump and two modifications used successfully to lift as much as 150 bbl of 4°API crude per day, with up to 70% of sand by volume. In combustion operations, wells producing combustion gases at high enough rates usually flow. Low bottomhole pressures may require the use of pumps, in which case gas anchors or some other type of downhole gas separators are desirable to achieve effective pump performance. Fig. 11.14, from a paper by Meldau and Lumpkin,[57] illustrates the advantage of using downhole separation to improve lifting performance (pump efficiency) in a combustion project in the Morichal field in Venezuela. Gas lifting is not efficient for viscous crudes, but steam lifting has potential. However, where steam is available, it also would be used to heat and stimulate production by means of cyclic steam injection. Production from cyclic steam injection usually flows at the outset and often can be maintained at acceptable levels by pumping or gas lifting. Where pumps are used, they sometimes are installed before the first steam injection, but they are left unseated so injected and produced fluids can be flowed through them until they are required for lifting.

Lifting production at elevated temperatures may present problems. Pumps may perform poorly, even where downhole gas separation is effective, because hot water may flash to steam on the intake pump stroke and cause vapor locking and reduced pump efficiency. Also, the pressure drop in the tubing when gas lift is used is difficult to anticipate because of the changes in temperature that occur as the fluids move toward the surface. The changes in pressure and temperature affect the viscosity of the crude, the relative amounts of steam and liquid water (through phase changes), and the flow regime (mist annular, transition to slug flow, etc.). As pointed out by Boberg *et al.*,[58] the performance of a gas-lifted or flowing well producing hot fluids is influenced by the total pressure drop in the tubing and flowlines.

Rod fall also has been improved by lifting through the annulus, with the rods in the tubing in contact with water under pressure. A downhole stuffing box just above a perforated tubing nipple isolates the tubing fluids. This system, reported by Ballard *et al.*,[59] is illustrated in Fig. 11.15. A similar one is discussed by Franco.[30]

Other methods used to help lift heavy crudes include the use of downhole heaters when crude viscosity can be reduced substantially, and emulsification of the crude to form low-viscosity readily pumpable oil-in-water emulsions.[54,60,61]

11.3 Surface Production Facilities

Separation

From the wellhead, the produced stream usually flows first to a separator or a free-water knockout vessel. This unit is designed to separate the free water and gas from the crude oil. Further processing may be necessary before the crude meets specified minimum sales (or refinery) requirements. The rate of separation of the phases is approximately proportional to their density difference and inversely proportional to the viscosity of the crude. The viscosity of heavy crudes may be reduced by heating, which increases the separation rate. This may require heating the separator if the production stream is not sufficiently hot. Density differences between water and crudes also are increased by heating, so that increased temperatures have a dual beneficial effect on separation. Another way to reduce the viscosity of the oil phase is to add diluents, either at the separator or in the well. The effect of diluents on density differences is usually small unless large concentrations are used. Another way to increase the density difference between the water and oil phases is to add salt to the produced water (but not if the water is to be treated for use in generating steam). Because of large differences between the liquid and gas densities, gas can be separated quickly. But for efficiency, the use of antifoamers may be found to be necessary.

The residence time of the production stream in the separator is another factor controlling the efficiency of free phase separation. In any case, simple separation as described is fairly effective on free fluids and coarse emulsions.

Dehydration-Demulsification

The crude stream from the primary separators is likely to contain water in emulsions in amounts too large to meet specifications. It then is passed through a second series of vessels to remove the remaining water. The water present at this stage is more difficult to separate because the droplets are dispersed more finely. To coalesce these droplets and ease their removal, a number of approaches[62-64] are used:

1. Chemicals (demulsifiers) are added to reduce the forces tending to stabilize the emulsion.
2. The rate of contact between droplets is increased to improve the rate of coalescence and eventual removal. This is accomplished by agitating the fluid, by forcing the fluids through packed beds of materials promoting spreading of water on their surfaces, and by creating electrical fields to induce dipole attraction between droplets.[65]
3. Diluents, chemicals, and heat are added to increase the density contrast between phases and to increase droplet mobility by reducing the viscosity of the crude.

Treatment vessels called heater treaters or wash tanks usually combine most of these approaches.[44,62,66] These vessels can be designed so that a single type can handle the production from most fields.

The degree of difficulty in removing water from crudes to acceptable levels is related to the stability of the emulsions. The stability of emulsions is controlled greatly by the properties of the crude. But emulsions produced in combustion operations are reported to be appreciably more difficult to break than those obtained in steam projects. Bertness[67] lists the following factors as contributing to the emulsion problems of combustion processes.

1. Low-temperature oxidation, which precedes initiation of a combustion front, adds to the surface activity of film-forming components of the crude. Oxidation of crude lowers interfacial tension, which can contribute to stability of emulsions.
2. The carbon dioxide generated by combustion may cause precipitation or growth of asphaltine colloids and resins in the crude oil.
3. The oxidized hydrocarbons, asphaltines, and resins bond to clays and silts and promote transport of these effective emulsifiers with flow of oil into the wellbore.
4. Iron sulfides are produced as corrosion products which stabilize water-in-oil emulsions.
5. Condensing steam ahead of a combustion front in the formation or in the producing facilities may be stabilized in micron and submicron droplets by the emulsifying agents present or generated in the crude oil. Lack of soluble salts in the water may aid stability of the interfacial film.
6. Acids generated in combustion processes form emulsifiers by reaction with the crude oil.
7. Increased energy in turbulent-flow gas lift of liquids creates emulsions which are stabilized by the multitude of emulsifying agents formed in the combustion process.

As a subclass of Item 7, it appears that inefficient pump action due to gas interference commonly leads to the formation of emulsions.

Disposal

As in any oilfield operation, solids, water, and gases must be disposed of properly. Typically, the wastewater stream itself can be expected to contain some solids, oil, and gas. Solids, which usually are contaminated with oil, include sand and silt sediments, inorganic scales, rust and complex metal-organic compounds, and sometimes organic solids.[37] The waste-gas stream may contain hydrocarbons, combustion products, and sulfur and nitrogen compounds. The only unusual disposal problems related to thermal operations are those resulting from the prevalence of hydrogen sulfide and from the considerable volumes of potential pollutants that must be handled. The treatment of these effluents varies markedly from one location to another, reflecting legal requirements and local practices.

For gases, condensation, incineration, and scrubbing are the usual procedures. Condensation may be cost effective where there are enough recoverable hydrocarbon liquids. Incineration is used to convert the hydrogen sulfide to sulfur dioxide, as well as to burn hydrocarbon gases. Burning the gas stream itself may not be possible or may not result in complete combustion of the components if the fuel value of the gas is too low. In such cases, additional fuel gas is added to the stream to attain the temperature necessary to complete combustion. Finally, scrubbing before venting to the atmosphere can reduce sulfur dioxide in the gas stream to acceptable levels. High stacks may be necessary to ensure an acceptable low ground-level concentration of contaminants. Where the volumes of the hydrocarbon gases are high enough, large separation plants may be built to recover them.

The method of wastewater treatment depends on the concentration of the contaminants and on the ultimate disposition of the water – whether to the surface or the subsurface. Subsurface disposal may require that the water be filtered, that oil be removed from it, and that chemicals such as bactericides be used to keep the injection wells from plugging. Furthermore, no effluent should be injected where there is the likelihood of contaminating sources of potable water. Oil and solids can be removed from a water stream by the use of filters,[68,69] flotation cells,[5] and retention basins. Hydrogen sulfide can be removed from the water by stripping with natural gas, by reacting it with oxygen in the presence of a catalyst,[70] and by frothing.[71,72] Wastewater sometimes is treated for use as feedwater for steam generators.[16] An example of the treating sequence used at Kern River field is reported by Smith.[13]

Solids usually are disposed of in pits or retaining basins. Depending on the nearness to habitation and sources of potable water and on the character of the solids, these repositories may be lined with materials to preclude or prevent drainage leakoff and rain runoff. Good practice is to cover these when they are full or out of use and to restore the land when the project is completed.

11.4 Surveillance

This section deals with surveillance operations – i.e., measurements aimed at monitoring the progress of the thermal operation in the reservoir. Measurements to monitor surface facilities, primarily for maintenance, are outside the scope of this monograph. Monitoring conditions related to health and safety is discussed briefly in Section 11.6.

The type and frequency of measurements vary with the type and purpose of the project. In a pilot project, where the purpose is to analyze the process to help decide about expansions, the tendency is to measure more than in a commercial project. But surveillance is important even in commercial projects.

Proper analysis of data is essential. When an inference is to be made from a measurement, it is essential to be aware of all phenomena that may affect the measured quantity. What is the cause of a high temperature in a pipe? When flow of hot fluids is anticipated, one may be ready to accept that flow as the explanation. But there may be no flow, and the high temperature may result from conductive heating. Or the flowing fluid may be relatively cold but be reacting and generating heat near the point where the measurement was made. An awareness of possible causes is essential for the proper analysis of apparently anomalous behavior.

Also, measurements provide not only direct but also indirect information. A wellhead temperature is direct information on the temperature in the wellhead at a producing well. But it may be used indirectly to infer bottomhole temperatures.

The accuracy and resolution of the instruments will affect the significance of the data and their interpretation. The condition and calibration of the instruments affect the reliability of the measurements. It is frustrating to have an instrument (or any other piece of equipment for that matter) malfunction over a critical period. Regular maintenance is essential.

Production and Injection Rates

Production rates are derived from the volumes produced or injected over a given time. It is important to recognize that in most cases, and especially in commercial projects, the liquid production is commingled from several wells before it is measured. It is customary to measure individual wells on a schedule – ranging from a few days to a few weeks – and then to allocate a fraction of the lease or block production to individual wells on the basis of their test performance. Such a procedure saves production equipment and is adequate to define trends in production rate at all wells. Where day-to-day information is required at a well, production facilities are provided for continuous use. It can be appreciated, then, that cumulative total volumes are known much better than individual well rates.

Where the output from one generator serves more than one well, there is no good way to determine the injection rate and quality of steam injected into each well. The causes for this difficulty are that steam contains both liquid and vapor, that the amount of liquid water increases as steam condenses because of heat losses, that it is difficult to measure two-phase flow rates, and that a header dividing a stream of steam to two or more wells may not divide both the vapor and the liquid equally.[73,74] However, it often is assumed that the quality of steam entering each branch off a header is the same. Then, if this assumption holds, orifice meters can be used to determine the steam mass flow rate into each branch and, thus, into each well. When the entire stream of steam is injected into a single well, the quality of the steam being injected at the wellhead may be calculated as described under Steam Generation in Section 11.1 or using other methods.[75,76]

Changes in the injection or production rate of a well (per unit pressure drop) are indicative of changes in the flow resistance, which may be caused by impairment or cleanup around the well, by fluid redistributions (such as oil banks or channeling) in the reservoir, or by mechanical problems, such as tubing, packer, or casing failure. Changes in the produced water/oil and gas/oil ratios are affected by channeling, relative permeability effects, and changes in bottomhole temperature and pressure.

In addition to the volumes and rates of each phase, the composition of each phase can impart significant information. Gas analyses provide information on the efficiency of combustion in the reservoir and on potential health and safety hazards. In steam drives and soaks, gas analyses provide information on the possible importance of gas expulsion as a production mechanism. Changes in the amount of light ends are indicative of cracking, distillation, and condensation phenomena. Changes in the composition of the produced water herald the arrival of injected water. The arrival of combustion and other fronts sometimes is preceded by changes in the requirements for breaking emulsions and separating fluids, in the color and smell of the produced stream, in the pH and salinity of the produced water, in the acid number and other properties of the produced crude, and in increased oxygen content.[27,77]

This discussion has considered events at only a single injector or producer. Differences in the performance of nearby wells may be attributed to (1) local differences in reservoir characteristics and initial fluid contents, (2) anisotropy in properties (perhaps a preferred fracture orientation), (3) effects of dip, (4) differences in well completion, operation, or impairment, and (5) differences in the location of fluid fronts in the reservoir.

The fluid measurements discussed here are not a complete listing of the information obtainable about the reservoir process. But they give a reasonable idea of what can be expected from typical measurements.

Temperatures

Surface and subsurface temperatures are measured in thermal operations to obtain (1) the amount of heat injected and the size of the heated reservoir volume, (2) information on the rate of advance and

TABLE 11.3 – TYPES OF THERMOCOUPLES, AND TEMPERATURES RANGES IN WHICH THEY ARE USED[102]

	Usual Temperature Ranges		Maximum Temperature	
Type	°C	°F	°C	°F
Chromel-alumel	– 200 to 1,200	– 300 to 2,200	1,350	2,450
Iron-constantan	– 200 to 750	– 300 to 1,400	1,000	1,800
Copper-constantan	– 200 to 300	– 300 to 570	600	1,100

distribution of heat in the reservoir (when measured downhole at other than an injection well), (3) information about stresses to which mechanical equipment is subjected, and (4) estimates of the thermal efficiency of steam generators and surface facilities.

Downhole temperatures seldom are measured continuously in commerical operations. In pilot and experimental field projects, especially in combustion operations, both continuous and spot readings often are obtained.

Both fixed and movable thermocouples are used to measure downhole temperatures. Thermistors seldom are used because of their temperature limitations. The range of use of typical thermocouple systems is given in Table 11.3. The more briefly a thermocouple is used at high temperatures, the longer it will last. Thermocouple wires may be sealed in a solid flexible metal tubing, typically ¼-in. OD, to keep out moisture and reactants. The thermocouple wires themselves are insulated from each other and from the sheath by a layer of packed, inert, electrical insulating powder such as magnesium oxide. The thermocouple junction sometimes is welded to the end of the metal sheath to improve the response time, but normally events move slowly enough that this is not required. To reduce costs, thermocouple extension wire may be used for the upper part of the thermocouple system in the unheated uphole section. The sheath itself is normally available in a wide choice of metals capable of withstanding most of the environments in downhole thermal operations. For ease of operation and to provide direct readings, temperature-compensated recorders generally are used.

A thermocouple is fixed when it is clamped and welded to a downhole tube. Temperature then is read as a function of time. Movable thermocouples can provide temperature profiles across the formation. They usually are run in a separate blind or open tubing string inside wells but also may be run in a cemented string. Downhole temperature measurements are affected by contact resistance and heat conduction of nearby tubing,[78] which tend to smear any sharp temperature profiles that might exist in the reservoir. Smearing reduces the significance of the temperature readings. Profiles also are smeared by any free convection of fluids in the well. Free convection in blind thermowells can be reduced by using special, high-viscosity, low-thermal-expansion fluids that decompose only slowly at the expected temperatures. Any significant amount of water in the tubing usually will result in temperatures equal to that of saturated steam at the prevailing downhole

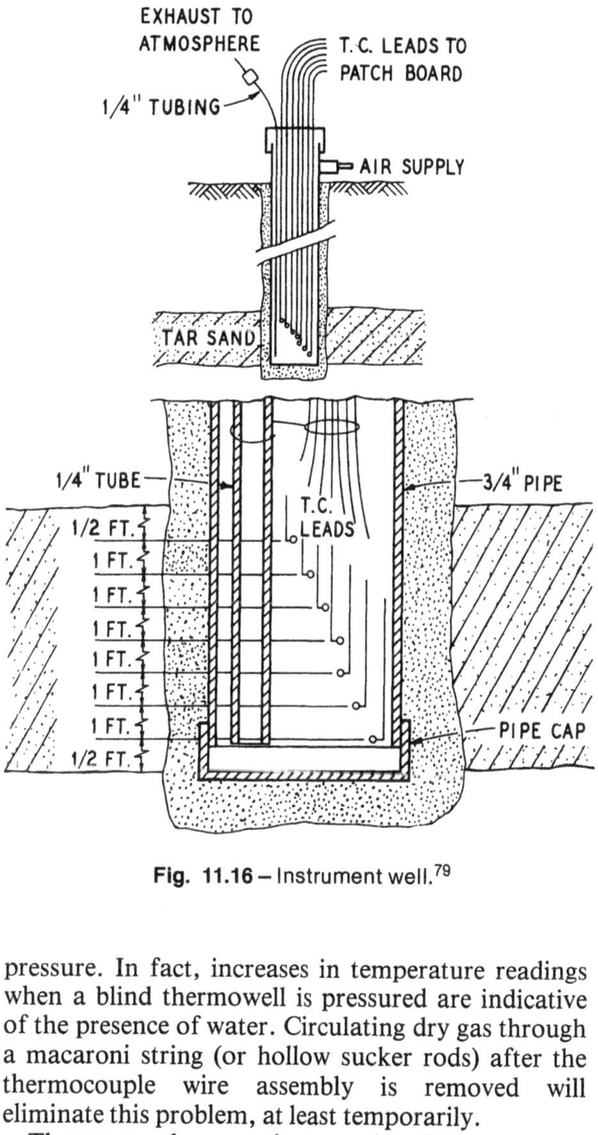

Fig. 11.16 – Instrument well.[79]

pressure. In fact, increases in temperature readings when a blind thermowell is pressured are indicative of the presence of water. Circulating dry gas through a macaroni string (or hollow sucker rods) after the thermocouple wire assembly is removed will eliminate this problem, at least temporarily.

Thermocouples sometimes are bundled to prevent their relative movement. Fig. 11.16 is the schematic of an observation well in which a ¼-in. open-ended tube was used to run a single thermocouple to obtain a continuous vertical traverse and a bundle of thermocouples was used in a capped and cemented pipe to record temperatures at fixed depths.[79] Thermocouples often are spring-loaded to improve thermal contact with the tubular string and are weighted to allow free fall.[20] Figs. 11.17 and 11.18 show schematic completions of production wells with capped thermowells inside[30] and outside casing,[80] respectively.

Vertical temperature profiles in active projects are excellent indicators of where the injection fluid is

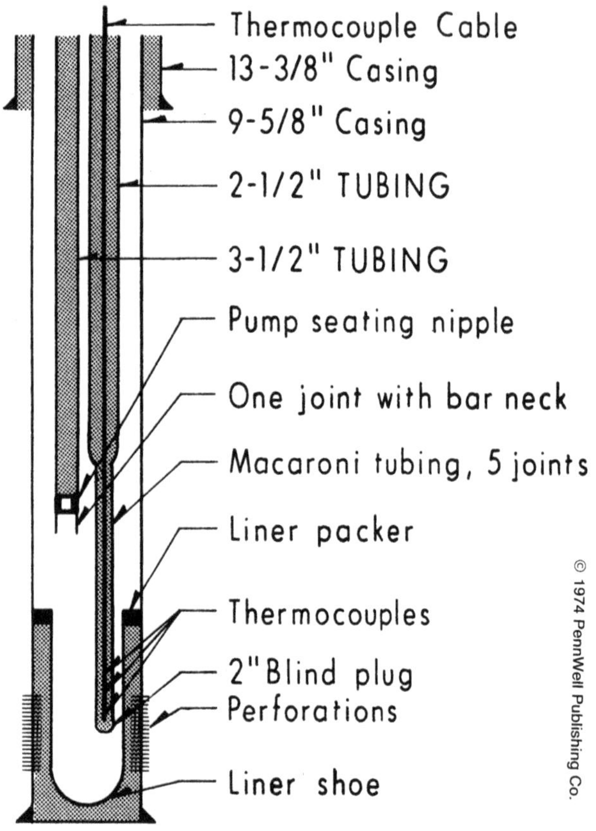

Fig. 11.17 – Completion of fireflood producing well with thermowell inside the casing.[30]

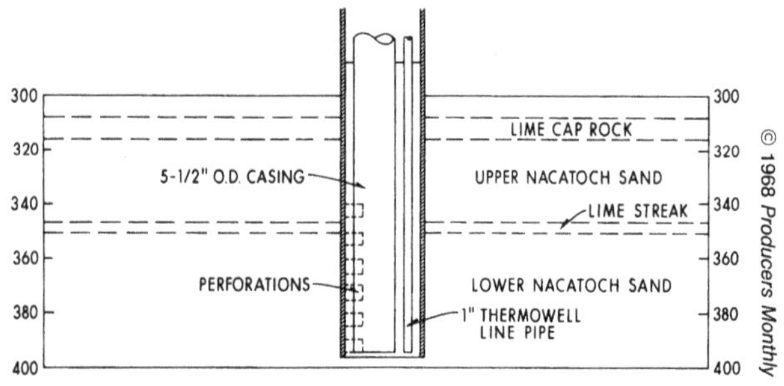

Fig. 11.18 – Completion method for a producer, using thermowell outside the casing.[80]

moving. They can indicate if gravity override is important in combustion and steam operations and reveal thief zones. Changes with time are particularly suggestive of the progress of the thermal operation in the reservoir. They also give the time of arrival of the heat front at various wells and, thus, indicate imbalances in the rate of heating among several directions and may suggest changes to promote a more balanced flood.

A less frequent use of temperature profiles is to determine the injection profile in combustion projects.[81]

Pressures

Pressure measurements are used during thermal operations to provide (1) information on the injectivity and productivity of wells and flow resistance in the reservoir, (2) confidence that equipment and wells are being operated at design pressure levels and within safe pressure limits, (3) information on the enthalpy of fluids, and (4) information about properties and conditions existing in the reservoir through the use of diagnostic well tests and continuous pressure observations.

Pressures are measured routinely at the wellhead and surface lines with Bourdon-type gauges. They also can be recorded on strip or circular charts. Pressures at downhole conditions in thermal operations are not measured routinely. At low pressures, a proved method for measuring downhole pressure is to use a small-diameter tubing run to the desired depth through the wellhead. In one method, a small slug of gas is injected down the small tubing and the stabilized pressure is noted. The process is repeated until the stabilized readings repeat, indicating that the gas fills all of the small tubing. The downhole pressure at the end of the tube is the stabilized pressure read at the surface plus a correction due to the weight of the gas column. At shallow depths, which correspond to low pressures, the correction is small. Because the gas in the tube is static, there are no pressure losses due to friction. Thus, the average density can be calculated quite accurately using the pressure-temperature-density properties of the gas. A temperature profile in the well (discussed in the preceding section) would improve the accuracy of the correction even more. The use of helium, which has about one-seventh the density of air, is indicated where it is desirable to reduce the numerical value of the correction term. Equipment for measuring downhole pressures based on this principle has been available commercially for some years.

Petrophysical Evaluation and Logging

Although it is generally inconvenient and even difficult to do so routinely, opportunities sometimes arise for evaluating the petrophysical properties of heated sections of the reservoir during a thermal operation. This is done either by obtaining cores or by using wireline logs.

Opportunities for doing this arise when wells must be replaced or new wells are required in or near the heated zone. Coring must be done under controlled conditions, which means that the interval must be cooled to safe operating levels. Fluid losses to the reservoir during coring may affect fluid saturations near the wellbore, and flushing also may affect the saturations in the cores.

Core information obtained in the course of a project may provide tangible evidence of burned or swept zones and of changes in fluid and rock properties. Sidewall core plugs taken during workover operations may provide similar information.[28]

The practical use of wireline logs is affected by the temperature limitations of the logging tools, which is about 350°F for the standard ones. However, there are a few high-temperature versions (to about 500°F) of most types of logging tools, and they can be special-ordered from the service companies. Although one can obtain logging tools that will operate to above 350°F, we recommend that the user inquire in detail about calibrations and interpretation procedures for high temperatures if at least normal accuracy is desired. The radioactive logs most commonly run are neutron logs, which provide estimates of the gas saturation profile near the wellbore. If the salinity of the formation water remains high (50,000 ppm and higher), meaningful oil saturation data can be obtained with a pulsed-neutron type log through casing. At lower salinities, special interpretation techniques must be used to distinguish signal from background noise. Carbon/oxygen logs, which also can be run through casing and are not affected by brine concentrations,[82] could provide information on hydrocarbon content. Resistivity logs, which require an open hole or the use of nonconducting pipe, normally cannot be run except when the formation is well consolidated and can be completed openhole or during the period between coring and completing replacement wells.[28] Fig. 11.19 compares the neutron log response taken during a steam drive with other log responses obtained before the project was started. Preferably, comparisons should be between logs of the same type.

Another type of log frequently run, especially in injection wells, is the spinner survey. This survey indicates the relative amounts of the injected stream entering an injection interval as a function of depth. Fig. 11.20 shows spinner surveys in a steam injection well completed separately into each of two zones.[83] But normally the significance of spinner surveys is highly questionable in determining the relative amounts of fluid entering several zones where downhole separation of liquids and vapors is likely. Radioactive and temperature surveys also have been used to determine injection and production[81] profiles. Radioactive-tracer surveys are now more common than spinner surveys.

Well Tests

A few well tests have been applied to find specific information about combustion operations. Van Poollen[84] reported the use of both injectivity and pressure falloff tests at the Fry project. In-

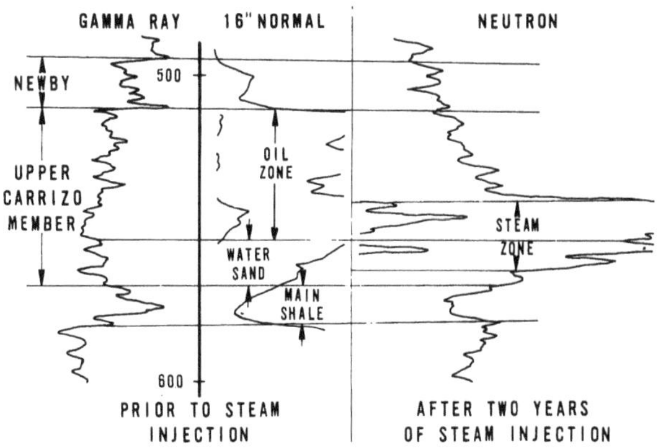

Fig. 11.19 – Steam zone delineation with neutron logging.[37]

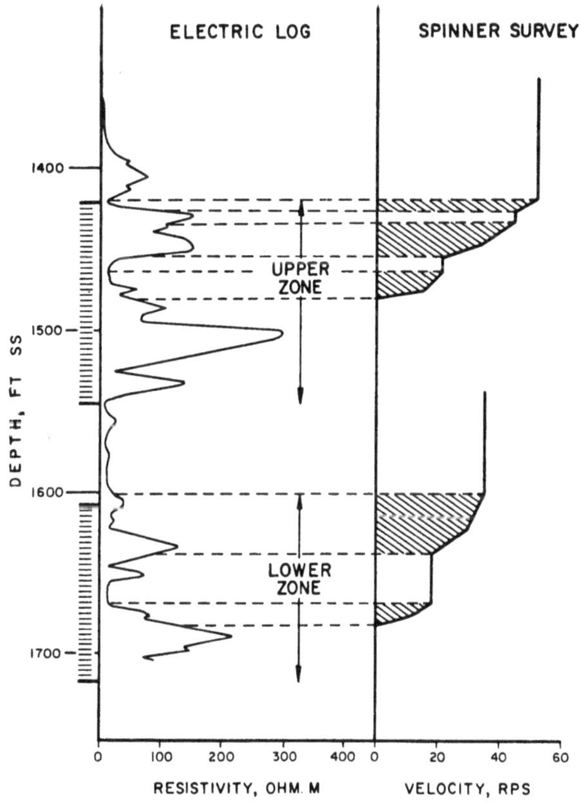

Fig. 11.20 – Typical steam injection profile.[83]

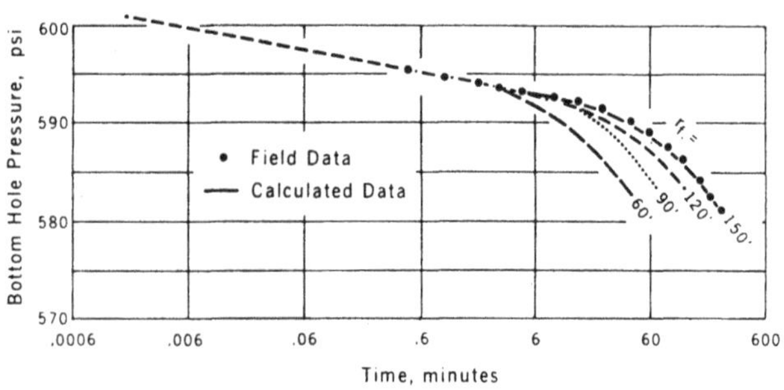

Fig. 11.21 – Comparison of calculated and measured pressure falloff curves.[85]

terpretation was based on a highly idealized model of the combustion process. The injectivity test was aimed at obtaining information about the impairment and reservoir transmissivity in the vicinity of an injection well. The pressure falloff tests were aimed at determining the static pressure and the distance to the combustion front as a function of time. Kazemi,[85] in a study of the Delaware-Childers pilot combustion project, introduced more realistic assumptions about combustion – assumptions that required intermediate numerical simulation calculations to interpret the results of pressure falloff tests. Fig. 11.21 shows the pressure falloff data obtained in this field test and that calculated for different positions of a radially cylindrical combustion front coaxial with the injection well. Interpretation of transient pressures from injection wells has recently been extended by Walsh *et al.*[86]

Well tests at hot production wells generally have been limited to attempts at determining conditions under which maximum oil production can be obtained. This type of test would be applicable only to wells in which the producing temperatures are high enough to flash water to steam if the bottom pressure is too low. This flashing tends to reduce the temperature and liquid permeabilities and to increase the crude viscosity. Thus, both the flow resistance and drawdown increase. The lifting conditions to obtain the best productivity are sought in this kind of well test. More conventional well tests[87,88] seldom are carried out in production wells that have responded thermally, probably because of the difficulties associated with properly carrying out and interpreting the test.

11.5 Operational Problems

The best solution to a problem is to prevent it from arising. Yet to ensure the prevention of any potential problem would require that a project not be carried out at all. Obviously, one must balance the risks and rewards to strike a compromise in the way a project is operated, always allowing for prudent operating practices and health, safety, and environmental factors.

This section discusses some of the more common problems in thermal projects, ranging from social ones, such as those related to environment and ecology to technical ones, such as tubing failure.

An integral part of many thermal projects is an automatic system for monitoring and displaying measured data, analyzing data to indicate the status of individual wells, scheduling production tests, sounding alarms, and even taking corrective action when malfunctions occur. This type of system, popularly known as SCAN (sample, control, and alarm network), is particularly warranted in projects having a large number of wells and pieces of equipment to be tended.[89]

Meeting Environmental and Ecological Needs

As responsible individuals and companies, we are vitally interested in maintaining and even improving the quality of our environment. The evolution of stringent legal requirements has added further incentive to develop low-cost processes for disposing of waste streams and refurbishing the surface of the land in a socially acceptable manner when a project is completed. Legal requirements and social acceptance vary widely from country to country and even within countries.

The two major sources of gas streams in thermal projects are the production wells and the smoke stacks of steam generators and air compressors. The streams may be treated separately or together, depending on the size of the project and the nature of the contaminants. The principal contaminants in these streams are hydrogen sulfide, carbon monoxide and hydrocarbons in the produced gas, and sulfur and nitrogen oxides in the stack gas. In 1965, Koch[90] reported that the contaminants in the produced gas from the West Newport combustion project were oxidized to carbon and sulfur dioxide in a gas-fired incinerator maintained at greater than 1,650°F. The resultant stream, including sulfur dioxide in concentrations of about 1%, then was vented to the atmosphere as "a colorless and odorless gas." It is now (1981) customary to pass the incinerated gas through a caustic scrubber (using sodium carbonate or sodium hydroxide) to remove the sulfur dioxide. Sulfur dioxide also has been scrubbed with oilfield-produced water.[91,92] Liquid scrubbers have the additional advantage of removing most of the particulates in the gas stream, especially when applied to the stack gas from oilfield steam generators. Several systems are now in use or being considered to remove sulfur compounds and other contaminants from the waste gas resulting from field operations.[93]

Casing gas produced from cyclic steam injection projects sometimes is collected and condensed,[69] yielding between 0.1 and 10 bbl of condensate per well per day. The noncondensable gas, which contains methane, sometimes is used as supplementary fuel for the steam generators. Also, steam produced along with the casing gas can be used to preheat the stream entering the first stage separations units or in other preheating applications.

Equal care is taken in disposing of waste water and produced solids.

Emulsions

No field project is known to have been abandoned because the produced emulsion could not be broken

TABLE 11.4 – FIELD A: STABILITY OF EMULSIONS FROM VARIOUS PRODUCING OPERATIONS[67]

Sample	Gravity (°API)	Days After Steam	Wellhead Temperature (°F)	Sample Temperature (°F)	Production Gross/Net (B/D)	Free Water (%)	Emulsion Cut (%)	Chemical Treatment (Qt./100 bbl)	Stability (V)
Well 1	13.5	7	225	175	160/60	0	60	None	480
								2	200
Well 2	14.3	21	190	138	160/102	0	36	None	520
								2	120
Well 3	13.9	36	175	138	180/140	0	22	None	540
								2	120
Well 4	13.6	55	150	130	300/78	34	40	None	440
								2	135
Well 5	13.7	197	110	110	20/9	25	30	None	400
								2	90
Production sump	13.6	–	–	85	–	–	32	None	600+

economically. Nevertheless, there are emulsions that require more effort to break than others. These are generally emulsions produced in combustion operations, although the stability of emulsions produced from steam projects also varies widely. Apparently, the character of the emulsion is affected not only by the properties of the crude but also by the operating environment.[67] Emulsions can be water-in-oil, oil-in-water, or both. And the discontinuous phase may itself be an emulsion.

The character of the emulsion produced from a well changes during production. For example, Table 11.4 shows the amount of emulsion and free water produced as a function of time from cyclic steam injection wells in a California field, together with the voltage required to break the emulsions. Note that the amount of emulsion tends to decrease with time and the free water tends to increase, whereas the stability of the emulsion (as measured by the voltage required for separation) decreases with time when a demulsifier is used but goes through a maximum when no additive is used. This means that the demulsification treatment should be altered during the production period to reflect the properties of the emulsion.

In some cases, this is indeed the procedure. Holke and Huebner[43] report that emulsions produced from the several wells in the hot-water stimulations at the Northeast Butterly Oil Creek Unit are significantly different in concentration and stability and that they were treated with demulsifying agents individually. Hall and Bowman[37] report a program for continuously monitoring the changing demulsification requirements in the steam drive at Slocum. At the Bellevue[94] combustion project, a "difficult emulsion coming from a single well can upset the entire treating system." Offending wells are treated individually. But individual wells seldom receive special treatment except to avoid upsets in the treating facilities or where resultant cost savings are important.

The rate at which emulsions can be broken affects the size of the surface facilities required to do the job. Obviously, there are economic incentives to reduce those facilities to a minimum. This can be done at the expense of using (1) high temperatures and solvents to reduce the crude viscosity and the separation time, (2) increased electric voltages, and (3) relatively expensive demulsifiers. The challenge is to achieve a compromise between facilities and operating costs, with some intangible value given for a system that will work trouble-free with little attention after the initial design and test period.

Foams, or gas emulsions, generally do not affect the acceptability of the crude directly but are difficult to dispose of and may hinder the proper operation of the dehydration units. They are likely to be found in combustion rather than steam projects. At Bellevue field[94] the foam is diverted to mist extractors, from which the liquid is gravity-fed to a settling pit. The liquid is tightly emulsified and contains sand, coke, iron sulfide, and other particles that tend to provide stability. Best results have been obtained by introducing an emulsion breaker downhole.

Sand Control and Production

Sand production is one of the most widespread and costly problems associated with thermal recovery projects. In any operation, sand and formation fines brought from the reservoir by the produced fluids may (1) impair production by filling the wellbore, (2) reduce production by preventing downhole pumps from operating properly, (3) require the replacement of equipment because of erosion damage, and (4) require costly workovers. In thermal operations, any well work (such as replacement of worn pumps) and workover would require cooling at least the immediate vicinity of the wellbore. The reduced temperatures may result in temporary reduction in oil production rate.

Fazio and Banderob[66] have explained how sand control techniques are selected before cyclic steam injection projects are initiated. Their study of Midway-Sunset field discusses the handling of existing wells, casing and liner completion practices in new wells, size of tubing, types of liner slots, the use of expansion joints, and some advantages and disadvantages of gravel-pack completions.

Reports have been published on subsurface pumps designed to handle relatively high sand content in the produced liquids.[45,56,80]

Although controlling sand is annoying and expensive, it may take years to find a reasonable method for doing so, especially where there is no experience from similar operations nearby. Holke and Huebner[43] describe a number of methods that were tried before an acceptable one was found for controlling sand in a hot-water stimulation project. Bleakley[39] reports similar efforts in cyclic steam injection operations in Cat Canyon.

Casing and Tubing Failure

Downhole casing and tubing in thermal projects may fail for a variety of reasons, including (1) buckling, (2) tension, (3) erosion, (4) corrosion, and (5) melting.

The avoidance of buckling, shearing, and tension failures (which usually occur while hot wells are cooling) through proper design and operational procedures has been discussed in Section 11.2. Liner failures due to compression and tension in cyclic steam injection operations have been reported by de Haan and van Lookeren.[95] Parted casings also have been reported.

Erosion failures are related to inadequate sand control and can be particularly severe in combustion operations. In such projects the gas enters the wellbore at low pressures and consequently high velocities. This is especially the case where the gas-producing interval is limited (e.g.,because of gravity override or where only a few perforations or slots are open to gas flow). The accepted preventive measure to reduce sand cutting is gravel packing. But sand cutting may remain a problem even when gravel packs are installed. In one field combustion project, Dietz[96] reported sand cutting to be more pronounced at gas rates above 700 Mscf/D. It is usually desirable to use liner and tubing made of abrasion-resistant steel opposite the production interval. Abrasion appears to aggravate corrosion, apparently by providing a clean surface where corrosion can occur as the corrosion products are removed.

Corrosion in thermal operations is generally most severe in combustion operations at production wells, especially at elevated temperatures and under erosive conditions. Corrosion field tests reported by Gates and Holmes[1] indicate that stainless steel 316 suffered much less weight loss than mild steel, which had a 73% weight loss over a 2-month period at the test conditions (which included temperatures maintained at 450°F or lower). But stainless steel 316 may not prove adequate everywhere because of its susceptibility to chloride attack. The choice of steel requires careful consideration. Where it is used, stainless steel tubing generally is restricted to the production interval because of its rather high cost. Not all operators use special steels, and the results have been mixed.[57,69,97,98] Corrosion is affected by a number of operating variables and conditions, including the carbon dioxide, oxygen, and hydrogen sulfide concentration of the effluent gas, the nature of the liquid stream, the temperature and pressure of the system, and rates of flow. Combustion reactions leading to corrosive environments have been discussed by Burger and Sahuquet.[99]

At injection wells, corrosion is generally significant only in wet combustion operations.[4,27,96] Some solutions include using separate strings for injecting air and water (or alternate injection of these fluids), removing moisture from the injected air, injecting oxygen-free water, and adding corrosion inhibitors to the water.

Corrosion is not generally a factor in steam injection projects. In rare cases, however, sulfide cracking may weaken the tubing to the extent it cannot sustain the changes in thermal stresses caused by cyclic steam injection.

Maximum temperatures in excess of 1,200°F (lower range of red heat) have been measured on a number of combustion projects, at both injectors and producers. At those temperatures, the mechanical strength of tubing is reduced greatly and often leads to failure. There have been reports of tubular steels melting in both injection and production wells in combustion projects.[1] This implies local temperatures well in excess of 2,000°F. At injection wells, these high temperatures invariably result from crude burning in the borehole during or even after ignition. As air is being injected, the crude flows into the borehole by gravity – especially in thick or multiple sand intervals and when the crude is hot. Preventive measures to reduce the likelihood of burning crude in the injection well include the following.

1. Artificially ignite the crude by means of a downhole heater. This may be difficult to achieve where spontaneous ignition is likely to occur and air injection is required before a project is started.
2. Limit the number of perforations and perforate through cemented casing to minimize gravity drainage.[49]
3. Limit the number of perforations at the top of the sand to minimize drainage. This approach appears useful where gravity override is expected to develop anyway.
4. Automatically inject cooling water (with or without air) on sensing a predetermined temperature level downhole.[49]
5. Inject steam to displace the bulk of the crude from near the wellbore and provide liquid water that would tend to keep temperatures to those of saturated steam.

High temperatures at production wells result from combustion in the reservoir and in the wellbore. In the latter instance, some oxygen must enter the wellbore. Since it is difficult to keep free oxygen and hot crude from reacting, the accepted solution is to inject water downhole to reduce the temperature and slow the reaction. If free oxygen is produced into a wellbore at a constant rate, the generation of heat downhole will be at a minimum when the oxygen content in the produced stream at the surface is at a maximum. It has been proposed that the rate at which cooling water is injected be adjusted to achieve this maximum concentration.[1,100] Some crude still may be burned downhole, but the temperature of the system is controlled. As can be appreciated, the calculated oxygen consumption efficiency for a field project may be somewhat high when based on surface measurements and where oxygen is reacted downhole.

Another operational problem is casing failure,

which could result, for example, in steam blowing out behind the casing and to the surface. Such casing failures may occur in old wells that are not cemented to the surface and that penetrate shallow depleted sands. Such steam blowouts can be difficult to bring under control. Adequate wellhead, pump, and tubing design can help retain control of the well in such circumstances. But casing failures must be corrected. Of course, inadvertent steam losses to shallow, depleted sands can reduce the overall thermal efficiency of a project severely.

Impairment

Mechanisms that impair flow are generally different at injection and production wells. At injection wells, flow may be impaired by a slow buildup of compressor lubricating oil, moisture in the compressed air, separation of liquids and vapors in steam injection wells, and particulates such as initial dirt on pipes, corrosion on tubing, scale loosened from steam generators, or loose fines from exposed overlying sands. In addition, feedwater going into heaters and steam generators may be filtered incompletely; perforations may fill up with sand and silt (this may be aggravated by pressure surges resulting from temporary adjustments in flow rates); inadequate sand control during shutdowns may cause the inflow of sand; and high temperatures, corrosion, or improper design or placement can cause a sand-control system to fail, allowing the inflow of sand.

Showalter and MacLean[49] discuss the impairment encountered in wells completed with limited-entry techniques to control the air injection profile in a combustion project. (See Fig. 11.22 for a schematic of the injection well completion.) But Gates *et al.*[48] did not report difficulties using a similar completion at South Belridge field. At Moco field, Gates and Sklar[46] report using "a slotted inner liner over the perforated interval to prevent sand from backflowing should equipment or piping fail." Generally, one would expect that a limited-entry completion would be more sensitive to impairment than other types. Where the method is used, each perforation generally is washed and tested.

Treatment is tailored to the source of impairment. Preventive measures include using condensing traps, other methods of separation, and filters[27] in air injection lines; cleaning tubing before injection is started; and, in wet combustion projects,[27,96] using inhibitors in the injection water. Acid, solvent washes,[27] and workovers may be necessary after impairment has become manifest. Acid washes must be planned carefully to avoid introducing undesirable side effects that might lead to impairment later.[49]

At production wells, impairment results primarily when particulates plug the reservoir near the well, the perforations or slots in the tubing, the gravel pack, or the wellbore. These particulates include organic precipitates,[37,67,96] fines loosened by flow, scale that results when water flashes to steam, precipitated carbonates, and coke resulting from high-temperature pyrolysis of crude. In the wellbore, loss of the sand-control measures resulting from corrosion or high temperatures may lead to a sandfill. Again, acid and solvent washes and workovers are the standard remedies. Preventive steps include properly designing and completing wells and keeping the well relatively cool in combustion operations. A high saturation bank of cold viscous oil near a producer also can cause a reduction in production rates. This decline in productivity often can be corrected by steam stimulation.

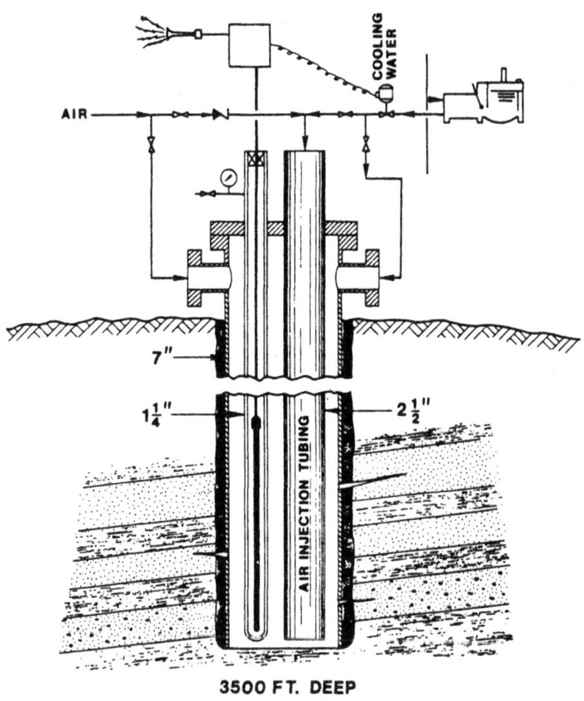

Fig. 11.22 – Schematic of the air injection wells, Stearns fireflood, Brea-Olinda field.[49]

A decline in production rate does not necessarily indicate flow impairment. The decline may arise from decreased lifting efficiency and change in three-phase relative permeabilities as the conditions in and near the wellbore change with time. This usually results when the producing temperature increases or the downhole producing pressure decreases and is accompanied by hot water flashing to steam either in the reservoir or in the pump.

In addition to treating and correcting the problem, conventional stimulation techniques such as hydraulic fracturing may be used.[30,77,101]

11.6 Health and Safety

The importance of health and safety, both to communities surrounding the projects and to operators, has been recognized in thermal operations from the start. In particular, the presence of high-pressure gas lines, some at temperatures higher than are found in conventional field operations, was recognized as a potential safety hazard. The presence of hydrogen sulfide – a matter that industry had had considerable experience with – also was recognized as a health and safety hazard. An awareness of potential health and safety hazards grew out of early tests, which were small and carried out at relatively low pressures. By the time high-pressure installations were ready to operate, previous experience and data published in the open literature made it possible to design such thermal projects adequately and prudently.

Preventive measures and fail-safe systems have been installed to (1) automatically cool wells[89] and wellheads that become too hot, (2) sound alarms when hydrogen sulfide and other contaminant concentrations reach unacceptable limits, (3) reduce hazards associated with high-pressure air compression through the careful choice of lubricants and lubricating adjustments,[28] and (4) regularly monitor surface equipment for signs of possible loss in integrity. Furthermore, it is common practice to provide training and safety instructions to the ever-changing personnel associated with thermal projects through "periodic safety meetings and project performance conferences attended by key operating personnel and engineers."[46]

Completion and operating guidelines have been offered by Gates and Holmes[1] to help maintain a safe combustion operation in unconsolidated sands, where production may continue long after the arrival of the combustion front. Their procedures include the following.

1. Use a gravel-pack to minimize abrasion.
2. Use stainless steel 316 for both liner and tubing opposite the productive interval to minimize hot corrosion.
3. Land tubing within 3 ft of bottom, after allowing for thermal expansion, to permit controlling high temperatures with cooling water.
4. Add cooling water to the casing continuously after an increase in surface temperature is observed.
5. Regulate cooling water rate so the oxygen content of the produced gas is at a maximum.
6. Replace the tubing at once if a hole is detected. A sudden increase in the gas/oil ratio is cause for testing the tubing for holes.
7. Halt the production of gas and liquids through the casing long before the well becomes hot.
8. Kill a hot well by pumping liquid into the casing rather than by venting gas off the casing.

Items 5 and 6 help reduce the likelihood that a well will be lost because of crude burning in the wellbore.

Special procedures are followed in preparing a hot well for a workover. These include procedures for killing the well,[1] for installing blowout preventers both to control pressures and to keep noxious gases off the workers,[90] and for ensuring that the mud does not lose its desired properties during the workover.[28]

References

1. Gates, C.F. and Holmes, B.G.: "Thermal Well Completion and Operation," *Proc.*, Seventh World Pet. Cong., Mexico City (1967) 419-429.
2. Giusti, L.E.: "Experiencias de la C.S.V. con la Inyección Alternada de Vapor en la Costa Bolívar, Estado Zulia," *Proc.*, Simposio Sobre Crudos Pesados, Inst. de Investigaciones Petroleras, U. del Zulia, Maracaibo, July 1-3, 1974.
3. Burkill, G.C.C. and Leal, A.J.: "Aspectos Operacionales de la Inyección Selectiva de Vapor en Pozos de la C.S.V. de la Costa Bolívar, Estado Zulia," *Proc.*, Simposio Sobre Crudos Pesados, Inst. de Investigaciones Petroleras, U. del Zulia, Maracaibo, July 1-3, 1974.
4. Rincón, A.: "Proyectos Pilotos de Recuperatión Térmica en La Costa Bolívar, Estado Zulia," *Proc.*, Simposio Sobre Crudos Pesados, Inst. de Investigaciones Petroleras, U. del Zulia, Maracaibo, July 1-3, 1974.
5. Noran, D.: "Enhanced Recovery Requires Special Equipment," *Oil and Gas J.* (July 12, 1976) 50-56.
6. McLennan, I.C.: "Raising of High Pressure Steam from Brackish Lake Maracaibo Water," *J. Inst. Petroleum* (Aug. 1968) 212-232.
7. Burns, W.C.: "Water Treatment for Once-Through Steam Generators," *J. Pet. Tech.* (April 1965) 417-421.
8. "Conditioning Water for Boilers," Nalco Chemical Co., Chicago (1962).
9. *Betz Handbook of Industrial Water Conditioning*, Betz Laboratories Inc., Trevose, PA (1962).
10. *Steam – Its Generation and Use*, 37th edition, The Babcock and Wilcox Co., New York City (1963).
11. Bradley, B.W. and Gatzke, L.K.: "Steam Flood Heater Scale and Corrosion," *J. Pet. Tech.* (Feb. 1975) 171-178.
12. Roberts, J.C. and Williams, J.W.: "Steam Injection, Oil Recovery, Techniques and Water Quality," *J. WPCF* (Aug. 1970) 1437-1445.
13. Smith, M.L.: "Waste Water Reclamation for Steam Generator Feed, Kern River Field, California," paper SPE 3689 presented at the 1971 SPE California Regional Meeting, Los Angeles, Nov. 4–5.
14. Elias, R. Jr., Johnstone, J.R., Krause, J.D., Scanlan, J.C., and Young, W.W.: "Steam Generation with High TDS Feedwater," paper SPE 8819 presented at the 1980 SPE California Regional Meeting, Los Angeles, April 9–11.
15. Fanaritis, J.P. and Kimmel, J.D.: "Review of Once-Through Steam Generators," *J. Pet. Tech.* (April 1965) 409-416.
16. Hull, R.J.: "The Thermosludge Water Treating and Steam Generation Process," *J. Pet. Tech.* (Dec. 1967) 1537-1540.
17. Wendt, R.E.: "Review of Stack Gas Scrubber Operation Experience on an Oil Field Steam Generator," paper SPE 7125 presented at the 1978 SPE California Regional Meeting, San Francisco, April 12–14.
18. Owens, M.E. and Bramley, B.G.: "Performance of Equipment Used in High-Pressure Steam Floods," *J. Pet. Tech.* (Dec. 1966) 1525-1531.
19. Trujano, D.N., Garza, B.T., and Lecona, S.C.: "How Ambient Conditions Affect Steam-Line Heat Losses," *Oil and Gas J.* (Jan. 21, 1974) 83-86.
20. Martin, W.L., Dew, J.N., Powers, M.L., and Steves, H.B.: "Results of a Tertiary Hot Waterflood in a Thin Sand Reservoir," *J. Pet. Tech.* (July 1968) 739-750; *Trans.*, AIME, **243**.
21. Bursell, C.G., Taggert, H.J., and Demirjian, H.A.: "Thermal Displacement Tests and Results, Kern River, California," *Prod. Monthly*, (Sept. 1966) 18-21.
22. Socorro, J.B. and Reid, T.B.: "Comportamiento de Equipos. Proyectos de estimulación cíclica con agua caliente – Campo Morichal, Estado Monagas," *Proc.*, Third Soc. Venezuela Eng. Pet. Tech. Petroleum Conference, Maracaibo (1972) **1**, 361-368.
23. Burke, R.G.: "Combustion Project is Making a Profit," *Oil and Gas J.*, (Jan. 18, 1965) 44-46.
24. *Engineering Data Book*, ninth edition, Gas Processors Suppliers Assn., Tulsa (1979).
25. *Compressed Air Handbook*, second edition, Compressed Air and Gas Inst., McGraw-Hill Book Co. Inc., New York City (1954).
26. Counihan, T.M.: "A Successful In-Situ Combustion Pilot in the Midway-Sunset Field, California," paper SPE 6525 presented at the 1977 SPE California Regional Meeting, Bakersfield, April 13–15.
27. Parrish, D.R., Pollock, C.B., and Craig, F.F. Jr.: "Evaluation of COFCAW as a Tertiary Recovery Method, Sloss Field, Nebraska," *J. Pet. Tech.* (June 1974) 676-686; *Trans.*, AIME, **257**.
28. Gates, C.F. and Ramey, H.J. Jr.: "Field Results of South Belridge Thermal Recovery Experiment," *Trans.*, AIME

(1958) **213**, 236-244.
29. Green, K.B.: "2-The Fireflood: Cox Penn Sand," *Oil and Gas J.*, (July 17, 1967) 66-69.
30. Franco, A.: "How MGO Handles Heavy Crude," *Oil and Gas J.*, (June 24, 1974) 105-109.
31. Leutwyler, K. and Bigelow, H.L.: "Temperature Effects on Subsurface Equipment in Steam Injection Systems," *J. Pet. Tech.* (Jan. 1965) 93-101; *Trans.*, AIME, **234**.
32. Willhite, G.P. and Dietrich, W.K.: "Design Criteria for Completion of Steam Injection Wells," *J. Pet. Tech.* (Jan. 1967) 15-21.
33. Bissoondatt, J.C.: "Casing Failure in Steam Stimulated Wells—A Case History of the Guapo Field," paper SPE 5951 presented at the 1976 SPE Trinidad and Tobago Section Conference, Trinidad, April 2-3.
34. U.S. Patent No. 3,654,691 (Oct. 20, 1969).
35. Penberthy, W.L. and Bayless, J.H.: "Silicate Foam Wellbore Insulation," *J. Pet. Tech.* (June 1974) 583-588.
36. Johnson, D.R. and Fox, R.L.: "Examination of Techniques for Thermally Efficient Delivery of Steam to Deep Reservoirs," paper SPE 8820 presented at the 1980 SPE/DOE Enhanced Oil Recovery Symposium, Tulsa, April 20-23.
37. Hall, A.L. and Bowman, R.W.: "Operation and Performance of Slocum Field Thermal Recovery Project," *Thermal Recovery Techniques*, Reprint Series, SPE, Dallas (1972) **10**, 144-157.
38. Schafer, J.C.: "Thermal Recovery in the Schoonebeek Oil Field–Fifteen Years of Experience," *Erdoel–Ergas-Feitschrift* (Oct. 1974) 372-379 (in English).
39. Bleakley, W.B.: "Steamed Wells Need Good Completions," *Oil and Gas J.* (April 4, 1966) 136-138.
40. Bott, R.C.: "Cyclic Steam Project in a Virgin Tar Reservoir," *J. Pet. Tech.* (May 1967) 585-591.
41. Saucier, R.J.: "Considerations in Gravel Pack Design," *J. Pet. Tech.* (Feb. 1974) 205-212; *Trans.*, AIME, **257**.
42. Coberly, C.J. and Wagner, E.M.: "Some Considerations in Selection and Installation of Gravel Packs for Oil Wells," *Pet. Tech.* (Aug. 1938) 1-20.
43. Holke, D.C. and Huebner, W.B.: "Thermal Stimulation and Mechanical Techniques Permit Increased Recovery from Unconsolidated Viscous Oil Reservoir," paper SPE 3671 presented at the 1971 SPE California Regional Meeting, Los Angeles, Nov. 4-5.
44. Reed, M.G.: "Gravel Pack and Formation Sandstone Dissolution During Steam Injection," *J. Pet. Tech.* (June 1980) 941-949.
45. Gates, C.F. and Brewer, S.W.: "Steam Injection Into the D and E Zones, Tulare Formation, South Belridge Field, Kern County, California," *J. Pet. Tech.* (March 1975) 343-348.
46. Gates, C.F. and Sklar, I.: "Combustion as a Primary Recovery Process–Midway Sunset Field," *J. Pet. Tech.* (Aug. 1971) 981-986; *Trans.*, AIME, **251**.
47. Suman, G.O. Jr.: *World Oil's Sand Control Handbook*, Gulf Publishing Co., Houston (1975).
48. Gates, C.F., Jung, K.D., and Surface, R.A.: "In-Situ Combustion in the Tulare Formation, South Belridge Field, Kern County, California," *J. Pet. Tech.* (May 1978) 798-806; *Trans.*, AIME, **265**.
49. Showalter, W.E. and MacLean, M.A.: "Fireflood at Brea-Olinda Field, Orange County, California," paper SPE 4763 presented at the 1974 SPE Improved Oil Recovery Symposium, Tulsa, April 22-24.
50. Anderson, G.W.: "New Methods Make Downhole Liquid Plugging Practical," *World Oil* (Feb. 1, 1978) 37-43.
51. Hess, P.H.: "Chemical Method for Formation Plugging," *J. Pet. Tech.* (May 1971) 559-564.
52. Felber, B.J., Dauben, D.L., and Marrs, R.E.: "Lignosulfonates for High-Temperature Plugging," Canadian Patent No. 1,041,900 (Nov. 7, 1978).
53. Fitch, J.P. and Minter, R.B.: "Chemical Diversion of Heat Will Improve Thermal Recovery," paper SPE 6172 presented at the 1976 SPE Annual Technical Conference and Exhibition, New Orleans, Oct. 3-6.
54. Kahn, S.A. and Marsden, S.S.: "Foam Proves Effective Way To Solve Production Problems," *Oil and Gas J.* (June 5, 1967) 126-129.
55. Skinner, W.C.: "A Quarter Century of Production Practices," *J. Pet. Tech.* (Dec. 1973) 1425-1431.
56. Vonde, T.R.: "Specialized Pumping Techniques Applied to a Very Low-Gravity, Sand-Laden Crude—Cat Canyon Field, California," *J. Pet. Tech.* (Sept. 1982) 1951-1957.
57. Meldau, R.F. and Lumpkin, W.B.: "Phillips Tests Methods to Improve Drawdown and Producing Rates in Venezuela Fire Flood," *Oil and Gas J.* (Aug. 12, 1974) 127-134.
58. Boberg, T.C., Penberthy, W.L. Jr., and Hagedorn, A.R.: "Calculating the Steam-Stimulated Performance of Gas-Lifted and Flowing Heavy-Oil Wells," *J. Pet. Tech.* (Oct. 1973) 1207-1215; *Trans.*, AIME, **255**.
59. Ballard, J.R., Lanfranchi, E.E., and Vanags, P.A.: "Thermal Recovery in the Venezuelan Heavy Oil Belt," *J. Cdn. Pet. Tech.* (April-June 1977) 22-27.
60. Simon, R. and Poynter, W.G.: "Downhole Emulsification for Improving Viscous Crude Production," *J. Pet. Tech.* (Dec. 1968) 1349-1353.
61. Strassner, J.E.: "Effects of pH on Interfacial Films and Stability of Crude Oil-Water Emulsions," *J. Pet. Tech.* (March 1968) 303-312; *Trans.*, AIME, **243**.
62. Lucas, R.N.: "Dehydration of Heavy Crudes by Electrical Means," paper SPE 1506 presented at the 1966 SPE Annual Meeting, Dallas, Oct. 2-5.
63. Bansbach, P.L.: "The How and Why of Emulsions," *Oil and Gas J.* (Sept. 7, 1970) 87-93.
64. Bansbach, P.L.: "Treating Emulsions Produced by Thermal Recovery Operations," paper SPE 1328 presented at the 1965 SPE California Regional Meeting, Bakersfield, Nov. 4-5.
65. Burris, D.R.: "Dual Polarity Oil Dehydration," *Pet. Eng.* (Aug. 1977) 30-41.
66. Fazio, P.J. and Banderob, L.D.: "Thermal Well Problems and Completion Techniques in the Midway-Sunset Field," *Prod. Monthly* (Nov. 1966) 10-15.
67. Bertness, T.A.: "Thermal Recovery: Principles and Practices of Oil Treatment," paper SPE 1266 presented at the 1965 SPE Annual Meeting, Denver, Oct. 3-6.
68. Evers, R.H.: "Mixed Media Filtration of Oily Waste Waters," *J. Pet. Tech.* (Feb. 1975) 157-163.
69. "Continuous Steam Injection, Heat Scavenging Spark Kern River," *Oil and Gas J.* (Jan. 27, 1975) 127-141.
70. Snavely, E.S. Jr. and Blount, F.E.: "Rates of Dissolved Oxygen with Scavengers in Sweet and Sour Brines," *Proc.*, NACE 25th Annual Conf. (1969) 397-404.
71. Bheda, M. and Wilson, D.B.: "A Foam Process for Treatment of Sour Water," *Water–1969*, Chem. Eng. Prog. Symposium Series (1969) 274.
72. Martin, J.D. and Levanas, L.D.: "Air Oxidation of Sulfide in Process Water," *Oil and Gas J.* (June 11, 1962) 184-187.
73. Fouda, A.E., Gregory, G.A., Rhodes, E., and Scott, D.S.: "Problems of Two-Phase Flow in Networks and Metering of Quality Mass Flow and Void Fraction," *The Oil Sands of Canada-Venezuela*, CIM (1977) Special Vol. **17**, 663-671.
74. Hong, K.C.: "Two-Phase Splitting at a Pipe Tee," *J. Pet. Tech.* (Feb. 1978) 290-296; *Trans.*, AIME, **265**.
75. Palm, J.W., Kirkpatrick, J.W., and Anderson, W.: "Determination of Steam Quality Using an Orifice Meter," *J. Pet. Tech.* (June 1968) 587-591.
76. Hodgkinson, R.J. and Hugli, A.D.: "Determination of Steam Quality Anywhere in the System," *The Oil Sands of Canada-Venezuela*, CIM (1977) Special Vol. **17**, 672-675.
77. Emery, L.W.: "Results from a Multi-Well Thermal-Recovery Test in Southeastern Kansas," *J. Pet. Tech.* (June 1962) 671-678; *Trans.*, AIME, **225**.
78. Closmann, P.J., Jones, E.R., and Vogel, E.A.: "Effect of a Heat-Conducting Well Casing on the Temperature Distribution in an Observation Well," *Trans.*, ASME, **101**, 20-27.
79. Trantham, J.C. and Marx, J.W.: "Bellamy Field Tests: Oil from Tar by Counterflow Underground Burning," *J. Pet. Tech.* (Jan. 1966) 109-115; *Trans.*, AIME, **237**.
80. Cato, B.W. and Frnka, W.A.: "Results of an In-Situ Combustion Pilot Project," *Prod. Monthly* (June 1968) 20-24.
81. Barber, R.M.: "Trix Liz Fireflood Looks Good," *Oil and Gas J.* (Jan. 1, 1973) 36-39.

82. Heflin, J.D., Lawrence, T.D., Oliver, D., and Koenn, L.: "California Applications of the Continuous Carbon/Oxygen Log," paper presented at 1977 API Meeting, Bakersfield, CA, Oct. 25-27.
83. de Haan, H.J. and Schenk, L.: "Performance Analysis of a Major Steam Drive Project in the Tia Juana Field, Western Venezuela," *J. Pet. Tech.* (Jan. 1969) 111-119; *Trans.*, AIME, **246**.
84. van Poollen, H.K.: "Transient Tests Find Fire Front in an In-Situ Combustion Project," *Oil and Gas J.* (Feb. 1, 1965) 78.
85. Kazemi, H.: "Locating a Burning Front by Pressure Transient Measurements," *J. Pet. Tech.* (Feb. 1966) 227-232; *Trans.*, AIME, **237**.
86. Walsh, J.W. Jr., Ramey, H.J. Jr., and Brigham, W.E.: "Thermal Injection Well Falloff Testing," paper SPE 10227 presented at the 1981 SPE Annual Technical Conference and Exhibition, San Antonio, Oct. 4-7.
87. Matthews, C.S. and Russell, D.G.: *Pressure Buildup and Flow Tests in Wells*, Monograph Series, SPE, Dallas (1967) **1**.
88. Earlougher, R.C. Jr.: *Advances in Well Test Analysis*, Monograph Series, SPE, Dallas (1977) **5**.
89. Shore, R.A.: "The Kern River SCAN Automation System—Sample, Control, and Alarm Network," paper SPE 4173 presented at the 1972 SPE California Regional Meeting, Bakersfield, Nov. 8-10.
90. Koch, R.L.: "Practical Use of Combustion Drive at West Newport Field," *Pet. Eng.* (Jan. 1965) 72-81.
91. Snavely, E.S. Jr. and Bertness, T.A.: "Removal of Sulfur Dioxide Emissions by Scrubbing with Oilfield-Produced Water," *J. Pet. Tech.* (Feb. 1975) 227-232.
92. Reyes, R.B. and Siemak, J.B.: "Comparison of Caustic Soda and Soda Ash as Chemical Media for SO_2 Recovery in Oilfield Steam Generators," *J. Pet. Tech.* (May 1980) 751-756.
93. van Ginneken, A.J.J. and Klein, J.P.: "The Desulphurization and Purification of Gases,' *Proc.*, Ninth World Pet. Cong., Tokyo (1975) **6**, 107-116.
94. "Getty Expands Bellevue Fire Flood," *Oil and Gas J.*, (Jan. 13, 1975) 45-49.
95. de Haan, M.J. and van Lookeren, J.: "Early Results of the First Large-Scale Steam Soak Project in the Tia Juana Field, Western Venezuela," *J. Pet. Tech.* (Jan. 1969) 101-110; *Trans.*, AIME, **246**.
96. Dietz, D.N.: "Wet Underground Combustion, State of the Art," *J. Pet. Tech.* (May 1970) 605-617; *Trans.*, AIME, **249**.
97. Hardy, W.C., Fletcher, P.B., Shepard, J.C., Dittman, E.W., and Zadow, D.W.: "In-Situ Combustion Performance in a Thin Reservoir Containing High-Gravity Oil," *J. Pet. Tech.* (Feb. 1972) 199-208; *Trans.*, AIME, **253**.
98. Parrish, D.R., Rausch, R.W., Beaver, K.W., and Wood, H.W.: "Underground Combustion in the Shannon Pool, Wyoming," *J. Pet. Tech.* (Feb. 1962) 197-205; *Trans.*, AIME, **225**.
99. Burger, J.G. and Sahuquet, B.C.: "Chemical Aspects of In-Situ Combustion – Heat of Combustion and Kinetics," *Soc. Pet. Eng. J.* (Oct. 1972) 410-422; *Trans.*, AIME, **253**.
100. Keller, H.H. and Couch, E.J.: "Well Cooling by Downhole Circulation of Water," *Soc. Pet. Eng. J.* (Dec. 1968) 405-412; *Trans.*, AIME, **243**.
101. Earlougher, R.C. Jr., Galloway, J.R., and Parsons, R.W.: "Performance of the Fry In-Situ Combustion Project," *J. Pet. Tech.* (May 1970) 551-557.
102. Roeser, W.F. and Wensel, H.T.: "Methods of Testing Thermocouples and Thermocouple Material," *J. Res. Natl. Bureau Standards* (March 1935) **14**, 247-282.

Chapter 12

Evaluation of Reservoirs for Thermal Recovery

It has been customary to attempt to define conditions – usually a limited list of reservoir and fluid properties and operating conditions – under which thermal recovery operations are likely to be commercially attractive. These conditions have been called by a variety of names, including screening criteria, screening guidelines, preferred criteria, and selection criteria. Any such set of criteria reflects its authors' opinions at a particular time and usually are affected by the current and local economic climate.

Although most available guidelines[1,2] attempt to include the effect of economic considerations in the technical variables, there are few technical considerations that in themselves limit the operability of thermal recovery processes. Important technical considerations are those that affect (1) the ability to generate heat within or inject heat into an oil-containing reservoir at efficient rates, (2) the ability to displace the oil, and (3) the ability to recover the oil, all in a controlled manner. It is difficult if not impossible to translate these broad considerations into specific values of properties or groups of properties that would limit the range of applicability of a thermal process. Some examples will help clarify this point.

In steam drives, the gravity of the crude plays no role in the technical considerations just listed except as it might affect plugging and, thus, the ability to maintain adequate communication between wells in reservoirs containing relatively heavy crudes. With regard to loss in injectivity or productivity, gravity override (or bypassing) of steam reduces the tendency of the formation to plug. Also, hydraulic fracturing, control of the injection temperature, and cyclic steam injection have been used successfully in thermal operations to avoid or minimize plugging. Thus, it seems impractical to place a limit on the range of API gravity of crudes to be considered for steam drive processes.

As another example, consider injectivity. Reduced ability to inject (or produce) fluids is a factor affecting the economics of all enhanced recovery processes, especially those requiring the injection of heated fluids. When the injection rate is reduced, the project life must be longer and more of the heat entering the formation is lost to the adjacent zones. This tends to reduce the oil displacement and production rates (i.e., the rate at which income is generated), which in turn reduces the economic attractiveness. But what may be an acceptable level of injectivity in one project could be economically disastrous in another.

Although guidelines giving parameter limits are useful in helping government bodies and oil companies decide how to allocate funds to support their general research and development efforts, they are of little use to an operator having a specific reservoir requiring either improved performance or special development. Each reservoir should be examined individually (at least briefly) as though there were no guidelines, especially where the reserves are great enough to support some engineering studies. A screening economic analysis for each likely process is the preferred method for forming initial decisions.

The first section of this chapter discusses, in an elementary way, the effect of capital expenditure and income histories on the present value of a venture. Then, it is shown how the income history may be estimated from the performance prediction model for the steam drive used in Chap. 7. Although it applies here to the steam drive process, this development is intended to demonstrate how performance predictions for other processes can be used to provide cash flow forecasts to help assess their economic attractiveness. The second part of the chapter presents general information on the sensitivity of income to certain physical properties and operating parameters. The need for sensitivity studies arises from uncertainties in the values of the fluid and reservoir properties, in the injectivities and productivities, in some aspect of the recovery process, etc. Because there are uncertainties, there is a risk in each investment decision.

Thus, the approach is to indicate some of the elements necessary to screen a potential venture, including an understanding of the possible effects of uncertainties on a parameter or group of parameters. This approach, which requires only a few hours to perform, should allow the engineer to make preliminary decisions or recommendations as to the merits of a prospective thermal project.

TABLE 12.1 – INCOME PATTERNS DISCUSSED IN SECTION 12.1 – IMPORTANCE OF CASH FLOW HISTORY

Year	Cash Outlays	Case A Revenue	Case B Revenue
0	$100,000	0	0
1	30,000	$120,000	$ 40,000
2	30,000	80,000	120,000
3	30,000	40,000	80,000
Total	$190,000	$240,000	$240,000

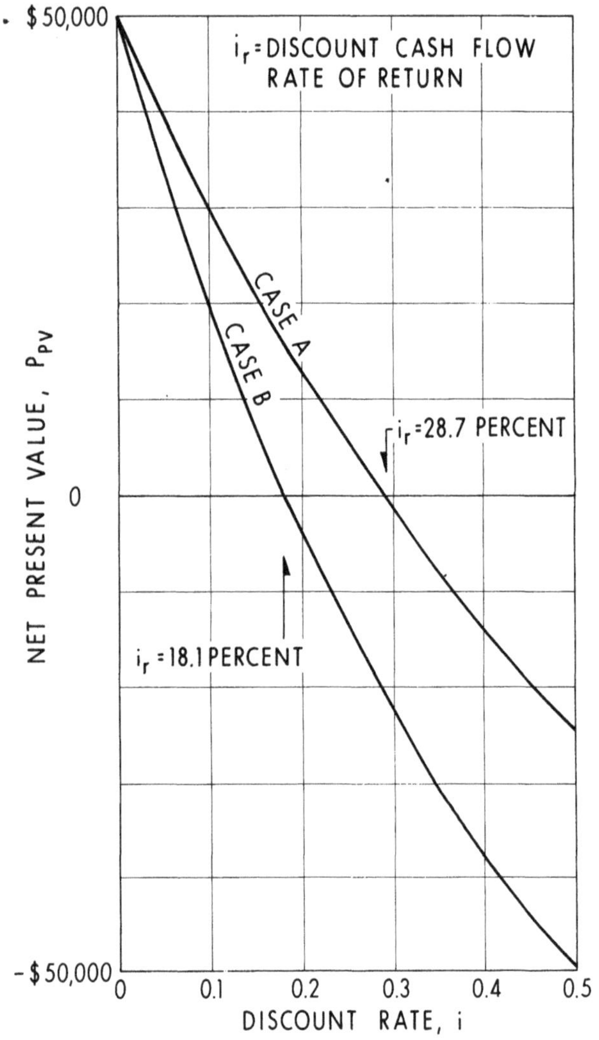

Fig. 12.1 – Significance of discount cash-flow rate of return, i_r.

12.1 Economics

There are many methods for measuring the economic attractiveness of production ventures.[3-5] In discussing the economics of thermal operations, the net present value method shall be used. It is assumed, as is customary, that a single discount rate i is applicable to *all* cash flows. For a net cash inflow after taxes, P_j, in the jth year after an initial investment of magnitude $P_i = P_0$, the net present value of an operation lasting n_y years is given by

$$P_{PV} = \sum_{j=0}^{n_y} P_j e^{-ij}, \quad \text{(12.1)}$$

when the discount rate is compounded continuously. Note that the P_j are positive when the net cash flow after taxes for the year is positive (inflow) and negative for an outflow of cash. The first term, corresponding to the initial investment, is always negative. But in many projects, the first several terms are negative.

The yearly after-tax cash inflow P_j is composed of four elements:

$$P_j = (V_j - O_j - C_j) f_{n,j}, \quad \text{(12.2)}$$

where V_j is the annual gross revenue after royalties and wellhead taxes, O_j is the annual operating cost, C_j is the annual capital outlay, and $f_{n,j}$ is the net fraction of the yearly cash inflow remaining after all taxes have been paid. The first two terms on the right side of Eq. 12.2 are the operating cash income, denoted by I_j:

$$I_j = V_j - O_j. \quad \text{(12.3)}$$

The annual gross revenue after royalties and wellhead taxes is proportional to the annual oil production:

$$V_j = V_{u,j} \Delta N_{p,j}, \quad \text{(12.4)}$$

where $V_{u,j}$ is the average unit crude price after royalties and wellhead taxes, and $\Delta N_{p,j}$ is the annual oil production.

Because any field project involves risks, the rate of return should be higher than that obtainable in relatively low-risk investments. The discounted cash-flow rate of return, denoted by i_r, is the value of the discount rate i in Eq. 12.1 that makes $P_{PV} = 0$. And i_r should be equal to or greater than the rate expected from other available investments.

Interest rates and rates of return are expressed as, for example, 10% per year, which commonly is shortened to 10%. In the equations, however, interest rates are expressed in fractional form, not as percentages. Thus, $i = 10\%$ would be expressed as 0.10/yr in Eq. 12.1.

Importance of Cash Flow History

Both the present value and the rate of return are affected by the oil production history. For example, suppose that the same ultimate production is obtained in two projects having the same operating and capital cost patterns, but with oil production patterns leading to revenues as indicated in the two cases shown in Table 12.1. The present value of each of these two cases is depicted in Fig. 12.1 as a function of the discount rate i. At any discount rate, the present value is higher for the case of high initial revenue (Case A) than for the case of deferred revenue (Case B). Also, the rate of return i_r is higher (28.7%) for the case of high initial revenue than for the other case (18.1%).

Cyclic steam injection processes, which are characterized by a high early production response, generally lead to high initial income patterns. On the other hand, interwell displacement processes, which are characterized by a delayed production response, generally lead to a revenue pattern in which the maximum revenue occurs later in the venture. Given identical investment schemes, cyclic steam injection is the preferred venture, since it gives a higher rate of return. Economic attractiveness, then, is the main reason for the popularity of cyclic steam injection as a thermal recovery process. It must be noted, however, that in most instances the cumulative recovery (or revenue) is higher in steam drives than in cyclic steam injection processes.

In general, the present value of a venture depends not only on the process but also on many reservoir and fluid properties and operating conditions. In any specific field where the depths of the wells and the area of development are known, these conditions are fixed. Other conditions – such as the process to be used, the average amount of oil in place at the start of the venture, whether existing producing wells must be used, and whether only new injection wells are to be drilled – also may be considered to be known by the operator. A suggested procedure for analyzing a venture is to obtain its present value at the applicable discount rate, using current estimates of the pertinent properties and conditions. Then, it is desirable to determine the sensitivity of the results on the input parameters, using values considered to be within the range of uncertainty. If necessary, properties or conditions having a significant influence on the results of the economic analysis can be targeted for further investigation and closer determination.

For example, the injection pressure of a steam drive may not be known with precise certainty, and since it affects the injection rate, the operating and capital costs, and the oil recovery rate, it is desirable to determine the economic sensitivity over the range of this variable. For Cases A and B, new production forecasts, cash flows, etc., were generated at pressure levels of 250, 500, and 750 psig, and the present value at a discount rate of 12% was calculated at each pressure level. As Fig. 12.2 shows, the present value in one case (Case A) is much less sensitive to variations in the injection pressure than it is in another (Case B). At low injection pressures, both cases exhibit low injection rates, increased project duration, large heat losses, and relatively low present values. Similarly, at high injection pressures, both cases have a high cost of steam generation, low reservoir volumes occupied per unit of steam, and relatively low present values. In both cases, there is a maximum present value at an injection pressure near 500 psig. The rate of return at an injection presure near 500 psig – the best estimate of the pressure required for the venture – is greater for Case B (33 vs. 27%). Also, for Case B the present value decreases rapidly when pressure varies from the 500 psig, whereas the present value of Case A is relatively insensitive to changes in pressure. If the pressure level were the only variable with significant uncertainty, it could be inferred that the economic risk is considerably higher in Case B than in Case A. Such risks always should be considered in making a decision about a venture.

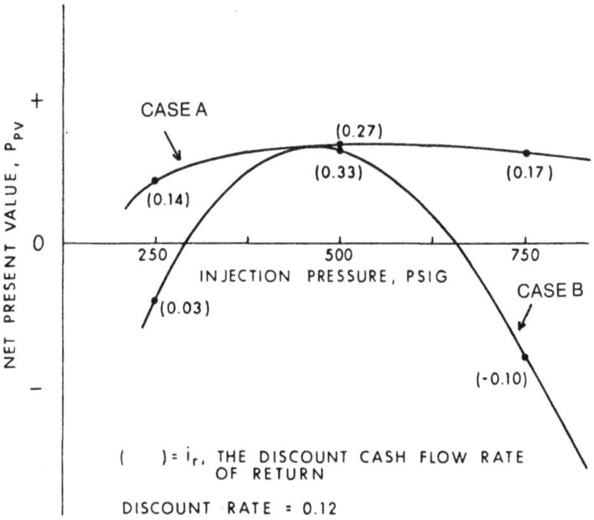

Fig. 12.2 – Sensitivity of net present value to injection pressure.

Estimating Gross Income History From Performance Prediction Models

The purpose of this subsection is twofold: to illustrate how the gross income history may be estimated from performance prediction models and to identify an operational variable and discuss how the gross income history is sensitive to it.

The steam drive process and the injection pressure are the process and the variable used to illustrate the procedures. It is to be understood that the procedures are equally applicable to any process and any variable (even an uncertain reservoir property).

To make a sensitivity analysis on the level of the injection pressure, one must be able to estimate how operating and capital costs and revenues will be affected by the injection pressure. If it is assumed that the operator can determine operating and capital costs as a function of injection pressure, the emphasis will be on indicating how the annual oil production (the project parameter that affects revenue; see Eq. 12.3) is affected by the injection pressure.

For simplicity's sake, the model of Marx and Langenheim[6] is used to calculate the volume of the steam zone.

The expression for the annual oil recovery from a five-spot-pattern steam drive in the jth year, obtained from Eqs. 7.10 through 7.17, 4.2, 4.3, 5.6, and 5.12, is

$$\Delta N_{p,j} = 0.25 F_o F_{\Delta h} i_S h_t^2 \Delta G_j / \alpha_S \text{ , } \quad \ldots\ldots\ (12.5)$$

where the steam injection rate in equivalent barrels of water per day is estimated from

$$i_s = \frac{3.54\times 10^{-3}(p_{inj}-p_p)\mathfrak{I}_{as}}{\ln(\sqrt{43{,}560A}/r_w)-0.964}. \quad \text{(12.6)}$$

$$F_o=(S_{oi}-S_{or})\phi h_n/h_t \quad \text{(12.7)}$$

is the fractional volume of oil per unit bulk volume of reservoir.

$$F_{\Delta h}=\frac{\rho_w(C_w\Delta T+f_sL_v)}{M_R\Delta T} \quad \text{(12.8)}$$

is the dimensionless ratio of the effective volumetric heat capacity of the injected steam to that of the reservoir steam zone.

$$\mathfrak{I}_{as}=(k_sh_n/\mu_s)_a \quad \text{(12.9)}$$

is the effective (apparent) transmissivity of the reservoir to the injected steam in md-ft/cp. (See Transmissivities in Section 12.2 for comments on the effective or apparent transmissivity $\mathfrak{I}_{as}$.)

$$\Delta G_j=G(j\,t_{D1})-G[(j-1)t_{D1}], \quad \text{(12.10)}$$

where the dimensionless function G, defined in Eq. 5.13, is a measure of the volume of the reservoir steam zone.

$$\Delta G_o\equiv 0. \quad \text{(12.11)}$$

$$t_{D1}=1{,}460\,\alpha_S/h_t^2 \quad \text{(12.12)}$$

is the value of the dimensionless time at 1 year (assuming the thermal properties are the same in the reservoir and contiguous formations). In these expressions,

- i_s = steam injection rate per injection well, B/D,
- h_t = gross reservoir thickness, ft,
- α_S = diffusivity of the surrounding formations, sq ft/D,
- $p_{inj}-p_p$ = the pressure difference between injection and production wells, psi,
- A = spacing per injection well, acres,
- r_w = wellbore radius, ft,
- $S_{oi}-S_{or}$ = the difference between initial and residual saturation,
- ϕ = porosity, fraction,
- h_n = net reservoir thickness, ft,
- ΔT = the difference between the injected steam and initial reservoir temperature, °F,
- f_s = the steam quality, fraction,
- L_v = latent heat of condensation of steam, Btu/lbm,
- k = reservoir permeability, md, and
- μ = fluid viscosity, cp.

The injection pressure affects the factor $F_{\Delta h}$ (given by Eq. 12.8) and appears explicitly in the expression for the injection rate (given by Eq. 12.6). Other factors, such as the apparent transmissivity to steam, also may be dependent on the injection pressure. When the effect of the injection pressure on the various parameters is known, the sensitivity of the calculated production response to the injection rate can be calculated.

The manner in which the sensitivity to injection pressure of a steam drive process can be estimated has been outlined on the basis of a specific model. Other sensitivity studies can be made in a similar manner for other parameters, for other recovery processes, and for different models of the steam drive process. Of the variables required to calculate the annual cash flow (Eq. 12.2), the annual oil recovery rate and sales price are the only project factors affecting revenues. One of the principal objectives of this monograph is to present methods for estimating the oil recovery from thermal processes. Methods for estimating the dependence of the oil recovery on the process used, on the reservoir and fluid properties, and on operating conditions are presented in the chapters discussing individual processes. It is important to recognize that the methods presented in these chapters are based on highly idealized models. Although only one model of a process is emphasized in any one chapter, other models developed and proposed in the literature also are referenced. Each would lead to slightly different conclusions with respect to the factors of interest. For these reasons, *good engineering judgment must be exercised in applying the available information* (whether presented in here or elsewhere) to estimate the annual production response. The uncertainties associated with the production-response estimates demand that the results of Eqs. 12.1 through 12.3 be used only as approximate or qualitative economic screening criteria.

Although the first part of this section was discussed in terms of a continuously compounded annual cash flow, obviously other time periods and compounding methods can be used.[3,4] Also, other economic criteria besides net present value and discounted cash-flow rate of return can be used in the screening process.[3,4]

Actual cost data in specific projects seldom are published for any recovery process, although there are exceptions.[7] The best discussion of the average cost components for thermal recovery processes and the impact of crude price on the extent of future thermal operations inthe U.S. is that published by the National Petroleum Council.[1] Sources of cost breakdowns, revenues, and production statistical data in the U.S. include the reports of the Joint Association Survey,[8] the U.S. Interior[9,10] and Commerce[11] departments, and agencies in the State of California.[12,13] Table 12.2 gives cost indices and other statistical data pertinent to the U.S. oil industry as reported in the *Petroleum Independent*.[14] Such information is useful in updating cost estimates.

Studies of specific interest to thermal recovery include those summarizing[15] the results of the National Petroleum Council[1] on the cost of generating steam[16-19] and those comparing relative merits of steam and combustion processes.[20-23] Considerations regarding the financing of commercial large-scale thermal projects also are receiving increased attention.[1,24,25]

12.2 Rock, Reservoir, and Fluid Properties

This section discusses parameters that affect success of thermal recovery projects, emphasizing how increases or decreases in the values of those parameters affect the overall behavior of the projects.

Recoverable Oil

The recoverable oil per unit bulk volume of reservoir is given by the quantity $\phi(S_o - S_{or})h_n/h_t$. Minimum acceptable values for this quantity may be estimated from the requirement that the fuel value of the oil displaced and recovered from a unit volume of the reservoir must be larger than the energy required to displace it. For steam displacement, such a balance leads to the following inequality (numbered, for convenience, as an equation):

$$(0.77)\phi(S_{oi} - S_{or})h_n/h_t > \frac{M_R\ (T_s - T_i)}{\rho_o \Delta h_F E_{h,s} E_c}, \qquad (12.12)$$

where

ϕ = porosity, fraction,
S_{oi} = initial oil saturation, fraction,
S_{or} = residual oil saturation, fraction,
h_n = net sand thickness, ft,
h_t = gross sand thickness, ft,
M_R = effective volumetric heat capacity of the steam zone, Btu/cu ft-°F,
T_s = steam injection temperature, °F,
T_i = original reservoir temperature, °F,
ρ_o = oil density, lbm/cu ft,
Δh_F = fuel value of the oil per unit mass, Btu/lbm,
E_c = fraction of the displaced oil that is produced at the end of the project, and
$E_{h,s}$ = fraction of the cumulative heat injected *into* the reservoir present in the steam zone at the end of the project. This quantity, defined in Eq. 7.12, is a strong function of the gross reservoir thickness.

The factor 0.77 represents the overall thermal efficiency of the generator, the surface lines, and the injection well. Here, it is considered to be constant.

Minimum values of the recoverable oil can be estimated by equating the left and right sides of Eq. 12.12. These values increase linearly with steam temperature and are inversely proportional to the fuel value of the oil, the heat efficiency $E_{h,s}$, and the capture factor E_c (the fraction of the displaced oil that is produced). For values of the properties given in Table 12.3,

$$[\phi(S_o - S_{or})h_n/h_t]_{min} = 0.074.$$

TABLE 12.2 – HISTORICAL COST INDICES OF INTEREST TO THE OIL INDUSTRY[14]

	PRICE INDICES 1967=100		WAGES ($ per hour)		PETROLEUM PRICE AND COST INDICES (Index Numbers - 1967 = 100)					EXPLORATION AND DRILLING ACTIVITY			
	Producer Price Index All Commod.	Consumer Price Ind. All Items	Crude Oil & Natural Gas Industry	All Mfg. Ind.	Crude Oil Prices	Hourly Wages Paid	Oil Field Machinery Prices	Oil Well Casing Prices	Line Pipe Prices	Rotary Rigs Active	Wildcat Wells Drilled	Total Wells Drilled	Total Ft. Drilled (Mil. Ft.)
1964	94.7	92.9	2.95	2.53	98.3	90.8	94.8	91.9	103.6	1,502	10,747	45,236	189.9
1965	96.6	94.5	3.03	2.61	98.2	93.2	95.2	96.2	97.9	1,388	9,466	41,423	181.5
1966	99.8	97.2	3.13	2.72	98.9	96.3	96.5	96.2	97.9	1,270	10,313	37,881	166.0
1967	100.0	100.0	3.25	2.83	100.0	100.0	100.0	100.0	100.0	1,134	8,878	33,818	144.7
1968	102.5	104.2	3.38	2.94	100.8	104.0	106.4	101.5	103.9	1,170	8,879	32,914	149.3
1969	106.5	109.8	3.59	3.19	105.2	110.5	112.7	104.5	109.2	1,195	9,701	34,053	160.9
1970	110.4	116.3	3.83	3.36	106.1	117.8	118.4	109.1	112.8	1,028	7,693	29,467	142.4
1971	114.0	121.3	4.16	3.57	113.2	128.0	122.6	120.7	124.7	975	6,922	27,300	128.3
1972	119.1	125.3	4.46	3.81	113.8	137.2	127.3	127.3	133.7	1,107	7,539	28,755	138.4
1973	134.7	133.1	4.80	4.06	126.0	147.7	133.2	133.2	139.1	1,194	7,466	27,602	138.9
1974	160.1	147.4	5.33	4.40	211.8	164.0	157.8	170.7	189.8	1,475	8,723	33,008	153.8
1975	174.9	161.2	6.05	4.80	245.7	186.2	196.3	211.5	227.8	1,660	9,214	39,097	178.5
1976	183.0	170.5	6.59	5.19	253.6	202.8	217.6	223.0	234.3	1,657	9,234	41,455	185.3
1977	194.2	181.5	7.13	5.63	274.2	219.4	236.5	244.3	260.5	2,001	9,961	46,479	215.0
1978	209.3	195.4	8.12	6.14	300.1	249.8	261.2	271.6	296.6	2,259	10,677	48,513	231.4
1978 July	210.6	196.7	8.15	6.13	301.9	250.8	261.5	271.1	297.2	2,307	5,958*	26,745*	128.6*
Aug.	210.4	u 197.8	8.02	6.16	302.7	246.8	263.1	279.2	302.8	2,325	6,856*	30,734*	147.3*
Sept.	212.3	u 199.3	8.22	6.28	305.7	252.9	265.3	279.2	302.8	2,332	7,784*	35,237*	168.8*
Oct.	215.0	u 200.9	8.13	6.32	305.7	250.2	269.9	279.2	302.8	2,346	8,713*	39,349*	188.8*
Nov.	215.7	u 202.0	8.14	6.38	310.5	250.5	272.5	279.2	302.8	2,356	9,572*	43,073*	206.9*
Dec.	217.4	u 202.9	8.27	6.47	312.4	254.5	273.8	279.2	302.8	2,286	10,677*	48,513*	231.4*
1979 Jan.	220.7	u 204.7	8.49	6.48	316.4	261.2	278.2	279.2	310.6	2,199	813	3,766	18.5
Feb.	223.9	u 207.1	8.72	6.50	322.2	268.3	280.6	279.2	310.6	2,064	1,589*	7,562*	36.8*
Mar.	226.4	u 209.1	8.74	6.55	324.4	268.9	280.2	279.2	310.6	1,971	2,486*	12,123*	59.9*
Apr.	229.7	u 211.5	8.85	6.53	325.8	272.3	281.7	279.2	310.6	1,943	3,092*	15,385*	76.3*
May	231.6	u 214.1	8.66	6.62	335.6	266.5	286.8	279.2	306.3	1,960	3,857*	18,947*	93.7*
June	233.5	u 216.6	8.66	6.66	356.3	266.5	288.6	279.2	301.9	1,999	4,652*	23,175*	113.5*
July	236.6	u 218.9	8.76	6.71	370.5	269.5	290.6	291.5	297.6	2,094	5,401*	27,023*	130.6*
Aug.	238.1	u 221.1	8.67	6.69	385.7	266.8	291.7	291.1	293.2	2,222	6,388*	31,310*	150.8*
Sept.	241.7	u 223.4	8.91	6.80	422.1	274.2	294.7	291.1	293.2	2,284	7,367*	36,080*	173.7*
Oct.	245.2	u 225.4	8.79	6.82	436.7	270.5	298.6	291.1	301.9	2,379	8,236*	41,370*	192.8*
Nov.	246.9	u 227.5		6.85	450.4		299.3	291.1	301.9	2,460	9,258*	46,163*	215.0*
Dec.													

*Cumulative u = all urban consumers.

NOTE: Other Supply includes Net Processing Gain, Unaccounted for Crude Oil, and Other Hydracarbons and Hydrogen Refinery Input.
SOURCE: Supply, demand, and ending stocks from Energy Information Administration, DOE. Price Indices, Wages, and Petroleum Price & Cost Indices from U.S. Bureau of Labor Statistics, except hourly wages paid calculated by IPAA. Rotary rigs active from Hughes Tool Co. Wells and footage drilled from American Association of Petroleum Geologists, American Petroleum Institute, and Oil and Gas Journal.

For combustion, a similar energy balance leads to the following inequality:

$$\phi(S_{oi}-S_{oF}) > \frac{7.7\times 10^3 a_R^* n_{CS} F_{comp}^{1/n_{CS}}}{\rho_o \Delta h_F E_{O_2} E_c}, \quad \ldots (12.13)$$

where

a_R^* = amount of injected air required to burn through a reservoir bulk volume and is a function of pressure (see Air Requirement in Section 8.2 and Air and Water Requirements in Section 8.5 for details),

E_{O_2} = oxygen consumption efficiency, fraction,

n_{CS} = number of compression stages,

F_{comp} = overall compression ratio, and

S_{oF} = oil saturation burned as fuel, fraction.

In obtaining this result from Eq. 11.7, it is assumed that interstage compressor cooling requires negligible energy [$F(n_{CS})=1$ in that equation] that the suction compressor temperature is 80°F and that the efficiency of the engine and compressor is 0.16 (typical of a diesel engine).

For values of the properties given in Table 12.4,

$[\phi(S_{oi}-S_{oF})]_{min}=0.066.$

The quantities $\phi(S_{oi} - S_{or})$ and $\phi(S_{oi} - S_{oF})$ determined in this manner are presented as $(\phi\Delta S_o)_{min}$ in Fig. 12.3, and are plotted vs. the injection pressure. Three values of $E_{h,s}$ (0.2, 0.25, and 0.5), and $h_n = h_t$, were used to calculate the steam results. For the values of the parameters chosen, the minimum recoverable oil initially in place satisfying the imposed energy requirement is larger for steam displacement than for combustion. Since results for the steam case should be multiplied by h_t/h_n when $h_t > h_n$, the difference between steam and combustion requirements is likely to be larger than indicated in Fig. 12.3.

As calculated, the requirements for steam displacement are affected strongly by the fraction of the injected heat within the steam zone. Although no such effect is explicit in the calculations for the combustion process, some effect would be introduced if the air injection rate were so low as to preclude the generation of heat at sufficiently high rates. As can be seen, the value of $E_{h,s}$ must be relatively high for steam displacement to be attractive at high steam pressures.

Notice that the minimum value of $\phi(S_{oi}-S_{or})$ for steamflooding, based on the energy criterion, may be larger or smaller than the generally accepted values of $(\phi S_{oi})_{min}=0.1$ (or 776 bbl/acre-ft).[1] On the other hand, the minimum value of $\phi(S_{oi}-S_{oF})$ for the combustion process generally corresponds to values less than $(\phi S_{oi})_{min}=0.1$, even for the relatively high air requirements of 400 scf to burn through a foot of reservoir. On the basis of these considerations alone, it is not possible to say that values of $\phi S_{oi}<0.1$ would be economically unattractive in practice.

TABLE 12.3 – PARAMETERS USED TO EVALUATE EQ. 12.12

M_R, Btu/cu-ft-°F	40
T_s-T_i, °F	200
ρ_o, lbm/cu ft	55
Δh_F, Btu/lbm	17,000
$E_{h,s}$	0.2
E_c	0.75

Note: The injection pressure (which affects the steam temperature, T_s) and the thermal efficiency, E_h, were varied to calculate the results presented in Fig. 12.3.

TABLE 12.4 – PARAMETERS USED TO EVALUATE EQ. 12.13

a_R^*, scf/cu ft	400
F_{comp}	200
n_{CS}	4
ρ_o, lbm/cu ft	55
E_{O_2}	1.0
E_c	0.75
Δh_F, Btu/lbm	17,000

Note: The injection pressure (which affects the overall compression ratio F_{comp} was varied to calculate the results presented in Fig. 12.3.

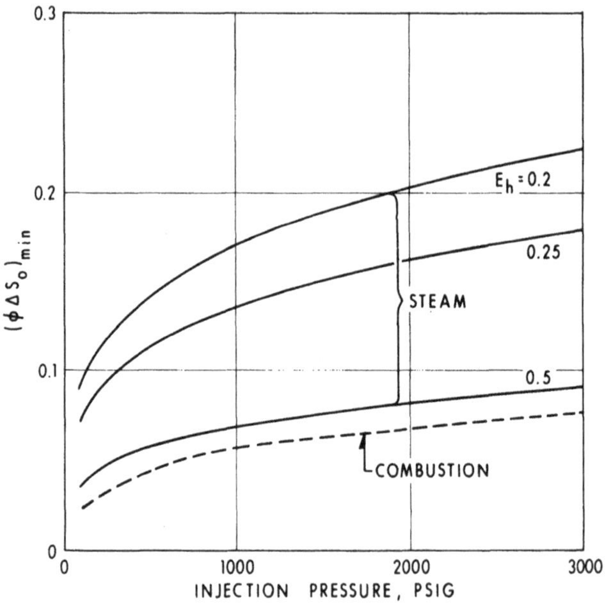

Fig. 12.3 – Minimum oil in place yielding a recovery that just meets surface fuel requirements (parameters used are given in Tables 12.3 and 12.4).

The net/gross sand interval ratio appearing in Eq. 12.12 can be interpreted as reducing either the effective porosity or the effective saturations of the gross reservoir sand. This factor should be used only where (1) the saturations are based on net sand and

(2) the net sand is present in the gross interval in such a way that the entire gross interval would be expected to be subjected to the process temperatures. Individual sand bodies separated by well-defined shale intervals, for example, should be treated separately if (1) only one sand interval is being treated thermally or (2) the intervening shales are so thick that their temperatures would be appreciably less than those of the treated intervals. In the latter case, each treated interval may be considered to have its own associated heat losses to adjacent beds (with some interaction possible), even though steam injection may be through a common well.

In combustion, the air requirement generally is determined in the laboratory under conditions representative of a net sand. To the extent that there is oil (or other fuel) that would burn in the nonpay part of the gross intervals, the left side of Eq. 12.13 should be multiplied by a factor less than one. However, this factor would not be as low as h_n/h_t unless the fuel consumption is expected to be uniform throughout the entire gross interval. If, on the other hand, the air requirement has been determined from a representative field test, no correction is necessary in Eq. 12.13.

The criteria used to develop minimum amounts of oil initially in place are only approximate. They provide only a plausible lower bound for the amount of oil that should be initially in place if each type of project is to break even on an energy balance basis, which is not necessarily the criterion for economic success.

Transmissivities

The effective (apparent) transmissivity[26] to fluid phase i, denoted by the symbol $\mathfrak{I}_{ai}$, is proportional to the k_ih product of the formation and inversely proportional to the fluid viscosity:

$$\mathfrak{I}_{ai} = \left(\frac{k_ih}{\mu_i}\right)_a . \qquad (12.14)$$

The fluid mobility k_i/μ_i is affected by saturation and temperature. Thus, the variables appearing in the right side of Eq. 12.14 are neither uniform throughout the reservoir (and even between wells) nor constant during the life of a project (except for the sand thickness h). Transmissivities are important because they affect the rates at which fluids can be injected, displaced, and produced. These, in turn, determine the life of a project and, thus, its economic attractiveness.

A minimum acceptable transmissivity to the oil phase may be estimated from Eq. 4.2 by using the maximum allowable pressure drop Δp and minimum acceptable average oil production rate q_o. Then, the effective oil transmissivity of the reservoir should be

$$\mathfrak{I}_{ao} > \langle 141.2(2\pi)\rangle \frac{F_G}{2\pi} \bigg/ \left(\frac{\Delta p}{q_o}\right)_{\max} , \qquad (12.15)$$

where F_G is a geometric factor associated with the location and spacing of wells (discussed in Section 4.4) and $(\Delta p/q_o)_{\max}$ represents the maximum flow resistance between the injection and production wells that the operator is willing to take.

The flow resistance between wells may be considered to be the sum of three contributions – one due to radial flow near the injector, another due to radial flow near the producer, and the third due to the flow resistance through the intervening space. In a reservoir of uniform properties and undamaged wells, these three contributions are approximately equal. Flow resistances near the wells can be reduced through conventional and thermal stimulation treatments – e.g., by hydraulically fracturing a well or by repeated application of cyclic steam injection. But even assuming the flow resistance around each well can be reduced to negligible levels, the total flow resistance between wells cannot be reduced much below a third of the prestimulation value. The part of the reservoir affecting this contribution to the flow resistance is halfway between the injector and the producer; thus, it may take considerable effort and time to affect its value through displacement processes. Even after the effect of the displacement process has been felt halfway between wells, the oil transmissivity is controlled primarily by the average temperature and saturation conditions downstream of the displacement front.

If the average temperature downstream of the displacement front were to be inferred from a frontal advance model of the process, it would be essentially equal to the initial reservoir temperature. Then, the oil transmissivity calculated from Eq. 12.14 could be much lower than the right side of Eq. 12.15 in the case of viscous crudes. With bypassing, however, the average temperature downstream of the displacement front can be significantly higher and the necessary transmissivity may be realizable.

Bypassing conditions, in fact, have been incorporated in the design of several thermal recovery operations. Examples include the use of horizontal fractures in the in-situ combustion projects at Lake Gregoire[27] in the Athabasca oil sands of Alberta and Kyrock, KY,[28] and in a steamflood in Loco field[29] in Oklahoma; the use of gravity override in combustion projects[30,31]; and the use of an underlying permeable water zone at Peace River[32] in Alberta.

There does not appear to be a well-defined lower limit for the oil transmissivity required for thermal recovery operations. The transmissivity of the injected fluid is also important as it affects the attainable injection rate. Again, flow resistance to the injected fluid may be decreased to acceptable levels through well stimulation treatments. In combustion processes, a noncondensable gas (usually air where spontaneous ignition is not considered a hazard) sometimes is injected for a few weeks or months where increasing the gas saturation before ignition is considered desirable. This usually results in improved injectivity. It is not uncommon for a process injection rate to be estimated from a field injectivity test. The attained rate then is based generally on the difference in pressure between the injector and the reservoir (and similarly for productivity tests). But the results of injectivity tests of short duration may be optimistic since they may not reflect adequately the influence of a well-developed oil bank in reducing the transmissivity to the other fluids. Also, natural and hydraulic fractures, gas caps and channels, and thief and bottom water have been used to provide the path through which the reservoir can be heated. Thus, the minimum matrix transmissivity to the injected fluids does not appear to have a well-defined value.

But although fractures and zones of relatively high permeability to the injected fluid may be useful for initially introducing heat into the reservoir at adequate rates, such zones may be a detriment to the efficiency of later operations. For example, steam and unreacted air might circulate through these high-permeability zones and be produced at unacceptably high rates.

Reservoir Thickness

Reservoir thickness is particularly important because it has a pronounced effect on the fraction of the introduced heat that remains in the reservoir – a quantity known as the heat efficiency. This effect is illustrated by using the heat efficiency E_h plotted as the lower curve of Fig. 5.3 vs. the dimensionless time

$$t_D = 4\alpha_S t / h_t^2, \quad \ldots\ldots\ldots\ldots\ldots\ldots\ldots (12.16)$$

where α_S is the thermal diffusivity of the surrounding formations and h_t is the gross reservoir thickness. In this expression for dimensionless time it is considered that the thermal properties of the reservoir and surrounding formations are equal.

As can be seen, a factor of one-half in the gross reservoir thickness h_t quadruples the values of the dimensionless time, which in turn may reduce the value of E_h significantly. For example, for $t=730$ days and $\alpha_S = 0.8$ sq ft/D, the heat efficiency would be 0.51 for a gross reservoir thickness of 40 ft but only 0.33 for one of 20 ft. These effects are even more pronounced on the fraction of the injected heat remaining in the reservoir within the steam zone ($E_{h,s}$) shown in Fig. 7.6, which is a measure of the volume of the steam zone in steam injection projects. For a given reservoir (which has a fixed thickness), the only way to increase the heat efficiency is to reduce the time t required to produce the oil from the zone between an injector and a producer. This generally can be done by increasing the injection rate (one also must be able to recover the oil at an increased rate) or by reducing the spacing. Each of these alternatives would cost money to implement but may be attractive compared with a slowly expanding steam front that results in low oil production rates and a long operating life.

The heat efficiency results given in Fig. 7.6 are based on the assumption that the temperature is uniform in a vertical direction throughout a gross interval h_t. In cases where the steam zone thickness is less than the gross reservoir thickness, such as where gravity override is dominant or where the steam is moving preferentially through a high-permeability zone, an average steam zone thickness is to be used for h_t in the expression for the dimensionless time t_D. Because the average steam zone thickness then is less than the gross thickness, the dimensionless time increases with increased bypassing, and $E_{h,s}$ and the volume of the steam zone decreases. Of course, the heat lost from the steam zone to the overlying and underlying formations would be heating any oil present in those formations and would not necessarily be wasted heat. The estimated values of $E_{h,s}$ would then be used in Eq. 12.12, and in the equations predicting production response, to determine if the process would be attractive.

Note that the heat efficiency (and, thus, the reservoir thickness) does not enter the combustion criteria directly. This is because the simple models for the combustion processes consider that the size of the burned volume is directly proportional to the volume of air reacted. In turn, the volume of air reacted is considered to be a weak function of the temperature levels and, thus, relatively insensitive to the vertical extent of the burned zone. For a first approximation this seems a reasonable approach – reasonably accurate and convenient to use.

Well Spacing

The geometrical factors F_G are not strong functions of the well spacing. But they decrease in value as the well spacing decreases. Together with the transmissivity and the allowed pressure drop between wells, they determine the well rates. In turn, injection rate and spacing affect both the life of the project and total heat losses to adjacent formations. The main effect of well spacing is on the cumulative production per well and on the project duration of an individual well, and to a lesser extent on the total flow resistance. One objective of any screening analysis is to determine an acceptable well spacing.

The main consideration regarding well spacing is sand continuity, whose determination may require the combined interpretation of geologists, stratigraphers, and logging and petrophysical experts. Sand continuity is vital in the selection of well spacing for processes involving interwell fluid displacement. Partial communication, as in "ratty" sands where only a fraction of the interval correlates between well locations, invariably leads to poor recoveries and poor effective transmissivities. Knowledge of the depositional setting of the sands may be crucial in establishing whether open intervals in nearby wells are part of the same blanket sand or are likely to be in unconnected and meandering river channels. Regional trends in the direction of river channels may result in an apparent anisotropy in reservoir permeability, with relatively good communication between wells in the same channels and poor or no communication in a transverse direction.

Displacement processes are not indicated where communication is poor over relatively short distances. Instead, cyclic steam injection or some other stimulation treatment that does not depend on interwell communication is preferred.

Another factor related to well spacing is the well pattern. Because of the extensive use of repeated well patterns (such as five-spots) in waterflooding, they were considered necessary in the early years of thermal recovery. Now, however, there is considerable evidence that the use of repeated regular well patterns is not always essential for the success of field projects.[30,31] The choice of well locations is important in reservoirs having significant dip or a strong lateral aquifer influx.

Other Considerations

In addition to the factors already discussed, there are a few other important considerations in selecting a reservoir for thermal projects. Foremost are any factors that may affect the control of the operation, especially where safety is involved.

Control means that the fluids are injected only into the target reservoir and that both displaced and injected fluids are collected efficiently and safely. The leakage of produced or injected fluids to the surface is an extreme example of loss of control.

Conditions that should be considered and evaluated carefully are injection pressures close to the fracture initiation pressure or close to the estimated pressure limits of well equipment (especially in old wells) and reservoirs where communication with nearby outcrops is likely. These could lead to fluid contamination of overlying formations, loss of production, loss of pressure, loss of control, and possibly both unexpected workover costs and early termination of the project.

Where displaced oil can bypass a producing well, it is generally desirable on technical grounds (but not necessarily justifiable economically) to fracture the well and to maintain it in unimpaired condition, especially where low-flow-resistance zones near the well compete for the flow of oil. Stimulation of producers by cyclic steam injection has long been recognized as an excellent method for reducing the resistance to flow into these wells.[8,31] Special attention should be paid to methods for avoiding or minimizing the resaturation of previously depleted zones or the saturation of previously clean sands (such as zones below a water/oil contact).

References

1. *Enhanced Oil Recovery*, National Petroleum Council (1976).
2. Geffen, T.M.: "Oil Production to Expect from Known Technology," *Oil and Gas J.* (May 7, 1973) 66-68, 73-76.
3. McCray, A.W.: *Petroleum Evaluations and Economic Decisions*, Prentice-Hall Inc., Englewood Cliffs, NJ (1975).
4. Bussey, L.E.: *The Economic Analysis of Industrial Projects*, Prentice-Hall Inc., Englewood Cliffs, NJ (1978).
5. Williams, R.L., Brown, S.L., and Ramey, H.J. Jr.: "Economic Appraisal of Thermal Drive Projects—A New Approach," paper SPE 9358 presented at the 1980 SPE Annual Technical Conference and Exhibition, Dallas, Sept. 21–24.
6. Marx, J.W. and Langenheim, R.N.: "Reservoir Heating by Hot Fluid Injection," *Trans.*, AIME (1959) **216**, 312-315.
7. Martin, W.L., Alexander, J.D., Dew, J.N., and Tynan, J.W.: "Thermal Recovery at North Tisdale Field, Wyoming," *J. Pet. Tech.* (May 1972) 606-616.
8. *Joint Association Survey of the U.S. Oil and Gas Producing Industry*, sponsored by the API, IPAA, and the Mid-Continent Oil and Gas Assn., published annually by API, Dallas.
9. *The Minerals Yearbook*, published annually by U.S. Dept. of the Interior, Washington, DC.
10. *Crude Petroleum and Petroleum Products Annual Petroleum Statements*, published annually by the USBM, Washington, DC.
11. *Statistical Abstracts of the United States*, published annually
12. "Annual Review of California Oil and Gas Production," Conservation Committee of California Oil Producers, Los Angeles.
13. "Summary of Operations of California Oil Fields," published annually by the California Div. of Oil and Gas, San Francisco.
14. "Vital Statistics," *Petroleum Independent* (Oct. 1980).
15. Kennedy, J.L.: "Study Analyzes Effect of Oil Price on Enhanced Recovery's Future Role," *Oil and Gas J.* (Dec. 27, 1976) 185-191.
16. Hampton, L.A.: "How Various Fuels Affect the Design and Operating Cost of Steam Generators," paper 6812 presented at the 19th Annual Petroleum Society of CIM Meeting, Calgary, May 7-10, 1968.
17. Friske, E.W.: "Costs of Steam Injection in the San Joaquin Valley," API paper 801-43i, presented at the Spring Meeting of the Pacific Coast Dist., API Div. of Production, Los Angeles, May 2-4, 1967.
18. Bartley, L.R. and Bullen, R.S.: "Calculate Your Steam Generation Costs," *Canadian Petroleum* (August 1967) 28-31.
19. Burns, J.A.: "A Review of Steam Soak Operations in California," *J. Pet. Tech.* (Jan. 1969) 25-34.
20. Root, P.J. and Wilson, L.A.: "A Cost Comparison of Reservoir Heating Using Steam or Air," *J. Pet. Tech.* (Feb. 1966) 233-239.
21. Simm, C.N.: "Improved Firefloods May Cut Steam's Advantages," *World Oil* (March 1972) 59-60, 62.
22. Chu, C.: "A Study of Fireflood Field Projects," *J. Pet. Tech.* (Feb. 1977) 111-120.
23. Doscher, T.M. and Ershaghi, I.: "Current Economic Appraisal of Steam and Combustion Drives," paper SPE 7073 presented at the 1978 SPE Improved Oil Recovery Symposium, Tulsa, April 17–19.
24. Sonosky, J.M. and Babcock, C.D.: "Financing Thermal Operations," *J. Pet. Tech.* (Sept. 1965) 999-1006.
25. Simpson, J.J.: "Financing Enhanced Recovery," *J. Pet. Tech.* (July 1977) 771-775.
26. Ramey, H.J. Jr.: "Commentary on the Terms 'Transmissibility' and 'Storage'," *J. Pet. Tech.* (March 1975) 294-295.
27. Giguere, R.J.: "An In-Situ Recovery Process for the Oil Sands of Alberta," paper presented at the 26th Canadian Chem. Eng. Conf., Toronto, Oct. 3-6, 1976.
28. Terwilliger, P.L.: "Fireflooding Shallow Tar Sands—A Case History," paper SPE 5568 presented at the 1975 SPE Annual Technical Conference and Exhibition, Dallas, Sept. 28–Oct. 1.
29. Wooten, R.W.: "Case History of a Successful Steamflood Project—Loco Field," paper SPE 7548 presented at the 1978 SPE Annual Technical Conference and Exhibition, Houston, Oct. 1-3
30. Gates, C.F. and Sklar, I.: "Combustion as a Primary Recovery Process – Midway Sunset Field," *J. Pet. Tech.* (Aug. 1971) 981-986; *Trans.*, AIME, **251**.
31. Gates, C.F., Jung, K.D., and Surface, R.A.: "In-Situ Combustion in the Tulare Formation, South Belridge Field, Kern County, California," *J. Pet. Tech.* (May 1978) 798-806; *Trans.*, AIME, **265**.
32. Dillabough, J.A. and Prats, M.: "Proposed Pilot Test for the Recovery of Crude Bitumen from the Peace River Tar Sands Deposit – Alberta, Canada," paper presented at the Simposio Sobre Crudos Pesados, Inst. de Investigaciones Petroleras, U. del Zulia, Maracaibo, July 1-3, 1974.

Chapter 13
Pilot Testing

13.1 Purpose of Pilots

Pilot projects are used in thermal recovery operations to test some critical part of the process under consideration. Stated another way, pilot projects are used to obtain answers to specific problems connected with the application of a process in a particular reservoir. The problem may be related to the reservoir, the process, the economics, or some interaction among these. In one case, for example, it may be necessary to show only that a combustion front can be propagated in a watered-out reservoir. In another case, it may be desirable to prove that a modification of a steam drive process actually works as predicted.

Some of the more common specific objectives of thermal pilot projects are to (1) determine vertical and areal coverage, (2) determine heat losses, (3) determine the floodability and continuity of a reservoir, (4) evaluate production problems, production rates, and recovery efficiency, (5) ascertain the rate of advance of a combustion or steam front and the production response times, (6) determine the amount of fuel burned and the air requirements in combustion operations, or the oil/steam ratios in steamfloods, (7) ascertain the remaining oil saturation in steam-swept zones, (8) determine injectivities and injection costs, and (9) determine the vertical and horizontal temperature distribution within the reservoir as an aid in interpreting the reservoir behavior.

As a generalized objective, a pilot is used to reduce the technical and economic risks of a proposed commercial operation. Pilots are precursors to commercial projects and are not necessarily profitable ventures in themselves.

There may be no need to conduct a pilot if the technology is sufficiently well developed. *If a pilot is considered necessary, the most important reasons for conducting it should be identified and ranked in order of importance.* In designing the pilot, this ranking can be used to advantage to obtain the necessary answers to the most pressing questions. Where critical information is needed to design and operate a costly commercial-size operation properly, it is always prudent to obtain field confirmation of the expected field performance and cost through a pilot test. Bypassing the pilot phase on such large-scale operations would be warranted only under the most unusual circumstances – e.g., if the pilot test cannot provide the necessary confirmation, if there is a need for immediate action, or if the reward-to-risk ratio is high.

13.2 Design Considerations

Very little has been written on the procedures for designing a pilot for any type of displacement process, with essentially none aimed at thermal recovery. That is not to say that the results of pilot projects have not been discussed frequently, for much has been written on their interpretation. What is lacking is a step-by-step description of how one goes about designing a pilot to meet specific objectives, apparently because it is difficult to develop an approach that would be applicable to all the situations that could arise in practice. This does not mean that the available results, even those discussed in terms of waterflooding and other displacement processes, cannot be used to advantage in the design of thermal pilots. Quite the opposite.

For general information on the design of pilot projects, Smith[1,2] and Craig[3] provide the best summaries, though they concern waterfloods. Faulkner[4] discusses the steps in designing and implementing a tertiary miscible-gas-drive pilot in west Texas. Many points examined in that paper have a counterpart in thermal projects and will serve as useful background.

In designing a pilot, the following questions are representative of those that may arise.

1. Is the site selected for the pilot representative of the commercial project?

2. Would the recovery mechanisms be present in the same proportions in both the pilot and the main project?

3. Would the production rates and the recovery efficiency of the pilot be comparable with that of the main project?

4. What data should be compiled to evaluate the pilot?

5. Is there a need for new wells?

These five questions are discussed in the following. The relative emphasis to be placed on each of these five questions, or any others that may arise, is determined by the order of importance of the reasons for conducting the pilot.

Pilot Location

In general, a pilot should be conducted, for the following reasons, in a portion of the reservoir representative of the field as a whole.

1. Net thickness, porosity, and oil saturation determine the oil in place and the amount of oil recovered.
2. Gross thickness is a measure of the amount of heat required to heat the oil.
3. Permeabilities affect the flow rates.
4. Initial gas caps, bottom water, and layering affect where the injected fluids flow in the reservoir.
5. The continuity of shales affects the ease with which fluids move normal to the depositional surfaces of the reservoir and the amount of gravity override that would occur.
6. The local degree of depletion affects the ease of injection.
7. The presence of free gas may facilitate the formation of an oil bank.

Whenever differences in any of these factors would affect significantly the nature of the response from the pilot area relative to that in the commercial operation, an assessment must be made of alternative locations, alternatives to the proposed pilot, or methods of correcting for the differences between the pilot and the proposed commercial operation.

When the primary purpose of a pilot is to test a new process (rather than to determine how an existing process works in a given field), it may be more important to select a very favorable portion of the reservoir to facilitate the evaluation of the process.

Pilot vs. Full-Scale Mechanisms

Just as it is desirable to run a pilot in a representative part of the reservoir, it is also desirable to run it so the prevailing mechanisms in the pilot are essentially the same as those expected in the commercial operation. Differences in mechanisms may have any number of causes, some connected with the reservoir properties and others with the operating variables. For example, consider the consequences of completing the injection well in the pilot in an atypically tight portion of the reservoir. Relatively high injection pressures would be required to deliver steam at a certain rate, which would increase the steam temperature. Furthermore, the tight reservoir would cause the ratio of viscous forces to buoyant forces to be fairly high and would tend to reduce the amount of steam override in the pilot. If the injection rate is reduced to obtain the injection temperature expected in the commercial operation, then the heat losses from the well would cause the downhole steam quality to be lower. And the heat losses to the adjacent formations would result in a somewhat smaller steam-swept zone per unit of heat injected, which could lead to poor pilot results.

In another instance, consider the consequences of having pilot producers completed near an edgewater boundary. Atypical encroachment and inflow of edge water may reduce the average temperature of the produced fluids, thus increasing the viscosity of the oil and adversely affecting its production rates. Alternatively, the presence of the nearby edge water may promote more pronounced advance of the displacement zone toward it, giving the impression that the reservoir is anisotropic. Or it may allow the displaced oil to move past the producers with relative ease, giving the impression of a poor sweep efficiency.

Spacing can affect production mechanisms greatly. Close spacing between wells often is desired in a pilot to obtain results early. But with close spacing, the viscous forces would be relatively small (for a given injection rate), thus minimizing override effects in steam and combustion drives relative to what could be expected in the commercial operation. (An example of the difference in behavior resulting from different ratios of buoyant forces to viscous forces is shown in Fig. 4.1.) In thermal operations, increased spacing results in increased heat losses to the adjacent formations, since it takes longer to flood wells that are widely spaced.

The advantages of carrying out a thermal pilot on close spacing must be weighed against the disadvantages. On the one hand, results can be obtained early; on the other, the fraction of the injected heat remaining in the reservoir at the end of the pilot project can be significantly larger than the corresponding amount in the expanded project. The resultant difference, if not recognized, could distort estimates of both the recovery performance and the economics of the expansion significantly.

It is important in designing the pilot, therefore, to be reasonably sure that the results can be scaled up even when the pilot entails mechanisms that are slightly different from those in the expansion. Normally, we can be quite confident of the predicted scaled-up performance when (1) the mechanisms are expected to be about the same in both cases, (2) the fluid and reservoir properties and operating variables are essentially the same in both cases, and (3) the actual pilot performance does not differ significantly from that which was predicted (or could have been predicted had the pilot or its preparation generated new information about reservoir properties).

Well Rates

Well performance can be a critical indicator in assessing the performance of a pilot. A damaged injection well, if taken as being representative of the behavior of other (unimpaired) wells, will suggest air-compression or steam-generation costs that are higher than necessary. Or it may indicate an injection rate too low for the process to be attractive and may result in a lost business opportunity that could have been identified through proper engineering. Conversely, a well that has been fractured inadvertently may indicate an injectivity that could not be attained in the commercial project without stimulating the wells. And what has been said of damaged and fractured wells applies to wells completed in portions of the reservoir that are tighter or more permeable than average.

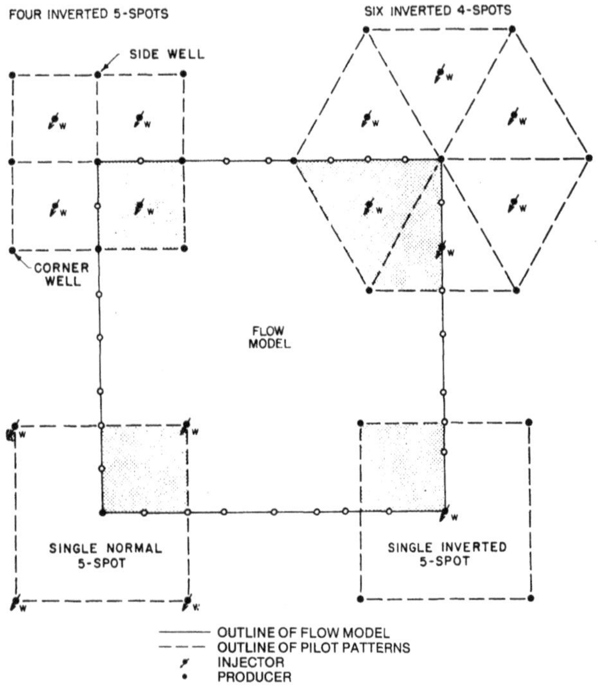

Fig. 13.1 – Idealized pilot patterns studied by Dalton *et al.*[6]

These effects are the same at the producer, but the psychological impact of oil in the tank is greater than just having poor or good injectivity. *A damaged producer can have disastrous consequences on the oil recovery rate.* The reason is that oil displaced by the injected fluid can be pushed past the producer, especially in depleted reservoirs containing significant gas saturations. This bypassing also can occur in fractured or stimulated wells, although to a lesser degree. Bypassing of oil is especially severe in producers placed at or near the periphery of the pilot area. For this reason, it is common to surround at least one producer with a number of injectors, leading to what is called "normal" patterns, such as those illustrated in Fig. 13.1.

The importance of the condition (and location) of wells is illustrated best by the wide range in the production performance reported for adjacent wells in cyclic steam injection projects. Because the condition of the production wells greatly affects the oil production rate, *there is merit in considering more than one producer per pilot.* When there is more than one producer, either the average production per well is taken as representative or the pilot performance is analyzed well by well to form a basis for predicting the behavior of the commercial development.

Data Gathering

The type of data to be obtained from a pilot, as well as the means used to interpret the results, follows directly from the primary purpose of the pilot. Injection and production rates, pressures, and temperatures often are measured routinely. Pressure and temperature measurements generally are obtained at the wellhead and corrected to downhole conditions. However, they sometimes are obtained at depth, where gauges or bubble tubes are used to measure pressure and thermocouples are used to measure temperature. The quality of the data desired, together with their value and the cost of acquiring them, ultimately determines the approach used.

Other types of data that often are measured include (1) concentrations of oxygen and other gases in samples from production and sampling wells, (2) changes in the composition of the produced crude and water, (3) vertical and horizontal distributions of saturations, pressure, temperatures, and compositions, (4) changes in rock properties, (5) changes in the viscosity and density of the crude, and the salinity and density of the water, (6) corrosion data from coupons in wells and surface lines, or from examination of tubing, and inhibitor performance, (7) lifting efficiency data obtained through pumping, flowing, or gaslift tests, (8) fuel consumption, and (9) costs of treating injected and produced fluids.

The type of wells from which data generally are taken, and the timing involved, are illustrated schematically in Table 13.1. Usually, observation and sampling wells are drilled and completed especially for their purpose. In Table 13.1, it is assumed that those wells provided core samples.

The frequency with which the data are taken varies widely. Continuous measurements seldom are made. More commonly, key items such as steam quality in steamfloods and production temperatures and oxygen concentratons in the produced gas of combustion projects are measured daily or weekly. Steam quality, since it is difficult to determine accurately (see Section 11.4), often is estimated. Total injection and production rates and cumulative volumes are measured daily. Because of the limitation of test facilities, rates from individual producers generally are measured every few days at best; the rest of the time, the total rate is measured for a group of producers and rates are allocated to each well on the basis of previous performance. Automation is used extensively in some pilot projects to improve control and efficiency of operations, reduce manpower requirements, and obtain frequent data in a systematic manner.

Are New Wells Needed?

Well requirements for pilot operations are similar to those for conventional projects (see Sections 11.2, 11.4, and 11.5). If anything, wells should be designed and selected even more carefully in pilot operations. A pilot implies lack of experience in the application of a process to a new field. It may even be a pilot of a new process. This lack of background alone should prompt the prudent operator to be especially careful on all aspects of the pilot, particularly the wells.

In thermal operations, the producing life of wells that get hot bears significantly on the success of the project. As already discussed in Chaps. 9 and 11, for example, wells intended for cyclic steam injection must be completed with special care. For this reason, the wells in cyclic steam injection projects are almost always new. Because a well is the smallest unit that can be tested and a cyclic steam injection project

TABLE 13.1 – TYPICAL LOCATION AND TIMING OF DATA GATHERING IN PILOTS

	Type of Well				
Data	Injector	Producer	Observer	Sampler	Core Hole
Temperature	B,D,A	B,D,A	B,D,A	–	A
Pressure	B,D,A	B,D,A	B,D,A	B,D,A	A
Flow rate	D	B,D	–	D (?)	–
Concentration[a]	D	D	–	D	–
Composition[b]	B,D,A	B,D,A	–	B,D,A	A
Distribution[c]	B,D,A	B,D,A	B,D,A	B,D,A	A
Rock properties[d]	?	?	B (?)	B (?)	A
Fluid properties[e]	–	B,D,A	–	B,D,A	A
Corrosion	D (?),A	D,A	A (?)	A (?)	–
Lifting efficiencies	–	D,A	–	–	–
Injectivity or productivity index	B,D,A (?)	B,D,A	B,D,A (?)	–	–

B = before pilot starts.
D = during pilot operation.
A = after pilot terminates.
[a] May include gas analysis, tracers, salinity, pH, and organic acid content.
[b] May include gas, crude, water, and rock composition before, during, and after the project.
[c] May include temperature, saturations, concentrations, and compositions.
[d] May include porosity, permeabilities, and mechanical properties.
[e] May include viscosity, API gravity, densities, and distillability.

involves a large number of individual wells, such a project is relatively easy to pilot and optimize.

Air injection wells do not get heated except near the ignition interval. An existing well with casing in good condition, with no leaks and with an interval on bottom that is still well cemented, is probably suitable as an air injector. In thick formations and when spontaneous ignition is used, appropriate steps should be taken to protect the casing in the pay interval against backburning. For steam injection, the casing in such a well would be likely either to buckle or to expand and rise above ground level, putting stress on surface connections unless such a rise was considered and accommodated in the design. If steam injection is shut down for any reason (and there are few operations with no shutdowns), the casing is placed in tension as it cools, often causing the casing to part at one or more joints. For hot-fluid injection, therefore, new wells generally are used.

The reverse is generally true at producers. Produced steam will be at low pressures and, consequently, relatively low temperatures. Temperature fluctuations at these lower levels rarely will cause failures in suitable cemented casings. Accordingly, old producers often serve well in steam and hot-fluid processes.

Producers in combustion operations require special attention for three reasons. One reason is that large amounts of heat may be released when oxygen breaks through into the well and produced oil is burned; the temperature may rise to levels high enough to melt or soften the casing. To alleviate such problems, the liners and bottom joints sometimes are made of metals capable of withstanding extremely high temperatures. The special liners may be installed in existing wells that are well cemented and in good condition, but the special metal bottom joints require that the well be new. Wells sometimes are protected by a monitoring/control system that detects sharp temperature rises or high temperature levels and automatically activates a cooling system. The interpretation of oxygen concentrations in the produced gas stream requires special attention since part of the oxygen entering the wellbore may be reacting with the crude downhole (see Casing and Tubing Failure in Section 11.5).

Another reason for giving special attention to producers is that they are subject to corrosion, which is likely to cause casing failure unless the casing is protected from the effluent stream and high temperatures. Although these problems also can be alleviated through the use of special metals, it is generally sufficient simply to use inhibitors and cooling water.

The third reason is abrasion, which results from the high velocities at which the combustion gases exit the formation and the well. Abrasion has been an important contributor to the failure of downhole, wellhead, and surface equipment. Sand control measures, pipe bends with large radii of curvature, and special materials are used to alleviate abrasion. All in all, the problems in combustion producers are often severe, especially when a producer is operated in a dry combustion project at high gas production rates. For these reasons, combustion producers often are drilled and completed especially for the purpose.

When the primary objective of the pilot is to test a new process (as opposed to determining how an existing process works in a particular field) every effort should be made to ensure that the wells are adequate. A well failure while testing a new process (whether failure is due to a poor choice of existing wells or to poorly designed and completed new wells) can set back development of the process for years.

13.3 Pilot Design

This section considers possible differences in the performance of pilot and full-scale projects and steps that might be taken to minimize, interpret, and predict these differences. There is a considerable

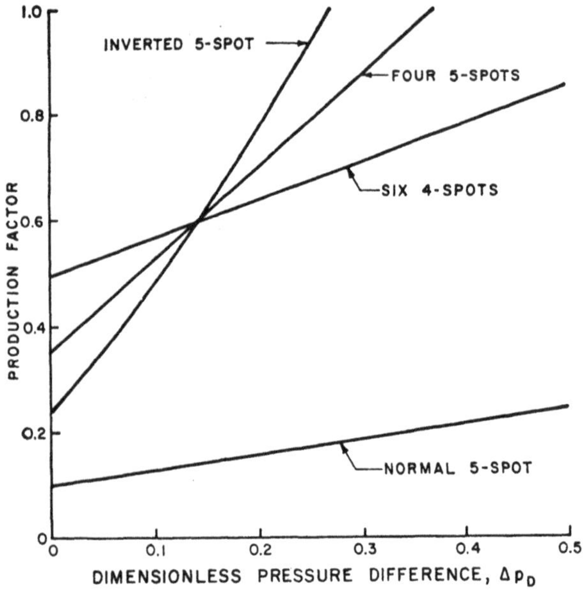

Fig. 13.2 – Effect of operating conditions on production factor.[6]

body of literature dealing with the differences in sweep efficiency and recovery between pilot and fully developed pattern floods, but there is little specifically directed at thermal recovery processes. The actual procedures for designing a pilot vary widely, again depending on the primary objectives of the pilot. Little planning is necessary when the primary purpose is, for example, to determine injectivity, but considerable simulation and analysis are required in support of a large and costly project aimed at extrapolating pilot performance to commercial performance. Here, the use of generalized well-pattern results and the use of physical and numerical simulation of a specific project are discussed.

Predictive Methods

Generalized results obtained over the years by means of potentiometric and other physical models, analytical results, and numerical simulation can serve as useful *guides* in planning pilot operations. Their limitations will be pointed out near the end of this section.

The effect of the number of wells (the size of the pilot) on the overall production performance has been studied by Rosenbaum and Matthews,[5] Dalton *et al.*,[6] and others. Dalton *et al.*[6] used potentiometric and flow models to study the effects of an initial free-gas saturation of 0.15 and of the relative pressures between the reservoir, the injector, and the producer on the waterflood production performance of the idealized pilot patterns shown in Fig. 13.1. The dimensionless pressure parameter Δp_D is defined as

$$\Delta p_D = \frac{p_i - p_{pdh}}{p_{idh} - p_i}, \quad \ldots\ldots\ldots\ldots\ldots\ldots (13.1)$$

and is used to show the effect of the relative pressures on the escape of oil outside the pattern area. Here, p_i represents the static reservoir pressure, p_{pdh} is the bottomhole producing pressure, and p_{idh} is the bottomhole injection pressure. The effect of the well pattern on the production factor, given by

$$F_p = \frac{1.5 \text{ PV of liquid produced}}{\text{liquid injected after production started, PV}}, \quad \ldots\ldots\ldots\ldots (13.2)$$

is illustrated in Fig. 13.2. This shows that a large number of producers per pattern is necessary if the liquid production is to be a substantial fraction of the liquid injection after fillup or when a substantial increase in liquid production first occurs. As the pressure difference between reservoir and producer increases, or that between injector and reservoir decreases, the production factor increases – i.e., less of the displaced liquids is pushed beyond the producers. With fully developed patterns, as much liquid would be produced as is injected after fillup, when shrinkage effects are negligible.

Fig. 13.3, from Rosenbaum and Matthews,[5] also shows that the production rate in an unconfined pilot is only a fraction of the injection rate. These results (which were obtained for $\Delta p_D = 0$, for a ratio of the side of a five-spot to the well radius equal to 3,600, for initial gas saturations different from those reported by Dalton *et al.*[6], for unit mobility ratio, and for swept zones extending beyond the pattern wells) show that the more patterns in the pilot, the closer the production rate comes to the injection rate.

Dalton *et al.*[6] report that "the area which supplies oil to the pilot producers divided by the basic pilot pattern area" – a value they call the areal recovery factor – is rather insensitive to the amount of liquid injected after breakthrough. Values reported in Fig. 13.4 illustrate that all pilot patterns produce from areas larger than that of the basic pattern at sufficiently large values of the dimensionless operating pressure Δp_D. The areal recovery and production factors of a normal five-spot pilot, with one center producer, are relatively insensitive to values of Δp_D.

The production rate of individual wells reported by Rosenbaum and Matthews[5] indicates that inner producers more closely approach the response of a fully developed pattern (Fig. 13.5).

The results presented in Figs. 13.1 through 13.5, though developed primarily for waterfloods, point out certain concepts that are applicable to pilot operations for any type of displacement process. Oil migration, the operating pressure parameter, the effect of an initial gas saturation, the effect of the number of patterns in a pilot, the number of producers in small pilots, the ratio of production rate to injection rate, the areal recovery factor, the importance of mobility ratio, and the distribution of production between sheltered inner wells and those on the periphery are factors that can be evaluated at least qualitatively in light of published results developed primarily for waterflooding.

Actually, the design and interpretation of thermal pilots are complicated further by possible differences between the temperature distributions in pilots and those in commercial operations and sometimes by relatively high (unfavorable) mobility ratios. Although personal experience and previously published results are two obvious sources of information on the design of a thermal pilot – and they may even provide rough estimates for some of the design parameters – there is no substitute for making studies specific to the proposed reservoir and process when sufficient information is available on *both*. For thermal recovery processes, there are very few published papers on the use of scaled physical models[7,8] or on numerical simulation studies[9] for pilot design.

Prats[7] reported results of scaled physical model laboratory experiments comparing the performance of steam drives in pilots and full-scale projects under conditions representative of the Peace River field in Alberta. Five sets of conditions and properties for the reservoir were considered, as shown in Table 13.2, but only Prototypes 2 and 6 were used in the pilot investigation. The seven-spot well pattern was chosen because of its relatively high ratio of producers to injectors, which is an important consideration in reservoirs where plugging by cold tar is likely to occur. The pilots considered were of either three or seven seven-spot patterns, with the same spacing as the full-scale projects. As shown in Fig. 13.6, pilots were considered to be affected by an aquifer, whereas the full-scale projects were not. A number of operating policies were considered for the steam drives, all aimed at reducing excessive production of steam, which tends to advance rapidly through the basal water zone (Table 13.2).

As Fig. 13.6c shows, the center well in the three seven-spot pilot is relatively sheltered from the aquifer inflow by a number of intervening production wells. In the seven seven-spot pilot (Fig. 13.6d), the inner ring of producers also is sheltered. Table 13.3 summarizes the comparisons between the sheltered wells, the total pilot, and the full-scale performance. Note that the ratio of producers to injectors is 4 1/3, 3 3/7, and 2:1 in the three seven-spot, seven seven-spot, and full-scale well patterns, respectively. The relatively large number of wells in the pilot is one of the reasons why the total oil/steam ration (labeled OSR in Table 13.3) in the pilot is about the same as that for the full-scale project. Recoveries per well are substantially lower in the pilots. Even the sheltered wells recover only about 70% of the production of a well in a full-scale development. Differences in performance between pilot and full-scale projects arise from the relatively small size of the pilot and include effects resulting from differences in the production-well/injection-well ratio, in drainage boundaries, and in aquifer influences on peripheral producers. The results of this study, which is only highlighted here, provided the technical basis for the pilot design at Peace River.

Another paper that discusses how to design some elements of a pilot for a steamflood project is that of Gomaa *et al.*[9] There are many publications that discuss *what* was done in pilot projects, but Gomaa *et al.*'s paper is one of the few known to describe some of the actual steps taken to design a thermal pilot. To quote, "Simulation results were used to

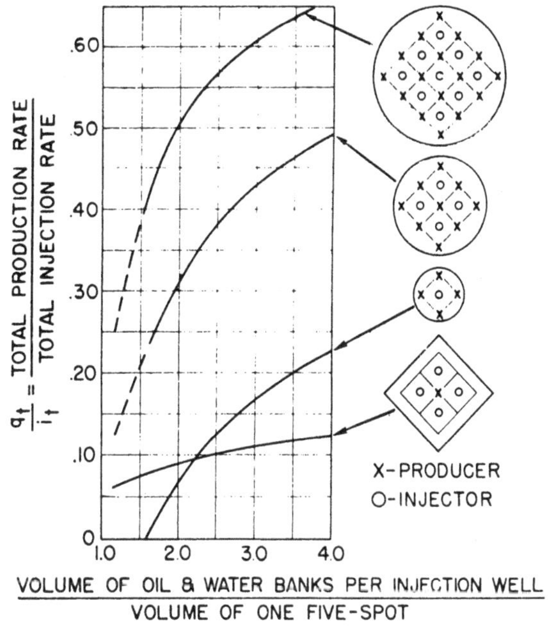

Fig. 13.3 – Effect of pilot pattern on total production rate for unit mobility ratio.[5]

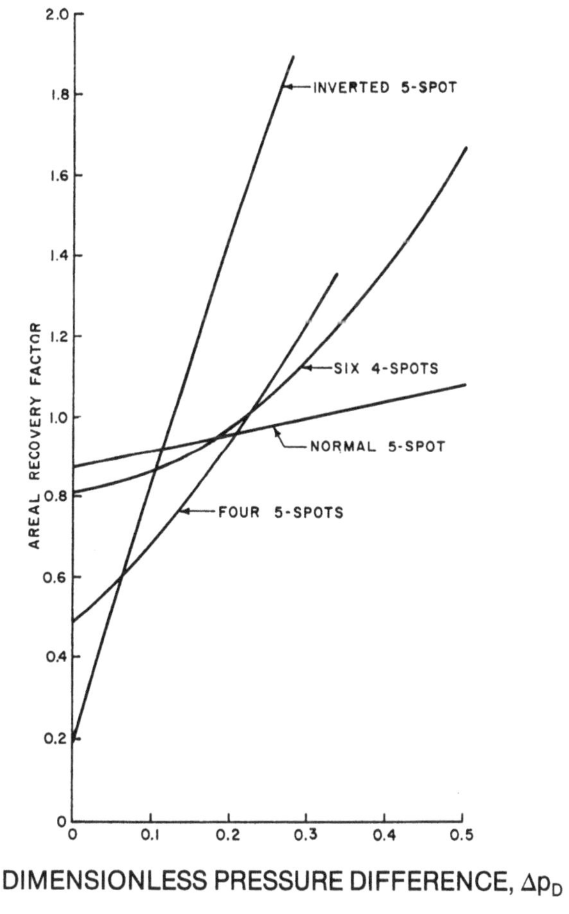

Fig. 13.4 – Effect of operating conditions on areal recovery factor.[6]

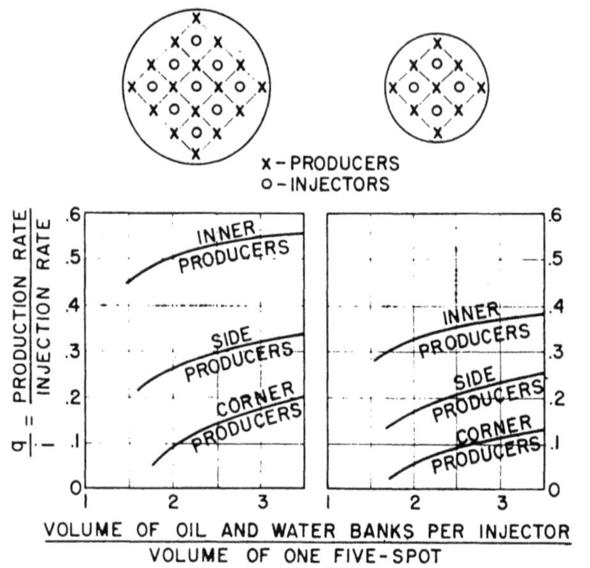

Fig. 13.5 – Rate response of individual producers for unit mobility ratio.[5]

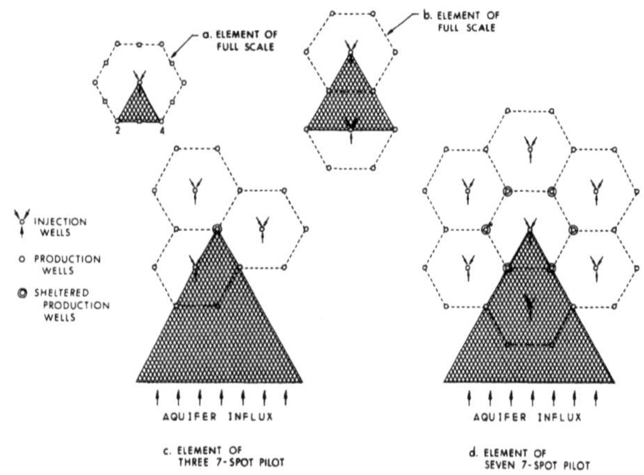

Fig. 13.6 – Modeled elements of symmetry.

determine injector completion intervals and to position three observation wells." Although the unconfined nature of the pilot was *not* taken into account in this study, the choice of steam injection rate and quality in the downdip and updip injectors was based on simulator results. The calculated preferential updip movement of the temperature and steam saturation is shown in Fig. 13.8. The downdip patterns received relatively large rates in anticipation of the upward migration of steam, and slightly higher steam quality was used in the downdip injectors. Two of the observation wells were located updip of an injector to take advantage of the preferential updip movement of heat and steam and to obtain the most information in a reasonable time. Calculations showed that steam zone advance updip would be approximately twice as fast as the one downdip.

Sensitivity to Process Variables

The preceding section discusses the various types of concepts, information, and model techniques that can be used in designing pilots. It states that results range from qualitative to the quantitative, depending not only on the models themselves but on whether they represent generalized and highly idealized situations or whether they specifically approximate the proposed pilot location and process. This section emphasizes sensitivity studies that develop information for improving the pilot results and their impact on the possible expansion to a commercial operation. The distinction between the two sections is one of emphasis, this one dealing specifically with pertinent features of the reservoir at the pilot location and with the process under consideration.

There is little in the literature that explains how operational variables are selected. The choice by Gomaa *et al.*[9] of steam injection intervals, rates, and quality and of the location of the observation wells in the Monarch sand of the Midway-Sunset field resulted from a sensitivity study using a thermal numerical simulator. Rock and fluid properties of the Monarch sand are given in Table 13.4. Although sensitivity studies can be made by varying more than one variable at a time, Gomaa *et al.* usually varied only one, as indicated in Table 13.5. The varied

TABLE 13.2 – PROTOTYPE PROPERTIES CONSIDERED IN PEACE RIVER STEAM DRIVE PILOT[7]

Prototype Identification	Thickness of Upper/Lower Layers (ft)	Permeability of Upper/Lower Layers (md)	Porosity of Upper/Lower Layers, (fraction)	Tar Saturations of Upper/Lower Layers (fraction)
2	60/30	345/1665	0.27/0.27	0.80/0.65 (0.75 avg)
3 (Homogeneous)	60/30	345/345	0.27/0.27	0.80/0.65 (0.75 avg)
4	51/13	345/1665	0.27/0.27	0.80/0.55 (0.75 avg)
5	36/9	345/1665	0.27/0.27	0.80/0.55 (0.75 avg)
6	74/16	215/1485	0.28/0.28	0.80/0.55 (0.775 avg)

Note: All wells have fractures. The lower layer coincides with the lower "water zone" in all prototypes except Prototype 6, where the "water zone" occupies the basal 9 ft. In the homogeneous Prototype 3, the basal water zone is 30 ft thick.

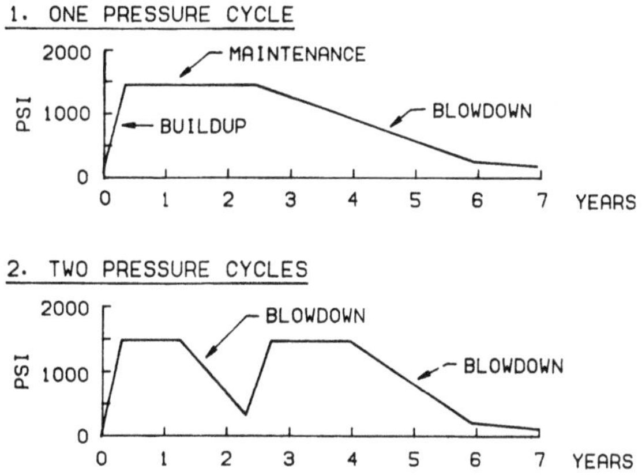

3. STEAM DRIVE - PRESET CONDITIONS

4. CONTROLLED STEAM DRIVE - RATES, TEMPERATURES AND/OR PRESSURES CONTROLLED TO CURTAIL EXCESSIVE HEAT PRODUCTION

Fig. 13.7 – Description of operating policies considered in Peace River steam drive pilot.

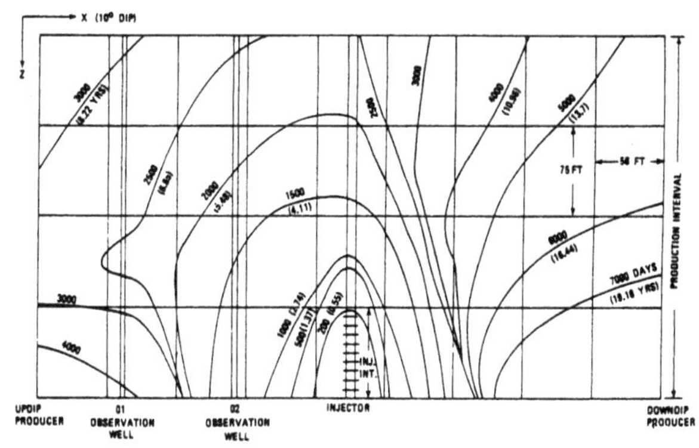

Fig. 13.8 – Simulation study of Monarch Sand, Midway Sunset field steamflood – 10% steam saturation profiles in a vertical cross section through updip and downdip producers and the injector.[9]

parameters included the relative permeabilities, bottomhole pressure, steam quality and injection rate, well spacing, sand thickness, productivity index, initial gas saturation, vertical permeability, and selected properties of a possible shale barrier. It is clear from the results presented in Table 13.5 that very low bottomhole producing pressures are desirable and that increasing the pattern size from 3.6 to 10 acres has a disastrous effect on the recovery performance unless the injection rate is proportionally increased.

It is also clear that sensitivity studies required quite a large number of simulations to investigate properly even the most pertinent parameters. For this reason, they generally are limited to projects where the potential is great enough to warrant the effort. In small projects, sensitivity studies rarely are conducted unless they and the pilot results can be applied in a number of analogous reservoirs. Factorial design procedures, which aim to reduce the number of runs required to ascertain sensitivity to operating variables, have been used only infrequently in thermal recovery processes, a notable exception being the study of Sawyer *et al.*[10] on wet combustion.

TABLE 13.3 – RELATIVE PERFORMANCE OF SHELTERED WELLS IN PEACE RIVER STEAM DRIVE PILOT[7]

Prototype, Spacing, Pilot Pattern	Operational Policy	Full-Scale Average Oil Recovery (PV)	Full-Scale Average Oil/Steam Ratio (bbl/bbl)	Pilot Average Oil Recovery (PV)	Pilot Average Oil/Steam Ratio (bbl/bbl)	Sheltered Well Average Oil Recovery (PV)	Sheltered Well Average Oil/Steam Ratio (bbl/bbl)	OSR Sheltered Wells / OSR Full-Scale*	Number of Pilot Runs	OSR Pilot / OSR Full-Scale**
Prototype 6										
7.0 acres										
Seven Seven-spot	1 cycle	0.45	0.21	0.44	0.20	0.34	0.15	0.71	5	0.95
Seven Seven-spot	2 cycle	0.48	0.23	0.45	0.20	0.37	0.16	0.70	7	0.87
Seven Seven-spot	Drive	0.48	0.24	0.41	0.21	0.33	0.17	0.71	3	0.88
Seven Seven-spot	Controlled drive	0.49	0.24	0.40	0.22	0.29	0.16	0.67	6	0.92
Prototype 2										
3.5 acres										
Three Seven-spot	1 cycle	0.45	0.20	0.46	0.14	0.48	0.15	0.75	2	0.70
Seven Seven-spot	1 cycle	0.45	0.20	0.41	0.17	0.31	0.13	0.66	9	0.85
Seven Seven-spot	Drive	0.59	0.17	0.28	0.18	0.22	0.14	0.82	2	1.06

Note: The oil/steam ratios of the sheltered well and pilot are in proportion to their oil recovery. OSR = oil/steam ratio.

*Defined as sheltered-well efficiency.

**Defined as pilot efficiency.

TABLE 13.4 – ROCK AND FLUID PROPERTIES OF MONARCH SAND, MIDWAY SUNSET FIELD[9]

Gross thickness, ft	0 to 600 (avg 350)
Net oil sand, ft	0 to 500 (avg 260)
Porosity under zero stress	0.327
Porosity under overburden stress	0.27
Permeability under zero stress, md	1,935
Permeability under overburden stress, md	520
Oil saturation in depleted zone	0.36
Oil saturation in undepleted zone	0.59
Water saturation	0.41
Oil content in undepleted zone, STB/acre-ft	1,236
Oil gravity, °API	14
Reservoir temperature, °F	105
Reservoir pressure, psia	75
Oil viscosity at reservoir temperature, cp	1,500
Oil viscosity at 300°F, cp	10

13.4 Evaluation of Pilot Performance

Depending on the primary objectives of the pilot, the evaluation of pilot results may be quite simple or it may involve many engineering and scientific disciplines using very sophisticated technology. Data Gathering in Section 13.2 discusses briefly the type of data that can be obtained in pilot projects. This section discusses some of the more common means of evaluating pilot projects: temperature observation wells, cores, well tests, and mathematical analyses.

Temperature Observation Wells

The location, number, and method of completing temperature observation wells, as well as the means used to acquire data, will depend on the primary objective of those wells. To obtain the temperature response early, the observation well must be relatively close to the heat source. To chart the rate of progress of the heat front, to determine areal anisotropy in the shape of the heated zone, and to determine the size and the vertical location of the heated zone within the reservoir, wells must be at varying distances. Analysis of the rate of growth of the burned volume in a combustion project, for example, may be used to provide information on the fuel burned. How frequently measurements should be taken depends on their purpose. Normally, data are taken more frequently (1) during start-up, (2) when the information is critical, (3) when anomalous behavior must be monitored or investigated, and (4) when there are rapid and important changes in temperature. Otherwise, temperatures are recorded only daily or even weekly. Examples of completions for temperature observation wells have been given and discussed in Chap. 11, under Surveillance of Operations. Note here that sometimes it is desirable to complete and equip the wells so temperatures can be monitored both above and below the injection interval. This is useful in determining heat losses and also is especially desirable in zones where the lateral continuity of bounding shales or the downhole integrity of the injection profile control system is questionable.

A very simple pilot was conducted in Mene Grande,[11] Venezuela in 1957 to test the steam drive process. It consisted of an existing production well, one newly drilled and completed injection well, and an observation/sampling well.[11] Steam injection started in March at 188 to 220 B/D, and steam temperatures corresponding to the subsurface pressures were measured at the observation/sampling well 50 ft away at a cumulative steam injection of 10,000 bbl. Calculation of radial steam-front advance based on the observed steam-swept interval (of about half the formation thickness between two packed-off shale stringers at the observation/sampling well) agreed well with the ob-

TABLE 13.5 – SIMULATION SENSITIVITY RESULTS FOR A FIVE-SPOT, MONARCH SAND, MIDWAY SUNSET FIELD[9]

Case No.	Parameter(s) Varied	Base Case Value	Oil Recovery (% OOIP) 2 Years	4 Years	6 Years	8 Years	Time of Steam Breakthrough (years)
1	Base Case	–	0.6	5.4	25.2	56.5	6.3
2	k_r curves (Ref. 20)	Set 2, Ref. 19	9.2	25.2	32.9	40.3	3.6
3	BHP = 60 psia	20 psia	0.5	3.3	13.5	40.8	7.0
4	Steam Quality = 0.6	0.7	0.5	3.3	15.2	45.7	7.0
5	Pattern area = 10 acres	3.6 acres	0.2	0.3	0.5	1.13	13.7
6	Pattern area = 10 acres Injection rate = 1,389 B/D	3.6 acres 500 B/D	0.5	6.0	27.2	53.8	6.0
7	Injection rate = 1,000 B/D	500 B/D	8.1	59.0	74.5	76.8	3.3
8	Sand Thickness = 150 ft Injection rate = 500 B/D	300 ft 500 B/D	6.0	54.5	73.0	78.0	3.1
9	Producer PI = 200 RB/D/psi	800 RB/D/psi	0.5	4.1	17.4	53.3	6.6
10	S_{gi} = 0.40 in top 60 ft	S_{gi} = 0	0.7	5.4	24.0	47.8	–
11	k_H – 1,900 md	950 md	0.4	5.1	31.3	46.5	5.4
	Permeability barrier*						
12a	k_V = 950, k_H = 1,900 md	950, 1,900	0.7	6.6	27.4	48.4	5.8
12b	k_V = 150, k_H = 300 md	950, 1,900	1.7	7.6	24.2	55.4	6.5
12c	k_V = 15, k_H = 30 md	950, 1,900	2.8	16.8	30.0	41.2	4.3
12d	k_V = 1.5, k_H = 3 md	950, 1,900	5.4	23.3	31.4	32.8	3.6

*Barrier considered is between 90 and 120 ft from the bottom of the 300 ft formation. Steam was injected in bottom 60 ft of formation in Case 12, as compared with 300 ft in the base case.

served arrival time. However, the steam front arrived at the producer earlier than predicted by the same model, suggesting increasing bypassing or channeling of steam with time. An increase in production was observed at a cumulative steam injection of 22,000 bbl, and oil production stabilized at a flowing rate of 125 B/D shortly afterward.

The lack of confinement in the pilot pattern did not allow an accurate determination of the swept zone or of ultimate recovery. But the well response proved that significant oil production coud be obtained by steamflooding. The test was expanded in the same area to four inverted five-spots on commercial spacing.

Pollock and Buxton[12] discuss steamflood pilot results from the Nugget reservoir of the Winkleman Dome field. Steam was injected into a 40-acre inverted five-spot, with an observation well 125 ft from the injector. With this large spacing, production increases observed after 6 months of steam injection were interpreted properly as being primarily due to the increased driving pressure resulting from steam injection rather than to steam displacement. Heat broke through at the observation well after about 457 days, at a cumulative heat injection of 25.2×10^9 Btu. Calculations of the steam front location using the Marx and Langenheim[13] method, an average injection rate of 2.3×10^6 Btu/hr, an average injection temperature of 550°F, and a swept thickness of 85 ft indicated the steam front should have advanced radially to only 68 ft. Further calculations indicated the observed heat breakthrough time was consistent with an average thickness of less than 5 ft swept by the steam. A temperature profile measured half a year later, and shown in Fig. 13.9, indicates that steam was present over only a limited interval at the top of the Nugget Sandstone.

The observation well was put on continuous production after heat breakthrough. Although the oil production rate was expected to be low, it stabilized at 45 B/D, flowing. This unexpectedly high level of production was encouraging and was interpreted as being the result of conduction heating of the oil below the steam zone. An adjacent inverted five-spot on a more realistic 10-acre spacing also indicated that 50 B/D of oil "could be obtained from wells responding to steam injection," and the operation was expanded by drilling enough wells "to complete development of six 10-acre inverted five-spot patterns."

These two cases exemplify field pilot tests in which responses at both observation and production wells were analyzed and interpreted to gain insight into reservoir behavior. This eventually led to expansion of the projects.

Core Holes

It is often desirable to determine the extent and distribution of the swept zone and of any oil bank formed during a thermal project. Information obtained during the course of the pilot sometimes is supplemented by means of post-mortem core holes. Core holes are used not only to recover samples that

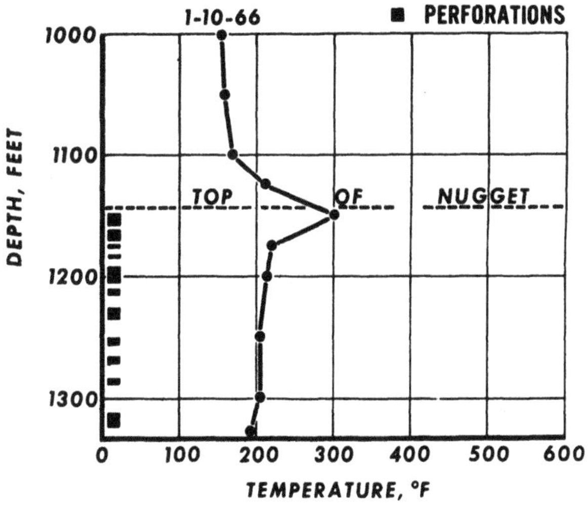

Fig. 13.9 – Temperature profile, Well 46, Winkelman Dome field.[12]

can be subjected to a variety of laboratory analyses but also for logging and well testing.

Coring after a thermal project requires special procedures. The temperature, pressure, and type of fluids in the portion of the reservoir to be cored must be considered, both to ensure control during drilling and coring operations and to reduce the amount of saturation redistribution just before and during coring. Although not essential in every case, it is often prudent to allow both the temperature and the pressure of the reservoir to drop before drilling into it, so as to maintain control of the well during coring. Should a well go out of control, a reservoir containing hot water at high temperature would generate large quantities of flowing steam. Any unused oxygen from a combustion pilot would ignite heated oil produced during a blowout. But apart from these safety considerations, blowouts would lead to loss of meaningful data from the core hole and even from the entire pilot, primarily because of the likelihood that (1) the swept zone would be resaturated with crude and water and (2) the formation interval in the well would exhibit saturations, temperatures, and liquid compositions unrepresentative of those existing at the time coring was begun.

Even when well control is maintained, fluid losses from the drilling mud may displace the heated oil both from the core and from the vicinity of the wellbore. This could lead to the conclusion that the pilot project was more efficient that it actually was. An estimate of the flushing resulting from coring operations can be made by using tracers in the water, by controlling its resistivity, and by comparing logging and extraction results.

Although these flushing effects would be reduced at lower temperatures, it is not always clear *how* best to achieve cooling in a pilot operation prior to coring. The quickest way to cool the reservoir is to inject unheated water, but that can cause additional displacement of oil from the test area. However, letting the reservoir cool by conduction would take

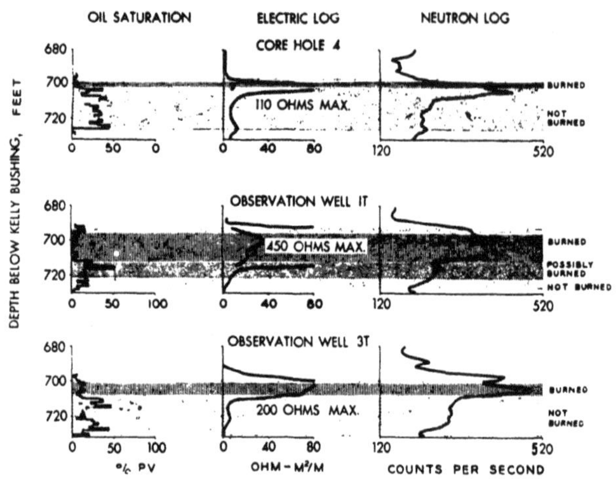

Fig. 13.10 – Post-thermal recovery coring and logging results from Wells CH-2, CH-3, and 2T, South Belridge field.[15]

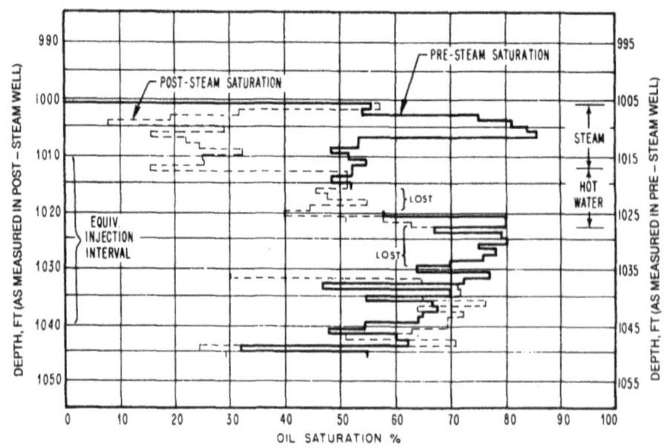

Fig. 13.11 – Comparison of core analyses of nearby wells, Inglewood field.[16]

too long. Moreover, the project volume can become resaturated because of the potential gradients resulting from local condensation of steam or because of gravity drainage from higher parts of the reservoir, and it can become desaturated by drainage from the project volume to lower parts of the reservoir. These flow movements, after the termination of the pilot but before coring, are aggravated in high-dip reservoirs and by pressure loss in the pilot area upon cooling, especially in steam zones. Interpretation of cores and boreholes, therefore, must take into consideration the *possibility* of (1) fluid migration during the time between the end of the pilot and the start of petrophysical studies and logging and (2) fluid displacement during the coring and drilling operation. Often, there is little fluid migration except in reservoirs of significant dip and where the reservoir has been allowed to cool over an extended time.

As discussed by Jorden and Campbell,[14] logging a well at elevated temperatures is not straightforward. Many types of logs are affected adversely by high temperatures, giving distorted responses that may lead to incorrect inferences as to the properties being measured; other types may be unusable altogether.

An early pilot in which the interpretation of the results was aided significantly by post-mortem coring was the in-situ combustion project at South Belridge.[15] The main conclusions of the three-hole coring program were that (1) the volume burned was at the top of the oil-bearing interval at each of the three locations and (2) there was a significant reduction in the average oil saturation (from 0.60 to 0.38) below the burned volume. A comparison of the core oil saturations and log responses is shown in Fig. 13.10.

Another case where coring has provided significant insight into the mechanisms of a pilot operation is reported by Blevins *et al.*[16] for a steam drive pilot at Inglewood. Comparisons of oil saturations before and after the pilot are shown in Figs. 2.5 and 13.11. As can be seen, the steam-swept zone was at the top of the sand, and the average oil saturation was reduced to from 0.20 to 0.25. Below the steam-swept zone, the average oil saturation was 0.42 in the core hole located 40 ft from the injector (Fig. 2.5) and 0.51 in the core hole located 135 ft from the injector (Fig. 13.11). These saturations compare with preflood values of 0.71 and 0.58 over the respective intervals in the two wells. These cores were taken in rubber-sleeve barrels using a low-loss, clay-based mud about 5 months after steam injection started. Steam injection was stopped during the coring. Subsequent production response indicates that "when injection stopped, the condensation of the vapor phase created a pressure sink and hot fluids originally displaced outside the first-line wells flowed back and were produced."[16] Coring results, thus, may reflect some resaturation.

One of the most complete accounts of the information obtainable from cores is given by Buxton and Pollock in the evaluation of a large-scale wet combustion project in a selected area of the Sloss field.[17] Buxton and Pollock not only describe the information obtained from the five core holes drilled after air injection was terminated but also discuss the considerations for their location. Data included foot-by-foot routine core analyses (porosity, permeability, and oil saturation), log analyses (compensated formation density, dual-induction laterolog, and temperature), photographs (black and white, color, and ultraviolet), visual examination, mineral analysis to indicate maximum temperatures to which samples had been subjected, microscopic examination of minerals (including scanning electron microscopy), and a drillstem test on one core hole. Of particular interest are the changes occurring in selected minerals that serve as temperature indicators, as shown in Table 13.6.

Sidewall samples from injection, production, or observation wells also provide information on changes in fluid content during the course of a pilot project.[15] They also may indicate the location of intervals where channeling occurs. Such samples, however, can be taken only in openhole completions,

TABLE 13.6 – MINERAL CHANGES WITH TEMPERATURE[17]

Temperature (°F)	Change
350	Glauconite, bright green pellets fade to grayish green
400	Glauconite changes color to gray or gray brown
480	10Å illite x-ray diffraction peak sharpens and mixed-layer minerals disappear
500 to 700	Chlorite starts to decompose, the 7.2Å and the 3.55Å peaks decrease, and the 14.5Å peak increases in intensity
700 to 800	Chlorite decomposes and becomes amorphous
950 to 1,000	Kaolinite decomposes

during drilling or during certain workover operations. Because of the probable flushing of the vicinity of the wellbore during drilling and workovers, sidewall samples obtained under those circumstances generally are aimed at determining the presence and concentration of nonmobile carbonaceous residues associated with cracking reactions in combustion operations. Even then, the interpretation of any carbonaceous residue in the samples must consider the possibility that they also could have resulted from combustion within the wellbore.

Well Tests

Well tests during or after a pilot project are the same as those carried out as part of surveillance operations and are discussed in Section 11.4.

Mathematical Analysis

Examples of mathematical analysis already have been given for the Mene Grande[11] and Winkleman Dome[12] fields. The latter used simple calculations based on the Marx and Langenheim[13] model to conclude that the steam zone was only a few feet thick – a conclusion confirmed by observed temperature profiles and subsequent production behavior.

Similar calculations have been made in combustion pilots and projects to determine the average reservoir thickness burned; the value is derived from the cumulative amount of air injected and the time at which breakthrough first is observed.

Field pilot results may be subject to different interpretations. In the South Belridge field[15] for example, a pronounced segregation of the injected air to the upper part of the pay was observed. Was this due primarily to buoyancy effects during the pilot, to higher absolute permeabilities at the top of the interval, to the presence of an initial gas saturation near the top of the interval, or to a combination of all three? Although a layered reservoir model without crossflow could account for the observed behavior within the accuracy of the data, there is no basis for assuming that vertical communication was poor. Years after the test was completed, Prats[18] was able to provide an alternative interpretation of the temperature response on the basis of complete gravity segregation of the injected air. The task of assessing the contribution of the possible causes for the observed behavior would have been easier had thermal numerical simulators been available when the pilot first was analyzed.

But even with the aid of numerical simulators, the data base may not be sufficient to provide unambiguous interpretations. The Sloss field wet combustion project is an example.[16] Although Buxton and Pollock conclude that the combustion front "moved through the most permeable part of the pay," they agree that "it cannot be unequivocably proved with the data" that the combustion front was not gravity controlled. They also recognize that gravity segregation of the injected water and air in the wellbore could have resulted in "air predominantly" entering "the top of the pay and water the bottom." This not only would have initiated gravity separation in the reservoir but also would have resulted in dry combustion near the top of the pay and no combustion at all near the bottom. However, even if some distribution of dry, wet, and completely quenched combustion did occur over the vertical extent of the reservoir, the project appears to have been a technical success.

The practical importance of resolving ambiguities and inconsistencies of pilot results is that it provides a firm basis for properly extending these results to other reservoirs or conditions, while it reduces the possibility of being confronted with an unexpected performance. In addition, with early resolution of such anomalies, reliable design and operating guidelines likely could be obtained from fewer and smaller pilot projects.

13.5 Prediction of the Expansion Performance

The last and most important step, and the reason for carrying out the pilot, is to use the information obtained during its evaluation to make a decision whether and how a commercial project should be undertaken. A pilot can be designed to give operational proof or operational experience and data about pattern size, rates, extent of gravity override, etc. A decision to proceed with a commercial project is based on a favorable assessment of the process and of its economic potential. Information derived during the pilot evaluation is used to determine or predict the production performance of the large-scale project under consideration. For example, the air required to burn a unit volume of reservoir rock in a combustion operation or the oil saturation remaining after steamflooding are some of the data used to predict the response of the expansion. Ideally, the prediction of the production performance of the full-scale operation considers the effect of the actual well location with respect to the character of the reservoir, the distribution of fluids within it, and the range of operational variables. Special care must be taken in predicting performance if it is thought that the displacement mechanisms in the expansion will be different from those in the pilot. *Prediction of full-scale performance on the basis of a thermal pilot is one of the toughest jobs in reservoir engineering.*

As stated at the beginning of this chapter, not all thermal projects warrant pilot testing. Key information for predictive purposes may be obtained from laboratory experiments, from experience in other fields, or from information obtained through numerical simulation. As with any other field operation, piloting a project must be justified. Piloting seldom is warranted in a small project. On the other hand, before large and costly projects, piloting can almost always be justified. This is especially true when some new modification is to be used or when the economic incentive is low and the risks are high. Of course, any new process that has the requisite economic and technical potential should be field tested on an appropriate scale. Pilots also can be used to choose between attractive competing processes, either thermal ones (e.g., combustion vs. steam drive or wet vs. dry combustion) or nonthermal ones (e.g., steam vs. cold-water drive).

References

1. Smith, C.R.: "Good Pilot Flood Design Boosts Field Project Profit," *World Oil* (Nov.-Dec. 1973; Jan.-July 1974) Parts 1 through 9.
2. Smith, C.R.: *Mechanics of Secondary Oil Recovery*, Reinhold Publishing Corp., New York City (1966).
3. Craig, F.F. Jr.: *The Reservoir Engineering Aspects of Waterflooding*, Monograph Series, Society of Petroleum Engineers of AIME, Dallas (1971) **3**, 97-100.
4. Faulkner, B.L.: "Reservoir Engineering Design of a Tertiary Miscible Gas Drive Pilot Project," paper SPE 5539 presented at the 1975 SPE Annual Technical Conference and Exhibition, Dallas, Sept. 28-Oct. 1.
5. Rosenbaum, M.J.F. and Matthews, C.S.: "Studies in Pilot Waterflooding," *Trans*, AIME (1959) **216**, 316-323.
6. Dalton, R.L. Jr., Rapoport, L.A., and Carpenter, C.W. Jr.: "Laboratory Studies of Pilot Waterfloods," *Trans.*, AIME (1960) **219**, 24-30.
7. Prats, M.: "Peace River Steam Drive Scaled Model Experiments," *The Oil Sands of Canada-Venezuela*, CIM (1977) Special Vol. **17**, 346-363.
8. Stegemeier, G.L., Laumbach, D.D., and Volek, C.W.: "Representing Steam Processes With Vacuum Models," *Soc. Pet. Eng. J.* (June 1980) 151-174.
9. Gomaa, E.E., Duerksen, J.H., and Woo, P.T.: "Designing a Steamflood Pilot in the Thick Monarch Sand of the Midway-Sunset Field," *J. Pet. Tech.* (Dec. 1977) 1559-1568.
10. Sawyer, D.N., Cobb, W.M., Stalkup, F.I., and Braun, P.H.: "Factorial Design Analysis of Wet-Combustion Drive," *Soc. Pet. Eng. J.* (Feb. 1974) 25-34.
11. Rincón, A.: "Proyectos Pilotos de Recuperación Térmica en la Costa Bolivar, Estado Zulia," *Proc.*, Simposio Sobre Crudos Pesados, U. del Zulia, Inst. de Investigaciones Petroleras, Maracaibo, Venezuela, July 1-3, 1974.
12. Pollock, C.B. and Buxton, T.S.: "Performance of a Forward Steam Project-Nugget Reservoir, Winkleman Dome Field, Wyoming," *J. Pet. Tech.* (Jan. 1969) 35-40.
13. Marx, J.W. and Langenheim, R.N.: "Reservoir Heating by Hot Fluid Injection," *Trans.*, AIME (1959) **216**, 312-315.
14. Jorden, J.R. Jr. and Campbell, F.L.: *Well Logging*, Monograph Series, Society of Petroleum Engineers of AIME, Dallas (in preparation).
15. Gates, G.F. and Ramey, H.J. Jr.: "Field Results of South Belridge Thermal Recovery Experiment," *Trans.*, AIME (1958) **213**, 236-244.
16. Blevins, T.R., Aseltine, R.J., and Kirk, R.S.: "Analysis of a Steam Drive Project, Inglewood Field, California," *J. Pet. Tech.* (Sept. 1969) 1141-1150.
17. Buxton, T.S. and Pollock, C.B.: "The Sloss COFCAW Project – Further Evaluation of Performance During and After Air Injection," *J. Pet. Tech.* (Dec. 1974) 1439-1448; *Trans.*, AIME, **257**.
18. Prats, M., Jones, R.F., and Truitt, N.E.: "In-Situ Combustion Away From Thin, Horizontal Gas Channels," *Soc. Pet. Eng. J.* (March 1968) 18-32; *Trans.*, AIME, **243**.
19. Coats, K.L., George, W.D., and Marcum, B.E.: "Three-Dimensional Simulation of Steam Flooding," *Soc. Pet. Eng. J.* (Dec. 1974) 573-592; *Trans.*, AIME, **257**.
20. Chu, C. and Trimble, A.E.: "Numerical Simulation of Steam Displacement – Field Applications," *J. Pet. Tech.* (June 1975) 765-776.

Chapter 14
Other Applications

Normally, we think of thermal recovery as applying to conventional crude-oil reservoirs; the crudes may be viscous, but they are crudes nevertheless. Yet one of the earliest in-situ field operations was aimed at recovering fuel from solid organic matter – coal. That was in the USSR in 1933.[1]

Thermal recovery also has been applied to oil shale. The interest in the solid organic matter in coal and oil shale as potential sources of fluid fuels has increased significantly in the U.S. since the 1973 oil embargo. That interest has encompassed both aboveground conversion and in-situ processes.

In the U.S., efforts at in-situ recovery from coal began at least as early as 1951, when attempts were made to retort coal electrically in a shallow coal seam at Gorgas, AL.[2] For oil shale, the field effort dates from at least 1953, when in-situ combustion was used in a hydraulically fractured shallow outcrop to recover shale oil.[3]

Currently, the thermal in-situ recovery of fluid fuels from coal and oil shale is being researched both in the laboratory and in the field, much of it with U.S. government participation or support. Specialized conferences[4-6] are held annually to discuss progress in both research and field projects. The Cameron Synthetic Fuels Report, referred to frequently in this chapter, is issued quarterly and includes a comprehensive summary of all matters pertaining to the development of in-situ recovery from oil shale and coal.

In addition to thermal processes applied to coal and oil shale, unusual sources of heat for the recovery of fuels, primarily from conventional oil reservoirs, are discussed in the last section of this chapter. Those sources include nuclear energy and electricity.

14.1 In-Situ Recovery From Coal

The most advanced in-situ method to convert coal to a fluid fuel, and the one currently being investigated intensively, is underground coal gasification. Reviews of the process are given by Elder,[1] Gregg *et al.*,[7] and Schrider and Whieldon.[8]

Heat for gasification is generated by reacting injected oxygen (usually as air, sometimes mixed with recycled products of gas) with the coal. The heat-generating reactions are idealized as

$$C + O_2 \rightleftharpoons CO_2 \quad \text{.......} (14.1)$$

and

$$4H + O_2 \rightleftharpoons 2H_2O, \quad \text{.......} (14.2)$$

as indicated schematically on the extreme right of Fig. 14.1. Since coal is not mixed with a rock matrix (i.e., the equivalent porosity approaches 100%), the heat generated increases the bulk temperature of the reacting coal surface (T_s) and vapor products (T_g) to very high temperature – high enough to form clinkers of the coal ash. Injection rates and oxygen concentrations in the injected stream may be adjusted to control temperatures. But high temperatures, in excess of about 1,600°F, are required to reduce carbon dioxide and water vapor. In the high-temperature reduction zone downstream of the combustion front, the reduction reactions are written as

$$CO_2 + C \rightleftharpoons 2CO \quad \text{.......} (14.3)$$

and

$$H_2O + C \rightleftharpoons H_2 + CO. \quad \text{.......} (14.4)$$

The latter reaction sometimes is referred to as the gasification reaction and is an important element in the overall process.

The hydrogen and carbon monoxide formed in the reduction zone commingle with the gaseous coal products (such as hydrogen, methane, and light hydrocarbon gases) formed in the retorting or carbonization zone to provide a fuel gas containing a relatively low fuel value per unit volume. Typical properties of the product gas from a coal gasification field test at Hanna[10] are given in Table 14.1.

There is evidence that the heating value of the product gas is affected greatly by the water/air ratio available to the process. Water seeping from the coal seam into the combustion zone (and vaporized near the reacting surface) is desirable to produce sufficient water vapor to contribute to the gasification reaction given by Eq. 14.4. If too little water enters the reaction zone, the fuel value tends to decrease. If too

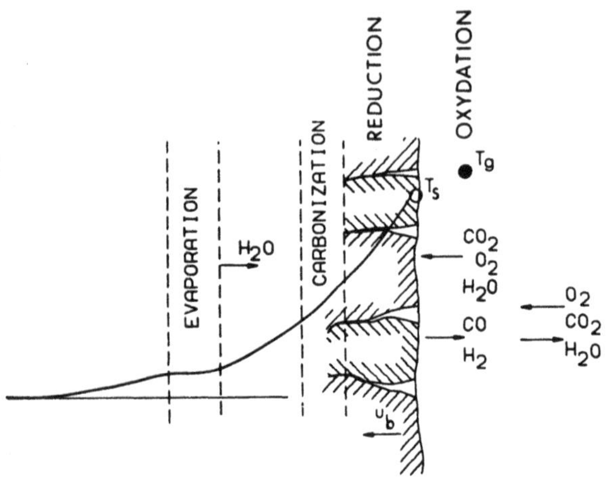

Fig. 14.1 – Model of coal face conditions during gasification.[9]

TABLE 14.1 – GAS COMPOSITION FROM PHASE I, HANNA II, IN WYOMING[10]

Component	Average Concentration (vol%)
H_2	17.3
N_2	51.0
CO	14.7
CO_2	12.4
CH_4	3.3
Ar	0.6
H_2S	0.1
C_2-C_4	0.6

Note: Gross heating value was 152 Btu/scf.

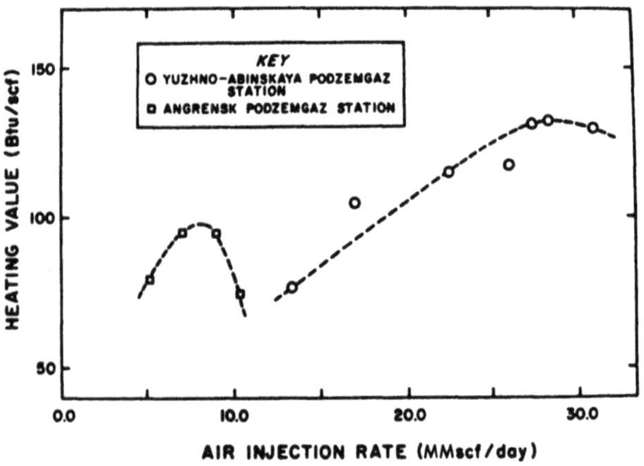

Fig. 14.2 – Effect of air injection rate on heating value of gas produced from underground coal gasification projects.[11]

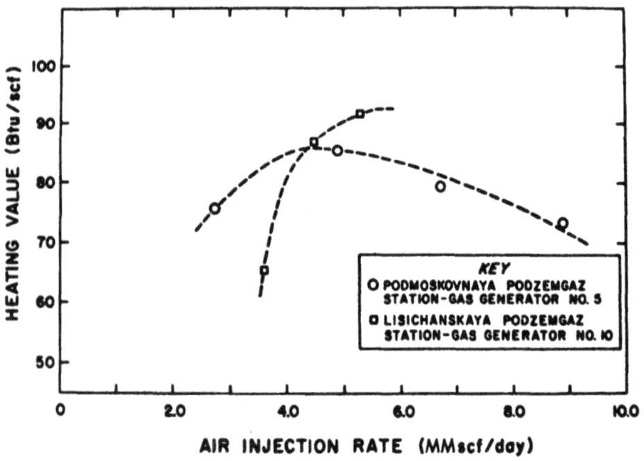

Fig. 14.3 – Effect of air injection rate on heating value of gas produced from underground coal gasification projects.[11]

much water enters the reaction zone, the associated cooling reduces the temperatures, and the loss in the efficiency of the gasification reaction leads to a product with low fuel value. Figs. 14.2 and 14.3, taken from Gunn *et al.*,[11] show how the fuel value of the product gas varies with the air injection rate in four USSR projects. Note that the amount of water seeping from the coal beds into the combustion zone is considered to be approximately constant (but different) in each case and that variations in the air injection rate are equivalent to variations in the water/air ratios entering the combustion zone.

Product gas fuel values are in the range of 50 to 200 Btu/scf, which would require upgrading to be an acceptable synthetic pipeline gas (usually referred to as SNG) of about 1,000 Btu/scf. Upgrading, which would include gas-shift and methanation reactions, would be carried out in surface plants in the presence of catalysts under controlled temperature and pressure conditions requiring cooling and purification of the gas product stream. In addition, it also would be necessary to remove water and organic liquids and compress the gas. Because of the rather high cost of making SNG from the low-Btu product gas obtained in air-fed coal gasification, alternative on-site or near-site uses are being considered. One possibility is to use the produced gas as fuel for electric-power plants.[12,13]

To implement a coal gasification project, there must be permeability or channels through which to inject the air and circulate and recover the product gas. The apparent injectivities of coal beds are in the same reange as those of oil and gas reservoirs. But there may be significant differences.[14,15] Because there is no rock matrix, coal is much more deformable than oil and gas reservoir rocks. Accordingly, coal permeabilities are fairly sensitive to compaction loads and tend eventually to creep, especially at moderate temperatures. Most coals are highly fractured, some will shrink and develop cracks upon loss of water, and all tend to expand when heated and to swell when contacted with hydrocarbon gases and liquids. These properties of coal make the flow resistance highly variable during a process and very different in various parts of the coal seam. Natural permeability is not relied upon in coal gasification. Concerted efforts are made to develop permeable paths in preferred orientations between wells and locations within seams to facilitate control of the process and improve its performance.

Permeable channels between wells have been obtained in a variety of ways, the more common being directional drilling and reverse combustion. Electrolinking, hydraulic fracturing, and explosive fracturing also have been tried, and the use of specially shaped charges, laser beams, and chemicals has been proposed.[16,17] "These channels are placed in the 'bottom' of the seam so that as gasification is carried out and the channel burns and widens, overlying coal falls into the channel which both feeds the coal to the combustion zone and provides a large amount of surface area for efficient reaction with the gases."[11] Fig. 14.4 shows a typical scheme developed in the USSR for in-situ coal gasification in steeply dipping coal seams. In very thin and essentially horizontal beds, coal would not fall into the channel, and widening of the channel is indicated. This is the basis for the longwall coal gasification process, which is illustrated schematically in Fig. 14.5. Parallel channels are made by drilling holes in the coal seam. Each channel can be operated as an injector/producer reactor. But a channel also can be used to inject air through the coal seam and recover product gas from an adjacent channel. Processes involving the flow of air (more generally, a reactant) through the coal bed sometimes are called filtration, percolation, and diffusion processes to distinguish them from a longwall reactor in which the direction of the main flow *stream* is parallel to the channel axis. Channels filled with coal and roof debris are associated with the stream process, but a rubbled coal bed formed explosively is considered to be a packed bed and is associated with filtration processes.[19] In-situ gasification of coal is affected by the coal properties (thickness, depth, water content, pressure, inclination, joint system, and permeability), the roof-collapse and leakoff properties of the adjacent strata, and the method used to achieve and maintain communication between wells. The importance of the coal properties cannot be overemphasized. Lowry,[20] for example, discusses coal classification based on volatile matter content, gross calorific value, reactivity, ash content, and caking, coking, and swelling properties. Another source of information on coal properties is the Synthetic Fuels Data Handbook.[21]

Recent or current field operations in the U.S. include process tests at Hanna and Hoe Creek in Wyoming and at Pricetown, WV.[8,22] The one at Hanna, operated by the Laramie Energy Research Center, makes use of vertical wells and permeability channels formed by reverse combustion and is known as the linked vertical wells (LVW) process. It is the most advanced of the three tests and appears suited for seams between 15 and 50 ft in thickness. Results from one of its phases are summarized in Table 14.1. The Hoe Creek project, operated by the Lawrence Livermore Natl. Laboratory, also makes use of vertical wells, but permeability was obtained initially by explosive fracturing. Because of the resulting permeable zone of broken coal, this is called the

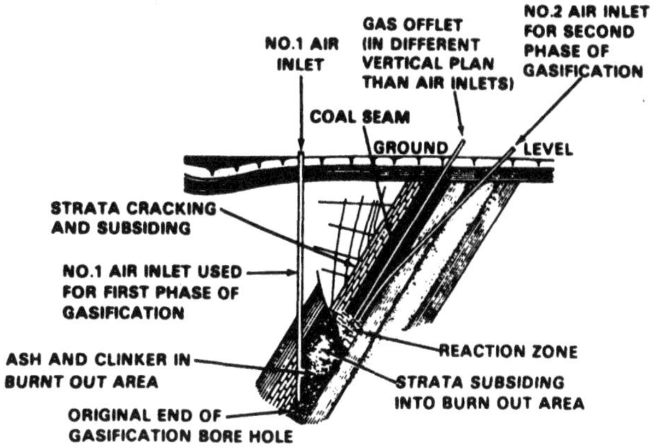

Fig. 14.4 – A steeply dipping bed concept of coal gasification.[19]

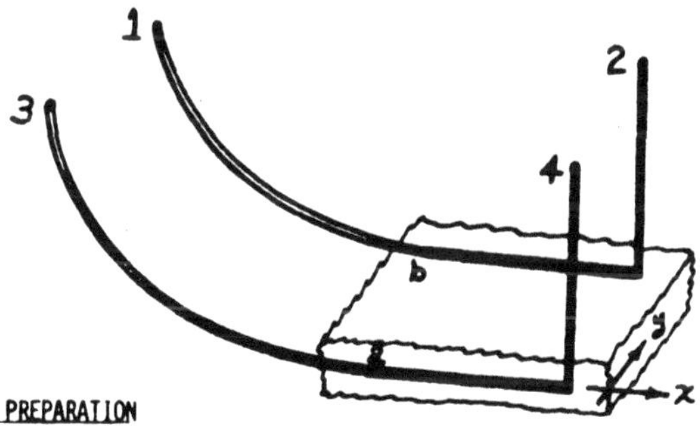

I. BED PREPARATION
A. PUMP WATER OUT OF 4
B. PRODUCE CH_4 OUT OF 4
" " " " 3 AND 4 (LOW P.P.)
C. PUMP WATER OUT OF 2
D. PRODUCE CH_4 OUT OF 2
" " " " 1 AND 2
" " " " 1 AND 1, AND 3 AND 4
E. INJECT AIR IN 1 OR 2 UNTIL O_2 ↑ AND CH_4 ↓ IN 3 AND 4
F. " " " " " " " H_2O CONTENT ↓ " " " "
G. MISCELLANEOUS PRESSURE INTERFERENCE AND OTHER TESTS BETWEEN STEP A AND G

II. WELLBORE IGNITION
A. DECREASE AIR RATE IN 1 AND 2 TO PRESET VALUE
B. VENT 3 WHILE HEATING 4
C. INJECT IN 3 AND REVERSE BURN FROM 4 TO (a)

III. BLOCK REVERSE BURN
A. WHEN IGNITION REACHES (a) CHANGE MODE
B. INJECT IN 1 AND 2 AND PRODUCE GASES OUT 3 AND 4

Fig. 14.5 – The MERC longwall generator development scheme (half-unit shown).[18]

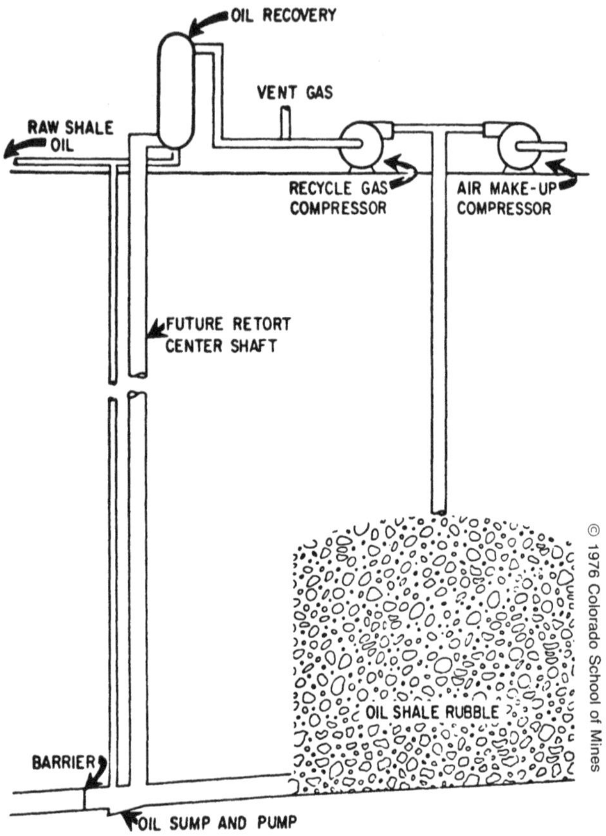

Fig. 14.6—Schematic representation of the retort operation in the MIS oil shale process.[29]

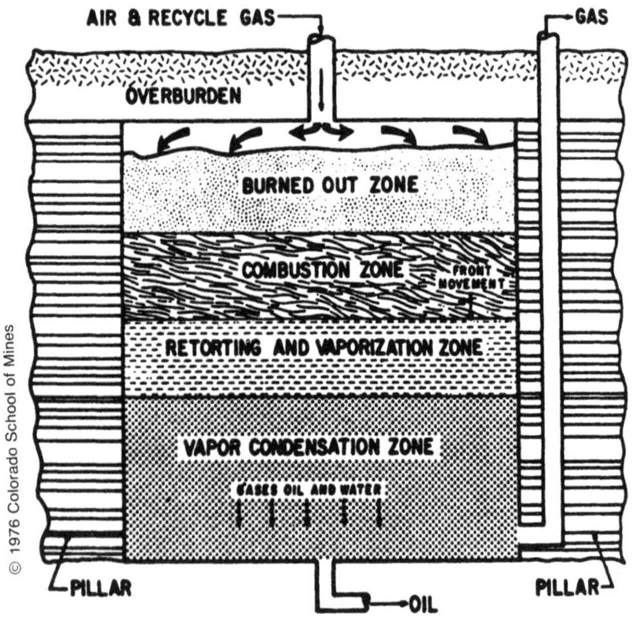

Fig. 14.7—Schematic representation of in-situ oil shale retorting.[29]

packed bed (PB) process—an approach that appears best suited for coal seams in excess of 50 ft. The produced gas at Hoe Creek had a heating value up to 150 Btu/scf. In more recent phases, reverse combustion and a horizontal channel have been used to control the movement of the combustion front, and gas has been produced with an average heating value exceeding 200 Btu/scf.[22] The test at Pricetown, operated by the Morgantown Energy Research Center, makes use of directionally drilled wells, as indicated in Fig. 14.5. This is known as the longwall generator (LWG) process, which appears best suited for coal seams less than 15 ft thick. Combustion at Pricetown was initiated in July 1979, and gas was produced with a heating value of about 130 Btu/scf.[22] Few descriptions had been published of coal gasification operations in the private sector of the U.S.[22,23] until the last few years.[6]

In addition to coal gasification processes, other in-situ processes for recovering coal products are being studied. A number of these, like liquefaction[19] and the formation of coal slurries by using chemicals[19,24,25] are likely to be combinations of thermal and chemical recovery processes.

Coal gasification is essentially a type of in-situ combustion. Wells are used to effect combustion of fuel in place, to generate the heat required for the processes, and to inject and produce fluids. These are the elements the technologies have in common. But there are also vast differences in the various processes, the most obvious of which is the gasification of the fuel in the case of coal.

14.2 In-Situ Recovery From Oil Shale

Another emerging technology is the use of thermal methods to recover fluid fuels from oil shale in situ. The oil shales of the Green River formation in Colorado, Utah, and Wyoming are the primary targets. Those oil shales are essentially impermeable. The organic matter in oil shale is kerogen, a solid that must be decomposed thermally to recover liquid and gaseous fuels. The Synthetic Fuels Data Handbook[21] is a good source of information on the properties of oil shale.

The most extensively tested recovery method, pioneered by Occidental Oil Shale Inc., "consists of three basic steps: (1) a limited amount of conventional mining, (2) blasting of the overlying oil shale to form the retort, and (3) retorting in place, normally using air and underground combustion."[26] Because the process entails mining technology to obtain the necessary fragmentation and permeability of the oil shale, and retorting technology similar to that used in surface processes,[21,27] it has been referred to as modified in-situ (MIS) processing. But because it makes use of wells to inject fluids and recover fuel from the subsurface, it also has all the elements of an oil recovery operation.

This process is a compromise between mining and in-situ recovery and offers some of the advantages and disadvantages of both. About 15 to 20% of the retort volume is mined and removed to provide access for the placement of explosive charges and to provide space for the bulking of the broken oil shale.

The latest blasting, firing, and fragmentation techniques then are used to achieve a permeable and rubbled zone of predetermined shape and volume.[28] When it is used where there are essentially no permeable channels in the oil shale, the retort can be operated at a low pressure as a sealed unit. The compressors are large, but because only low pressures are required, they need not be very powerful. The modified in-situ process is attractive because only small amounts of solids need be removed in the mining phase, the resulting retort is large, the required process pressure is low, and relatively little water is needed (an important factor in the oil shale deposits of the western U.S.).

A schematic representation of the overall operation is shown in Fig. 14.6. Fig. 14.7 illustrates the retorting of the oil shale and the recovery of the shale oil resulting from downward combustion. Recycled gas is used to control the temperature levels and the convective heat transfer in the retort. Field testing started in 1972 with relatively small retorts. A retort measuring 120×120×270 ft in height was evaluated in 1976.[30] The predicted and observed production responses are compared in Fig. 14.8 for the first 1.5 months of operation, before "the retort conditions were changed significantly."[31] Ultimate oil recovery, not shown in the figure, amounted to 30,000 bbl.[30] Properties of the shale oil recovered from two smaller retorts are given in Table 14.2. A postfragmentation vertical section of Retort 6, the latest to be evaluated, is shown in Fig. 14.9. Over 55,000 bbl of shale oil have been produced from this retort.[22] And other companies are testing variations of the MIS process.[22]

In other in-situ processes field tested by both industry and government since 1953,[32] the necessary permeable paths have been created by hydraulic fracturing, by detonating explosives from wellbores, and by electrolinking.[33] Until recently, these approaches had not resulted in the large volumes of broken or exposed oil shale desirable for retorting operations. However, current efforts aim at overcoming this drawback by explosively fragmenting large volumes of oil shale solely through wells.[22,34,35] One of these projects is in the Antrim shale of Michigan.[35]

The retorting in all the operations mentioned thus far has been carried out by in-situ combustion. Dougan,[36] however, injected heated methane (>800°F) into naturally fractured oil shale zones in Colorado to recover oil from wells as far as 500 ft apart.[26] Losses of the injected gas through the fracture system apparently affected the viability of the process, which currently is being further field tested and evaluated with the injection of superheated steam.[22,37] Steam also has been injected to dissolve salts that are interspersed intimately with oil shale in certain deep parts of the Piceance Creek basin and, thus, develop the required permeability and particle size.[38] The high-temperature steam retorting (up to 620°F) was carried out concurrently with the solution mining. This field test disclosed severe corrosion and generation of fines. The time required to convert kerogen to fluids at various

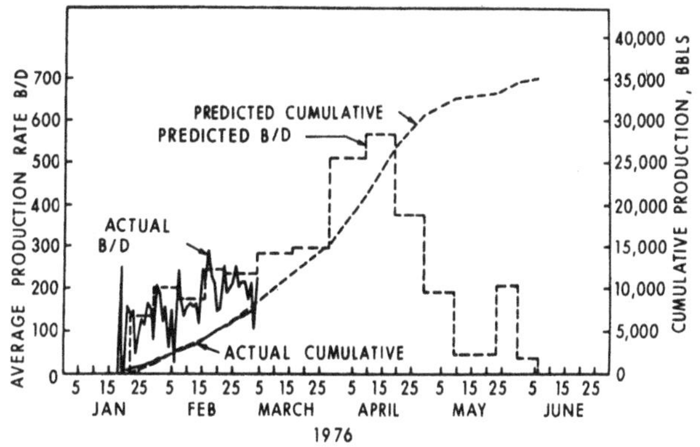

Fig. 14.8 – Predicted and actual shale oil production.[29]

TABLE 14.2 – SHALE PROPERTIES AND COMPOSITION[29]

Properties of Shale Oil From Retort 2E	
API gravity, 60/60, °F	24.3
Viscosity, 100°F, SUS	116
Pour point, °F	65
Flash point, °F	241
Properties of Shale Oil From Retort 3E	
Fire point, °F	258
Nitrogen, wt%	1.50
Sulfur, wt%	0.71
Carbon, wt%	84.86
Hydrogen, wt%	11.80
Shale Oil Composition (vol%)	
Naphtha, IBP to 400°F	4.6
Light distillate, 400 to 500°F	25.4
Light gas oil, 600 to 800°F	45.0
Heavy gas oil, 600 to 1,000°F	20.0
Residuum, >1,000°F	5.0

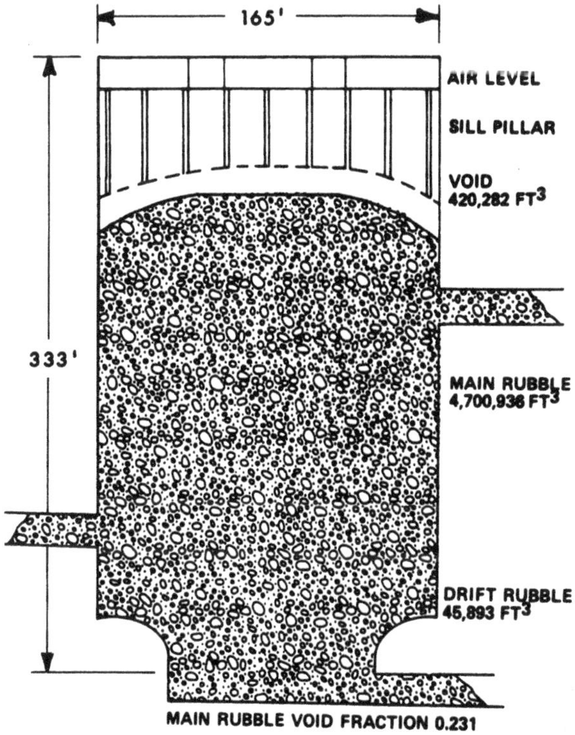

Fig. 14.9 – Section view of Retort 6 after blasting and fragmentation.[28]

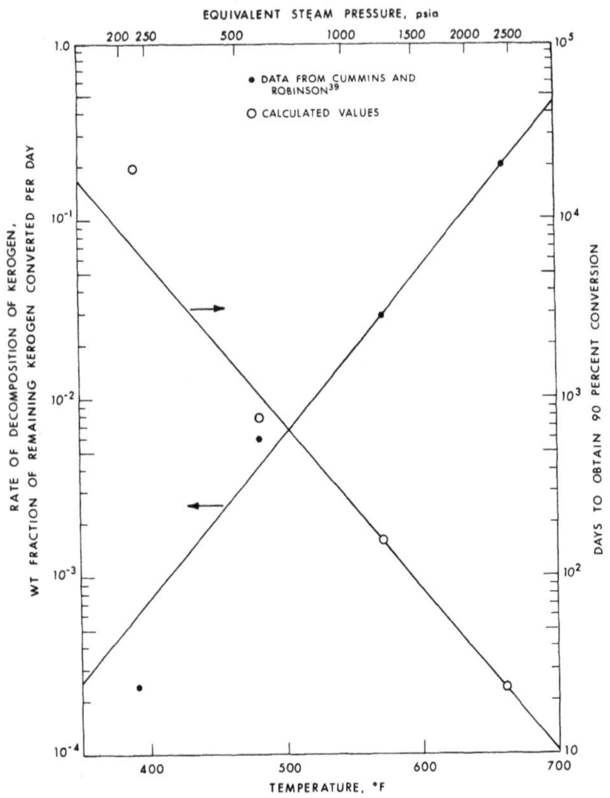

Fig. 14.10 – Days required to obtain 90% conversion of kerogen at temperatures of 350 to 700°F.

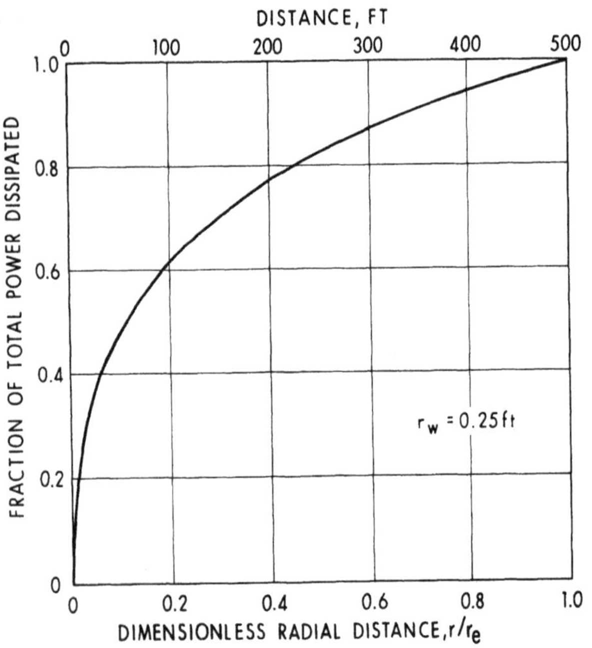

Fig. 14.11 – Dissipated electric power as a function of distance from the wellbore.

TABLE 14.3 – ANALOGY BETWEEN FLUID FLOW AND ELECTRIC CONDUCTION

Quantity	Fluid FLow	Electric Conduction
Flux	u	u_e
Rate	q	I
Driving force	p or Φ	V_e
Resistivity	λ^{-1}	σ^{-1}
Resistance	$\frac{1}{J}=\frac{\Delta p}{q}$	$R_e=\frac{\Delta V_e}{I}$
Power	q^2/J	$P=I^2R_e$ $=I\Delta V_e$ $=(\Delta V_e)^2/R_e$

temperatures can be estimated from Fig. 14.10. At 620°F, 90% can be converted in about 55 days.

14.3 Other Methods of Heating Reservoirs

Thus far, this chapter has discussed fairly conventional means of extracting fluid fuels from solid raw materials such as coal and oil shale; now some unconventional methods to heat reservoirs will be examined. The methods are unconventional in that they do not entail the injection of fluids to introduce heat into the reservoir. Of the two agents involved – electricity and nuclear energy – only electricity actually has been used in oil reservoirs.

Electric Heating

The dissipation of electric power in a reservoir causes it to heat – i.e., to increase in temperature. Power can be dissipated either through resistance or through absorption. Resistance heating occurs when an electric current passes through a resistive element under a voltage gradient, which is governed by laws that are analogous to those that govern the flow of fluids in porous media. Because of this analogy, it is possible to use the methods presented in other chapters to obtain results applicable to the flow of direct currents. The analogy between fluid and electric fluxes, driving forces, resistances, and other properties are given in Table 14.3.

For example, the electric resistance to a radial current from a wellbore confined to a reservoir of electrical conductivity σ can be found from the resistance to fluid flow given in Table 14.3 and Eq. 1.2 to be

$$R_e=\frac{1}{2\pi h\sigma}\ln(r_e/r_w) \;, \quad \ldots\ldots\ldots\ldots\ldots\ldots (14.5)$$

and the electric voltage is

$$V=V_{e,w}-\Delta V_e\frac{\ln(r/r_w)}{\ln(r_e/r_w)}, \quad \ldots\ldots\ldots\ldots (14.6)$$

where ΔV_e is the voltage drop between r_w and r_e, and $V_{e,w}$ is the voltage at $r=r_w$. From these expressions, the power dissipated between the wellbore and the reservoir volume a distance r from its axis is

$$P(r)=2\pi\sigma h(\Delta V_e)^2\frac{\ln^2(r/r_w)}{\ln^3(r_e/r_w)}. \quad \ldots\ldots\ldots (14.7)$$

The total power dissipated to the external radius is

$$P(r_e)=(\Delta V_e)^2 2\pi\sigma h/\ln(r_e/r_w), \quad (14.8)$$

$$=(\Delta V_e)^2/R_e, \quad (14.9)$$

which is the familiar form for electric resistance heating. The plot of the power dissipated between the wellbore and a radius, r, given by Eq. 14.7 and shown in Fig. 14.11, indicates that most of the power is dissipated near the wellbore. Since power dissipation is the rate of energy released in the formation in the form of heat, it is clear that electric heating would cause the highest temperatures to be developed near the wellbore.[40]

Because of the near-well heating effect and the absence of an externally applied driving force to produce fluids, except possibly for electro-osmosis, electric heating is considered to be primarily a stimulation treatment.

Of course the fluids flowing toward the wellbore carry sensible heat, so that the effectively heated zone near the wellbore is somewhat smaller than would be estimated solely on the basis of power dissipation. Also, contrary to the assumptions used in developing Fig. 4.11, it is necessary to keep in mind (1) that the electrical conductivity of shales and water-bearing sands is appreciably higher than that of highly saturated oil-containing rocks, (2) that the electric currents will not be radial, and (3) that formation electrical resistivities are affected by temperatures and saturations. Although the near-wellbore heating effect is real, the quantitative distribution of the power dissipation with distance from the wellbore given by Fig. 14.11 must be considered strictly illustrative. It should be noted from the first form of the power dissipation expression in Table 14.3 ($P=I^2R_e$) that little power is dissipated where the electrical conductivity is high, even when the currents are large.

Alternating current has been used in at least three widely separated field tests.[41] They consisted of single-well tests completed in reservoirs containing asphaltic crudes with gravities between 8 and 14°API and a high pour point (93°F). Maximum depth was 3,362 ft. A crude-oil production rate as high as 186 B/D was reported in one test, but it is not known if this rate represents sustained production or what the production would have been in the absence of the treatment. In another test, the crude-oil production rate stabilized at 17 B/D, about 300% above that of the average field producer.

Fig. 14.12 is a schematic representation of the electrothermic process. Fig. 14.13 illustrates some special completion techniques associated with the process, including the use of the tubing as a conductor, the use of downhole electrical insulators, and the location of disc electrodes filled with steel shot and/or graphite pellets relative to the zones of high electrical resistivity as determined from electric logs.

Direct[42] and high-frequency alternating[43,44] currents also have been proposed for heating oil-containing reservoirs and oil shales, but few field tests appear to have been reported.[45] Both direct and low-frequency alternating currents dissipate power through resistance losses. Where impedance power losses are small, usually at low frequencies, Table 14.3 also may be used to *estimate* alternating current

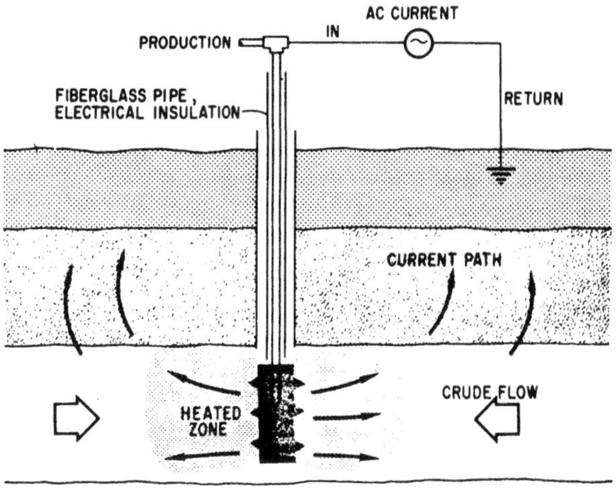

Fig. 14.12 – The electrothermic process.[41]

Electrothermic process uses continuous flow of single-phase AC current to heat formation around wellbore. Viscosity of heavy crude is reduced, improving flow. Current flows through steel tubing into special electrode formed by packing steel shot into openhole notched formation. Tubing is electrically insulated with concentric string of fiberglass pipe. Current disperses into the earth beyond the heating zone and returns to the surface to complete the circuit. Heat results from resistance of semiconductive reservoir fluids in the area where the intensity of the current is greatest.[41]

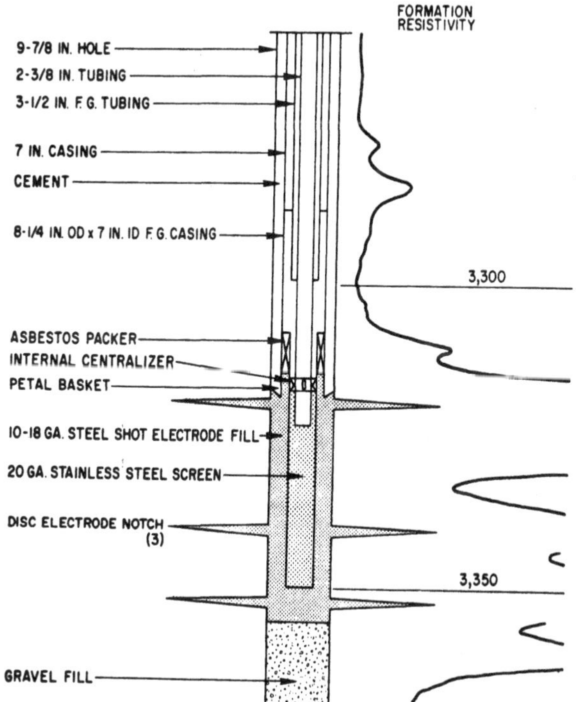

Fig. 14.13 – Special completion technique for electrothermic process.[41]

Special completion method uses steel casing with several joints of fiberglass casing on bottom. Open hole is notched and packed with steel shot. A steel screen is washed into the shot and latched to the fiberglass casing with an asbestos packer. Fiberglass pipe, hung separately in the wellhead, isolates production tubing, which is run into the electrode screen and electrically connected with an internal centralizing device. Conventional pumping equipment can be used. Resistivity curve illustrates zone that will be heated by AC current. In this example, current returns through the steel casing into surface.[41]

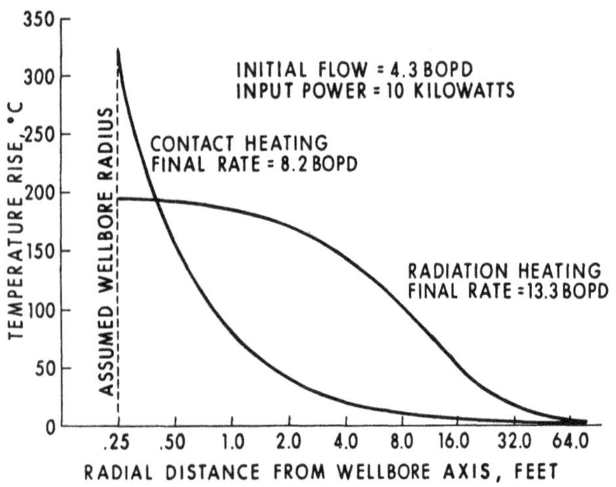

Fig. 14.14 – Comparison of steady-state temperatures.[43]

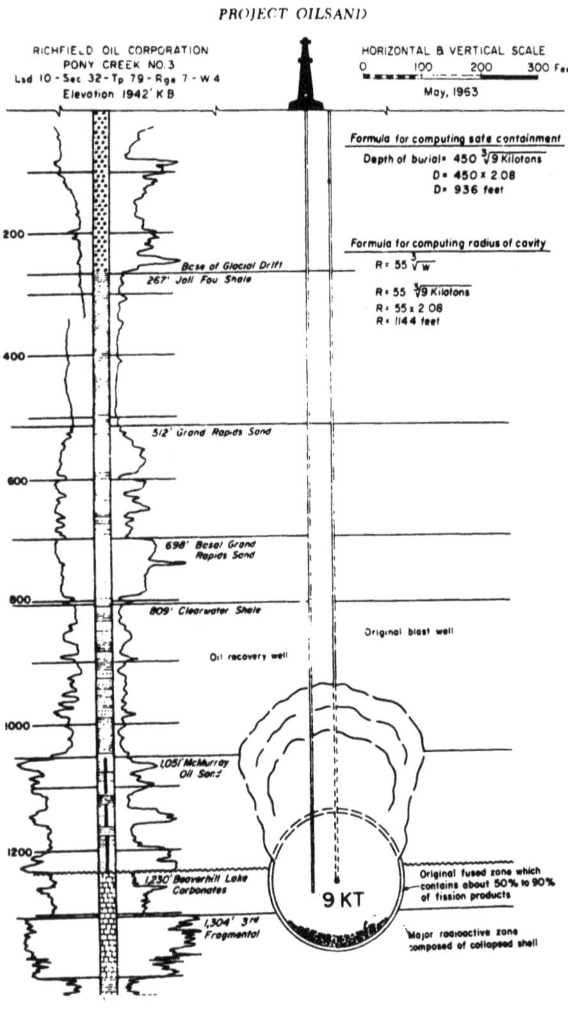

Fig. 14.15 – Proposed nuclear test for bitumen recovery.[47]

needs. High-frequency electromagnetic power is dissipated by absorption, where the power radiated across a plane surface at a distance, l, from a reference surface can be represented by

$$P = P_r e^{-\alpha l}, \qquad (14.10)$$

where P_r is the power radiated across the reference plane surface. α is the power absorption coefficient, a function of the dielectric properties of the fluid-filled rock and of the frequency. Dielectric properties of a fluid-filled formation depend on the type and amounts of fluids present and on the frequency. Comparisons of theoretical steady-state temperature distributions and steady-state radial flow production response for radial energy transfer by electromagnetic heating and by a downhole heater (in the absence of heat losses to adjacent formations) are shown in Fig. 14.14; the electromagnetic system is more efficient that the use of bottomhole heaters. There are no available comparisons between electromagnetic and resistive reservoir heating.

Heating by dissipating electric power directly in the reservoir has some advantages.

1. Heat is generated instantaneously throughout the reservoir elements. There is no wait for energy to be convected by hot fluids or conducted through the reservoir, though it may take some time for the reservoir temperature to increase significantly halfway between wells.

2. The generation of heat, which does not rely on fluid injection, is essentially independent of the reservoir's fluid permeability and of injection pressure limitations.

Thus, the method may be of advantage in applications where the fraction of the injected heat remaining in the reservoir would be too low because of long heating times resulting from low rates of heat and fluid injection.

Electric heating methods also have their disadvantages.

1. Electric power is expensive compared with conventional heating sources used in thermal projects; thus, the methods are economically attractive only where the increase in thermal (energy) efficiency is great enough to offset the high cost of the electric energy.

2. The operating life of the currently available downhole equipment is inadequate for trouble-free operation; equipment needs further development.

Nuclear Energy

Nuclear energy for thermal recovery has not been field tested, though at least three different types of applications have been proposed.

In one application, a contained nuclear explosion would be detonated underground in or slightly below a thick column of the target reservoir, which could be either oil shale[47] or tar sands. Here, the primary purpose of the detonation is to fracture and fragment the resource, thus making it permeable and suitable for in-situ recovery. Some temperature increase also would accompany the detonation, but it would have little effect except possibly in the tar sands. The detonation of repeated shots in the same fragmented volume to increase the temperature to more satisfactory levels has been proposed, but the use of nuclear energy for the recovery of fuels has not been approved by any government. A schematic of the single-shot process is shown in Fig. 14.15.

Other proposed uses of nuclear energy include (1) reducing the oil viscosity by making use of heat liberated from radioactive wastes stored in adjacent formations and (2) building a subsurface nuclear-powered steam generator to inject steam directly into the reservoirs.

References

1. Elder, J.L.: "The Underground Gasification of Coal," *Chemistry of Coal Utilization, Supplementary Volume*, H.H. Lowry (ed.), John Wiley & Sons Inc., New York City (1963).
2. Dowd, J.J., Elder, J.L., Capp, J.P., and Cohen, P.: "Field Scale Experiments in Underground Gasification of Coal at Gorgas, Alabama. Use of Electrolinking-Carbonization as Means of In-Situ Preparation," RI 5357, USBM (1957).
3. Grant, B.F.: "Retorting Oil Shale Underground – Problems and Possibilities," *Quarterly*, Colorado School of Mines, Golden (July 1974) 39-46.
4. *Proc.*, Eleventh Oil Shale Symposium, J.H. Gary (ed.), Colorado School of Mines Press, Golden (1978).
5. Abstracts of Third Annual Oil Shale Conversion Symposium, *Cameron Synthetic Fuels Report*, Cameron Engineers Inc., Denver (March 1980) **17**, No. 1.
6. *Proc.*, Sixth Annual Underground Coal Gasification Symposium, Laramie Energy Technology Center and Williams Brothers Engineering Co., Shangri-La, OK (1980).
7. Gregg, D.W., Hill, R.W., and Olness, D.U.: "An Overview of the Soviet Effort in Underground Coal Gasification," UCRL-52004, Lawrence Livermore Natl. Laboratory, Livermore, CA (Jan. 29, 1976).
8. Schrider, L.A. and Whieldon, C.E.: "Underground Coal Gasification – A Status Report," *J. Pet. Tech.* (Sept. 1977) 1179-1185.
9. Stewart, I.McC. and Wall, T.F.: "Reaction Rate Analysis of Borehole 'In-Situ' Gasification Systems," paper presented at 16th Intl. Symposium on Combustion, The Combustion Inst. (1977).
10. Fischer, D.D., Grandenburg, C.F., King, S.B., Boyd, R.M., and Hutchinson, H.L.: "Status of the Linked Vertical Well Process in Underground Coal Gasification," *Proc.*, Second Annual Underground Coal Gasification Symposium, Morgantown Energy Research Center, U.S. DOE, Morgantown, WV (1976) MERC/SP-76/3.
11. Gunn, R.D., Gregg, D.W., and Whitman, D.L.: "A Theoretical Analysis of Soviet In-Situ Coal Gasification Field Tests," *Proc.*, Second Annual Underground Coal Gasification Symposium, Morgantown Energy Research Center, U.S. DOE, Morgantown, WV (1976) MERC/SP-76/3.
12. *Synthetic Fuels*, Cameron Engineers Inc., Denver (March 1978) 4-2.
13. Stewart, R.E.D.: "Utilization of Low BTU Gas From In-Situ Coal Gasification for On-Site Power Generation," *Proc.*, Second Annual Underground Coal Gasification Symposium, Morgantown Energy Research Center, U.S. DOE, Morgantown, WV (1976) MERC/SP-76/3.
14. Dabbous, M.K., Reznik, A.A., Taber, J.J., and Fulton, P.F.: "The Permeability of Coal to Gas and Water," *Soc. Pet. Eng. J.* (Dec. 1974) 563-572; *Trans.*, AIME, **257**.
15. Reznik, A.A., Dabbous, M.K., Fulton, P.F., and Taber, J.J.: "Air-Water Relative Permeability Studies of Pittsburgh and Pocahontas Coals," *Soc. Pet. Eng. J.* (Dec. 1974) 556-562; *Trans.*, AIME, **257**.
16. *Proc.*, Second Annual Underground Coal Gasification Symposium, L.Z. Schuck (ed.), Morgantown Energy Research Center, U.S. DOE, Morgantown, WV (1976) MERC/SP-76/3.
17. *Synthetic Fuels*, Cameron Engineers Inc., Denver (June 1977) 4-11 and 4-14.
18. Schuck, L.A., Pasini, J. III, Martin, J.W., and Bisset, L.A.: "Status of the MERC In-Situ Gasification of Eastern U.S. Coals Project," *Proc.*, Second Annual Underground Coal Gasification Symposium, Morgantown Energy Research Center, U.S. DOE, Morgantown, WV (1976) MERC/SP-76/3.
19. Wieber, P.R. and Sikri, A.P.: "The 1976 ERDA In-Situ Coal Gasification Program," *Proc.*, Second Annual Underground Coal Gasification Symposium, Morgantown Energy Research Center, U.S. DOE, Morgantown, WV (1976) MERC/SP-76/3.
20. *Chemistry of Coal Utilization – Supplementary Volume*, H.H. Lowry (ed.), John Wiley & Sons Inc., New York City (1963).
21. *Synthetic Fuels Data Handbook*, Cameron Engineers Inc., Denver (1975).
22. *Cameron Synthetic Fuels Report*, (June 1980) **17**, No. 2; for oil shale: 2-31 through 2-34; for coal: 4-70 through 4-73.
23. Raimondi, P., Terwilliger, P.L., and Wilson, L.A. Jr.: "A Field Test of Underground Combustion of Coal," *J. Pet. Tech.* (Jan. 1975) 35-41.
24. Datta, R.: "An Ammonia Injection Scheme for Fracturing and Permeability Enhancement of Coal Beds," *Proc.*, Second Annual Underground Coal Gasification Symposium, Morgantown Energy Research Center, U.S. DOE, Morgantown, WV (1976) MERC/SP-76/3.
25. Drinkard, G., Prats, M., and O'Brien, S.M.: "Coal Disaggregation by Basic Aqueous Solution for Slurry Recovery," U.S. Patent No. 4,023,193 (June 1977).
26. Ridley, R.D.: "In-Situ Processing of Oil Shale," *Quarterly*, Colorado School of Mines, Golden (April 1974) 21-24.
27. Matzick, A. and Dannenberg, R.O.: "USBM Gas Combustion Retort for Oil Shale – A Study of Process Variable," RI 5545, USBM (1960).
28. Ricketts, T.E.: "Occidental's Retort 6 Rubblizing and Rock Fragmentation Program," *Proc.*, 13th Oil Shale Symposium, Colorado School of Mines, Golden (1980) 46-61.
29. McCarthy, H.E. and Cha, C.Y.: "OXY Modified In-Situ Process Development and Update," *Quarterly*, Colorado School of Mines, Golden (April 1976).
30. *Synthetic Fuels*, Cameron Engineers Inc., Denver (March 1977) 2-11 and 2-12.
31. *Synthetic Fuels*, Cameron Engineers Inc., Denver (June 1976) 2-54 through 2-57.
32. Burwell, E.L., Sterner, T.E., and Carpenter, H.C.: "In-Situ Retorting of Oil Shale," RI 7783, USBM (1973).
33. Campbell, G.G., Scott, W.G., and Miller, J.S.: "Evaluation of Oil Shale Fracturing Tests Near Rock Spring, Wyoming," RI 7397, USBM (1970).
34. Lekas, M.A.: "Progress Report on the Geokinetics Horizontal In Situ Retorting Process," *Proc.*, 12th Oil Shale Symposium, Colorado School of Mines, Golden (1979) 228-236.
35. McNamara, P.H., Peil, C.A., and Washington, L.J.: "Characterization, Fracturing, and True In Situ Retorting in the Antrim Shale of Michigan," *Proc.*, 12th Oil Shale Symposium, Colorado School of Mines, Golden (1979) 353-365.
36. Dougan, P.M., Reynolds, F.S., and Root, P.J.: "The Potential for In-Situ Retorting of Oil Shale in the Piceance Creek Basin of Northwestern Colorado," *Quarterly*, Colorado School of Mines, Golden (Oct. 1971) 57-72.
37. *Synthetic Fuels*, Cameron Engineers Inc., Denver (Sept. 1977) 2-30 and 2-31.
38. Prats, M., Closmann, P.J., Ireson, A.T., and Drinkard, G.: "Soluble-Salt Processes for In-Situ Recovery of Hydrocarbons From Oil Shale," *J. Pet. Tech.* (Sept. 1977) 1078-1088.
39. Cummins, J.J. and Robinson, W.E.: "Thermal Degradation of Green River Kerogen at 150°C to 350°C," RI 7620, LERC, USBM (1972).
40. Todd, J.C. and Howell, E.P.: "Numerical Simulation of In Situ Electrical Heating To Increase Oil Mobility," *Oil Sands of Canada-Venezuela*, CIM (1977) Special Vol. **17**, 19.
41. "AC Current Heats Heavy Oil for Extra Recovery," *World Oil* (May 1970) 83-86.
42. Bell, C.W. and Titus, C.H.: "Electro-Thermal Process for Promoting Oil Recovery," Canada Patent 932,657 (Aug. 28, 1973).

43. Abernethy, E.R.: "Production Increase of Heavy Oil by Electromagnetic Heating," *J. Cdn. Pet. Tech.* (July-Sept. 1976) 91-97.
44. Bridges, J.E., Taflone, A., and Snow, R.H.: "Net Energy Recoveries for the In-Situ Dielectric Heating of Oil Shale," *Proc.*, 11th Oil Shale Symposium, Colorado School of Mines, Golden (1978) 311-330.
45. "Petro-Canada Uses Electricity Before Steam," *Enhanced Recovery Week* (Sept. 14, 1981) 2.
46. Lombard, D.B. and Carpenter, H.C.: "Recovering Oil by Retorting a Nuclear Chimney in Oil Shale," *J. Pet. Tech.* (June 1967) 727-734.
47. Natland, M.L.: "Project Oil Sand," *Athabasca Oil Sands: A Collection of Papers*, M.A. Carrigy (ed.), Research Council of Alberta, Edmonton, Canada (1963) 143-155.

Chapter 15

Status and Potential of Thermal Recovery

This chapter summarizes the state of the art regarding thermal recovery processes, discusses some of the major current problems and recognized research needs, touches on the impact of thermal recovery on society, and comments on the potential importance of the processes.

15.1 State of the Art

One way to assess the current importance of thermal recovery processes is to examine the extent of field operations, not only in an absolute sense but also relative to the other enhanced recovery processes, excluding waterflooding.

The only known complete statistics that are readily available on thermal recovery operations are those for California.[1,2] Fig. 15.1, which shows the total production increase in California due to thermal recovery, is based on Ref. 1. The incremental production rate of 199,000 B/D in 1979 amounted to 20% of the state's total production and has been growing steadily since 1964.

The most readily available information on the general extent of thermal recovery applications is that provided every 2 years or so by *Oil and Gas Journal*[3-6] as part of their worldwide survey on enhanced recovery processes; the latest survey covers 1977 through 1979.[6] But the most exhaustive and up-to-date presentation on worldwide enhanced oil recovery projects, including thermal ones, are those of Dafter.[7,8]

For 1981, Table 15.1 provides information on the thermal and enhanced recovery production rates in the U.S., Venezuela, Canada, and the rest of the world. Of the 468,000 B/D obtained in the U.S. by enhanced recovery methods, some 83%, or 388,000 B/D, was produced by thermal methods. The thermal recovery production from Venezuela, which comes primarily from the region near the east shore of Lake Maracaibo, amounted to 163,000 B/D.

Note that total thermal production accounts for about 73% of the total 915,000 B/D worldwide production from enhanced recovery processes (waterfloods excluded).

Of the estimated 388,000 B/D thermal production in the U.S., only about 20,000 B/D is from combustion processes. The bulk of the production, some 204,000 B/D, is from cyclic steam injection, with the remaining 164,000 B/D coming from steam drives.

Table 15.2 lists 1981 data for some of the largest projects for each type of recovery process. Of these, the largest reported production is for the steam processes – even the largest of the combustion projects is small compared with those using steam. Kernridge's Belridge project used both cyclic and continuous steam injection.

Even though combustion, being the older form, had some advantage early in its development and application, it has fallen into disfavor by comparison with steam processes. Fig. 15.2 shows the number of steam drives, cyclic steam injection projects, and combustion projects in the U.S. as reported by *Oil and Gas Journal*[3] in 1970, 1973, 1975, and 1977 and inferred from the 1979 tables.[6] Since the production of crude (income) for cyclic steam injection begins soon after a project is started, it is not difficult to see why this process often is favored over drives. Steam drives increased sharply in 1975, partly as a result of the economic pressures arising from the oil embargo of 1973. The number of combustion projects leveled at about 20 from 1973-79, but has increased to 34 in 1981. The corresponding numbers for steam drives and cyclic steam injection are 80 and 77.[8]

15.2 Current Problems and Research

The main problems facing thermal recovery are economic – i.e., it must compete with alternative processes. This competition must be met even in the face of the greatly increased cost of fuel used at the surface and the increased costs of meeting environmental constraints. These, together with the cost of treating water for steam generation, appear to be the only costs that place a specific burden on thermal recovery operations. Special metallurgy and fairly frequent well repairs may contribute to the cost of some projects.

Also affecting the economics of thermal recovery is the fact that it is frequently applied to heavy crudes, which usually are high in sulfur. Both the low API gravity and the high sulfur content of the average thermally produced oil reduce its value and, thus, the income from the operation.

These economic pressures lead to a continuous search for more efficient, cost-effective methods of meeting environmental constraints and operational

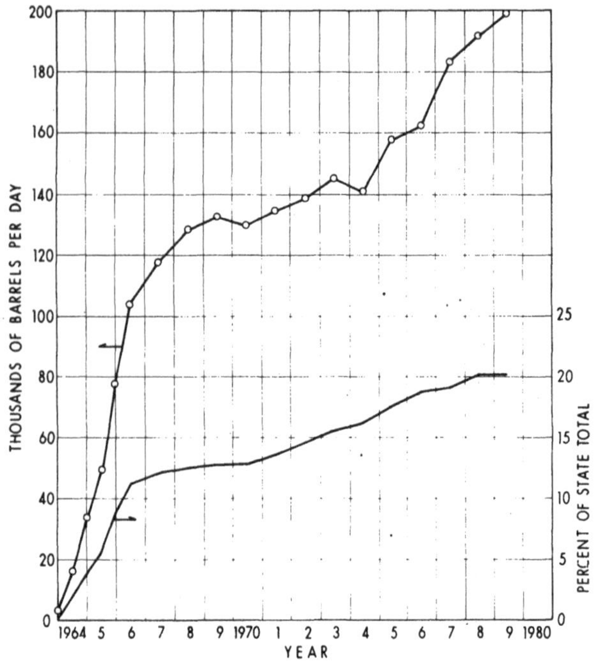

Fig. 15.1 – Total Production increase in California due to thermal recovery processes.[1]

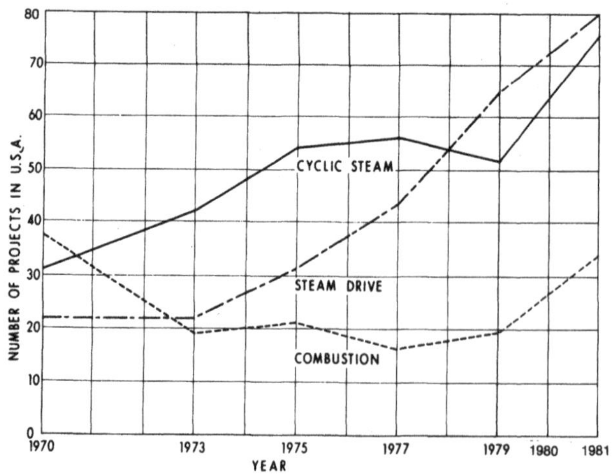

Fig. 15.2 – Number of thermal recovery projects in the U.S.[3,7]

needs in a socially acceptable manner. Continuing efforts in that direction are being devoted to developing methods of removing particulates, hydrocarbons, sulfur, and noxious compounds from waste gas streams in thermal projects; to improving metallurgy and corrosion inhibitors; to improving demulsification procedures and chemicals; to developing more cost-effective steam generation equipment; to reducing heat losses from wells and surface lines; and to developing low-cost water treatments and ways to use cheaper fuels such as coal to fire steam generators.

All those efforts are associated with improving the hardware or procedures used outside the reservoir. This is not to say there are no problems associated with the reservoir processes themselves. Even though, as we have pointed out, thermal recovery processes are used widely, it is important to increase the oil production per unit energy spent. The last significant improvement in that area was wet combustion. Current efforts include controlling mobility by using foams to improve sweep efficiencies in reservoirs where bypassing is excessive,[9,10] using other chemicals to improve displacement efficiency in steam drives,[10,11] and ascertaining the potential of oxygen injection in combustion projects to enhance the beneficial effects of CO_2 while reducing compression costs and the adverse effects of the noncondensable N_2.[12,13] Improving the effectiveness of thermal recovery processes currently is the goal of both the operating and research branches of the oil industry, and is often financially supported by governments.

From a microscopic point of view, much remains to be learned, for example, about the efficient displacement resulting from a steam condensation front (not the fact that it occurs, but why and how); about the effect of bypassing on the fuel burnt along the trailing combustion surface; about the way the continued condensation of light ends affects the displacement processes; about the role of noncondensable gases; about the flow of foam and bubbly liquids; and about the importance of the rock matrix, multiphase relative permeabilities, and many other phenomena. This lack of a complete understanding of many facets of thermal recovery certainly does not preclude its use. But additional knowledge surely will lead to improvements.

Experience has proved the usefulness of fairly simple and reasonably accurate methods for predicting overall reservoir performance. Today, with the availability of reservoir simulators, engineers are less dependent on simple models in dealing with complex reservoirs or situations. But we need better ways to represent properly how chemical reactions, fluid bypassing, and mixing affect macroscopic (grid-block level) behavior, and we need to enhance the cost effectiveness and general capabilities of thermal reservoir simulators.

It appears that the processes and their use can be improved through a fuller understanding of the phenomena occurring in the reservoir during thermal recovery. Such improvements likely would result in higher recoveries, reduced costs, extended conditions under which the processes can be commercially applied, or a combination of all these.

15.3 Social Impact

The most beneficial impact of thermal oil is its contribution to the supply of needed liquid fuels in a competitive market. Although the amount of thermal oil produced is small compared with the total amount of oil produced worldwide, thermal oil is an important part of the economy wherever there is a developed industry. Venezuela and California, for example, are two areas where thermal production currently affects the local economy. The effect of any one industry on the economic well-being of an area of course depends on the extent of other industries. At a very local level – e.g., a small town – the effect of thermal production may be critical to the financial

TABLE 15.1 – ENHANCED OIL PRODUCTION FROM ACTIVE PROJECTS, 1981[8]

	All Enhanced Recovery Methods		Thermal Recovery Methods	
Country	Rate (1,000 B/D)	Mean Percent of Total Rate	Rate (1,000 B/D)	Mean Percent of All Methods
U.S.A.	468	51	388	83
Venezuela	163	18	163	100
Canada	89	10	37	42
Others	195	21	83	43
Totals	915	100	671	73

Note: Average values have been used for the production rates where the source gave a range.

solvency and future of the community.

Associated with any human activity, industrial or otherwise, there are peripheral consequences affecting the quality of life in the community. The generation and disposal of waste products and thermal, visual, and noise pollution are familiar factors affecting society. Eliminating or ameliorating factors such as these imposes constraints on the use of thermal recovery and increases its cost. The oil industry usually not only meets the minimum legal requirements but also responds to the current and future needs of the community.

In a broader sense, thermal recovery also may affect society through the consequences of the technology on certain global balances such as the CO_2 and energy balances. There is concern that expanded synfuel programs (both at the surface and in situ) could increase the CO_2 content of the earth's atmosphere, which in turn might create worldwide climatic changes.[14] The energy ratio, which for our purpose is a measure of the amount of energy extracted in the form of fuel per unit of energy consumed, is also important. The social concern is to avoid implementing energy extraction processes that have an energy ratio of less than unity, since such efforts would be self-defeating from an energy-conservation point of view. Although all thermal recovery processes are energy-intensive and several interpretations of how to evaluate numerically the energy ratio have been proposed and used, their energy ratio nonetheless is significantly larger than unity,[15-17] although on the average it is lower than for other recovery processes. Energy ratios and global balances have become popular tools for deriving information on which to base decisions affecting the development of enhanced oil recovery and synthetic fuel processes.[17-21]

TABLE 15.2 – LARGEST ACTIVE THERMAL RECOVERY FIELD PROJECTS, 1981[8]

Operator	Field	Average Extra Oil Rate (B/D)
Steam Drive		
Getty	Kern River, CA	52,250
Caltex	Duri, Indonesia	40,000
Texaco	San Ardo, CA	22,500
Shell	Mount Poso, CA	20,000
Maraven	Tia Juana Este (M-6), Venezuela	15,000
Cyclic Steam Injection		
Getty	Kern River, CA	52,250
Maraven	Lagunillas (T-6), Venezuela	40,850
Caltex	Duri, Indonesia	22,000
Santa Fe Energy	Midway-Sunset, CA	20,000
Combustion		
Mobil	Battrum No. 1, Saskatchewan	2,900
Cities Service	Bellevue, LA	2,800
Combination Process		
Kernridge	Belridge, CA	45,000

15.4 Potential Importance

The potential of thermal recovery processes is much

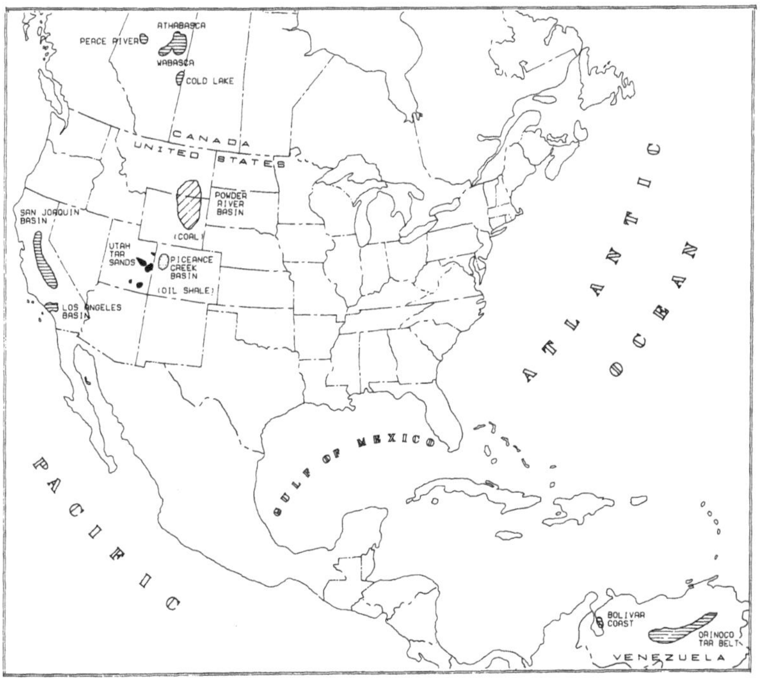

Fig. 15.3 – Location of selected major resources amenable to thermal recovery.

greater than the current production would suggest. One reason is that there are large accumulations of heavy crudes (some of which are called tar sands) amenable to thermal recovery in certain parts of the world, especially in western Canada and central Venezuela (see Fig. 15.3). The amount of crude in place in those two areas is estimated at about 950×10^9 bbl for Canada[19] and 750×10^9 bbl for Venezuela[22]; only a fraction of those amounts ultimately will be proved as reserves. (For comparison, the proved reserves for the world as of the end of 1975 were 567×10^9 bbl.)[23] Thermal recovery is expected to play a dominant role in the production of those reserves.

In Alberta, Imperial Oil Ltd. announced plans to produce up to 141,000 B/D of bitumen from the Cold Lake area by 1985.[24] The cost of the proposed 25-year operation, which appears to be based (at least initially) on cyclic steam injection, is estimated to be $7 billion.[25] Shell Canada Resources Ltd. and Shell Explorer Ltd. announced plans to use a modified steam drive/pressure cycle process to produce bitumen from Peace River in commercial quantities. The pilot phase of this project,[26] which is a joint venture of the two companies and the Alberta Oil Sands Technology and Research Authority (AOSTRA), is currently under way.[27] These are only two of the announced long-range major efforts at recovering bitumen from the accumulations of heavy oil in western Canada.

In 1977, there were 16 in-situ projects in operation, under construction, or being planned for these accumulations, five of which were joint operations with AOSTRA.[19] Fifteen are thermal recovery projects and included cyclic steam injection, steam drive, and wet and dry combustion, supplemented in some cases by the use of solvents and electricity. In 1981, the total number of active thermal recovery projects in Canada was reported as 44.[8] Obviously, the main effort in thermal recovery in western Canada is just beginning.

In Venezuela, there are a number of thermal recovery projects planned for the Lake Maracaibo area. The cyclic steam injection effort is scheduled to be increased substantially. The M-6 Tia Juana steam drive project,[28,29] now being piloted on a large scale (current production is about 15,000 B/D of extra oil), will result in increased application of the steam drive process in steam-soaked areas, with potentials well in excess of 125,000 B/D. But the largest accumulation of heavy oil in Venezuela is the Orinoco Heavy Oil Belt, with oil in place considered to be $1{,}000 \times 10^9$ bbl.[30] Lagoven S.A. is scheduled to produce 150,000 B/D from the Cerro Negro area of the belt by 1988, and Meneven S.A. plans to produce 75,000 B/D by the same year from the Hamaca area of the oil belt, with 50,000 B/D to be available as early as 1983.[31] The greatest impact of thermal recovery in Venezuela is yet to come.

In the U.S., the analogous heavy crude accumulations are already under thermal production in California. The potential of thermal recovery operations in California is still great; it is merely dwarfed by the immense potential in other countries. Other U.S. deposits of heavy oil, known as oil sands or tar sands, have resources amounting to about 30×10^9 bbl; the six accumulations that account for over 90% of this amount are in Utah.[32] The largest of the six is the Tar Sand Triangle area, with tar in place estimated at between 12.5 and 16.0×10^9 bbl. Properties of these six accumulations are given in Table 15.3. Characteristically, they have no significant amounts of reservoir energy and relatively little crude per unit bulk volume (less than 800 bbl/acre-ft, except for the Tar Sand Triangle which has 1,080 bbl/acre-ft). Certain areas within each accumulation may have significantly more crude per unit bulk volume than the reported values, which are averages. Field tests of steam drive have been made at Sunnyside[33] and at Northwest Asphalt Ridge[34] by the industry and the government, respectively. Reverse combustion also has been investigated at Northeast Asphalt Ridge.[35] However, a significant fraction of these resources may be more amenable to surface mining than to in-situ recovery.

Another reason for the great potential of thermal

TABLE 15.3 – UTAH TAR SAND CHARACTERISTICS[32]

Deposit	Number and Type of Samples*	Porosity	Permeability (md)	Bitumen Saturation	Water Saturation	Range of Estimated Resources (10^6 bbl)
Tar Sand Triangle**	29 S	0.200	207	0.063†	–	12,500 to 16,000
	14 C	0.197	788	0.707	–	
P.R. Spring	1,038 C	0.250	1,510	0.425	0.030	4,000 to 4,500
Sunnyside	129 C	0.213	729	0.448	–	3,500 to 4,000
Circle Cliffs	6 C	0.123	228	0.177	–	1,000 to 1,300
Asphalt Ridge	120 C	0.196	497	0.514	0.027	1,000 to 1,200
N.W. Asphalt Ridge	1,087 C	0.228	603	0.452	0.202	
Hill Creek	203 C	0.202	325	0.297	0.021	300 to 1,160
						22,300 to 28,160

*S = surface; C = core.
**A large fraction of the resource is present at low oil saturations. In many cases, the zones of high oil saturations are separated by significant intervals of low saturations.
†Total liquid saturation present by weight.

recovery is that it can be used to displace and recover light crudes. Except possibly where there is insufficient fuel for combustion, or where steam condensate causes formation damage, there are few known *technical* reasons for not considering thermal processes for the recovery of any kind of crude. For example, *depth* is an economic rather than a technical factor, for we have the technology for delivering some form of heat to great depths. Thermal recovery generally has not been used in light-oil reservoirs because other processes (such as waterflooding) yield a higher rate of return on investment. The oil remaining after waterflooding now is considered a target for a third try – the so-called tertiary recovery processes. We say "so-called" because the term tertiary, like primary and secondary, actually indicates only the sequential position of a recovery process relative to the other processes that are applied in the lifespan of a field. The process itself, in one form or another, may be used earlier or later in the sequence of processes, depending on the circumstances. For example, combustion still may be considered by some to be a secondary recovery process – i.e., to be used after primary production. But it has been used to assist and supplement the primary drive production at Midway-Sunset[36] and as a tertiary process after a waterflood.[37] It would be more accurate and of benefit to the general understanding if the term enhanced recovery were consistently used instead of primary, secondary, and tertiary recovery. These comments apply not only to thermal recovery, but to all other forms of enhanced recovery as well.

The fact that some of the other enhanced recovery processes have not reached the maturity of thermal recovery, the pressures that exist to increase oil production in the U.S., and the recognition that thermal processes may be used today to supplement primary methods for increasing U.S. oil production should accelerate the early commercial use of thermal recovery in light-crude reservoirs. In such applications, the production rates are increased because the total field operating life is reduced and less oil is left in the reservoir at the end of the project. The number of light-oil reservoirs in which thermal recovery may be economically attractive is probably small, although this may change with experience.

15.5 Frontier Areas

Besides the conventional uses of thermal processes discussed in the preceding sections, there are two major emerging uses. These are to recover shale oil and coal products in situ. An in-situ recovery process is defined as one in which wells are used to inject fluids and to bring the hydrocarbons to the surface, either as a fluid or in a slurry. The modified in-situ process pioneered by Occidental Oil Co. for the recovery of shale oil falls within this definition, even though the underground combustion retorts are prepared mechanically using mining technology.[38] In-situ coal gasification, such as that being field tested at Hoe Creek, WY, by the Lawrence Livermore Natl. Laboratory and at Hanna, WY, by the U.S. DOE,[39] are also frontier applications of thermal recovery. There is less experience in the use of this technology in the U.S. than in the USSR, where several large demonstration-size in-situ coal gasification projects have been operated for a number of years.[40]

Hydrocarbons have been extracted from oil shale and coal primarily by the combustion process, in which temperatures are high enough to pyrolyze the kerogen and coal quickly and to cause reactions between the coal, water, and combustion products to yield a low-Btu gas. But the injection of hot gas and steam to recover shale oil also has been field tested.[41,42]

In most of the oil shale resources, there is essentially no permeability – i.e., no possibility to inject or circulate fluids as the deposits occur in nature. In coal, there is usually some permeability, but it is inadequate and is sometimes highly sensitive to temperature and environment. The principal technical problems to be solved in developing in-situ processes for these resources are how to create and control a process volume having an adequate bulk permeability and (for oil shale) how to fragment the resource volume into small, reasonably uniform pieces. Mining and rubbling, hydraulic and explosive fracturing, and salt dissolution and thermal rubbling are some of the methods that have been or are being tried in oil shales to achieve adequate communication.

The resources associated with oil shale and coal are very large. In the Powder River basin alone, the equivalent oil (on a Btu basis) in coal seams more than 50 ft thick and more than 1,000 ft below the surface is 900×10^9 bbl.[43] In the Piceance Creek basin, oil shale assaying at 25 gal/ton and higher contains 450×10^9 bbl of shale oil by Fischer assay.[44] It should be noted that some of these resources can be mined. The principal holdup to developing these resources is economics, which is affected by a lack of sufficient water near the resources and by expensive technology (including that required to meet environmental constraints). There is, in addition, a lack of unanimity among private groups and government agencies on how (and even whether) to allow development.

References

1. *Annual Review of California Oil and Gas Production,* Conservation Committee of California Oil Producers, Los Angeles (yearly through 1979).
2. *Summary of Operations of California Oil Fields,* California Div. of Oil and Gas, San Francisco (yearly since 1915).
3. Noran, D.: "Growth Marks Enhanced Oil Recovery," *Oil and Gas J.* (March 27, 1978) 113-140.
4. Bleakely, W.B.: "Journal Survey Shows Recovery Projects Up," *Oil and Gas J.* (March 25, 1974) 69-76.
5. Noran, D. and Franco, A.: "Production Reports," *Oil and Gas J.* (April 5, 1977) 106-138.
6. Matheny, S.L. Jr.: "EOR Methods Help Ultimate Recovery," *Oil and Gas J.* (March 31, 1980) 79-124.
7. Dafter, R.: "Scraping the Barrel – The Worldwide Potential for Enhanced Oil Recovery," *Financial Times,* London (1980).
8. Dafter, R.: "Winning More Oil," *Financial Times*, London (1981).
9. Elson, T.D. and Marsden, S.S.: "The Effectiveness of Foaming Agents at Elevated Temperatures over Extended Periods of

Time,'' paper SPE 7116 presented at the 1978 SPE California Regional Meeting, San Francisco, April 12–14.

10. Kuuskraa, V.A., Hammershaimb, E.C., and Doscher, T.M.: ''Improved Steam Flooding by Use of Additives,'' U.S. DOE, Oakland, CA (Dec. 1979).
11. Robinson, R.J., Bursell, C.G., and Restine, J.L.: ''A Caustic Steamflood—Kern River Field, California,'' paper SPE 6523 presented at the 1977 SPE California Regional Meeting, Bakersfield, April 13–15.
12. Cady, G.V. and Moss, J.T.: ''The Design and Installation of an Oxygen Supported Combustion Project in the Lindbergh Field, Alberta, Canada,'' paper presented at the 1981 World Oil and Gas Show, Dallas, Dec. 14-17.
13. Moss, J.T. and Cady, G.V.: ''Laboratory Investigation of the Oxygen Combustion Process for Heavy Oil Recovery,'' paper SPE 10706 presented at the 1982 SPE California Regional Meeting, San Francisco, March 24–26.
14. ''Panel Says Synfuels Pose No CO_2 Hazard, for Now,'' *Science* (Aug. 31, 1979) 884.
15. Clark, C.E. and Varisco, D.C.: ''Net Energy and Oil Shale,'' *Proc.,* Eighth Oil Shale Symposium, Colorado School of Mines, Golden (July 1975) **70**, No. 3, 3-19.
16. Melcher, A.G.: ''Net Energy Analysis: An Energy Balance Study of Western Fossil Fuels,'' *Proc.,* Ninth Oil Shale Symposium, Colorado School of Mines, Golden (July 1976) **71**, No. 3, 11-31.
17. *Energy Study,* Oregon Office of Energy Research and Planning, Office of the Governor, State of Oregon, Salem (1974) 150-180.
18. *Energy Alternatives: A Comparative Analysis,* The Science and Public Policy Program, U. of Oklahoma, Norman (1975) Chap. 15; U.S. Govt. Printing Office No. 041-011-00025-4.
19. Nicholls, J.H. and Luhning, R.W.: ''Heavy Oil Sand In-Situ Pilot Plants in Alberta (Past and Present),'' *J. Cdn. Pet. Tech.* (July-Sept. 1977) 50-61.
20. Berry, K.L.: ''Conceptual Design of Combined In-Situ and Surface Retorting of Oil Shale,'' *Proc.,* Eleventh Oil Shale Symposium, Colorado School of Mines, Golden (1978) **73**, 176-183.
21. Bridges, J.E., Taflove, A., and Snow, R.H.: ''Net Energy Recoveries for the In-Situ Dielectric Heating of Oil Shale,'' *Proc.,* 11th Oil Shale Symposium, Colorado School of Mines, Golden (1978) **73**, 311-330.
22. ''Venezuela – Hay Que Echarle Mano al Orinoco,'' *Petróleo Intl.* (Aug. 1977) 17-19.
23. ''Twentieth Century Petroleum Statistics,'' DeGolyer and MacNaughton, Dallas (1976).
24. ''Imperial Poised to Launch Project at Cold Lake Heavy Oil Region,'' *Oilweek* (Nov. 21, 1977) 16, 17.
25. *Cameron Synthetic Fuels Report,* Cameron Engineers Inc., Denver (Sept. 1980) **17**, No. 3.
26. Dillabough, J.A. and Prats, M.: ''Proposed Pilot Test for the Recovery of Crude Bitumen from the Peace River Tar Sands Deposit – Alberta, Canada,'' *Proc.,* Simposio Sobre Crudos Pesados, Inst. de Investigaciones Petroleras, U. del Zulia, Maracaibo (1974).
27. ''Shell In-Situ Pilot Project Opens,'' *Oilweek* (Dec. 3, 1979) 3.
28. Herrera, A.J.: ''M-6 Steam Drive Project Design and Implementation,'' *J. Cdn. Pet. Tech.* (July-Sept. 1977) 62-71.
29. Schenk, L.: ''Analysis of the Early Performance of the M-6 Steam-Drive Project, Venezuela,'' paper SPE 10710 presented at the 1982 SPE/DOE Enhanced Oil Recovery Symposium, Tulsa, April 5-8.
30. Fiorillo, G.: ''Exploration of the Orinoco Oil Belt – Review and General Strategy,'' paper UNITAR/CF10/V/3 presented at the 11th Intl. Conference on Heavy Crude and Tar Sands, Caracas, Feb. 7-17, 1982.
31. ''Enhanced Recovery Report – Canadians, Venezuelans Exchange View at Edmonton Oil Symposium,'' *Oilweek* (June 20, 1977) **9**, 12-32.
32. Glassett, J.M. and Glassett, J.A.: ''The Production of Oil from Intermountain West Tar Sands Deposits,'' BUMines-OFR-92-76, USBM (1976).
33. Thurber, J.L. and Welbourn, M.E.: ''How Shell Attempted to Unlock Utah Tar Sands,'' *Pet. Eng.* (Nov. 1977) 31, 34, 38, and 42.
34. Johnson, L.A. Jr., Fahy, L.T., Romanowski, L.J., and Thomas, K.P.: ''An Evaluation of a Steamflood Experiment in a Utah Tar Sand Deposit,'' *J. Pet. Tech.* (May 1982) 1119–1126.
35. Land, C.S., Cupps, C.Q., Marchant, L.C., and Carlson, F.M.: ''Field Test of Reverse-Combustion Oil Recovery from a Utah Tar Sand,'' *J. Cdn. Pet. Tech.* (April-June 1977) 34-38.
36. Gates, C.F., and Sklar, I.: ''Combustion – A Primary Recovery Process, Moco Zone Reservoir, Midway Sunset Field, Kern County, California,'' *J. Pet. Tech.* (Aug. 1971) 981-986; *Trans.,* AIME, **251**.
37. Parrish, D.R., Pollock, C.B., Ness, N.L., and Craig, F.F. Jr.: ''A Tertiary COFCAW Pilot Test in the Sloss Field, Nebraska,'' *J. Pet. Tech.* (June 1974) 981-986; *Trans.,* AIME, **257**.
38. Ridley, R.D.: ''In-Situ Processing of Oil Shale,'' *Quarterly of the Colorado School of Mines* (April 1974) **69**, No. 2, 21-24.
39. Schrider, L.A. and Whieldon, C.E., Jr.: ''Underground Coal Gasification – A Status Report,'' *J. Pet. Tech.* (Sept. 1977) 1179-1185.
40. Gregg, D.W., Hill, R.W., and Olness, D.U.: ''An Overview of the Soviet Effort in Underground Gasification of Coal,'' UCRL-52004, ERDA Lawrence Livermore Natl. Laboratory (Jan. 1976).
41. Dougan, J.L.: ''Recovery of Petroleum Products from Oil Shale,'' U.S. Patent No. 3,241,611 (March 22, 1966).
42. Prats, M., Closmann, P.J., Ireson, A.T., and Drinkard, G.: ''Soluble-Salt Processes for In-Situ Recovery of Hydrocarbons from Oil Shale,'' *J. Pet. Tech.* (Sept. 1977) 1078-1088.
43. Roupert, R.C., Choate, R., Cohen, S., Lee, A.A., Lent, J., and Spraul, J.R.: ''Energy Extraction from Coal In-Situ – A Five-Year Plan,'' T1D27203, prepared by TRW Systems, available from Natl. Technical Information Service (1976).
44. Duncan, D.C. and Swanson, V.E.: ''Organic-Rich Shale of the U.S. and World Land Areas,'' USGS Circular 523, Washington, DC (1965).

Appendix A

Selected Conversion Factors

Conversion factors are necessary because there is no single system of units used by everyone in the oil industry. However, it appears that the time is near[1] when the SI system of units (Le Système International d'Unités) will be adopted throughout most of the world, including the U.S.

There are seven base quantities used in the SI metric system, each having a separate unit and denoted by a separate symbol, as shown in Table A-1. These base quantities may be combined to form derived units. The derived units used in this monograph are listed in Table A-2.

Base and derived SI units are multiplied by prefixes to modify and denote the magnitude of the quantities. These SI unit prefixes, and some comments on their use, are given in Table A-3.

The Nomenclature lists conversion factors for the units and combination of units used in this monograph; Table A-4 lists additional useful conversion factors. (For a fuller explanation of the SI metric system and a more exhaustive list of conversion factors, see Ref. 1.) In all cases, the conversions are purposely unidirectional: from customary units to SI units. While this does not prevent one from finding the inverse, it should discourage the continued use of customary units.

The importance of developing a sense of the magnitude of the SI quantities cannot be overemphasized. To this end, Table A-5 is offered as a "memory jogger" or guide to arrive at the "metric ball park" relative to customary units. Table A-5 is *not* a conversion table.

TABLE A-1 – SI BASE QUANTITIES

Base Quantity	SI Unit	SI Unit Symbol
Length	Meter	m
Mass	Kilogram	kg
Time	Second	s
Thermodynamic Temperature	Kelvin	K
Amount of Substance	Mole	mol
Electric Current	Ampere	A
Luminous Intensity	Candela	cd

Reference

1. "The SI Metric System of Units and SPE's Tentative Metric Standard," *J. Pet. Tech.* (Dec. 1977) 1575-1611; also available as pamphlet from SPE, Dallas (1979).

TABLE A-2 – SELECTED SI DERIVED UNITS

Quantity	Unit	Symbol	Definition
Acceleration	Meter per second squared	–	m/s^2
Area	Square meter	–	m^2
Density	Kilogram per cubic meter	–	kg/m^3
Energy	Joule	J	N·m
Force	Newton	N	$kg \cdot m/s^2$
Power	Watt	W	J/s
Pressure	Pascal	Pa	N/m^2
Quantity of heat	Joule	J	N·m
Radiant flux	Watt	W	J/s
Specific heat	Joule per kilogram kelvin	–	J/kg·K
Stress	Pascal	Pa	N/m^2
Thermal conductivity	Watt per meter kelvin	–	W/m·K
Velocity	Meter per second	–	m/s
Viscosity, dynamic	Pascal-second	–	Pa·s
Viscosity, kinematic	Square meter per second	–	m^2/s
Volume	Cubic meter	–	m^3
Work	Joule	J	N·m
Diffusivity (any)	Square meter per second	–	m^2/s

TABLE A-3 – SI UNIT PREFIXES

Factor	Prefix	Symbol
10^{18}	exa	E
10^{15}	peta	P
10^{12}	tera	T
10^9	giga	G
10^6	mega	M
10^3	kilo	k
10^2	hecto	h
10	deka	da
10^{-1}	deci	d
10^{-2}	centi	c
10^{-3}	milli	m
10^{-6}	micro	μ
10^{-9}	nano	n
10^{-12}	pico	p
10^{-15}	femto	f
10^{-18}	atto	a

Notes: Double or multiple prefixes are not to be used. For example, use GJ (gigajoule), not kMJ; use 6.68 kg, not 6 kg 680 g.

Multiple units appearing in the same line are separated by a product dot or parenthesis. For example, mK means 10^{-3}K, but m·K means (meter)(Kelvin).

A unit *and its prefix* are raised to the power expressed by the exponent – e.g., $1\ cm^3 = (10^{-2}\ m)^3 = 10^{-6}\ m^3$, $1\ cm^3 \neq 1\ c(m^3) \neq 10^{-2}\ m^3$.

TABLE A-4 – SUPPLEMENTAL CONVERSION FACTORS

Quantity	Customary Units	Metric Unit	Conversion Factor (Multiply value in customary units by factor to obtain value in metric units)
Amount of substance	scf (60 F, 1 atm)	kmol	1.195×10^{-3}
Pressure	atm (760 mm Hg, 14.696 psi)	kPa	1.013×10^{2}
Velocity	mph	km/h	1.609×10^{0}
Viscosity, dynamic	lbf-s/sq ft	Pa·s	4.788×10^{1}
	lbm/ft-s	Pa·s	1.488×10^{0}
Viscosity, kinematic	sq ft/s	mm^2/s	9.290×10^{4}

TABLE A-5 – "MEMORY JOGGER" FOR METRIC UNITS

Customary Unit	"Ballpark" Metric Values; Do *Not* Use As Conversion Factors	
Acre	4000	square meters
	0.4	hectare
Barrel	0.16	cubic meters
British thermal unit	1000	joules
British thermal unit per pound-mass	2300	joules per kilogram
	2.3	kilojoules per kilogram
Calorie	4	joules
Centipoise	1*	millipascal-second
Centistokes	1*	square millimeter per second
Darcy	1	square micrometer
Degree Fahrenheit (temperature *difference*)	0.5	kelvin
Dyne per centimeter	1*	millinewton per meter
Foot	30	centimeter
	0.3	meter
Cubic foot (cu ft)	0.03	cubic meter
Cubic foot per pound-mass (cu ft/lbm)	0.06	cubic meter per kilogram
Square foot (sq ft)	0.1	square meter
Foot per minute	0.3	meter per minute
	5	millimeters per second
Foot-pound-force	1.4	joules
Foot-pound-force per minute	0.02	watt
Foot-pound-force per second	1.4	watts
Horsepower	750	watts (¾ kilowatt)
Inch	2.5	centimeters
Kilowatthour	3.6*	megajoules
Mile	1.6	kilometers
Ounce (avoirdupois)	28	grams
Ounce (fluid)	30	cubic centimeters
Pound-force	4.5	newtons
Pound-force per square inch (pressure, psi)	7	kilopascals
	0.07	bar
Pound-mass	0.5	kilogram
Pound-mass per cubic foot	16	kilograms per cubic meter
Section	260	hectares
	2.6	milion square meters
	2.6	square kilometers
Ton, long (2240 pounds-mass)	1000	kilograms
Ton, metric (tonne)	1000*	kilograms
Ton, short	900	kilograms

*Exact equivalents.

Appendix B

Data and Properties of Materials

Rock and fluid properties of general interest in petroleum engineering have been summarized and reported in previously published monographs.[1-3] Properties of interest in thermal recovery operations have been reported by Farouq Ali.[4] The summaries of properties in this appendix incorporate essentially all such material presented by Farouq Ali[4] and Earlougher.[3] (The entire text of Appendix D of Ref. 3 has been incorporated here, much of it verbatim with appropriate changes in figure and reference numbers.) In addition, attempts have been made to extend the temperature range of some of their correlations, as well as to provide information specific to thermal recovery. In summary, this appendix edits and supplements the information provided in Refs. 3 and 4, emphasizing aspects that bear on the subject of this monograph.

The quality of the data used in thermal recovery operations is generally best when measured on the actual materials (fluid and rocks) to be used under simulated operational conditions. Even when the materials are on hand, equipment may not be available to make the desired measurements at the simulated operational conditions. Where appropriate, measurements may be made at less severe conditions and then extrapolated to the desired ones. More commonly, rock and fluid samples are not available from the reservoir of interest, in which case samples from reservoirs considered similar may be measured. In other cases, properties may be estimated on the basis of experience, from available correlations, or from published data on the same or similar material.

This appendix is organized so that to obtain information on a particular property (e.g., of water), it is necessary first to locate the section emphasizing the desired property and then to locate the subsection dealing with water.

Sec. B.1 treats density and PVT properties of crudes and water and the densities of minerals and fluid-saturated rocks. Sec. B.2 discusses isothermal compressibilities of pore volumes, liquids, and gases. Sec. B.3 presents information in the isobaric thermal coefficient of volume expansion for liquids, gases, and solids. Sec. B.4 discusses methods and correlations for estimating the viscosity of crudes, crude components, water, and gases. Sec. B.5 presents information on the enthalpy and specific heat of steam and water, crude oil and its fractions, and reservoir rocks. Sec. B.6 considers thermal conductivities of liquids, gases, and reservoir rocks and other solids. Sec. B.7 briefly discusses thermal diffusivity and how this property can be estimated from thermal conductivities and densities. Sec. B.8 provides correlations for estimating coefficients of heat transfer in tubular goods. Sec. B.9 introduces data on standard tubing and casing. And Sec. B.10 gives sources of information on *K*-values and other miscellaneous properties.

B.1. Density and Other PVT Properties

Crude Oils

Though the title of this subsection is Crude Oils, some of the information deals with properties of pure hydrocarbons and a few nonhydrocarbon gases and how their individual properties are combined to obtain properties of their mixture. Strictly speaking, a hydrocarbon contains only hydrogen and carbon in its molecular structure, whereas crude oil components (especially the viscous crudes of interest in thermal operations) more often than not contain additional elements, usually from the following series: sulfur, oxygen, nitrogen, vanadium, nickel, and other heavy metals. These additional elements generally are found in the denser components of the crude oil, the ones more likely to remain liquid at high temperatures. And even where crude properties are used in the correlations, no two crudes have the same composition. Accordingly, this Appendix refers to the original articles for a clearer understanding of the limitations of the correlations.

To ensure accuracy and to include realism in reservoir engineering applications, one should obtain and use carefully measured data. These usually are obtained in laboratories. Where accurate data are critical, it is both poor economics and poor engineering to resist obtaining good laboratory data simply because correlations are available.

Table B.1 gives physical properties of methane through decane and some other compounds commonly associated with petroleum reservoirs. More

TABLE B.1 – PHYSICAL PROPERTIES OF HYDROCARBONS AND SELECTED COMPOUNDS[5]

Constituent	Molecular Weight	Normal Boiling Point (°F)	Normal Boiling Point (°R)	Liquid Density (lbm/cu ft)	Gas Density at 60°F, 1 atm (lbm/cu ft)	Critical Temperature (°R)	Critical Pressure (psia)
Methane, CH_4	16.04	−258.7	201.0	19*	0.04228	343.1	668
Ethane, C_2H_6	30.07	−127.5	332.2	22.22**	0.07924	549.8	708
Propane, C_3H_8	44.10	−43.7	416.0	31.66**	0.1162	665.7	616
Isobutane, C_4H_{10}	58.12	10.9	470.6	35.12**	0.1531	734.7	529
n-butane, C_4H_{10}	58.12	31.1	490.8	36.44**	0.1531	765.3	551
Isopentane, C_5H_{12}	72.15	82.1	541.8	38.96	–	828.8	490
n-pentane, C_5H_{12}	72.15	96.9	556.6	39.35	–	845.4	489
n-hexane, C_6H_{14}	86.18	155.7	615.4	41.41	–	913.4	437
n-heptane, C_7H_{16}	100.20	209.2	668.9	42.92	–	972.5	397
n-octane, C_8H_{18}	114.23	258.2	717.9	44.08	–	1,023.9	361
n-nonane, C_9H_{20}	128.26	303.5	763.2	45.01	–	1,070.4	332
n-decane, $C_{10}H_{22}$	142.29	345.5	805.2	45.79	–	1,112	304
Nitrogen, N_2	28.01	−320.4	139.3	–	0.07380	227.3	493
Air (O_2+N_2)	28.96	−317.6	142.1	–	0.07630	238.4	547
Carbon dioxide, CO_2	44.01	−109.3	350.4	51.5**	0.1160	547.6	1,071
Hydrogen sulfide, H_2S	34.08	−76.6	383.1	49.3**	0.08977	672.4	1,306
Water	18.02	212.0	671.7	62.37	–	1,365.3	3,208
Hydrogen, H_2	2.02	−423.0	36.7	–	0.005313	59.9	181
Oxygen, O_2	32.00	−297.4	162.3	–	0.08432	278.6	737
Carbon monoxide, CO	28.01	−313.6	146.1	–	0.07380	240	507

*Apparent density in liquid phase.
**Density at saturation pressure.

exhaustive data are given in Refs. 5 and 6. Such information can be used to estimate some properties of hydrocarbon mixtures.

The pseudocritical temperature T_{pc} and pseudocritical pressure p_{pc} of a mixture are used in many correlations and equations in this appendix. If the composition of the mixture is known, these quantities may be estimated from

$$T_{pc} = \sum_{j=1}^{n_C} y_j T_{cj} \quad \text{(B.1)}$$

and

$$p_{pc} = \sum_{j=1}^{n_C} y_j p_{cj}, \quad \text{(B.2)}$$

where

n_C = number of components in the mixture,
y_j = mole fraction of component j,
T_{cj} = critical temperature of component j, °R, and
p_{cj} = critical pressure of component j, psia.

If the system composition is not known, Figs. B.1 through B.3 may be used to estimate T_{pc} and p_{pc}.

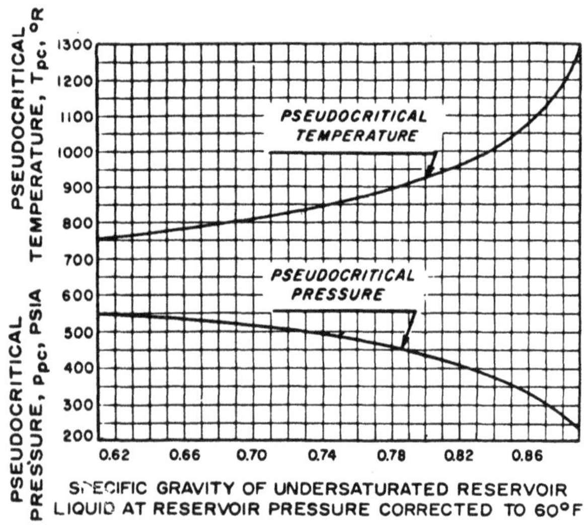

Fig. B.1 – Approximate correlation of liquid pseudocritical pressure and temperature with specific gravity.[7]

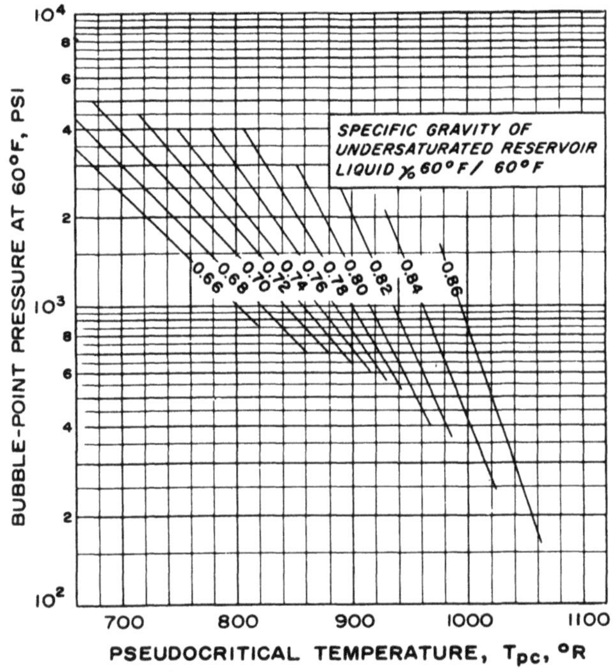

Fig. B.2 – Correlation of liquid pseudocritical temperature with specific gravity and bubble point.[7]

Fig. B.1 provides a way to estimate those quantities for undersaturated oil at reservoir pressure; the oil specific gravity corrected to 60°F (the value normally reported) is used. "The specific gravity of a petroleum oil and of mixtures of petroleum products with other substances is the ratio of the weight of a given volume of the material at a temperature of 60°F . . . to the weight of an equal volume of distilled water at the same temperature, both weights being corrected for the buoyancy of air."[8] This often is referred to as γ_o at 60°F/60°F. Sometimes the weight of the oil is not determined at 60°F, in which case the specific gravity is either corrected to 60°F to give γ_o at 60°F/60°F, or it is reported as γ_o at temperature. The pressure at which the liquids are weighed is not specified. In Fig. B.1, the specific gravity at reservoir pressure corrected to 60°F is the ratio of the oil density at reservoir pressure and temperature, corrected to 60°F, to the density of distilled water at 60°F and 1 atm. Fig. B.2 applies to bubble-point liquids, again using the specific gravity corrected to 60°F. The bubble-point pressure at 60°F should be determined in the laboratory.* Fig. B.3

*If only the value at reservoir temperature is known, Fig. B.5 may be used to estimate the 60°F value by going vertically upward from the bubble-point pressure to reservoir temperature, horizontally left to 60°F, and vertically downward to the estimated bubble-point pressure.

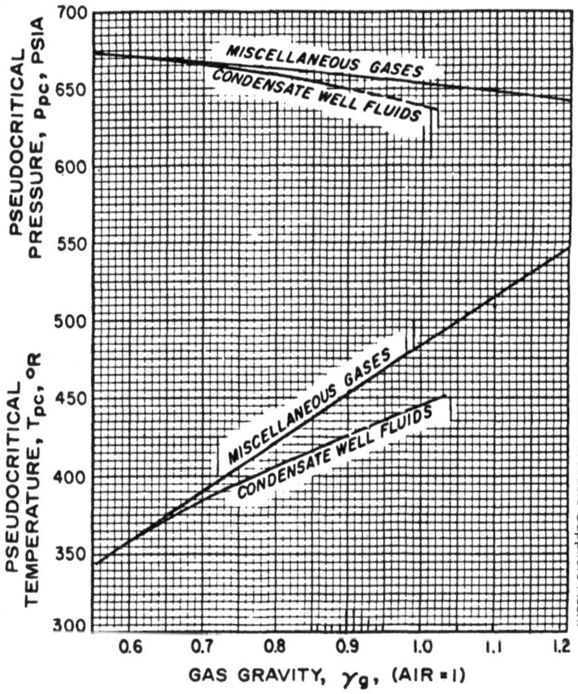

Fig. B.3 – Correlation of pseudocritical properties of condensate well fluids and miscellaneous natural gas with fluid gravity.[9]

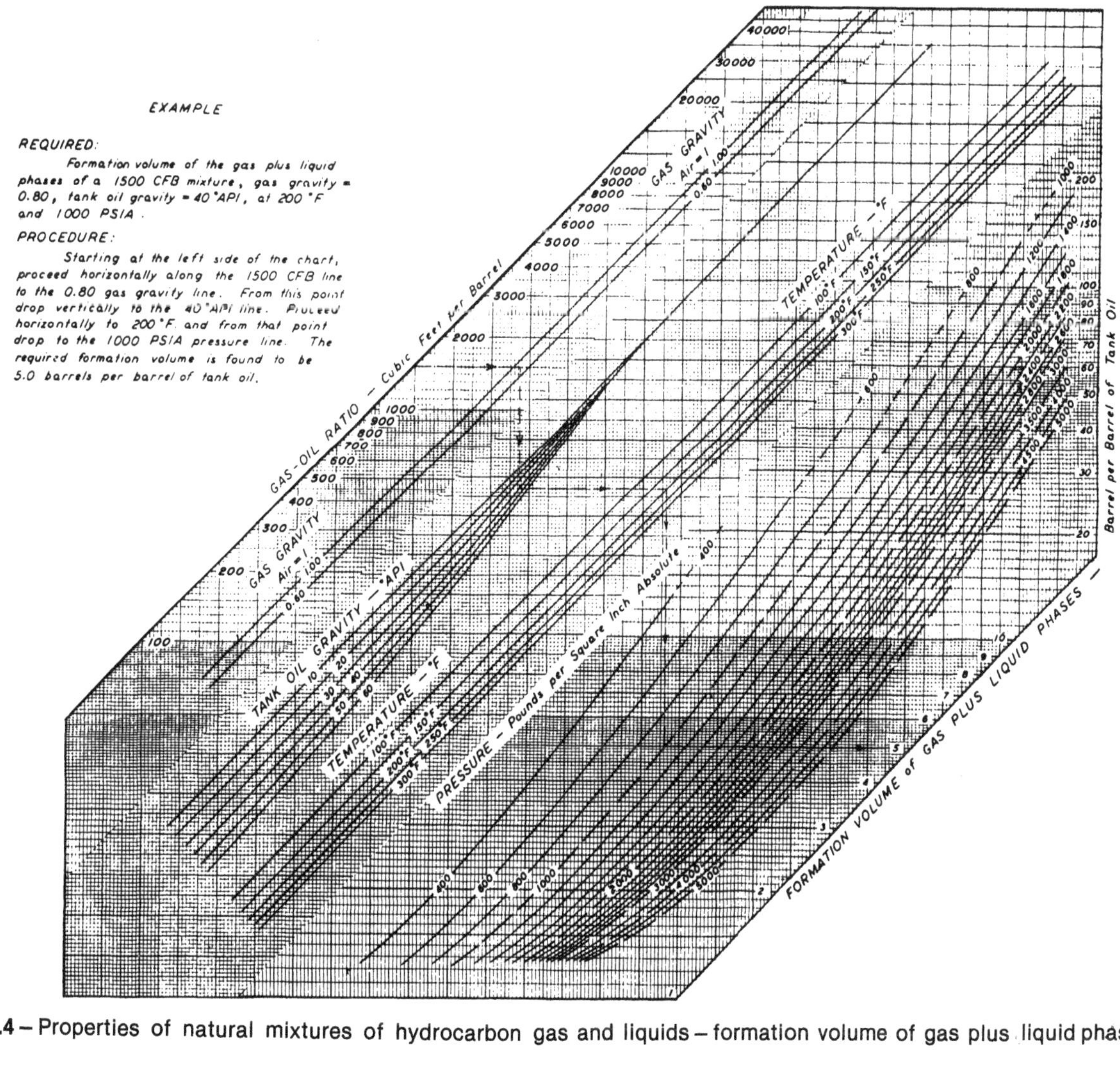

Fig. B.4 – Properties of natural mixtures of hydrocarbon gas and liquids – formation volume of gas plus liquid phase.[12]

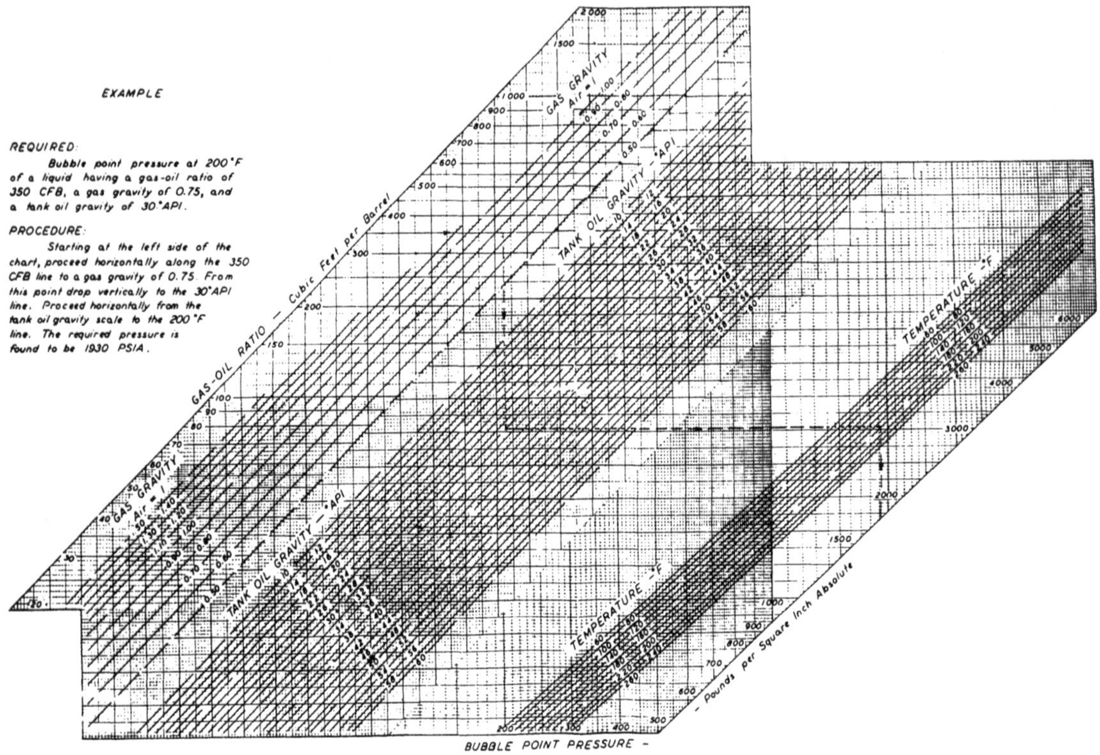

Fig. B.5 – Properties of natural mixtures of hydrocarbon gas and liquids – bubble-point pressure.[12]

applies to condensate well fluids and natural gases; to use Fig. B.3 the gas gravity must be known.

T_{pc} and p_{pc} normally are used to estimate the pseudoreduced temperature and pressure:

$$T_{pr}=\frac{T_{ab}}{T_{pc}}, \quad \text{(B.3)}$$

and

$$p_{pr}=\frac{p_{ab}}{p_{pc}}, \quad \text{(B.4)}$$

where

T_{ab} = temperature of interest, °R, and
p_{ab} = pressure of interest, psia.

Note in Eqs. B.1 through B.4 that both temperature and pressure must be absolute.

Since many correlations in this appendix use specific gravity or API gravity, it is worthwhile to restate the relationships between those two quantities:

$$°\text{API}=\frac{141.5}{\gamma_o}-131.5. \quad \text{(B.5)}$$

In Eq. B.5, γ_o is the specific gravity at 60°F/60°F. If the API gravity is reported at other than 60°F, it may be corrected to 60°F using the technique described in Ref. 8. (In Ref. 8, Table 5 is used for hydrometer measurements at other than 60°F; Table 7 allows the correction of volume at a given temperature to volume at 60°F.)

Figs. B.4 through B.6 are Standing's[10] correlations for properties of mixtures of hydrocarbon gases and liquids. Examples of their use are shown in the figures. Standing's correlations are based mainly on the properties of California crude oils. Cronquist[11] gives correlations that may be useful for U.S. gulf coast oils.

Fig. B.4 is a correlation to estimate the formation volume factor of the crude liquid and gas phases measured at stock-tank conditions, which depend on the total GOR (free and dissolved), the gas specific gravity (referred to air), the tank oil gravity in °API, and the ambient pressure and temperature in the reservoir. Fig. B.5 is a correlation for estimating the bubble-point pressure p_b (at a desired temperature T in °F up to 260°F) of a liquid when its solution GOR R_s, the specific gravity of the dissolved gas γ_g, and the API gravity of the stock-tank oil are known. Standing[12] gives

$$p_b=18.2\left[\left(\frac{R_s}{\gamma_g}\right)^{0.83}10^{(T/1100-°\text{API}/80.0)}-1.4\right] \quad \text{(B.6)}$$

as the relationship on which Fig. B.5 is based.

Fig. B.6 is a correlation to estimate the oil formation volume factor when the solution GOR, dissolved gas specific gravity, API gravity of the stock-tank oil, and reservoir temperature are known. The equation on which this figure is based is[12]

$$B_o=0.9759+12\times10^{-5}\left(R_s\sqrt{\frac{\gamma_g}{\gamma_o}}+1.25T\right)^{1.2}. \quad \text{(B.7)}$$

Fig. B.7 is the well known chart of real gas deviation factor for natural gases. Pseudoreduced properties may be estimated from Eqs. B.1 through B.4.

The gas formation volume factor B_g is given by

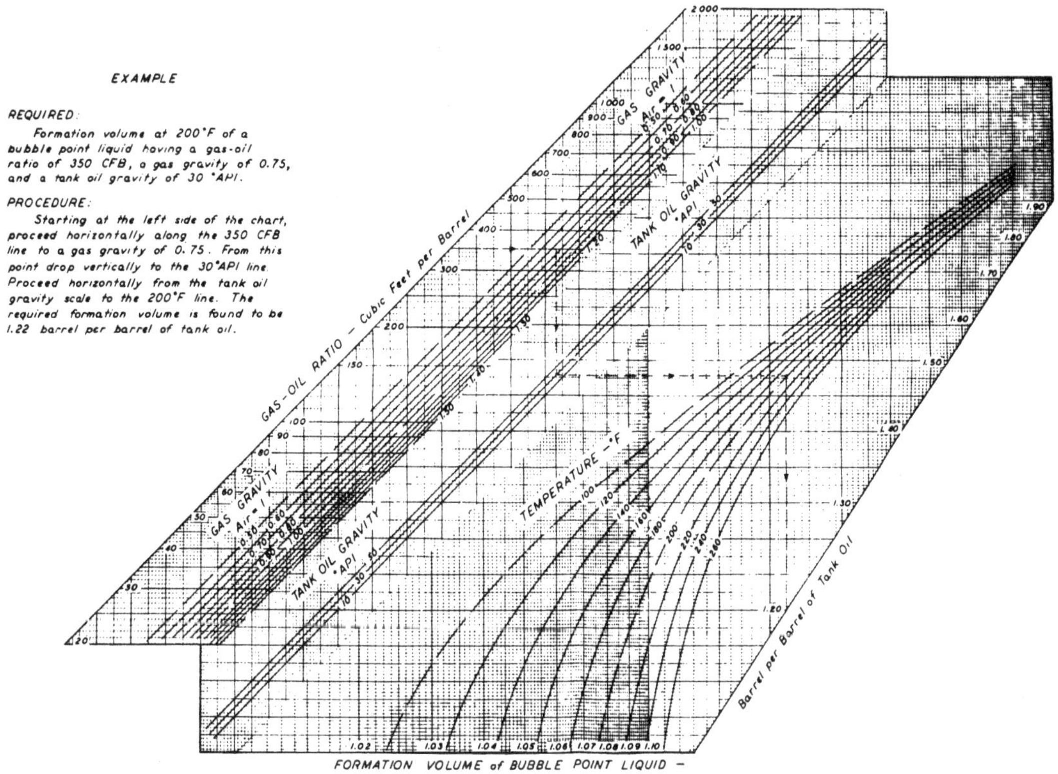

Fig. B.6 – Properties of natural mixtures of hydrocarbon gas and liquids – formation volume of bubble-point liquids.[12]

$$B_g = \langle 178.1\ rb/\text{Mscf} \rangle \frac{p_{sc}}{T_{sc}} \frac{T_{R,ab}}{p_{R,ab}} z, \quad \ldots\ldots\ (B.8)$$

where the deviation factor z may be estimated from Fig. B.7, and $T_{R,ab}$ and $p_{R,ab}$ are the absolute reservoir temperature and pressure.

Water

The water formation volume factor, including dissolved gases, may be estimated from Fig. B.8. Table B.2 presents information[15,16] on the specific volume of the saturated liquid and vapor phases of solute-free water for several values of temperature and pressure. Densities are the reciprocal of the specific volumes. Fig. B.9 gives specific volumes for superheated steam.

Fig. B.10 shows the relationship between saturation pressure and temperature for water obtained from the entries in Table B.2. The approximate expression

$$T_s = 1 + 115.1 p_s^{\,0.225},$$

where p_s is in psia and T_s is in °F, gives saturation temperatures with less than 0.9% error for pressures ranging from 9 to 3,206.2 psia. When used inversely to calculate the saturation pressure, the error is less than 4%.

Fluid-Saturated Rocks

The density of fluid-saturated rock is the volumetric average of the densities of its constituents:

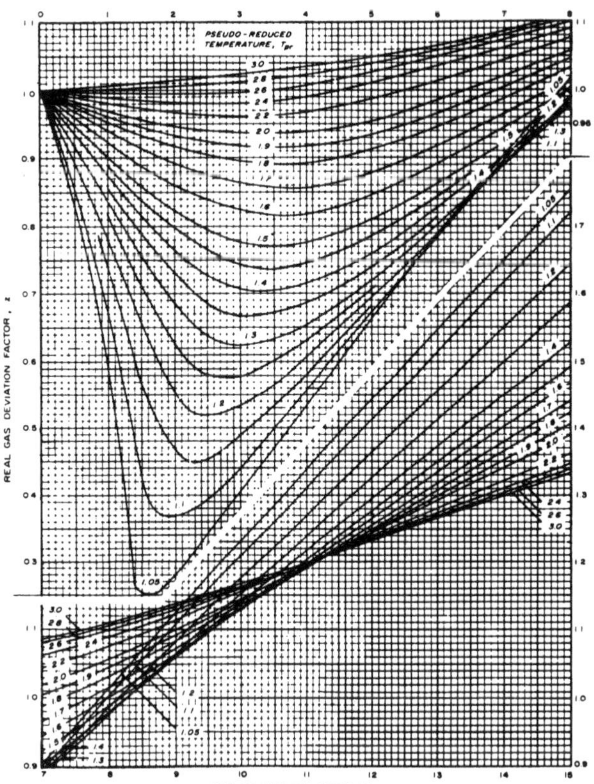

Fig. B.7 – Real-gas deviation factor for natural gases as a function of pseudoreduced pressure and temperature.[13]

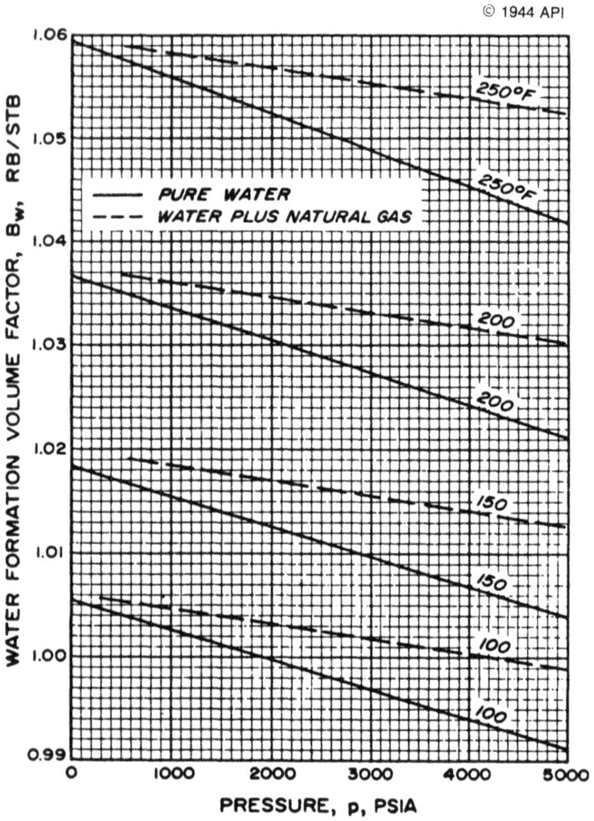

Fig. B.8 – Formation volume factor of pure water and a mixture of natural gas and water.

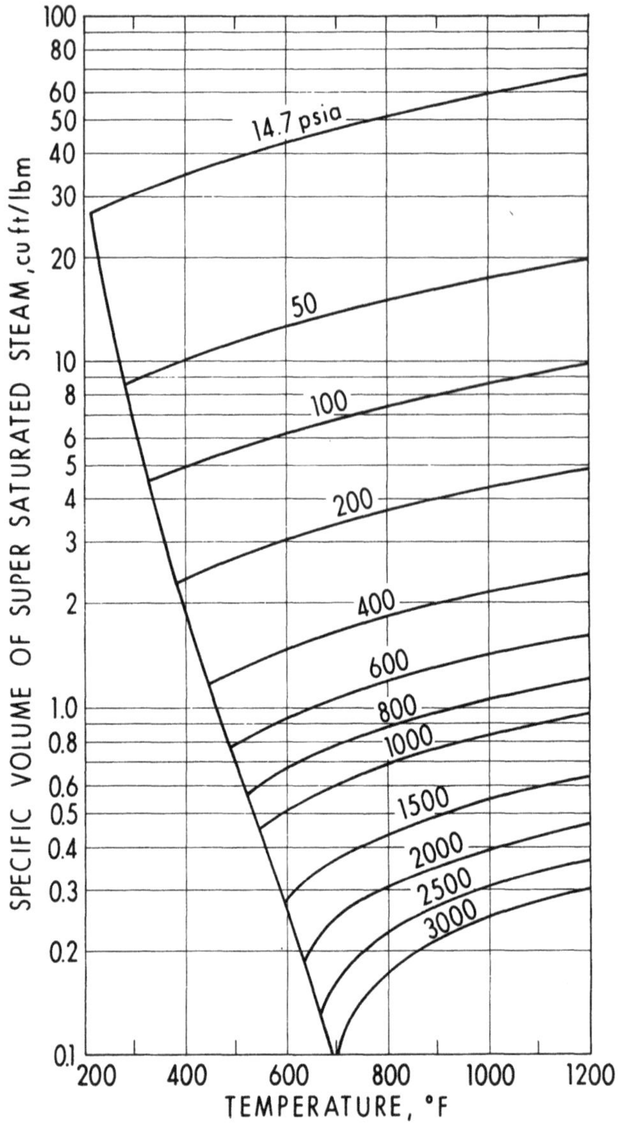

Fig. B.9 – Effect of temperature on specific volume of superheated steam.

$$\rho=\rho_\sigma(1-\phi)+\phi(S_o\rho_o+S_w\rho_w+S_g\rho_g). \quad \text{(B.10)}$$

Fluid densities have been discussed earlier. Table B.3 presents the densities of selected common minerals found in reservoir rocks, measured at room temperatures. The densities given for clays reflect uncertainties about the amount of water present. It is to be understood that the solids density ρ_σ used in Eq. B.10 is itself a volumetric average of the solid constituents of the rock matrix.

B.2 Compressibility

Pore Volume

The isothermal formation (rock pore volume) compressibility generally is defined as

$$c_f=\frac{1}{V_p}\left(\frac{\partial V_p}{\partial p}\right)_T. \quad \text{(B.11)}$$

The subscript T indicates that the partial derivative is taken at constant temperature. The subscript frequently is omitted, since all compressibilities used in this monograph are isothermal compressibilities. Formation compressibility is defined so that it is a positive number. Thus, Eq. B.11 indicates that as fluid pressure decreases, the pore volume decreases. That occurs because the confining lithostatic pressure is essentially constant while the reservoir is depleted, thus causing compression of the rock.

Several authors have attempted to correlate formation compressibility with various physical parameters. The correlations of Hall[20] and van der Knaap[21] have been used extensively in the petroleum literature. Newman[22] has shown that those correlations apply to only a limited range of reservoir rocks. Fig. B.11 shows data for compressibility of limestone samples superimposed on both van der Knaap's and Hall's limestone correlations. In Fig. B.11, and in other figures in this section, the lithostatic pressure is defined as the pressure obtained by multiplying reservoir depth by 1 psi/ft.

TABLE B.2 – PROPERTIES OF SATURATED WATER

Temperature (°F)	Absolute Pressure (psia)	Specific Volume (cu ft/lbm) Saturated Liquid	Saturated Vapor
101.76	1	0.01614	333.79
150	3.716	0.01634	97.20
162.25	5	0.01641	73.600
193.21	10	0.01659	38.462
200	11.525	0.01663	33.67
212	14.696	0.01672	26.83
250	29.82	0.01700	13.841
300	67.01	0.01745	6.471
327.83	100	0.01774	4.433
350	134.62	0.01799	3.342
358.43	150	0.01809	3.016
381.82	200	0.01839	2.288
400	247.25	0.01864	1.8632
400.97	250	0.01866	1.8431
417.33	300	0.01890	1.5426
420	308.82	0.01894	1.4995
431.71	350	0.01912	1.3255
440	381.59	0.01926	1.2166
460	466.97	0.0196	0.9941
467.00	500	0.0197	0.9274
480	566.12	0.0200	0.8172
486.21	600	0.0201	0.7695
500	680.80	0.0204	0.6748
520	812.68	0.0209	0.5591
540	962.80	0.0214	0.4647
544.56	1000	0.0216	0.4456
560	1133.4	0.0221	0.3869
580	1326.1	0.0228	0.3217
596.20	1500	0.0235	0.2765
600	1543.2	0.0236	0.2668
620	1787.0	0.0247	0.2202
635.78	2000	0.0257	0.1879
640	2060.3	0.0260	0.1799
660	2366.2	0.0278	0.1445
680	2708.4	0.0305	0.1115
695.37	3000	0.0346	0.0849
700	3094.1	0.0369	0.0758
705.34*	3206.2*	0.0541	0.0541

*Critical properties.

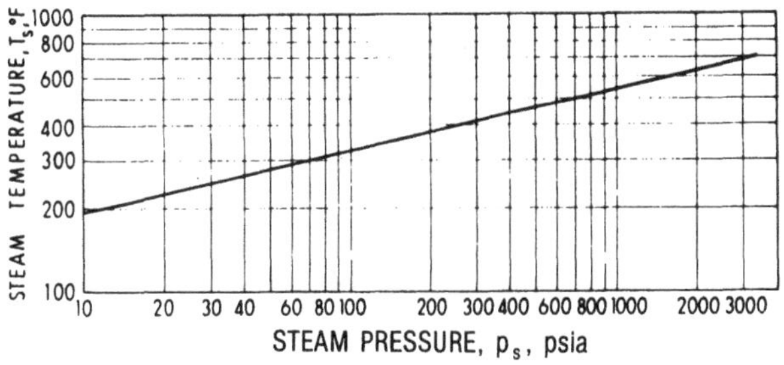

Fig. B.10 – Saturation pressure and temperature of steam.

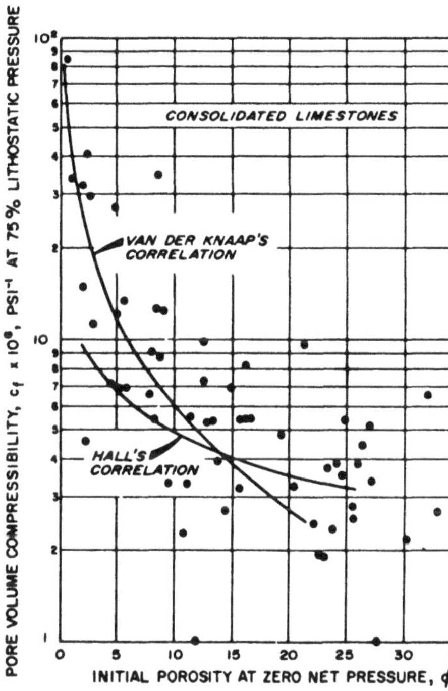

Fig. B.11 – Pore-volume compressibility at 75% lithostatic pressure vs. initial sample porosity for limestone.[22]

TABLE B.3 – DENSITY OF SELECTED MINERALS

Mineral	Formula	Density (lbm/cu ft)	Reference, Appendix B
Calcite	$CaCO_3$	169	16
Dolomite	$CaMg(CO_3)_2$	179	16
Siderite	$FeCO_3$	246	16
Aragonite	$CaCO_3$	183	16
Anhydrite	$CaSO_4$	185	16
Gypsum	$CaSO_4 \cdot 2H_2O$	145	16
Halite	NaCl	135	16
Quartz	SiO_2	165	16
Pyrite	FeS_2	313	16
Feldspars			
Albite	$NaAlSi_3O_8$	163	16
Microcline	$KAlSi_3O_8$	160	18
Clays*			
Kaolinite	$Al_2Si_2O_5(OH)_4$	162-167	18,19
Illite	$K_{0.6}Mg_{0.25}Al_{2.3}Si_{3.5}O_{10}(OH)_2$	165-168	18,19
Montmorillonite**	$(Mg,Ca,Na_2,K_2)_{0.17}Al_{2.33}Si_{3.67}O_{10}(OH)_2$	137-168	18,19
Chlorite	$Mg_5Al_2Si_3O_{10}(OH)_8$	162-185	18,19
Micas			
Biotite	$K(Mg,Fe)_3(AlSi_3O_{10})(OH)_2$	168-193	18
Muscovite	$KAl_2(AlSi_3O_{10})(OH)_2$	177	16

*Formulas for clays are those reported by Helgeson.[17]

**(Mg,Ca, . . .) Any combination of the elements in the parentheses may be present.

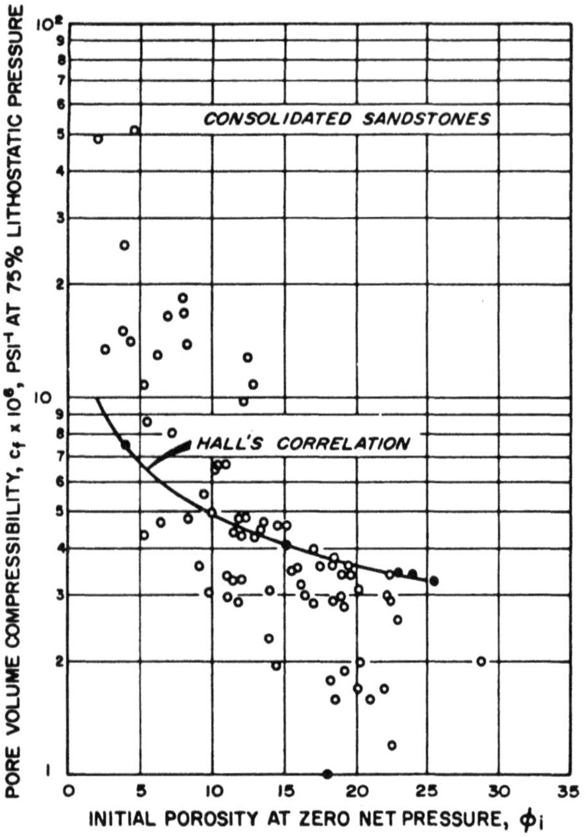

Fig. B.12 – Pore-volume compressibility at 75% lithostatic pressure vs. initial sample porosity for consolidated sandstones.[22]

Figs. B.12 through B.14 compare Newman's and other data with Hall's sandstone correlation. In preparing the three figures, Newman[22] used the following criteria.

1. Consolidated samples consisted of hard rocks (thin edges could not be broken off by hand).
2. Friable samples could be cut into cylinders but the edges could be broken off by hand.
3. Unconsolidated samples could fall apart under their own weight unless they had undergone special treatment, such as freezing.

As can be seen in the three figures, no correlation would provide a good description of the large suite of samples studied. It is apparent from Fig. B.13 that there is no correlation at all for the friable samples. Fig. B.14 indicates that if there is any correlation for unconsolidated samples, the trend may be opposite the trend for consolidated samples (Fig. B.12).

Unfortunately, Figs. B.11 through B.14 lead to only one conclusion: formation compressibility should be measured for the reservoir being studied. At best, correlations can be expected to give only order-of-magnitude estimates.

Oil Phase

The isothermal compressibility of an undersaturated oil (oil above the bubble point) is defined as

$$c_o = -\frac{1}{V_o}\left(\frac{\partial V_o}{\partial p}\right)_T = \frac{1}{\rho_o}\left(\frac{\partial \rho_o}{\partial p}\right)_T$$

$$= -\frac{1}{B_o}\left(\frac{\partial B_o}{\partial p}\right)_T . \qquad \text{(B.12)}$$

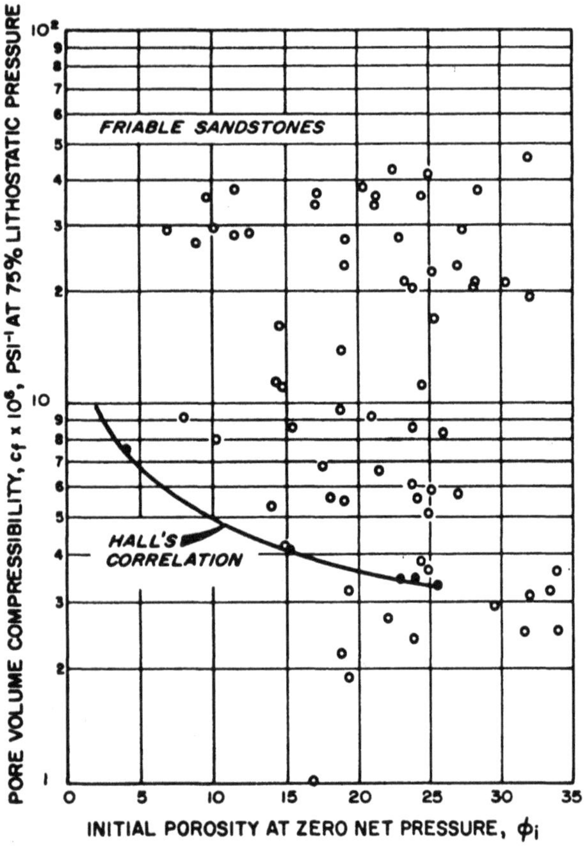

Fig. B.13 – Pore-volume compressibility at 75% lithostatic pressure vs. initial sample porosity for friable sandstones.[22]

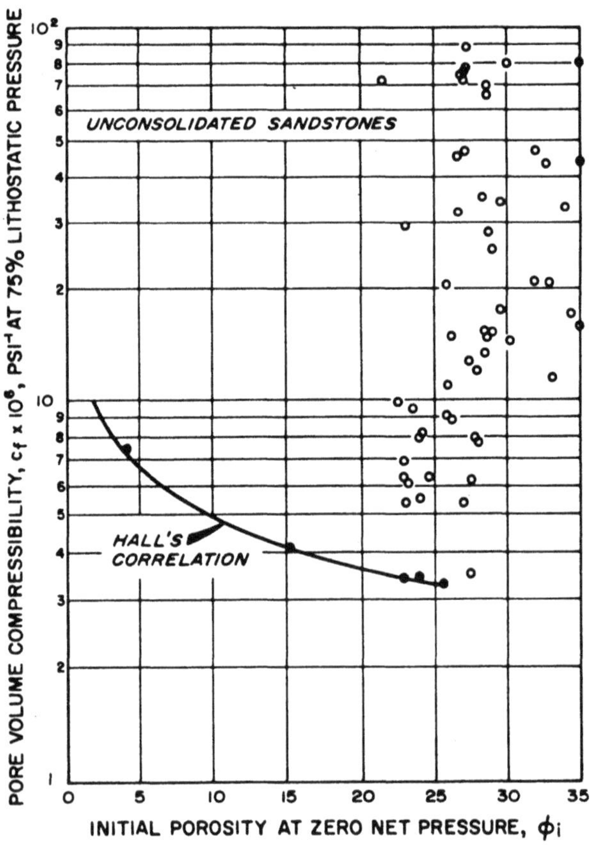

Fig. B.14 – Pore-volume compressibility at 75% lithostatic pressure vs. initial sample porosity for unconsolidated sandstones.[22]

Since the volume of an undersaturated liquid decreases as the pressure increases, c_o is positive.

Generally, oil compressibility should be computed from laboratory PVT data for the oil existing in the subject reservoir. The final equality in Eq. B.12 is useful for calculating c_o from such data. For some reservoir crudes, c_o is essentially constant above the bubble point, while in others it varies with pressure.

If laboratory data are not available, Trube's[7] correlation for the compressibility of an undersaturated oil (Fig. B.15) may be used. It is necessary to estimate T_{pr} and p_{pr} from Fig. B.1 or Fig. B.2. The pseudoreduced compressibility c_{pr} is read from Fig. B.15 and the oil compressibility is estimated from

$$c_o = \frac{c_{pr}}{p_{pc}}. \quad \text{(B.13)}$$

Below the bubble point, dissolved gas affects the compressibilities. Thus, an *apparent* oil compressibility for the region below the bubble point where oil volume increases with pressure as a result of gas going into solution is defined by

$$c_{oa} = -\frac{1}{B_o}\frac{\partial B_o}{\partial p} + \frac{B_g}{B_o}\frac{\partial R_s}{\partial p}. \quad \text{(B.14)}$$

Note that Eq. B.14 reduces to Eq. B.12 above the bubble point when R_s is constant with pressure. Special care must be exercised in evaluating Eq. B.14, since B_g is calculated in a great variety of units. If available, laboratory data should be used to estimate c_{oa}; otherwise, correlations may be used, but with caution. When using correlations, the $\partial R_s/\partial p$ factor in Eq. B.14 may be estimated from Fig. B.16 or from

$$\frac{\partial R_s}{\partial p} \simeq \frac{R_s}{(0.83p+21.75)}. \quad \text{(B.15)}$$

Eq. B.15 and Fig. B.16 are from Ramey[23] and are based on Standing's data.[10] The gas formation volume factor may be estimated from Eq. B.8 where the z factor is estimated from Fig. B.7. The factor $\partial B_o/\partial p$ in Eq. B.14 may be estimated from

$$\frac{\partial B_o}{\partial p} \simeq \frac{\partial R_s}{\partial p} \cdot \frac{\partial B_o}{\partial R_s}, \quad \text{(B.16)}$$

where the first factor on the right side is from Eq. B.15 or Fig. B.16 and the second factor on the right side is from Fig. B.17. To use Fig. B.17, oil and gas gravities must be known. The oil formation volume factor B_o may be estimated from Standing's correlation (Fig. B.6) or from Eq. B.7. It should be obvious that oil compressibilities vary significantly with the reservoir conditions.

Water

The water compressibility is defined in a similar manner to the oil compressibility (Eq. B.12). The compressibility of water or brine *without* any solution gas is estimated from Figs. B.18 through B.21. Linear interpolation may be used for intermediate pressures and salinities.

To estimate the compressibility of undersaturated water or brine (i.e., with solution gas), Long and Chierici[24] recommend using

$$c_w = (c_w)_{0,n}[1+0.0088\times 10^{-Kn}(R_{sw})], \quad \text{(B.17)}$$

where

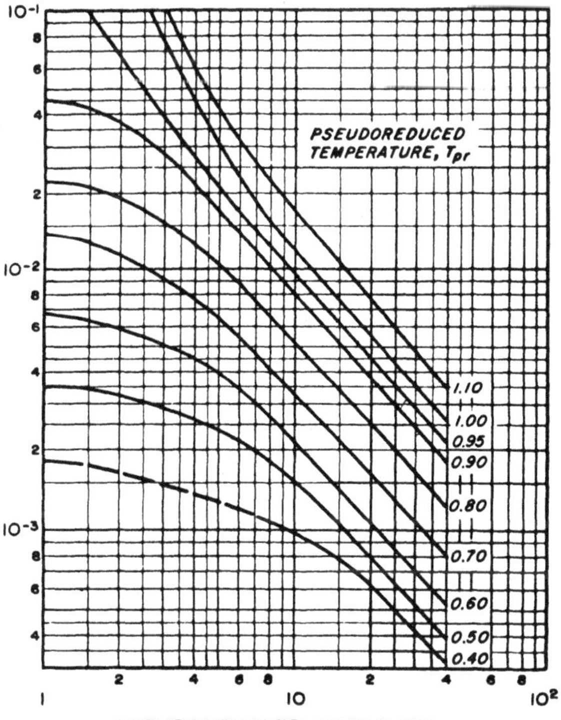

Fig. B.15 – Correlation of pseudoreduced compressibility for an undersaturated oil.[7]

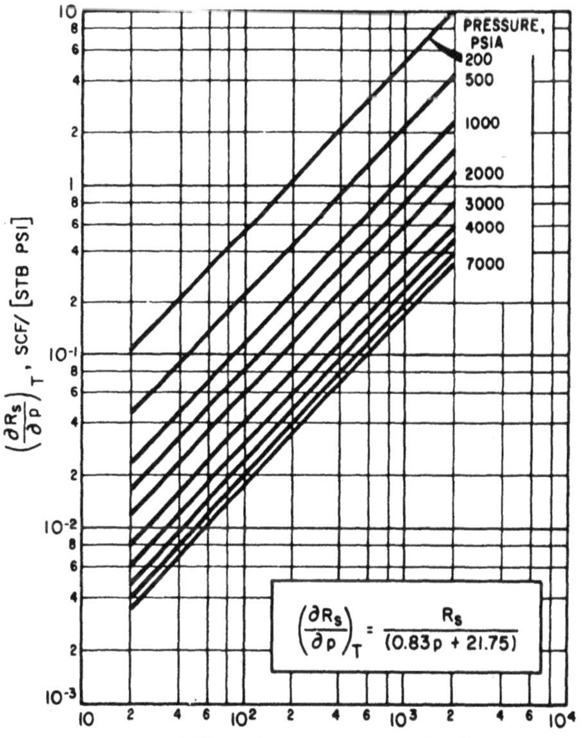

Fig. B.16 – Change of gas in solution in oil with pressure vs. gas in solution.[23]

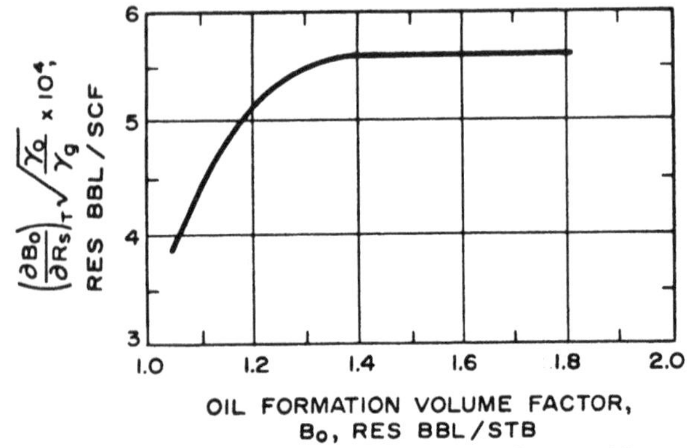

Fig. B.17 – Change of oil formation volume factor with gas in solution vs. oil formation volume factor.[23]

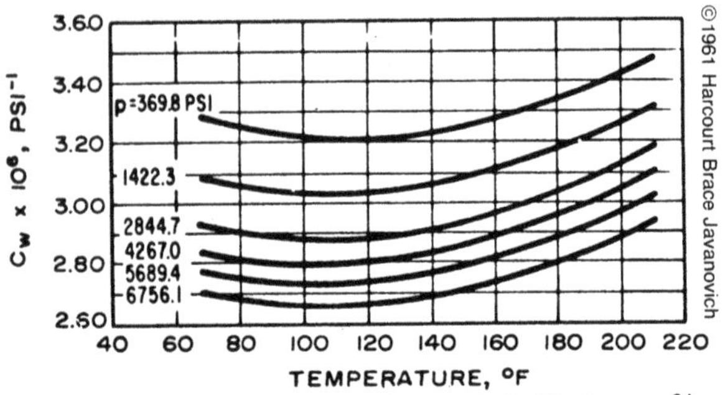

Fig. B.18 – Average compressibility of distilled water.[24]

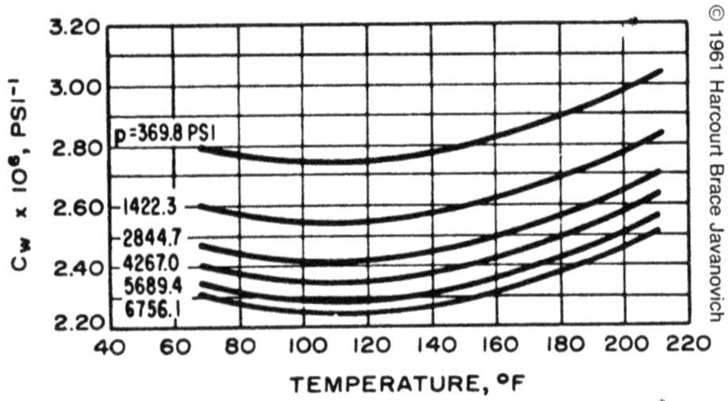

Fig. B.19 – Average compressibility of 100,000 ppm of NaCl in distilled water.[24]

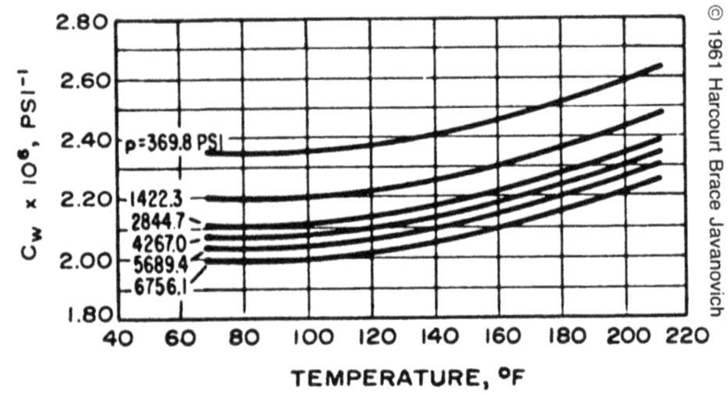

Fig. B.20 – Average compressibility of 200,000 ppm of NaCl in distilled water.[24]

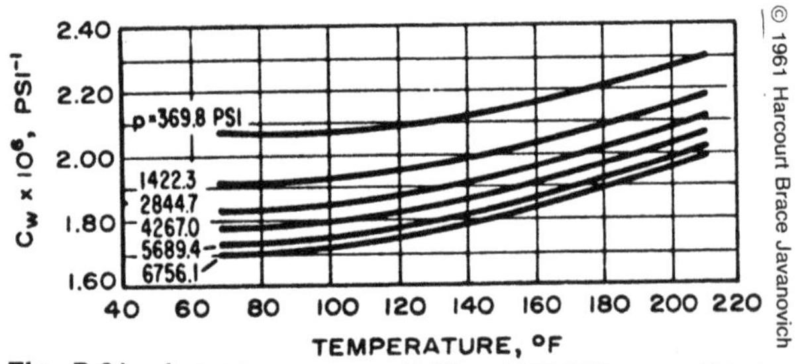

Fig. B.21 – Average compressibility of 300,000 ppm of NaCl in distilled water.[24]

c_w = compressibility of an undersaturated brine containing solution gas and n gram-equivalents of dissolved solids, psi^{-1},

$(c_w)_{0,n}$ = compressibility of a *gas-free* brine containing n gram-equivalents of dissolved solids, psi^{-1}, from Figs. B.18 through B.21,

n = dissolved solids (ppm) ÷ 58,443, concentration of dissolved solids, gram-equivalents/L,

K = Secenov's coefficient, obtained at reservoir temperature from Fig. B.22, and

R_{sw} = gas solubility in distilled water at the required pressure and temperature, from Fig. B.23, scf/bbl.

An alternative approach to estimating the compressibility of undersaturated water is to use Fig. B.24 to estimate the water compressibility at reservoir temperature, pressure, and solution gas/water ratio. Fig. B.25 is used to estimate the solution gas/water ratio as a function of temperature, pressure, and salinity.

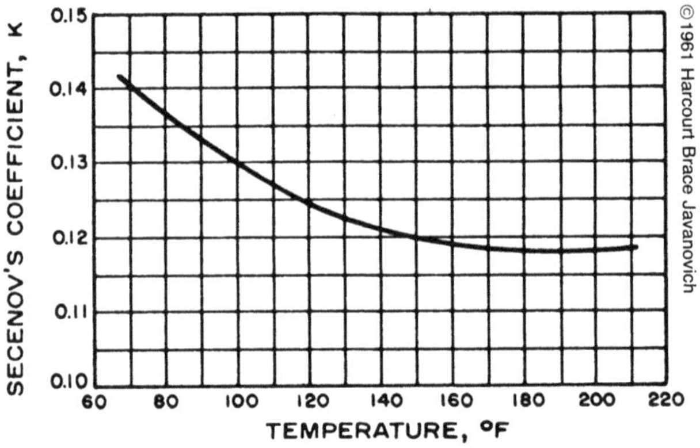

Fig. B.22 – Secenov's coefficient for methane used in Eq. B.17.[24]

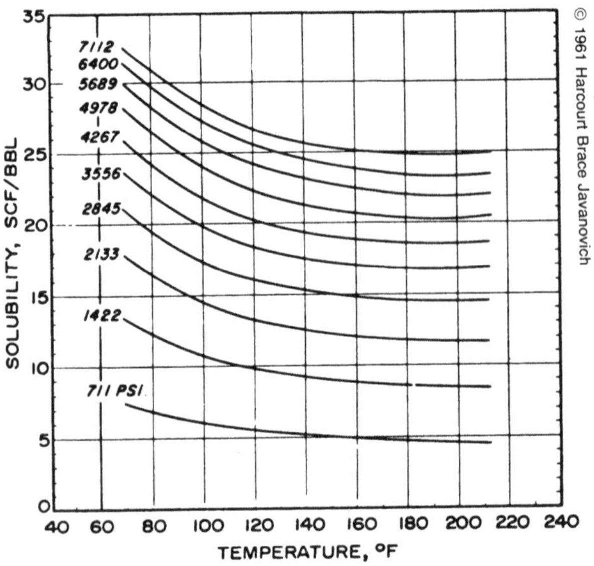

Fig. B.23 – Solubility of methane in distilled water.[24]

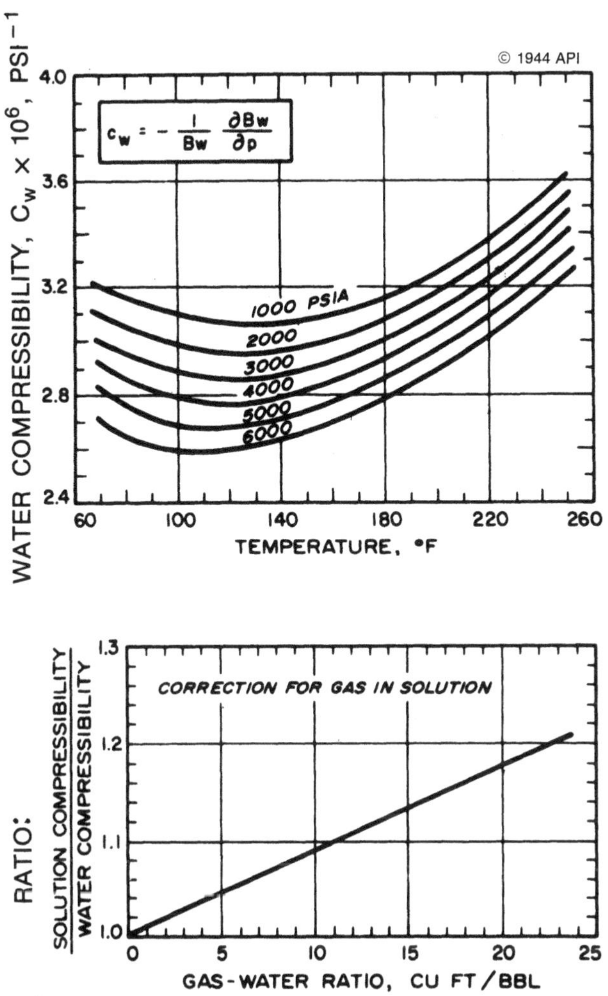

Fig. B.24 – Effect of dissolved gas on water compressibility.[14]

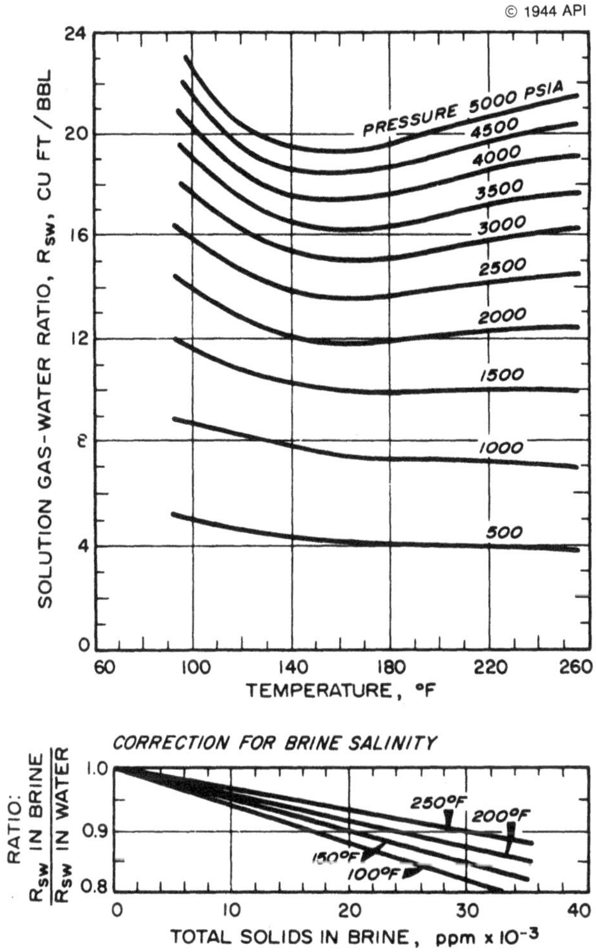

Fig. B.25 – Solubility of natural gas in water.[14]

The apparent compressibility of water below the bubble point is given by

$$c_{wa} = -\frac{1}{B_w}\frac{\partial B_w}{\partial p} + \frac{B_g}{B_w}\frac{\partial R_{sw}}{\partial p}. \quad \text{(B.18)}$$

Again, it is best to compute c_{wa} from PVT analyses if they are available. However, since they seldom are available, correlations often must be used. The $\partial R_{sw}/\partial p$ term may be approximated from Fig. B.26, and B_w may be approximated from Fig. B.8. B_g is estimated with Eq. B.8. The first term on the right side of Eq. B.18 must be estimated from Fig. B.24 or Eq. B.17.

Gas

Isothermal gas compressibility is defined in a similar manner to oil compressibility (Eq. B.12). The gas equivalent of Eq. B.12 may be written using the real gas deviation factor z:

$$c_g = \frac{1}{p} - \frac{1}{z}\left(\frac{\partial z}{\partial p}\right)_T. \quad \text{(B.19)}$$

If pseudoreduced pressures and temperatures are introduced into Eq. B.19, the isothermal gas compressibility may be written as

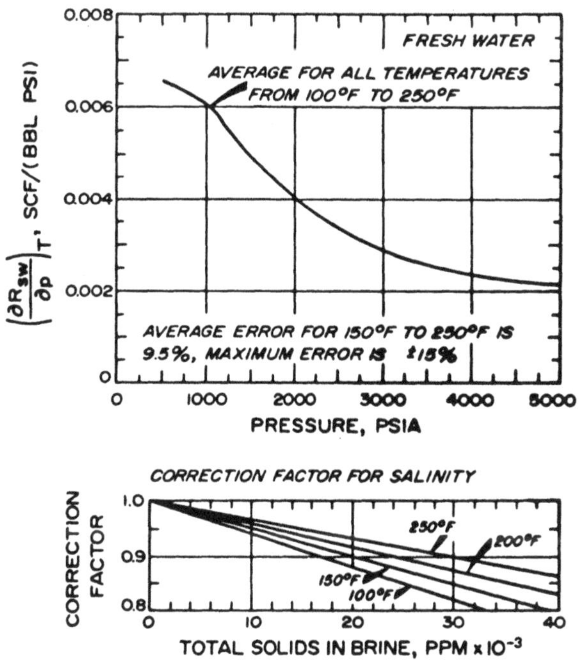

Fig. B.26 – Change of natural gas in solution in formation water with pressure vs. pressure. Multiply $(\partial R_{sw}/\partial p)_T$ by the correction factor to get result for brine.[23]

$$c_g = \frac{1}{p_{pc}}\left[\frac{1}{p_{pr}} - \frac{1}{z}\left(\frac{\partial z}{\partial p_{pr}}\right)\right]_{T_{pr}} \quad \text{(B.20)}$$

The z-factor chart (Fig. B.7) may be used directly to estimate the derivative term for Eq. B.20.

Gas compressibility also may be estimated from the pseudoreduced compressibility correlation shown in Figs. B.27 and B.28. The pseudoreduced compressibility is read from one of those figures, and the gas compressibility is computed from

$$c_g = \frac{c_{pr}}{p_{pc}}. \quad \text{(B.21)}$$

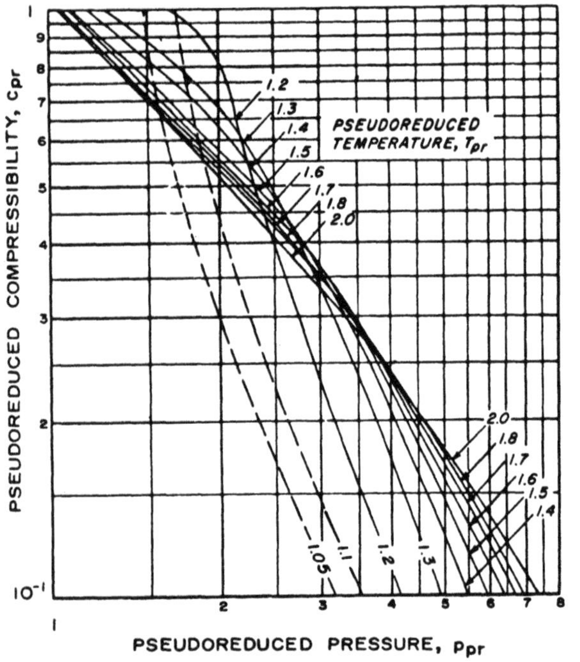

Fig. B.27 – Correlation of pseudoreduced compressibility for natural gases.[25]

B.3 Thermal Expansion

Crude Oils

The isobaric thermal expansion of an undersaturated oil (oil above the bubble point) is defined as

$$\beta_o = -\frac{1}{\rho_o}\left(\frac{\partial \rho_o}{\partial T}\right)_p = \frac{1}{V_o}\left(\frac{\partial V_o}{\partial T}\right)_p$$

$$= \frac{1}{B_o}\left(\frac{\partial B_o}{\partial T}\right)_p. \quad \text{(B.22)}$$

Since the density of an undersaturated liquid decreases as the temperature increases, β_o is positive. Volumetric thermal expansion coefficients β_o under constant pressure are used to estimate the effect of temperature changes on the density of crudes by means of the approximate relation

$$\rho_o = \rho_{or}[1 - \beta_o(T - T_r)], \quad \text{(B.23)}$$

where the subscript r is used to identify a reference condition. For crudes containing dissolved gases under pressure, Fig. B.29 provides values of

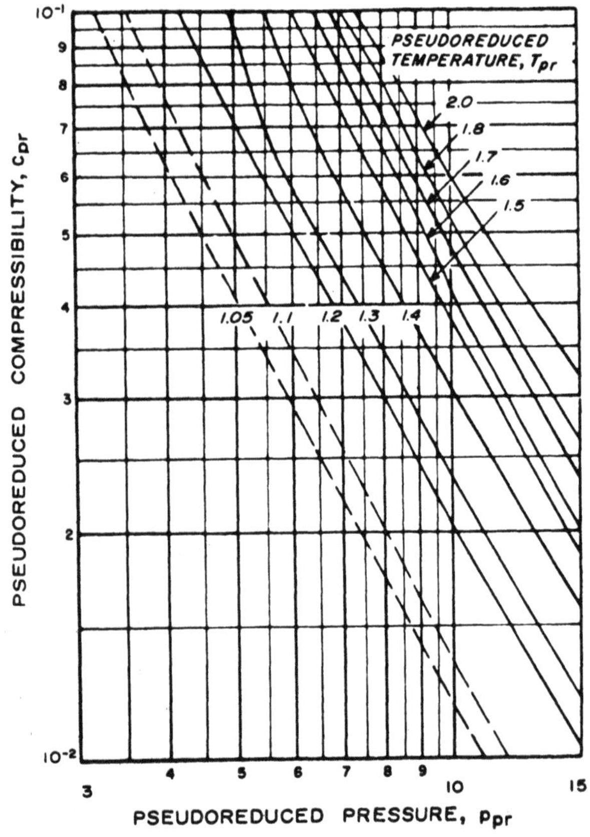

Fig. B.28 – Correlation of pseudoreduced compressibility for natural gases.[25]

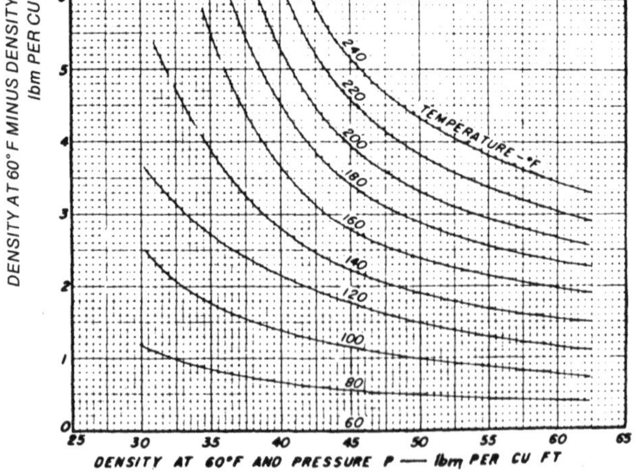

Fig. B.29 – Density correction for thermal expansion of liquids.[12]

$$\Delta\rho_{op} = (\rho_{or} - \rho_o) = \beta_o \rho_{or} [T - T_r] \quad \ldots\ldots (B.24)$$

vs. ρ_{or}, the reference density. The reference temperature is 60°F, and the pressure is arbitrary but constant. The correlation

$$\Delta\rho_{op} = \left(0.0133 + 152.4\rho_{or}^{\;-2.45}\right)(T-60) - [8.1(10^{-6}) - 0.0622(10^{-0.0764\rho_{or}})] \cdot (T-60)^2 \quad \ldots\ldots (B.25)$$

has a reported[12] standard deviation for 50 samples of no more than 3.4%. Thermal expansion coefficients calculated from

$$\beta_o = \frac{\Delta\rho_{op}}{\rho_{or}(T-T_r)} \quad \ldots\ldots (B.26)$$

(using the correlation given by Eq. B.25) are given in Fig. B.30. Fig. B.30 also gives the thermal expansion coefficients listed in Table 7 of Ref. 8, which is considered to be applicable at atmospheric pressures.

Fig. B.31 gives the approximate specific gravity of petroleum fractions of several API gravities as a function of temperature up to 600°F. The negative secant slopes of these curves can also be converted to thermal expansion coefficients using Eq. B.26.

Farouq Ali[4] proposed using a β value of 5×10^{-4} $°F^{-1}$ where there are no other data available for the volumetric thermal expansion of oil. Edminster[6] presents correlations for estimating the thermal expansion coefficients of petroleum liquids in terms of their viscosities, API gravity, density, UOP K factor, Jessup's modulus, and ASTM average boiling point.

Water

The isobaric thermal coefficient of volume expansion of water is defined in a similar manner to that of oil (given by Eq. B.22). The thermal expansion coefficient for pure water may be deduced from values of specific volumes for the compressed liquid reported in steam tables.[15,16] The specific volume of a substance is its reciprocal density. Given the reduction in specific volume Δv_w at any temperature and pressure from that at saturation conditions, we can find the specific volume at any temperature for a constant pressure from the relation

$$v_w(T,p) = v_w(T_s,p_s) - \Delta v_w. \quad \ldots\ldots (B.27)$$

Values of the specific volume at saturation conditions, $v_w(T_s,p_s)$, are given in Table B.2. And values of the correction Δv_w vs. temperature are given in Fig. B.32 for several values of pressure.

The effect of dissolved solids and gases on the thermal coefficient of volume expansion of water is probably small; however, we could find no information on it.

Gases

The isobaric thermal coefficient of volume expansion for gas is defined analogously to that for oil (given by Eq. B.22). For gases, the expression for β_g may be written using the real gas deviation factor z as

$$\beta_g = \frac{1}{T} + \frac{1}{z}\left(\frac{\partial z}{\partial T}\right)_p. \quad \ldots\ldots (B.28)$$

In terms of pseudoreduced temperatures and pressures,

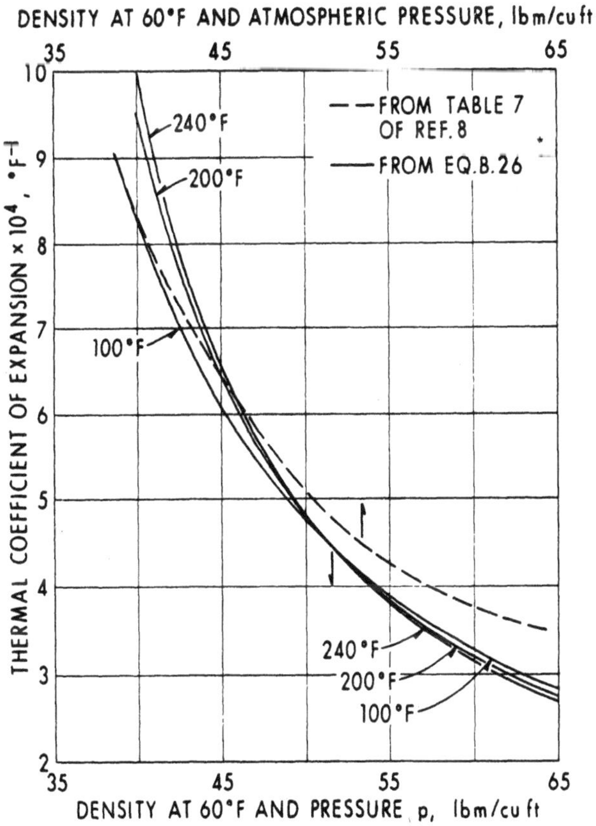

Fig. B.30 – Volume thermal expansion coefficients of crudes and crude fractions.

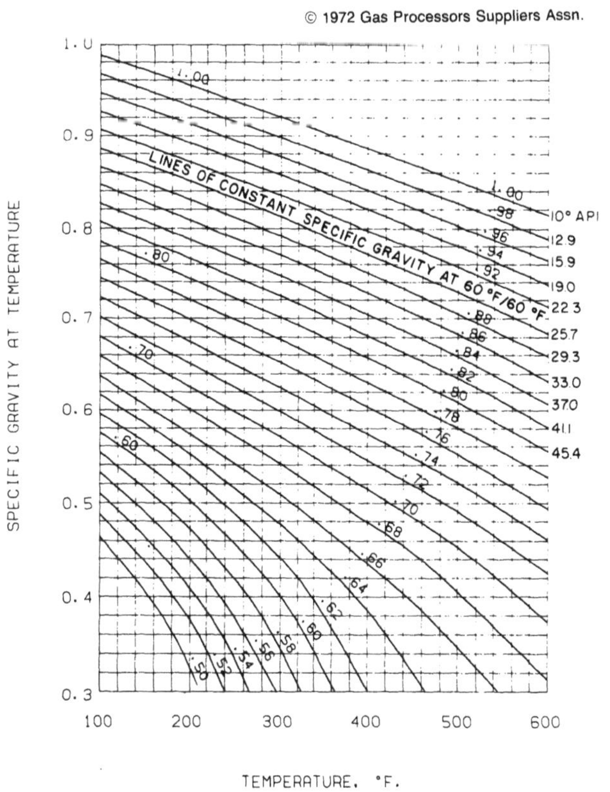

Fig. B.31 – Approximate specific gravity of petroleum fractions.[5]

TABLE B.4 – VOLUMETRIC THERMAL EXPANSION OF SOLIDS[18]

Material	Expansion (%) Relative to Volume at 20°C (68°F) at Temperature of				
	100°C (212°F)	200°C (392°F)	400°C (752°F)	600°C (1112°F)	800°C (1472°F)
Elements and Metals					
Aluminum	0.562	1.317	3.049	5.115	
Graphite	0.193	0.450	1.018	1.596	2.221
Copper	0.425	0.976	2.110	3.338	4.661
Iron steel (1.2% C)	0.27	0.60	1.47	2.43	3.45
Minerals					
Halite (NaCl)	0.963	2.288	5.256	8.932	
Pyrite (FeS_2)	0.219	0.529	1.291		
Aragonite ($CaCO_3$)	0.36	1.00	2.48		
Calcite ($CaCO_3$)	0.105	0.285	0.765	1.395	
Barite ($BaSO_4$)	0.434	1.023	2.381		
Quartz (SiO_2)	0.36	0.78	1.87	4.54	4.43
Feldspars	~0.1	~0.3	~0.8	~1.1	~1.5

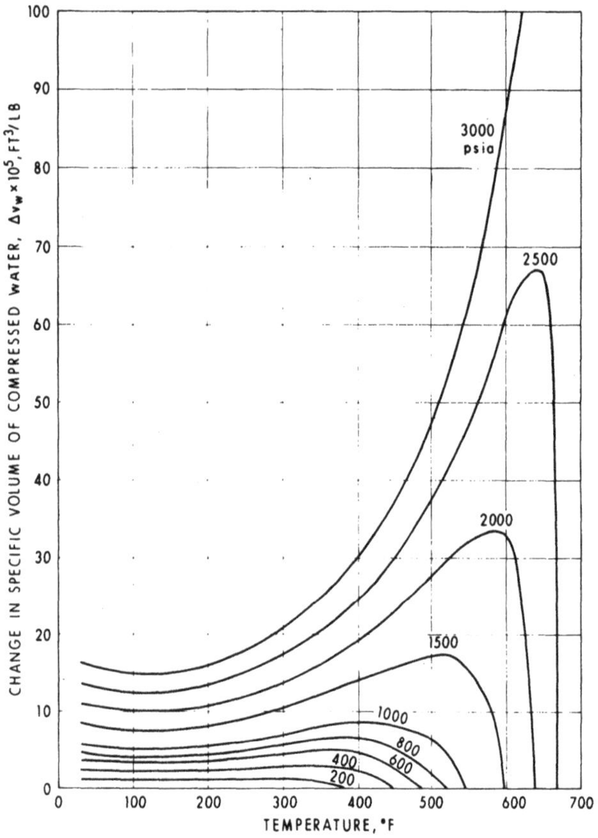

Fig. B.32 – Specific volume of water as a function of temperature.

TABLE B.5 – VOLUME EXPANSION COEFFICIENTS OF ROCKS[18]

Rock Types	Average Volume Expansion Coefficient $\beta \times 10^5$ (°F^{-1})
Granites and rhyolites	4±2
Andesites and diorites	4±1
Basalts, gabbros, and diabases	2.9±0.5
Sandstones	5.4±1
Quartzites	5.9
Limestones	4±2
Marbles	4±1
Slates	5±0.5

Temperature range 68° to 392°F (20° to 200°C).

$$\beta_g = \frac{1}{T_{pc}}\left[\frac{1}{T_{pr}} + \frac{1}{z}\left(\frac{\partial z}{\partial T_{pr}}\right)_{p_{pr}}\right], \quad \text{........ (B.29)}$$

where the z factor and its derivative may be estimated from Fig. B.7.

For superheated steam, the specific volume (reciprocal density) is plotted vs. temperature in Fig. B.9 for several values of pressure.

Solids

The effect of temperature on the volume expansion of solids is generally much smaller than on that of liquids. Table B.4 gives values of the percent volume expansion from a reference temperature of 68°F for selected solids, including some elements, metals, and minerals. Table B.5 gives average values of the thermal coefficient of volume expansion at low temperatures for several types of rock. No data were given on the porosity ranges of the rocks.

B.4 Viscosity

Crude Oils

Farouq Ali[4] presents a good discussion on the estimation, correlation, and measurement of oil viscosity at high temperatures.

The standard way to plot viscosity vs. temperature is on ASTM Viscosity Temperature Charts for Liquid Petroleum Products, which for Newtonian fluids generally yield a straight line for the kinematic viscosity (μ/ρ). These charts are available in several

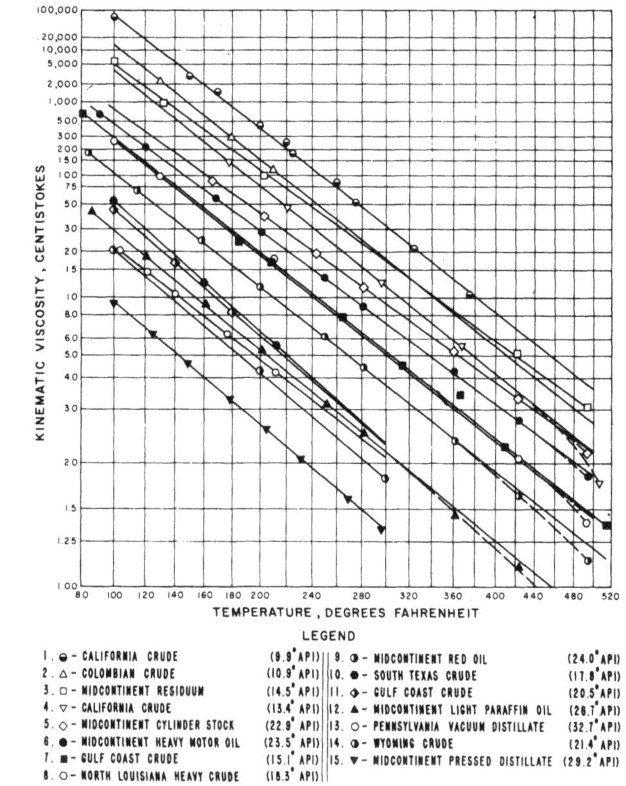

Fig. B.33 – Viscosity of gas-free oils vs. temperature.[26]

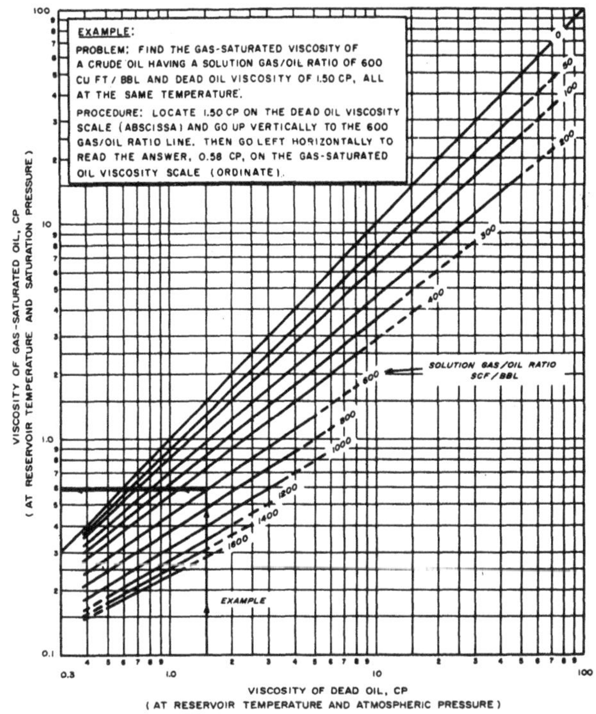

Fig. B.34 – Viscosity of gas-saturated crude oil at reservoir temperature and pressure (dead-oil viscosity from laboratory data or Fig. B.37).[27]

sizes and ranges of viscosity and temperature. An example of such a plot for gas-free oils at atmospheric pressure is given in Fig. B.33. With values of the kinematic viscosity at two different temperatures it is possible to extrapolate or interpolate to other temperature values.

The straight lines in these plots are based on a mathematical relation between viscosity and temperature that allows its calculation at any temperature T (in °R) from the expression

$$\mu/\rho=\nu=(\nu_1+0.8)^{10^{-x}}-0.8, \quad \text{(B.30)}$$

where

$$x=\frac{\log(T/T_1)}{\log(T_2/T_1)}\{\log[\log(\nu_1+0.8)] - \log[\log(\nu_2+0.8)]\}, \quad \text{(B.31)}$$

and ν_1 and ν_2 are the known values of the kinematic viscosity at the temperatures T_1 and T_2, respectively. The exponent x is always positive and leads to the result that the kinematic viscosity approaches a value of 0.2 as the temperature increases to large values *for all crudes*. This result is, of course, a consequence of the mathematical relationship used to prepare the ASTM charts. It is important to note that for most crudes, the kinematic viscosity of 600°F is at most a few centistokes. Because the density of the crude is not a strong function of temperature, plots such as those shown in Fig. B.33 often are made of the dynamic viscosity μ (usually called viscosity and measured in centipoise) vs. temperature. Such a plot

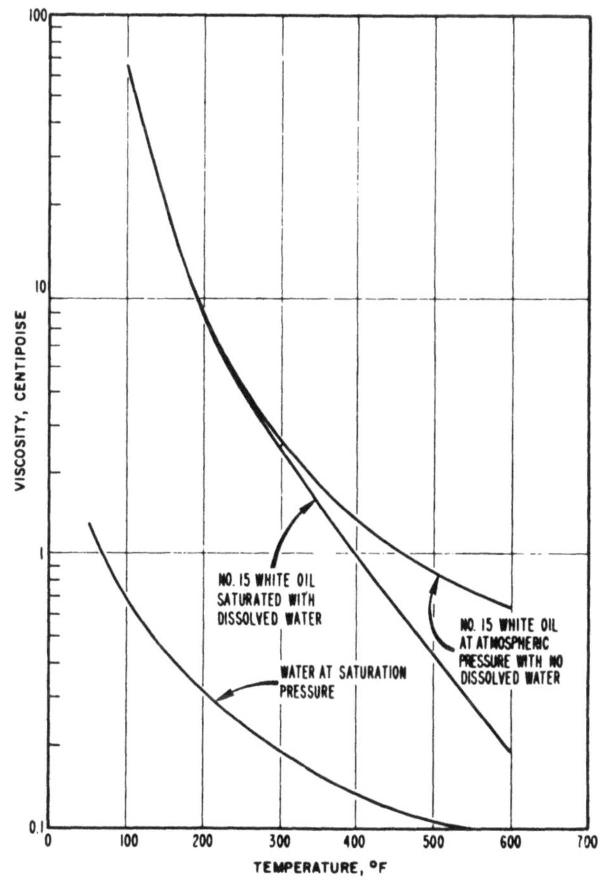

Fig. B.35 – Effect of dissolved water on viscosity of a white oil.[28]

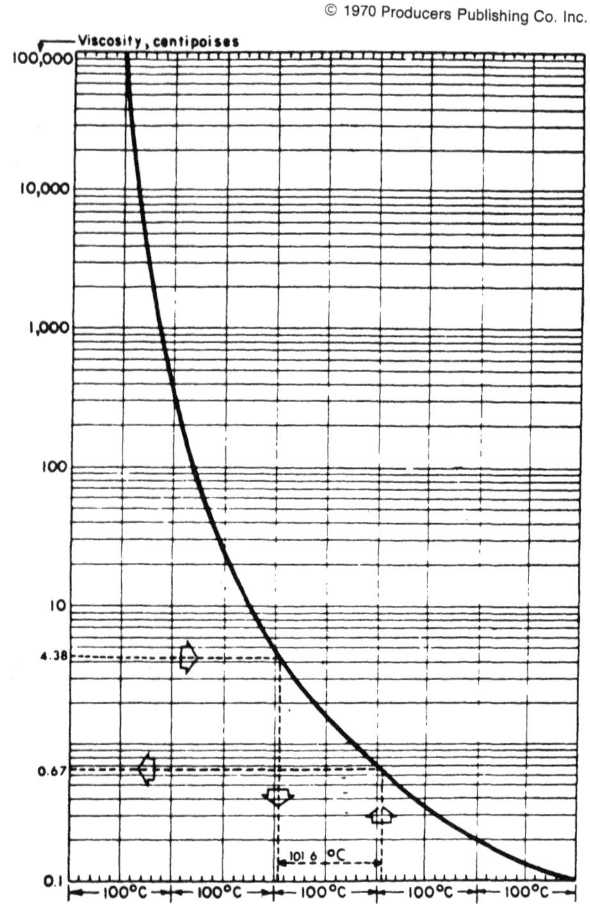

Fig. B.36 – Generalized viscosity/temperature correlation for use when only one value of the viscosity is known.[4]

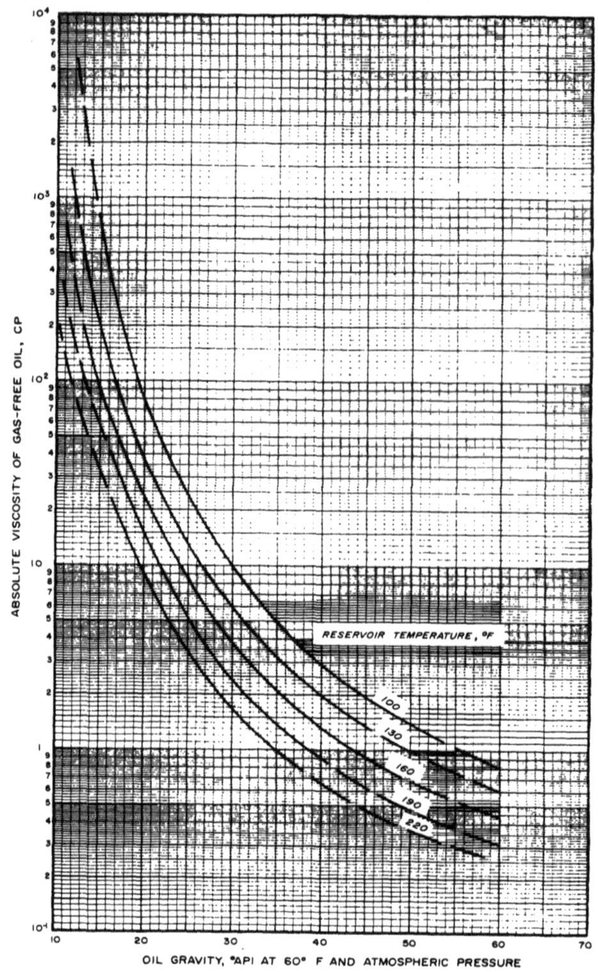

Fig. B.37 – Dead-oil viscosity at reservoir temperature and atmospheric pressure.[30]

is often useful (errors are generally less than 25% over practical temperature ranges) in providing adequate estimates of crude viscosities at other than the measured temperatures. The same can be said for the use of Eqs. B.30 and B.31.

At high temperatures, crude oils tend to lose light ends if the sample is under inadequate pressure. When this happens, the viscosity values will be higher than would have been expected by extrapolation from measurements at lower temperatures. Of course, other intrinsic properties of the crude could cause those (or any other) deviations from linearity. For example, high-temperature viscosity values that are lower than extrapolated values could be due to low-temperature viscosities that are unusually high because of the presence of highly structured liquids such as asphaltenes and waxes or because of any non-Newtonian behavior.

The loss of light components is known to increase the viscosity of crudes, and can be seen in Fig. B.34. Fig. B.35 shows that dissolved water seems to reduce the viscosity of the subject oil by a factor of about three at high temperatures and has essentially no effect at low temperatures.

Fig. B.36 is of particular interest because it allows the estimation of crude viscosities at any temperature when only one set of viscosity-temperature data is known. Farouq Ali[4] stated, "An example of the use of the graph is given in [Fig. B.36]. Suppose that the viscosity of the Bradford crude is required at a temperature of 255°F, given the viscosity of 4.38 centipoises at a temperature of 72°F. Locate the point representing 4.38 cp on the viscosity scale, and proceed horizontally to the curve. Proceed vertically downward to the base line, and measure off a distance representing the difference of 255° and 72°F in degrees centigrade. This is given by

$$(255-72)\times\frac{5}{9}=101.6°\text{C}.$$

(Note that each major division on the base line represents 100°C.) Now proceed vertically upward to the curve and read the corresponding viscosity. This is seen to be 0.67 cp. The experimental value at 255°F was found to be 0.836 cp, thus giving an error of about −20 percent."

If no viscosity data are available, but the API gravity of the crude is known, the crude viscosity as a function of temperature may be estimated from the correlation given in Fig. B.37. Viscosity values obtained from this figure may be plotted on ASTM graph paper and the best straight line extrapolated to

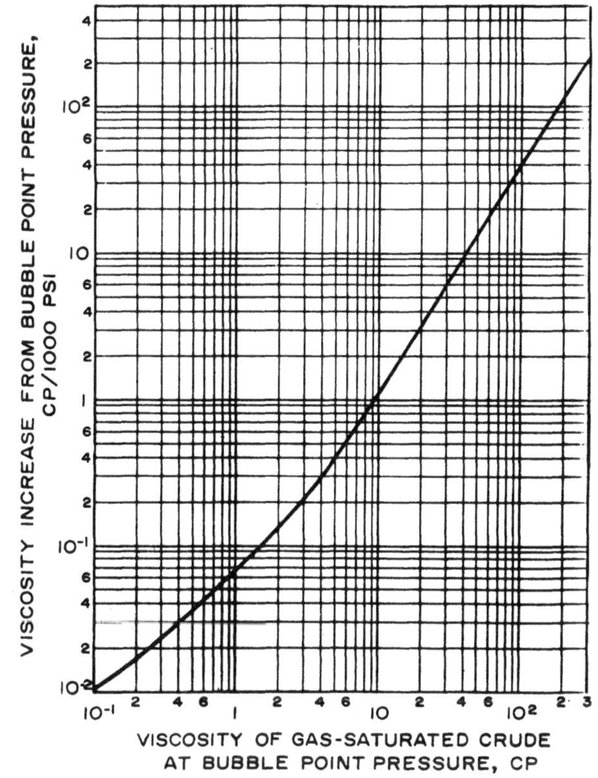

Fig. B.38 – Rate of increase of oil viscosity above bubble-point pressure.[30]

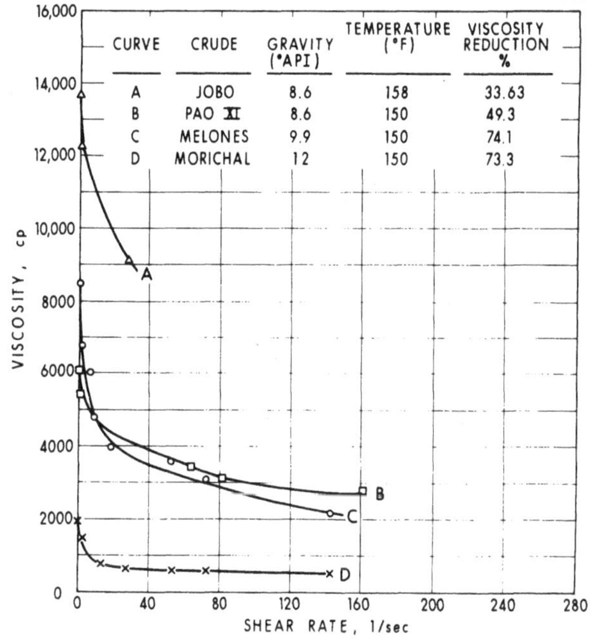

Fig. B.39 – Effect of shear rate in the viscosity of crude oils from the Orinoco oil belt.[31]

the desired temperature range. Alternatively, two data pair (say those at 100 and 200°F) may be used in Eqs. B.30 and B.31 to obtain the desired viscosity at a different temperature. Yet another method is to use Fig. B.37 and correct for solution gas as indicated in Fig. B.34.

It is not known to what temperatures these procedures would yield viscosities within a prescribed error limit. All procedures essentially assume that the crude oil does not change in character either through cracking or distillation effects at high temperatures or through the development of suspended solids or an internal structure (e.g., waxes) at low temperatures, and that the crude is Newtonian in character.

To estimate oil viscosity above the bubble-point pressure, use Fig. B.38. The figure shows the increase in viscosity above the bubble-point pressure, expressed in cp/1,000 psi. It is based on a small amount of data and should be used only as a rough guide.

Some crudes exhibit non-Newtonian behavior. This simply means that the viscosity is dependent on the fluid velocity. The rheological properties of such crudes must be measured in the laboratory. No correlations are known for estimating the effective viscosity of such crudes. Fig. B.39 presents examples of the effect of shear rate on the crude viscosity, adapted from Rojas *et al.*[31] Such crudes generally exhibit Newtonian behavior at higher temperatures.[31,32]

Gases

Fig. B.40 is one of the simplest hydrocarbon gas

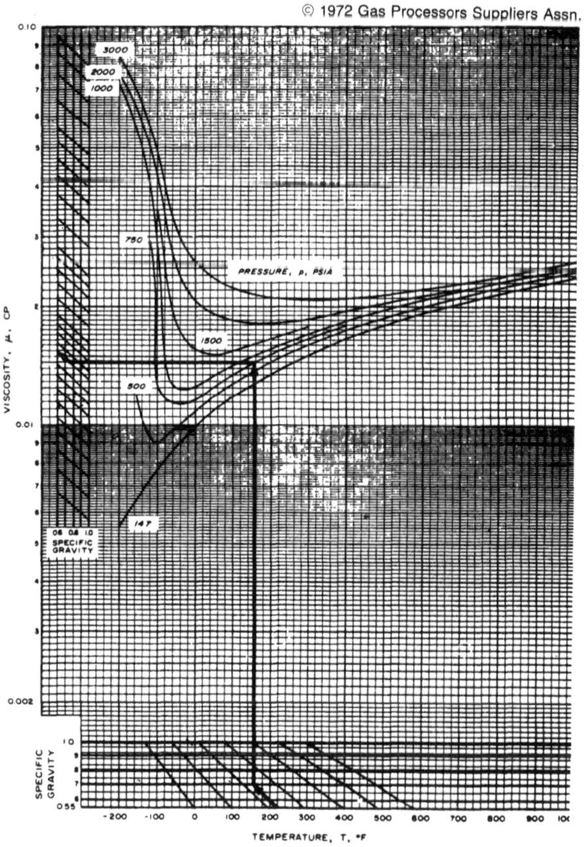

Fig. B.40 – Viscosity of hydrocarbon gases.[5]

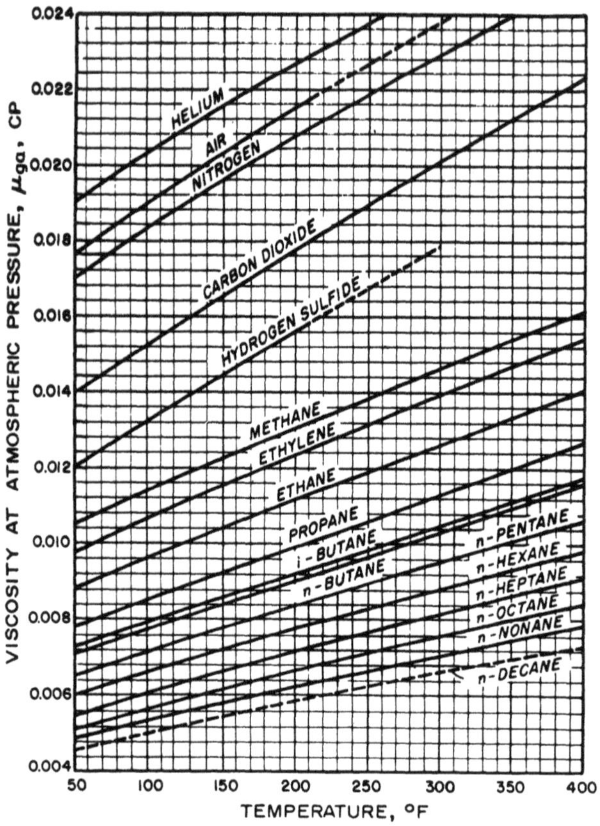

Fig. B.41 – Viscosity of pure hydrocarbon gases at 1 atm.[33]

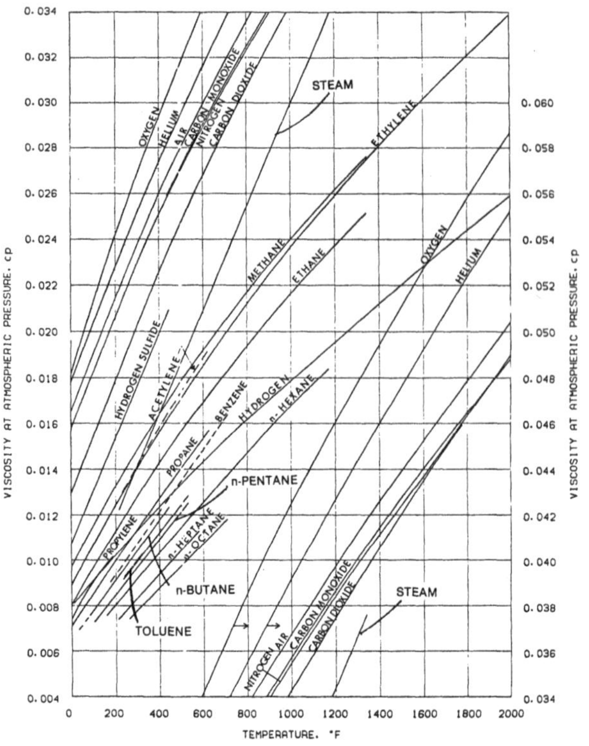

Fig. B.42 – Viscosity of gases at 1 atm.[34]

viscosity correlations available.[5] That figure gives gas viscosity as a function of gas gravity, pressure, and temperature. Its use is illustrated by the arrows. For a 0.7-gravity gas at 750 psia and 220°F, viscosity is 0.0158 cp.

Carr and colleagues[33] present a widely used method for estimating natural gas viscosity; it requires knowledge of the gas composition and of the viscosity of each component at atmospheric pressure and reservoir temperature. The viscosity of the mixture at atmospheric pressure is estimated from

$$\mu_{ga}=\frac{\sum_{j=1}^{n_C} y_j\mu_j\sqrt{M_j}}{\sum_{j=1}^{n_C} y_j\sqrt{M_j}}, \quad \text{(B.32)}$$

where

μ_{ga} = viscosity of the gas mixture at the desired temperature and *atmospheric* pressure, cp,
n_C = number of components in the gas,
y_j = mole fraction of component j,
μ_j = viscosity of component j at the desired temperature and *atmospheric* pressure, obtained from Figs. B.41 or B.42, and
M_j = molecular weight of component j (Table B.1).

Viscosity for many gaseous components is shown in Figs. B.41 and B.42 at 1 atm and various temperatures. If the gas composition is not known, Fig. B.43 may be used with the gas molecular weight to estimate the gas viscosity at reservoir temperature and atmospheric pressure. Molecular weight is related to gas gravity by

$$M=28.97\,\gamma_g. \quad \text{(B.33)}$$

The gas viscosity at reservoir pressure is estimated by determining the ratio μ_g/μ_{ga} at the appropriate temperature and pressure from Fig. B.44 or Fig. B.45. Then, that ratio is applied to μ_{ga} computed from Eq. B.32 or Fig. B.43. The pseudoreduced temperatures and pressures for use in Figs. B.44 and B.45 are estimated from Eqs. B.1 through B.4 or from Fig. B.3.

The kinematic viscosity of steam at the saturation pressure is shown in Fig. B.46 as a function of temperature. The dynamic viscosity of water vapor is shown in Fig. B.47 as a function of temperature for several pressures.

Water

Fig. B.46 shows the kinematic viscosity (ν) of water at saturated conditions vs. temperature. The dynamic viscosity (μ) vs. temperature is shown in Fig. B.48 for several pressures. These graphs correspond to pure water.

Fig. B.49 provides a means for estimating water viscosity as a function of salinity and temperature; a pressure correction is included. There are no provisions in Fig. B.49 for modifying the viscosity of

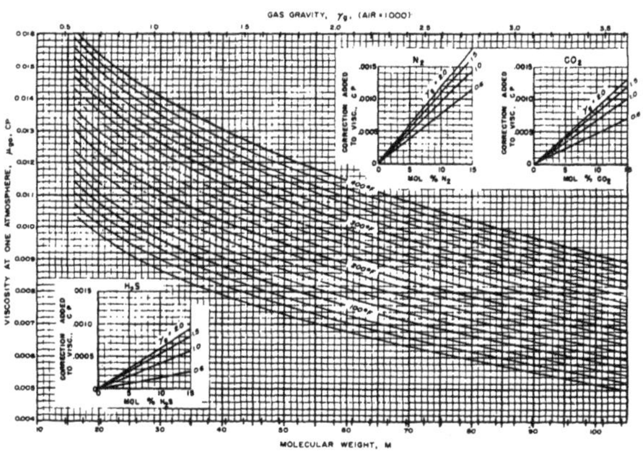

Fig. B.43 – Viscosity of natural gases at 1 atm.[33]

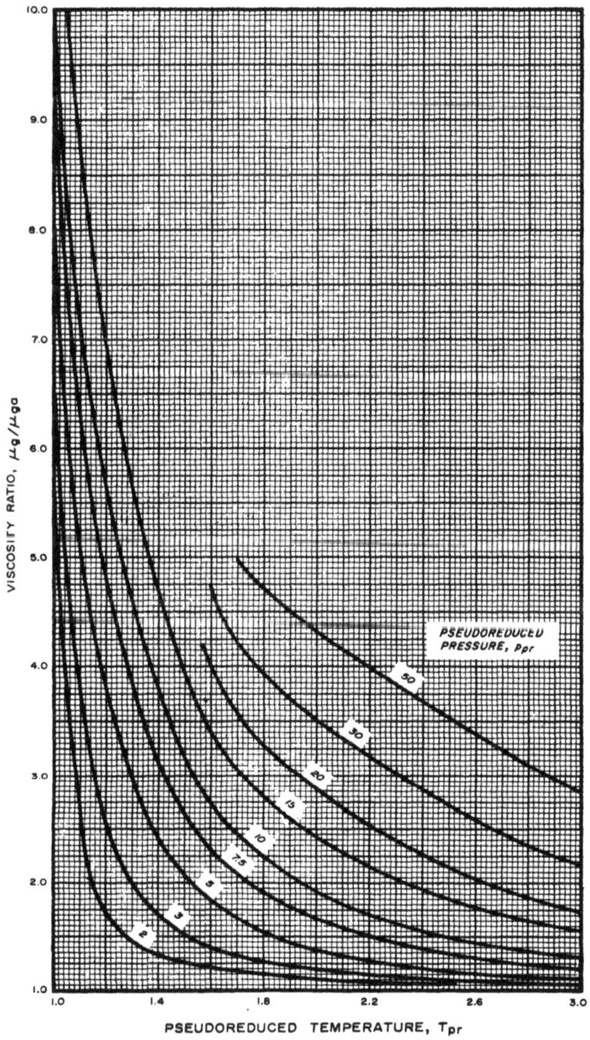

Fig. B.44 – Effect of temperature and pressure on gas viscosity (μ_{ga} is estimated from Eq. B.32 or Fig. B.43).[33]

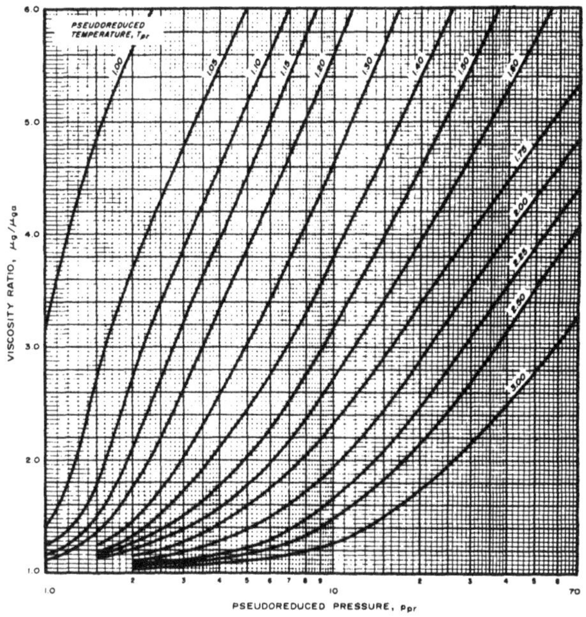

Fig. B.45 – Effect of pressure and temperature on gas viscosity (μ_{ga} is estimated from Eq. B.32 or Fig. B.43).[33]

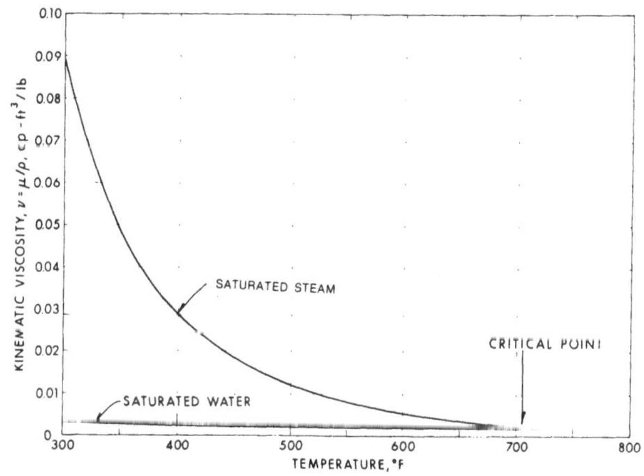

Fig. B.46 – Kinematic viscosity of steam and water as a function of temperature.

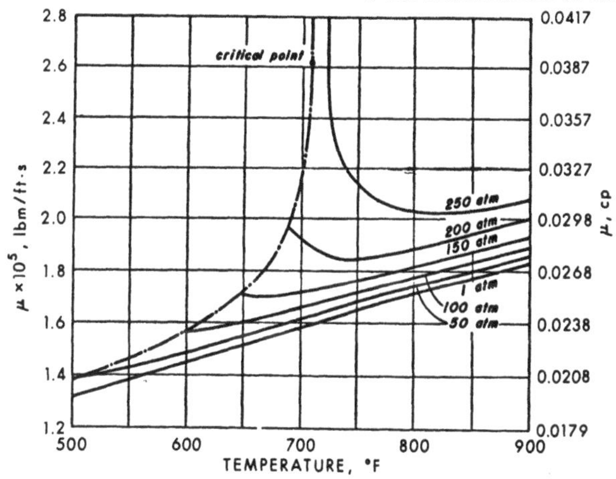

Fig. B.47 – Dynamic viscosity μ of water vapor.[35]

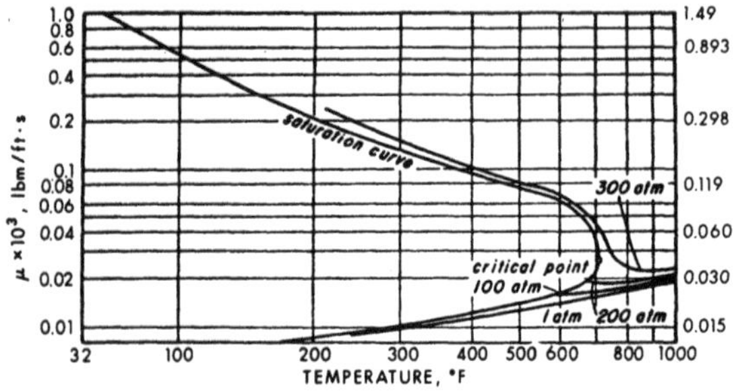

Fig. B.48 – Dynamic viscosity μ of water and water vapor.[35]

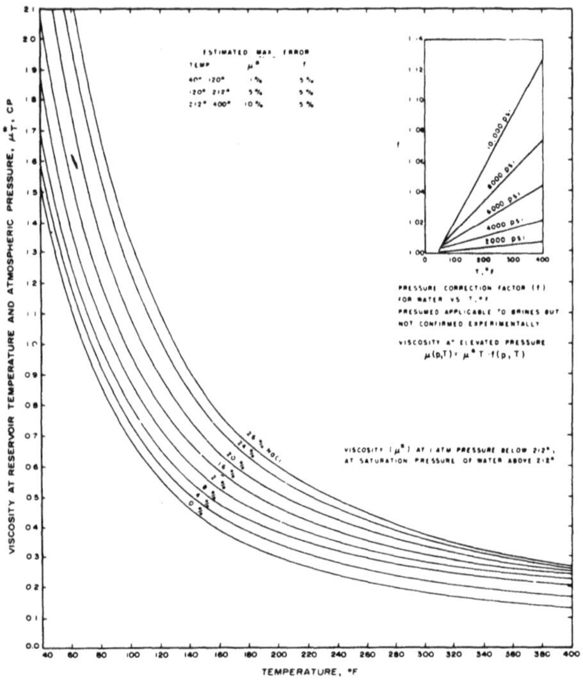

Fig. B.49 – Water viscosity at various salinities and temperatures.[1]

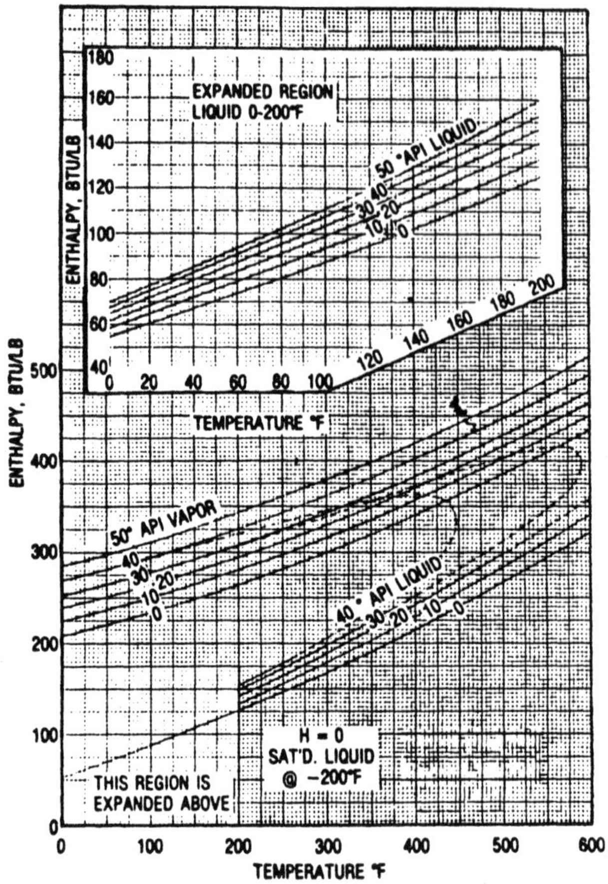

Fig. B.50 – Enthalpy of petroleum fractions ($K_w = 10$, T = 0 to 600°F).[36]

water as a function of gas in solution. As for oil, it is best to measure water viscosity as a function of pressure at reservoir temperature. The water viscosity in the reservoir should be measured with the dissolved gas saturant and salinity in the reservoir.

B.5 Enthalpy and Heat Capacity

Crude Oils and Fractions

Enthalpy of crude oil fractions has been investigated and reported by Kesler and Lee[36] and others[5,6,37,38] for both liquids and vapors. The correlations of Kesler and Lee involve obtaining base values of the enthalpy at zero pressure (denoted by h^o) and then correcting these for pressure effects according to the relation

$$h = h^o - \Delta h. \quad \text{(B.34)}$$

The base enthalpies h^o are given in Figs. B.50 through 57 and require information on the API gravity and the Watson characterization factor of the crude (or crude fraction), K_w. K_w is defined as follows.

$$K_w = (\bar{T}_b)^{1/3}/\gamma_o, \quad \text{(B.35)}$$

where $\bar{T}_b$ is the cubic average boiling point and γ_o is the specific gravity.

The pressure correction Δh is found thus:

$$\Delta h = \left\{\left[(\Delta h/RT_{pc})^{(0)} + \omega(\Delta h/RT_{pc})^{(1)}\right] \cdot \frac{RT_{pc}}{M}\right\}_{p_r} - \left\{\left[(\Delta h/RT_{pc})^{(0)} + \omega(\Delta h/RT_{pc})^{(1)}\right]\frac{RT_{pc}}{M}\right\}_{p_r=1}, \quad \text{(B.36)}$$

where $(\Delta h/RT_{pc})^{(0)}$ is found in Fig. B.58, $(\Delta h/RT_{pc})^{(1)}$ is found in Figs. B.59 and B.60, and the quantities T_{pc}, p_{pc}, ω, and RT_{pc}/M are found in Fig. B.61. The second term of Eq. B.36 is omitted when the reduced temperature is less than 0.8.

The corrections for pressure given in Figs. B.58 through B.60 depend on the reduced temperature and

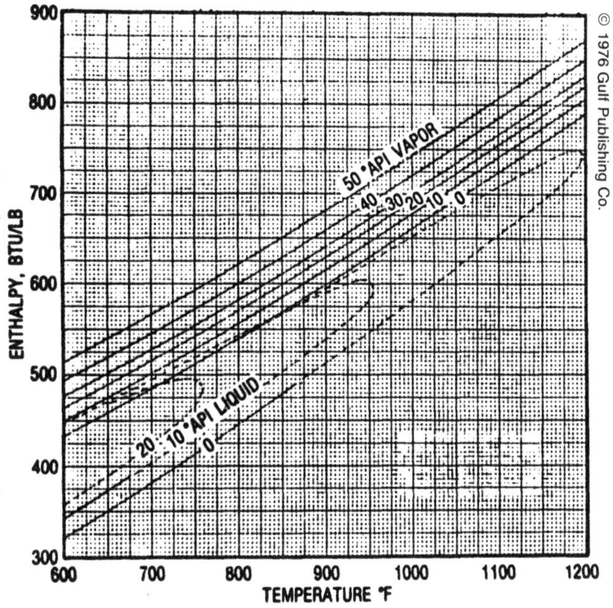

Fig. B.51 – Enthalpy of petroleum fractions ($K_w = 10$, $T = 600$ to $1{,}200°F$).[36]

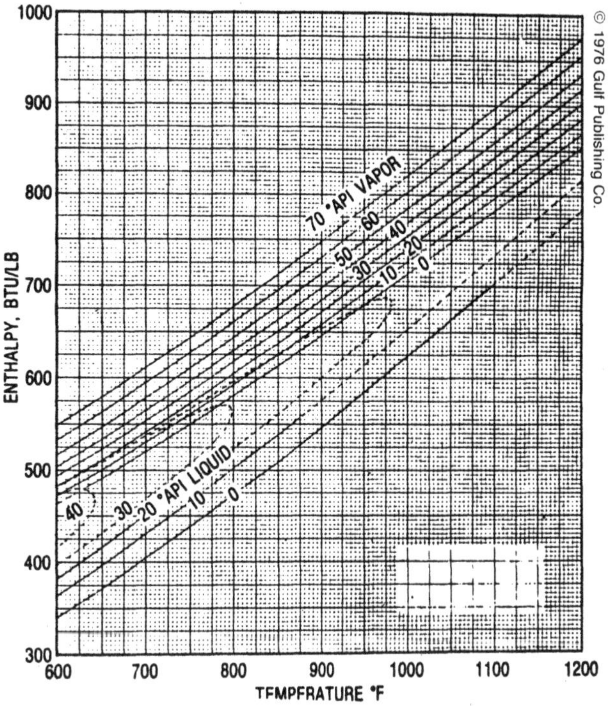

Fig. B.53 – Enthalpy of petroleum fractions ($K_w = 11$, $T = 600$ to $1{,}200°F$).[36]

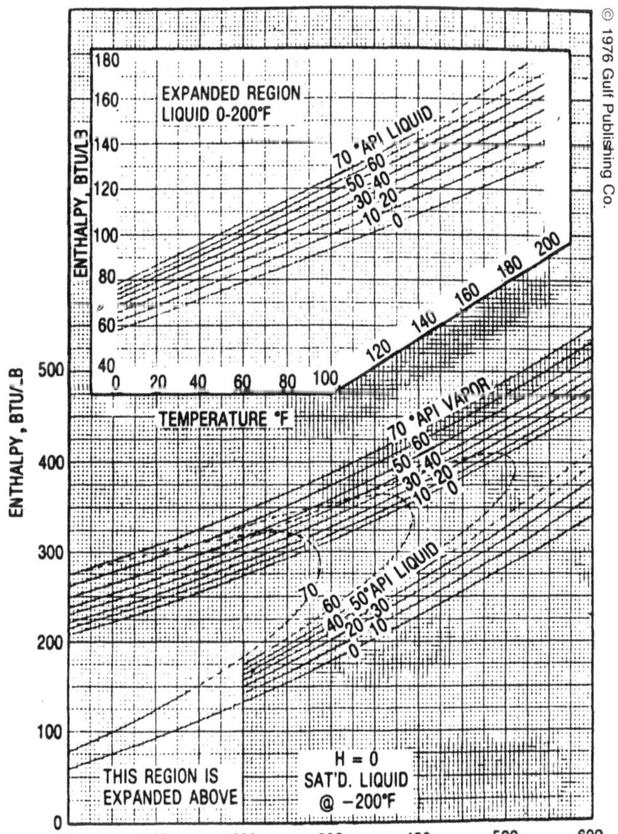

Fig. B.52 – Enthalpy of petroleum fractions ($K_w = 11$, $T = 0$ to $600°F$).[36]

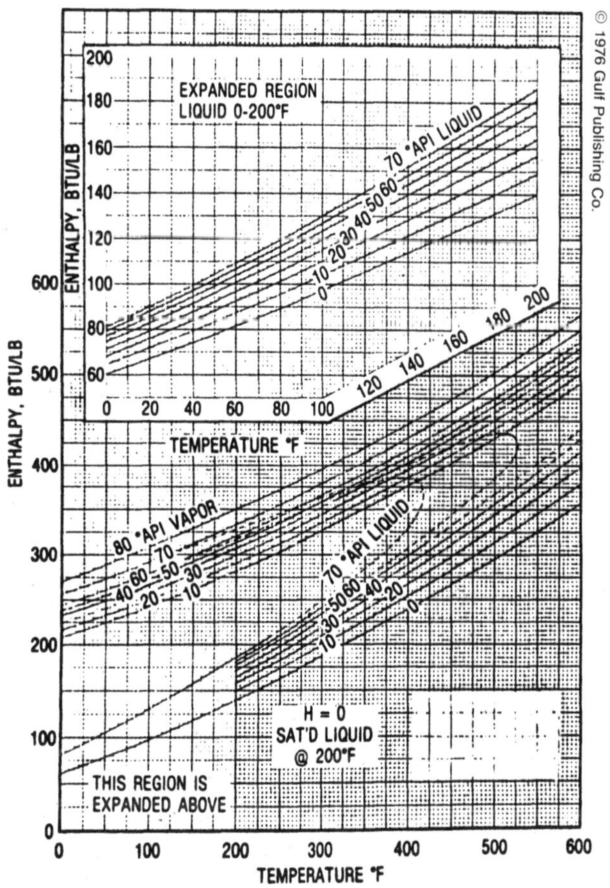

Fig. B.54 – Enthalpy of petroleum fractions ($K_w = 11.8$, $T = 0$ to $600°F$).[36]

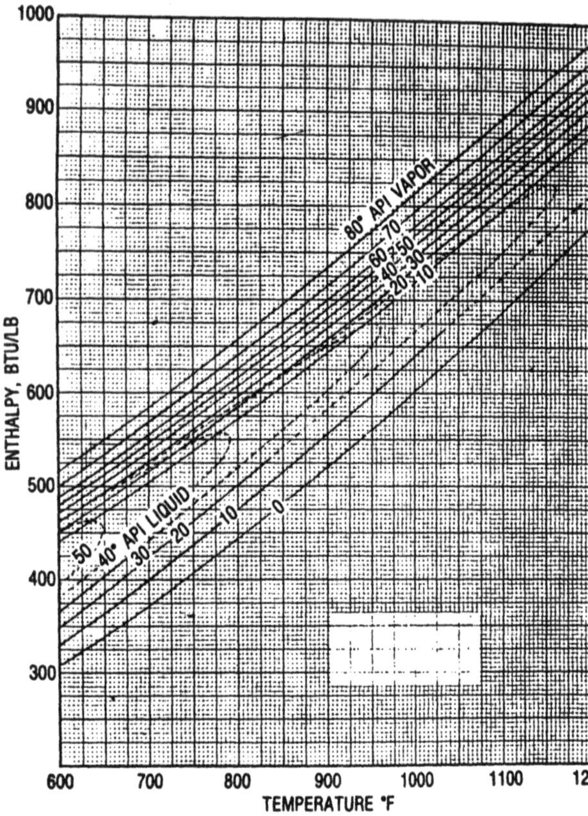

Fig. B.55 – Enthalpy of petroleum fractions ($K_w = 11.8$, $T = 600$ to 1,200°F).[36]

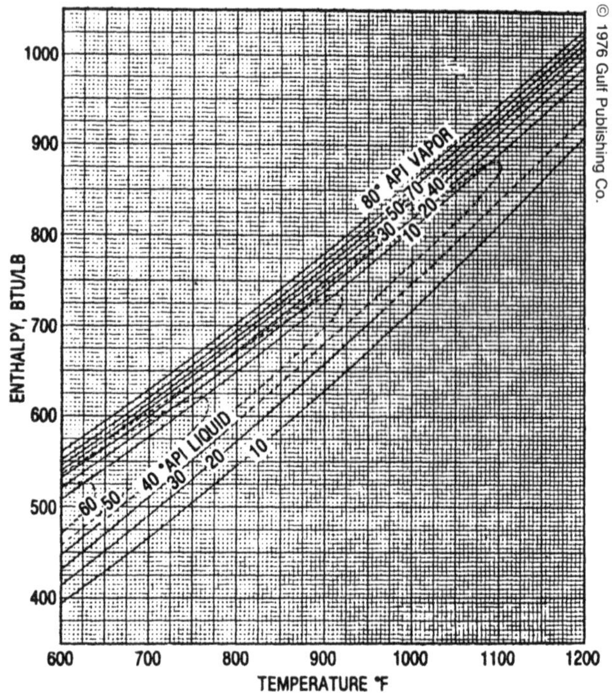

Fig. B.57 – Enthalpy of petroleum fractions ($K_w = 12.5$, $T = 600$ to 1,200°F).[36]

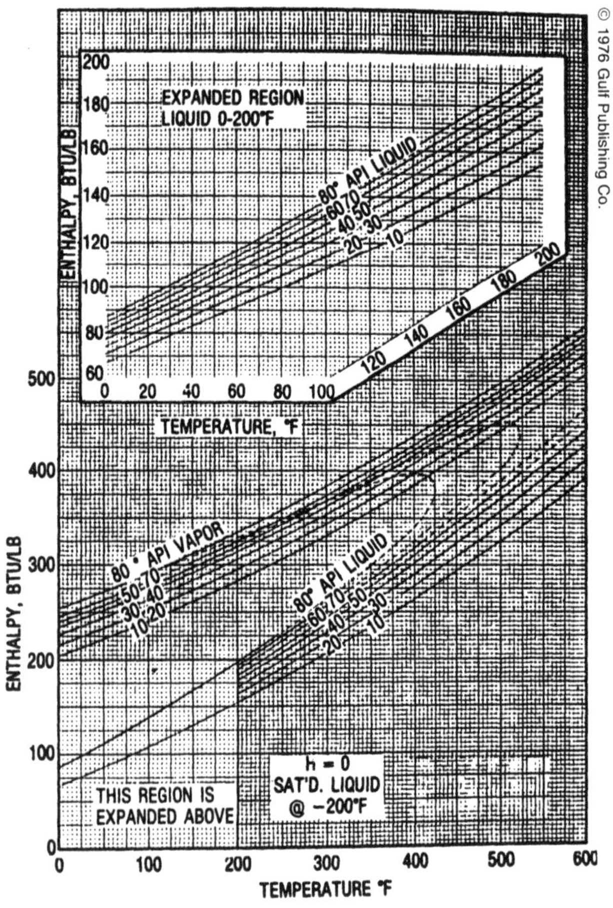

Fig. B.56 – Enthalpy of petroleum fractions ($K_w = 12.5$, $T = 0$ to 600°F).[36]

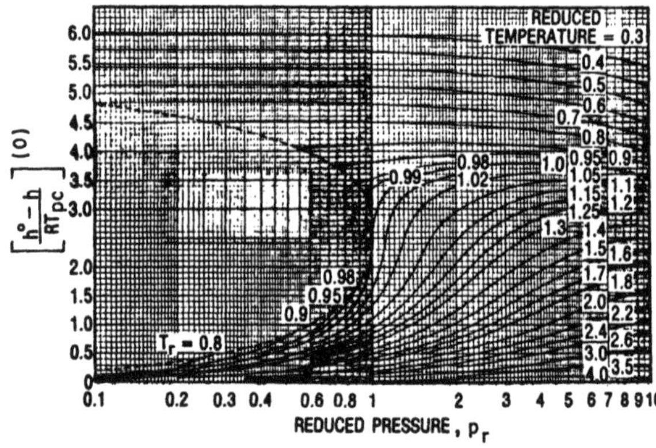

Fig. B.58 – Enthalpy departure (from zero pressure) simple fluid function.[36]

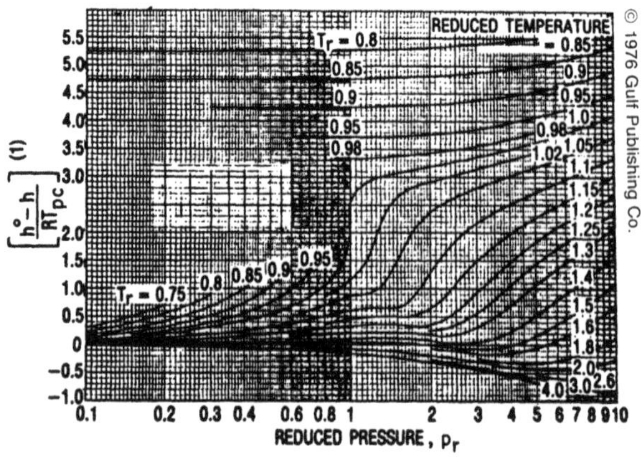

Fig. B.59 – Enthalpy departure (from zero pressure) deviation function (high-temperature part).[36]

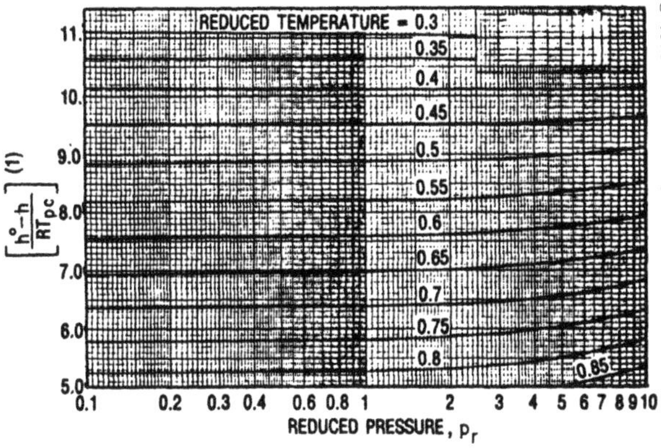

Fig. B.60 – Enthalpy departure (from zero pressure) deviation function (low-temperature part).[36]

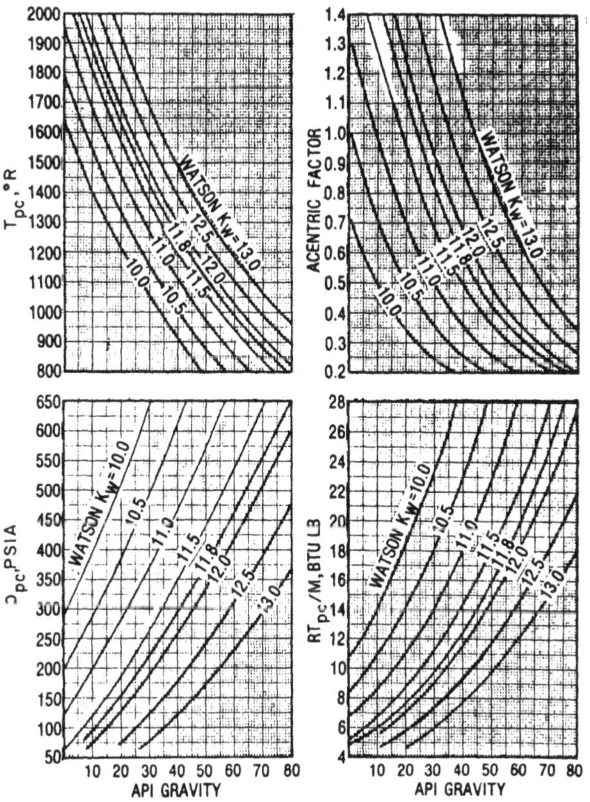

Fig. B.61 – T_{pc}, p_{pc}, ω, and RT_{pc}/M as functions of API gravity and Watson's characterization factor K_w.[36]

pressure defined by Eqs. B.3 and B.4, and are independent of the Watson characterization factor K_w and acentric factor ω. It appears that if they are not available from measurements or other sources,[5] the values of K_w and ω may be estimated from Fig. B.61 when the API gravity and the pseudocritical temperature or pressure of the crude are known. For additional details, refer to the original article of Kesler and Lee.[36] The base enthalpies h^o are referred to a value $h^o = 0$ at $T = -200$°F. Enthalpies calculated by this procedure take into account changes in the isobaric specific heat C_o with temperature and of the latent heat of condensation and evaporation of the crude fractions. Average specific

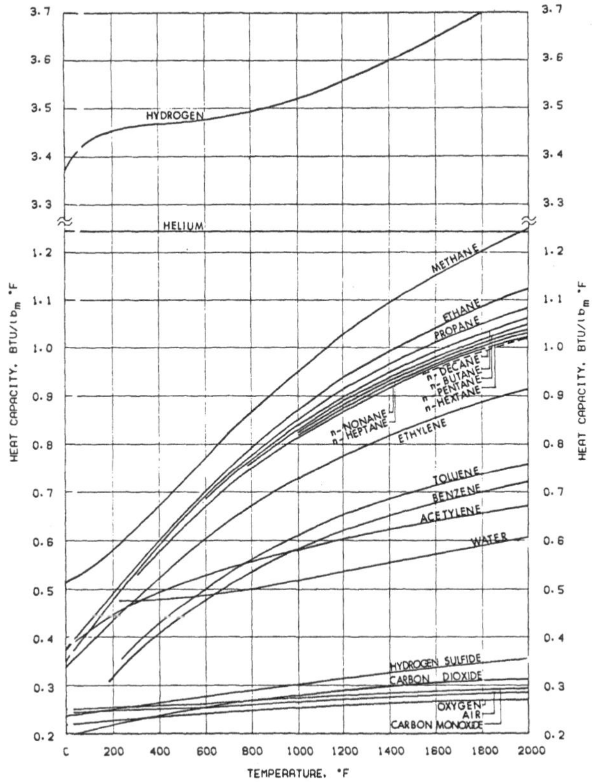

Fig. B.62 – Heat capacity of gases at 1 atm. c_{ga}, Btu/lbm-°F.[40]

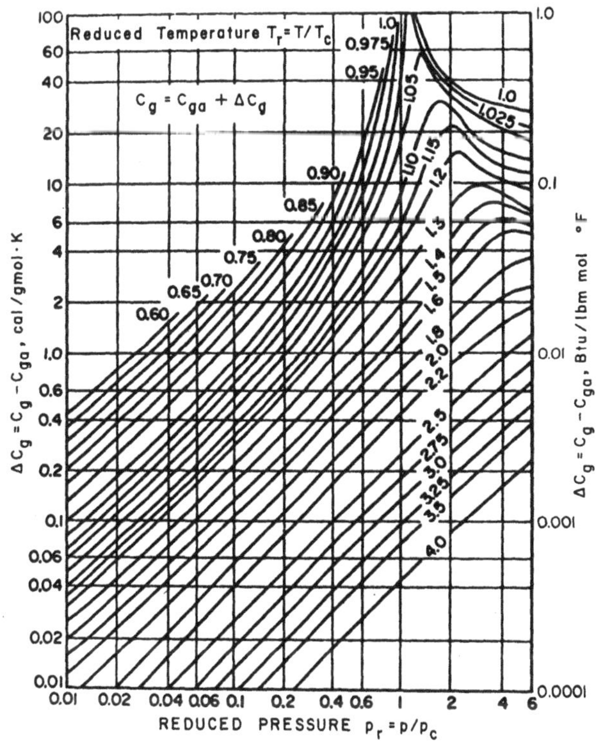

Fig. B.63 – Pressure correction to the molar heat capacity of gases.[29]

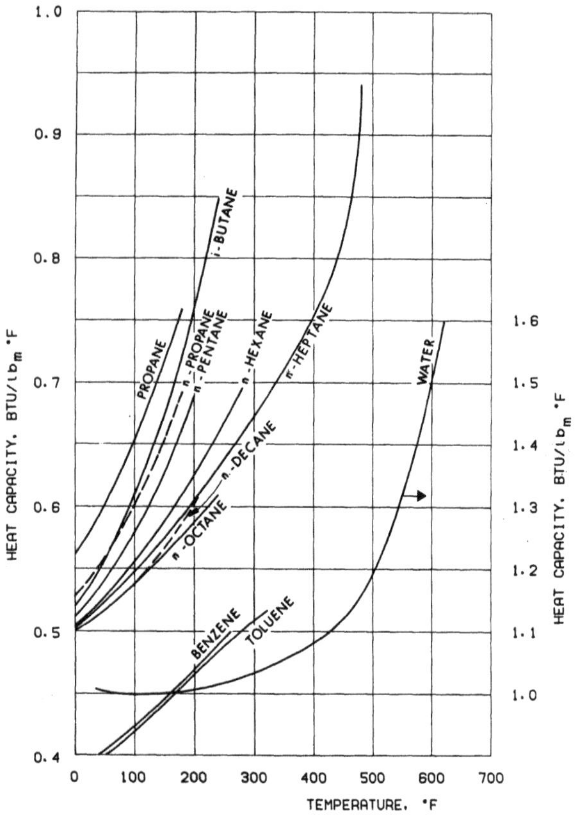

Fig. B.64 – Heat capacity of saturated pure liquids.[40]

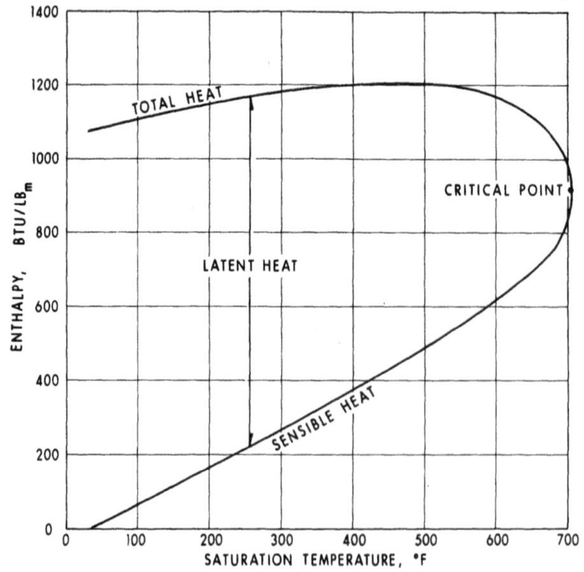

Fig. B.66 – Enthalpies of water and steam as a function of the saturation temperatures (reference temperature is 32°F).

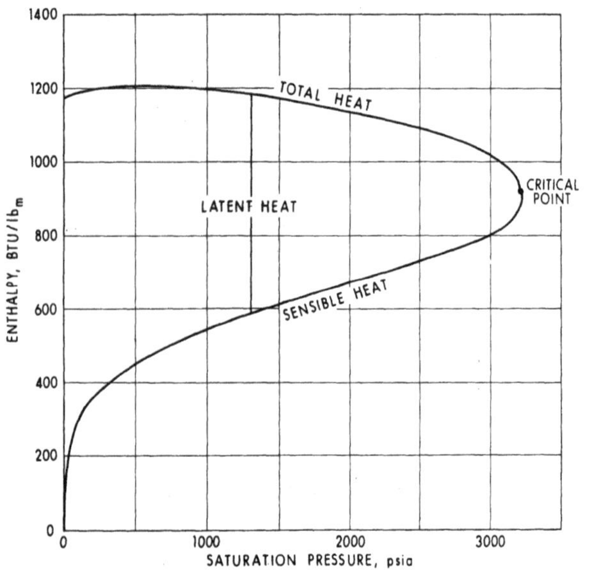

Fig. B.65 – Enthalpies of water and steam as a function of the saturation pressure (reference temperature is 32°F).

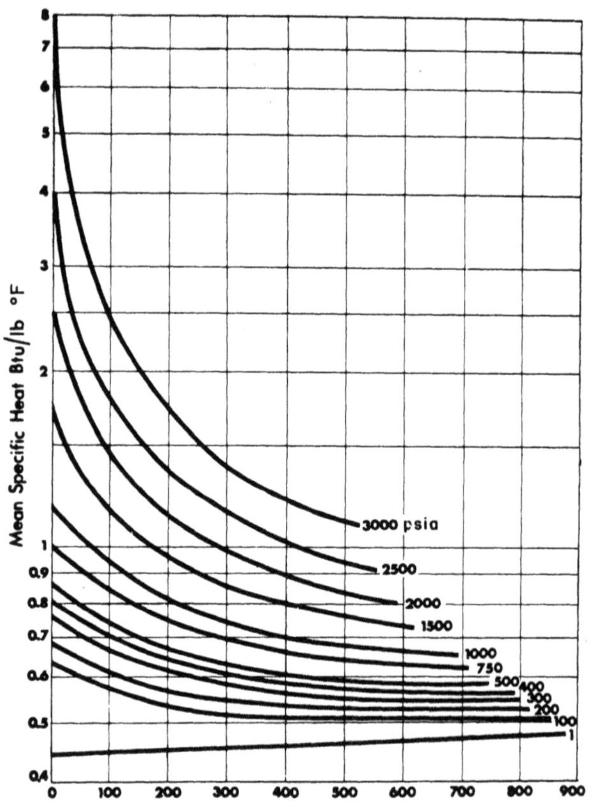

Fig. B.67 – Mean specific heat of superheated steam. Heat added (Btu/lbm) = mean specific heat × superheat (°F).[15]

heats $\overline{C}_o$ of the liquid and vapor phases can be estimated by taking the chord secant of the enthalpies over the desired temperature range:

$$\overline{C}_o = \frac{h(T_1) - h(T_2)}{T_1 - T_2}. \quad \text{(B.37)}$$

When only the density or API gravity of the crude is known, its heat capacity at initial reservoir temperatures may be estimated from Gambill's[39] relation:

$$C_o = (0.388 + 0.00045T)/\sqrt{\gamma_o}, \quad \text{(B.38)}$$

where T is less than about 300°F.

Heat capacities of selected hydrocarbon and nonhydrocarbon gases and vapors at atmospheric pressure are given in Fig. B.62, which can be

TABLE B.6 – ENTHALPY OF WATER AND STEAM AT SATURATED CONDITIONS

Temperature (°F)	Absolute Pressure (psia)	Enthalpy (Btu/lbm) Liquid h_w	Enthalpy (Btu/lbm) Vapor h_s	Latent Heat L_v (Btu/lbm)
101.76	1	69.72	1,105.2	1,035.5
150	3.716	117.87	1,125.7	1,007.8
162.25	5	130.13	1,130.8	1,000.7
193.21	10	161.17	1,143.3	982.1
200	11.525	167.99	1,145.8	977.8
212	14.696	180.07	1,150.4	970.3
250	29.82	218.48	1,163.8	945.3
300	67.01	269.60	1,179.7	910.1
327.83	100	298.43	1,187.3	888.9
350	134.62	321.64	1,192.6	871.0
358.43	150	330.53	1,194.4	863.9
381.82	200	355.40	1,198.7	843.3
400	247.25	374.97	1,201.2	826.2
400.97	250	376.04	1,201.4	825.4
417.33	300	393.85	1,203.2	809.3
420	308.82	396.78	1,203.5	806.7
431.71	350	409.70	1,204.4	794.7
440	381.59	418.91	1,204.8	785.9
460	466.97	441.42	1,205.0	763.6
467.00	500	449.40	1,204.9	755.5
480	566.12	464.37	1,204.2	739.8
486.21	600	471.59	1,203.6	732.0
500	680.80	487.80	1,202.0	714.2
520	812.68	511.9	1,198.4	686.5
540	962.80	536.6	1,193.3	656.7
544.56	1,000	542.4	1,191.9	649.5
560	1,133.4	562.2	1,186.3	624.1
580	1,326.1	588.9	1,177.0	588.1
596.20	1,500	611.4	1,167.7	556.3
600	1,543.2	616.8	1,165.2	548.4
620	1,787.0	646.5	1,150.2	503.7
635.78	2,000	671.7	1,135.2	463.5
640	2,060.3	678.7	1,130.7	452.0
660	2,366.2	714.4	1,104.9	390.5
680	2,708.4	757.2	1,067.2	310.0
695.37	3,000	802.6	1,019.3	216.7
700	3,094.1	823.9	995.6	171.7
705.34*	3,206.2*	910.3	910.3	0

*Critical properties.

corrected for pressure by means of Fig. B.63. Heat capacities of saturated liquids are given in Fig. B.64. Enthalpies of water (both liquid and vapor) are discussed in the next section.

Water and Steam

The word "water" commonly is used in at least two ways. In one, water means the chemical H_2O. In the other, water means liquid water or the liquid phase of H_2O. Here, water is used to mean the latter – the liquid manifestation of H_2O. Table B.6 gives the enthalpy of water and steam at the saturation pressure and temperature. The difference between these two quantities represents the heat required to convert H_2O from liquid to its vapor and is called the latent heat of evaporation:

$$L_v = h_s - h_w. \quad \text{(B.39)}$$

Enthalpies of water and steam at saturation pressures and temperatures are given in Figs. B.65 and B.66.

At temperatures higher than its boiling point, steam is said to be superheated. The additional enthalpy (above its saturation value, given in Table B.6) of superheated steam is found by multiplying the mean specific heat given in Fig. B.67 by the excess temperature above the saturation value T_s. This product is called superheat:

$$\Delta h_s = \bar{C}_s(T - T_s), \quad T \geq T_s. \quad \text{(B.40)}$$

Note that the mean specific heat of steam at atmospheric pressure over a temperature range indicated in Fig. B.67 is somewhat lower than the actual specific heat of superheated steam at a given temperature shown in Fig. B.62. The specific heat of saturated water is included in Fig. B.64.

Reservoir Rocks and Minerals

Figs. B.68 and B.69 and Table B.7 provide data reported by Somerton[41] on the heat capacity of dry reservoir rocks.

Fig. B.69 also shows the calculated heat capacity corresponding to the porosities of Samples 1 and 8 of Table B.8. Those calculated heat capacities are based on reported heat capacities of pure quartz and pure calcite. They agree rather well with the experimental values of the two samples. This is an example of the rule that the heat capacity of a mixture is the mass average of that of its constituents. For fluid-saturated rocks, the heat capacity thus may be estimated from

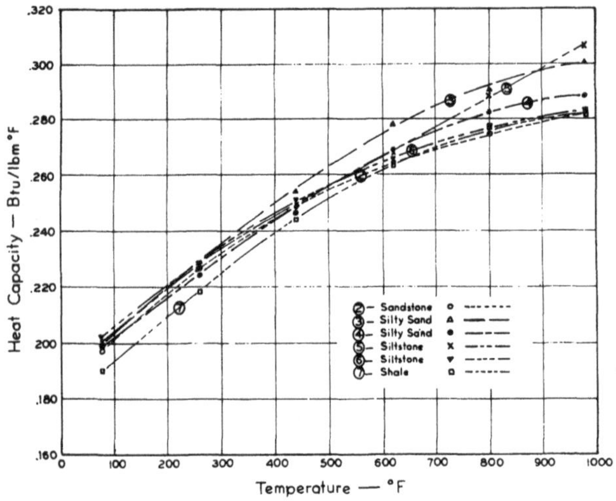

Fig. B.68 – Heat capacities of some reservoir rocks.[41]

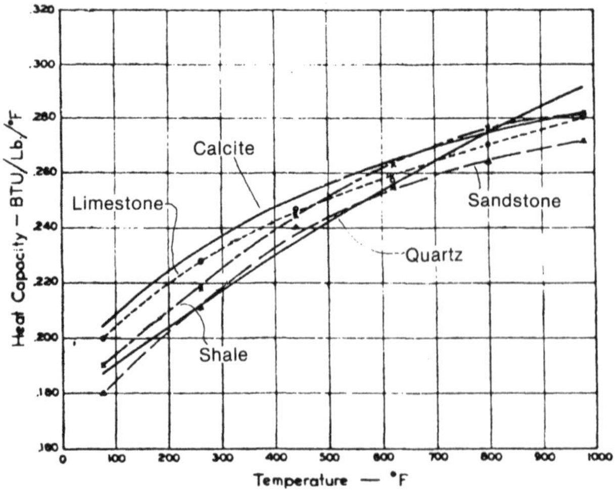

Fig. B.69 – Comparison of calculated heat capacities of reservoir rocks and principal constituents.[41]

$$C_R = M_R / \rho, \quad \ldots\ldots\ldots\ldots\ldots\ldots\ldots\ldots (B.41)$$

where M_R is given by Eq. 5.3 of the text and the average density is given by Eq. B.10.

Table B.7 gives the calculated heat capacities for the first eight samples in Table B.8 when those samples are saturated with methane or water at 620°F and several pressures between 14.7 and 3,000 psia. Note that the water is in the liquid state only at the highest pressure.

Helgeson[17] provides coefficients to calculate isobaric heat capacities of several minerals – including calcite, dolomite, anhydrite, micas, and clays in cal/mol-°K – as a function of temperature. Touloukian *et al.*[42] provide both tables and graphs of isobaric heat capacities in cal/g-°K for nonmetallic solids, including elements, oxides, sulfides, carbonates, sulfates, and other oxygen compounds.

Table B.9 provides thermophysical properties of selected metals and alloys, and Table B.10 provides them for nonmetals. Included in the properties are isobaric heat capacities.

B.6 Thermal Conductivity

Liquids

Thermal conductivities of saturated organic liquids decrease with increasing temperature. Values for some pure organic liquids are plotted in Fig. B.70. For petroleum fractions and hydrocarbon mixtures, Cragoe[45] proposes the relation

TABLE B.7 – CALCULATED HEAT CAPACITIES OF FLUID-SATURATED ROCKS[41]
(heat capacity at 620°F and pressure indicated – Btu/cu ft-°F)

	14.7 psia			500 psia		1,500 psia		3,000 psia	
Sample	Dry	Methane	Water	Methane	Water	Methane	Water	Methane	Water
1. Sandstone	34.0	34.0	34.0	34.1	34.1	34.3	34.9	34.6	45.9
2. Sandstone	32.9	32.9	32.9	33.0	33.0	33.3	34.0	33.7	47.6
3. Silty Sand	35.6	35.6	35.6	35.7	35.7	36.0	36.5	36.3	49.1
4. Silty Sand	33.5	33.5	33.5	33.5	33.6	33.9	34.5	34.3	48.3
5. Siltstone	32.0	32.0	32.0	32.1	32.1	32.5	33.2	32.9	48.7
6. Siltstone	33.6	33.6	33.6	33.7	33.7	34.0	34.6	34.3	47.7
7. Shale	39.6	39.6	39.6	39.6	39.6	39.7	40.0	39.9	40.1
8. Limestone	35.4	35.4	35.4	35.5	35.5	35.7	36.2	36.0	46.8

TABLE B.8 – DESCRIPTION OF TEST SAMPLES[41]

			Principal Minerals		
Sample	Description	Porosity	Quartz (%)	Clay Mineral	Other
1. Sandstone	Well-consolidated, medium-coarse grain	0.196	80	Trace kaolinite	Trace pyrite
2. Sandstone	Poorly-consolidated, medium-fine grain	0.273	40	Illite(?)	Feldspars
3. Silty sand	Poorly-consolidated, poorly sorted	0.207	20	Kaolinite type	Feldspars
4. Silty sand	Medium hard, poorly sorted	0.225	20	Kaolinite type	Feldspars
5. Siltstone	Medium hard, broken	0.296	20	Kaolinite type	Feldspars
6. Siltstone	Hard	0.199	25	Illite	Feldspars
7. Shale	Hard, laminated	0.071	40	Illite-kaolinite	....
8. Limestone	Granular, uniform texture	0.186	.....	....	Calcium carbonate
9. Sand	Unconsolidated, fine-grained	0.38	100	....	....
10. Sand	Unconsolidated, coarse-grained	0.34	100	....	....

TABLE B.9 – THERMOPHYSICAL PROPERTIES OF SELECTED METALS AND ALLOYS[43]

Metals and Alloys	Density ρ (lbm/cu ft)	Isobaric Heat Capacity C (Btu/lbm-°F)	Thermal Conductivity λ (Btu/D-ft-°F)			Thermal Diffusivity α (sq ft/D)
	68°F	68°F	68°F	212°F	1112°F	68°F
Aluminum, pure	169	0.214	2,830	2,860		87.96
Brass (70% Cu, 30% Zn)	532	0.092	1,500	1,800		31.73
Constantin (60% Cu, 40 Ni)	557	0.098	310	310		5.69
Copper, pure	559	0.0915	5,350	5,260	4,900	104.5
Iron						
Pure	493	0.108	1,000	940	550	18.8
Cast (C~4%)	454	0.10	720			16.0
Wrought (C<0.5%)	490	0.11	820	790	500	15.2
Lead, pure	710	0.031	480	463		22.2
Magnesium, pure	109	0.242	2,400	2,300		90.29
Molybdenum	638	0.060	1,700	1,600	1,500	49.78
Nickel						
Pure (99.9%)	556	0.1065	1,200	1,200		21.1
Impure (99.2%)	556	0.106	960	890	770	16.2
Silver, pure	657	0.056	5,800	5,760		158.4
Steel, mild, 1% C	487	0.113	600	600	460	10.8
Stainless steel (18 Cr, 8 Ni)	488	0.11	230	240	310	4.13
Tin, pure	456	0.054	890	820		36.12
Tungsten	1208	0.032	2,300	2,100	1,600	58.32
Zinc, pure	446	0.092	1,560	1,500		38.18
		516				

$$\lambda_o = 1.62[1-3(T-32)10^{-4}]/\gamma_o \quad \ldots\ldots\ (B.43)$$

to estimate the thermal conductivity in Btu/D-ft-°F. For $0.78<\gamma_o<0.95$ and $32°F<T<392°F$, Eq. B.43 gave "average and maximum errors of 12 and 39 percent."[29]

The ratio of two liquid thermal conductivities at the same temperature but different pressures is given by Lenoir[46] as

$$(\lambda_1/\lambda_2)=(\epsilon_1/\epsilon_2), \quad \ldots\ldots\ (B.42)$$

where ϵ is given in Fig. B.71. Values for organic liquids are not very sensitive to changes in pressure, as can be seen from Fig. B.71, except at low values of the reduced pressure and high values of the reduced temperature. Comparing measured values with predicted values[46] based on Fig. B.71 revealed an arithmetic deviation of 1.6% with a maximum deviation of 4.2%.

The thermal conductivity of saturated water, also shown in Fig. B.70, is several times larger than that of pure hydrocarbons and exhibits a maximum at about 260°F.

TABLE B.10 – THERMOPHYSICAL PROPERTIES OF SELECTED NONMETALLIC MATERIALS[40]

Material	Temperature T (°F)	Density ρ (lbm/cu ft)	Isobaric Heat Capacity C (Btu/lbm-°F)	Thermal Conductivity λ (Btu/D-ft-°F)	Thermal Diffusivity α (sq ft/D)
Aerogel, silica	100	5.3	0.205	0.31	0.29
Asbestos	32	36	0.25	2.1	0.24
	800	36		3.1	
Brick					
Common	68	100	0.20	2.4 to 4.8	0.2 to 0.5
Fire clay	1472	145	0.23	19	0.58
Bakelite	68	79.5	0.38	3.2	0.11
Concrete	68	119 to 144	0.21	11 to 19	0.45 to 0.65
Corkboard	100	10	0.4	0.60	0.1
Diatomaceous earth, powdered	100	14	0.21	0.72	0.2
Fiber insulating board	100	14.3		0.58	
Glass, window	68	162	0.16	12	0.48
Glass wool					
Fine	100	1.5		0.74	
Packed	100	6.0		0.52	
Ice	32	57	0.46	30.7	1.2
Magnesia, 85%	100	17		0.94	
Marble	68	156-169	0.193	38	1.3
Paper				1.8	
Rock wool	100	12		0.55	
Rubber, hard	32	74.8	0.48	2.1	0.058
Wood, oak, ⊥ to grain	70	51	0.57	2.9	0.1
Wood, oak, ∥ to grain	70	51	0.57	5.5	0.17

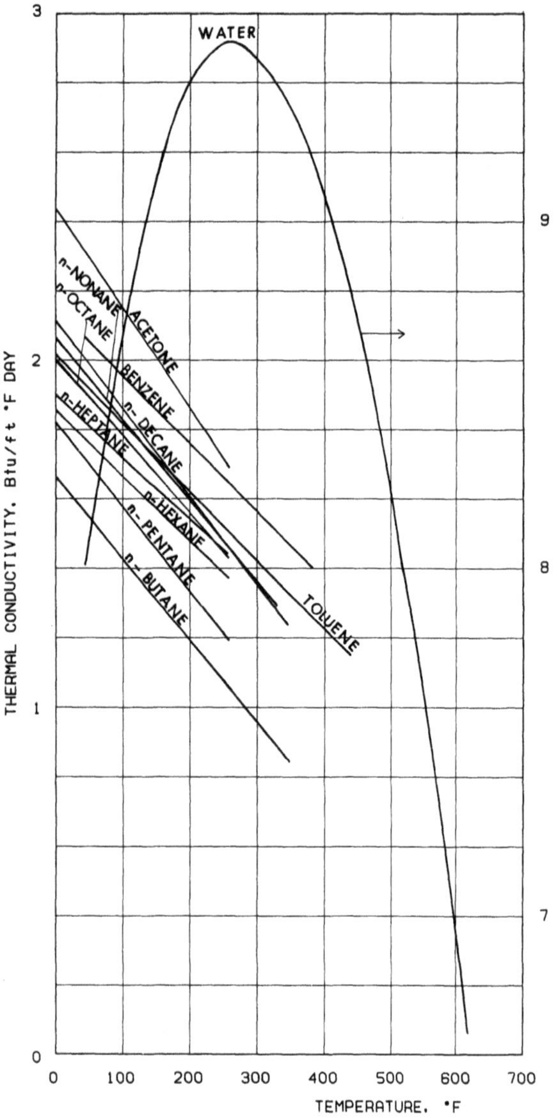

Fig. B.70 – Thermal conductivity of saturated pure liquids.[44]

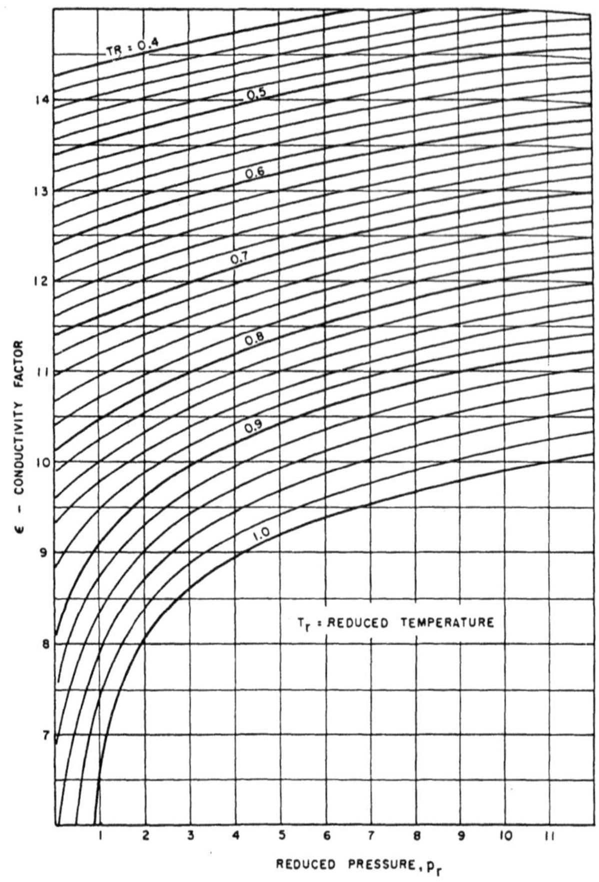

Fig. B.71 – Generalized chart for liquid thermal conductivities and effect of pressure.[46]

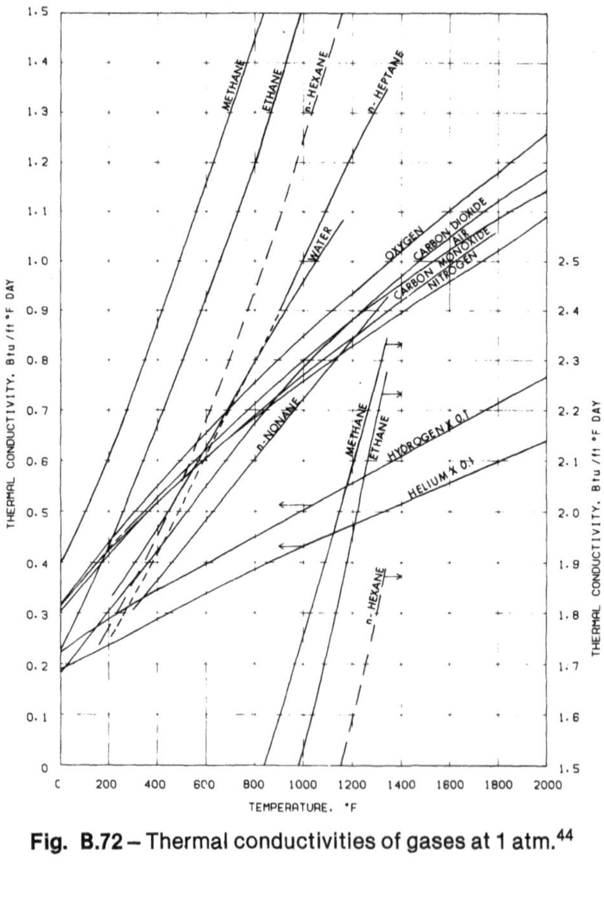

Fig. B.72 – Thermal conductivities of gases at 1 atm.[44]

Gases

Thermal conductivities of gases at atmospheric pressure are plotted vs. temperature in Figs. B.72 and B.73. Ref. 44 reviews available methods for estimating the thermal conductivity of mixtures of gases ($\bar{\lambda}$) from those of constituents (λ_j) and suggests that the following relation be used "because of its simplicity and moderate reliability" for "approximate engineering calculations":

$$\bar{\lambda} = \frac{1}{2}\left\{ \sum_{j=1}^{n_c} y_j \lambda_j + \left[\sum_{j=1}^{n_c} \left(\frac{y_j}{\lambda_j} \right) \right]^{-1} \right\}, \quad \ldots\ldots (B.44)$$

where y_j is the mole fraction of the jth gas component.

Fig. B.74 provides information on the effect of elevated pressures on the thermal conductivity of selected gases.

Fig. B.75 gives the thermal conductivity of superheated steam vs. temperature at several values of pressure and for the saturated vapor.

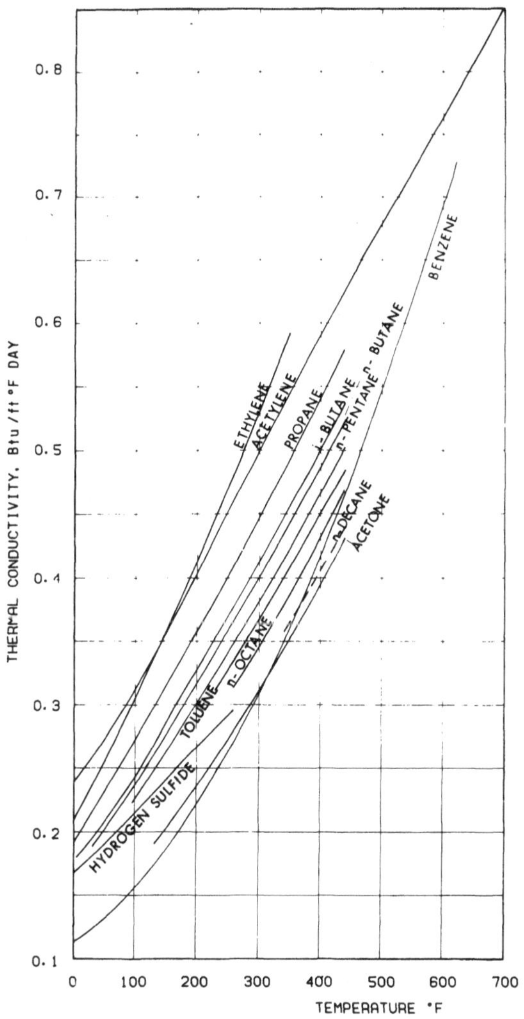

Fig. B.73 – Thermal conductivities of gases at 1 atm.[44]

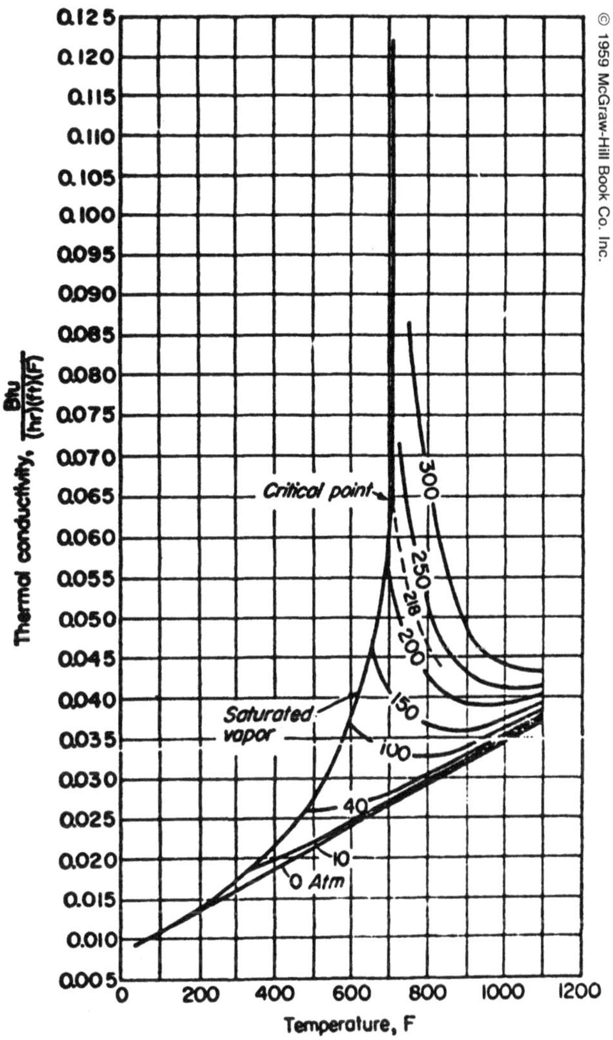

Fig. B.75 – Dependence of thermal conductivity of water vapor on pressure and temperature.[35]

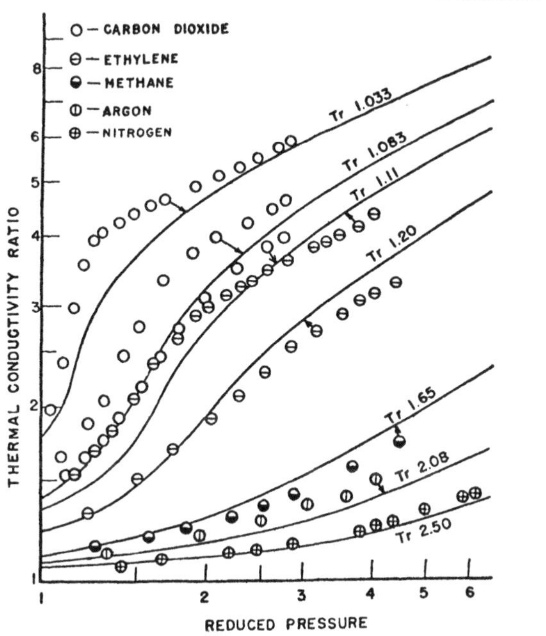

Fig. B.74 – Comparison of measured values of thermal conductivity with correlation by Comings and Nathan [*Ind. Eng. Chem.* (1947) **39**, 964].[47]

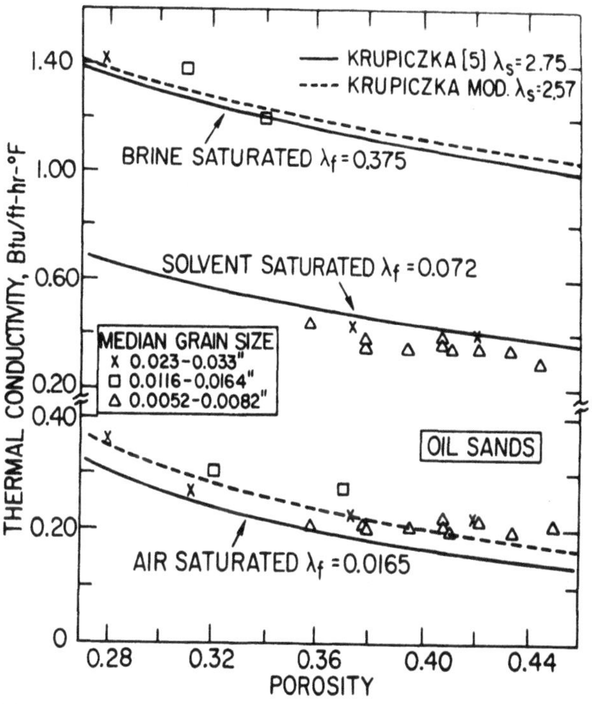

Fig. B.76 – Thermal conductivity of unconsolidated oil sands.[48]

TABLE B.11 – THERMAL CHARACTERISTICS OF TEST SAMPLES[50]

Sample*	Bulk Density (lbm/cu ft)		Thermal Diffusivity Unsteady State (sq ft/D at 200°F)		Thermal Diffusivity Steady State (sq ft/D)		Thermal Conductivity Unsteady State (Btu/D-ft-°F at 200°F)		Thermal Reactions
	Initial	Repeat	Initial	Repeat	90°F	275°F	Initial	Repeat	
Bandera SS	134.2	131.8	0.816	0.660	0.895	0.660	2.16	16.6	α−β quartz 1,015°F, clay
Berea SS	134.8	126.0	0.821	0.581	0.948	0.665	21.8	15.7	α−β quartz 1,020°F, clay
Boise SS	118.9	116.0	0.833	0.492	0.694	0.624	19.5	11.8	α−β quartz 1,040°F, clay
Limestone	140.2	78.5**	0.780	0.497	0.780	0.643	21.7	13.9	$CaCO_3 \rightarrow CaO + CO_2$, 1,530°F
CaO**	78.5	–	0.924	–	–	–			
Shale	137.1	128.2	0.936	0.516	0.950	0.698	26.2	13.9	α−β quartz 1,045°F, clay
Rock Salt	135.0	128.0	2.95	–	1.7	1.5	8.30	8.30	Dehydration, 864°F
Tuffaceous SS	115.3	107.2	0.444	–	0.504	–	9.55	9.55	α−β quartz 1,035°F

*SS = sandstone.
**After reaction.

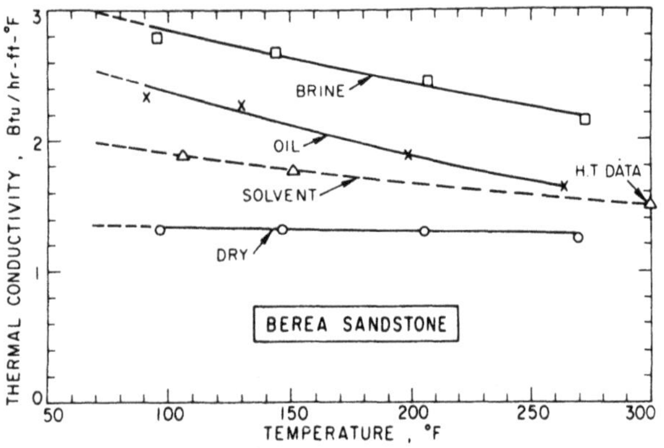

Fig. B.77 – Thermal conductivity of Berea sandstone – effects of temperature and fluid saturant.[49]

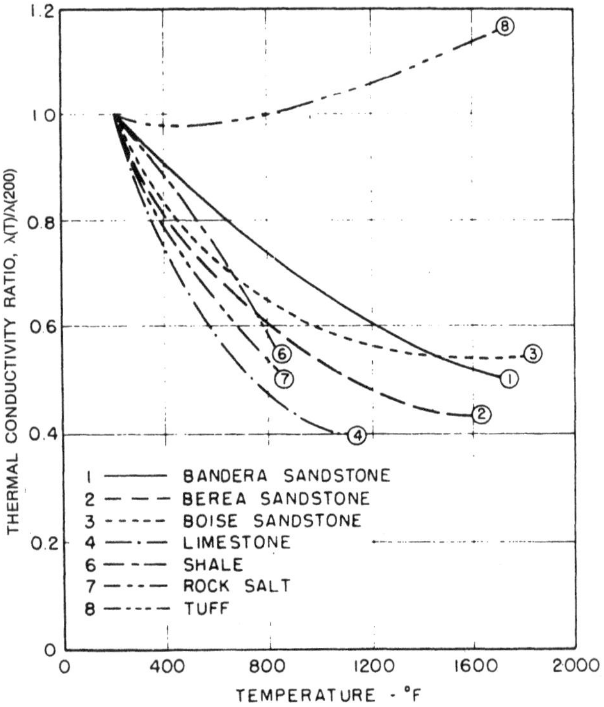

Fig. B.78 – Thermal conductivity relative to that at 200°F – samples subjected to increasing temperatures.[50]

Reservoir Rocks and Other Solids

Somerton and colleagues[48,49] have reported extensively on the thermal conductivity of reservoir rocks, including the effects of fluid and mineral content, particle size, temperature, and pressure. For unconsolidated quartzitic sands saturated with water and oil, the thermal conductivity at 125°F is given by

$$\lambda_R = 0.735 - 1.30\phi + 0.390\lambda_m\sqrt{S_w}, \quad \ldots\ldots \text{(B.45)}$$

where the thermal conductivity of the minerals may be estimated from

$$\lambda_m = 4.45 f_q + 1.65(1 - f_q) \quad \ldots\ldots\ldots\ldots \text{(B.46)}$$

for quartzitic sands, unless the mineral thermal conductivity can be estimated from its composition. f_q is the fractional volume of quartz in the sand. Eqs. B.45 and B.46 are correlations based on samples having porosities between 0.28 and 0.37 and fluids and minerals corresponding to those in Kern River oil sands. In these expressions, no distinction is made about the fluids other than the brine.

The effect of temperature is given by

$$\lambda_R(T) = \lambda_R - 1.28 \times 10^{-3}(T - 125)(\lambda_R - 0.82), \quad \ldots\ldots\ldots\ldots \text{(B.47)}$$

where λ_R is given by Eq. B.45. And the effect of pressure, which is generally rather minor, is estimated from

$$\Delta\lambda_R = \Delta p \times 10^{-5}(0.50\rho_b\phi + 5.75\phi - 0.37k^{0.10} + 0.12F), \quad \ldots\ldots\ldots \text{(B.48)}$$

where k is the permeability in millidarcies, F is the formation resistivity factor, and ρ_b is the bulk density in grams per cubic centimeter. Values of $\Delta\lambda_R$ are added to the value of λ_R at the base pressure. Typical effects of fluid content and porosity on the thermal conductivity of unconsolidated oil sands are shown by the data points in Fig. B.76.

For consolidated sands, Anand *et al.*[49] give the relation

$$\frac{\lambda_{R,l}}{\lambda_{R,d}} = 1.00 + 0.3\left(\frac{\lambda_l}{\lambda_a} - 1.0\right)^{1/3} +$$

TABLE B.12 – DESCRIPTION OF SAMPLES USED IN TABLE B.11[50]

Sample*	Description**	Principal Minerals: Quartz	Feldspar	Other	Porosity
Bandera SS	Well-consolidated, VFG	35%	25%	Calcite, clay	0.200
Berea SS	Well-consolidated, FG	65%	10%	Calcite, sericite, clay	0.205
Boise SS	Well-consolidated, MG	40%	35%	Clay, sericite	0.265
Limestone	Small vugs, MCG	–	–	Ca CO_3	0.186
Shale	Hard, laminated VFG	50%	–	Clay, iron oxides, biotite	0.170
Rock salt	Crystalline	–	–	Halite	0.010
Tuffaceous SS	Well-consolidated, large to VFG	10%	60%	Clay, pumice lapilli, calcite	0.280

*SS = sandstone.
**VFG = very fine grained, FG = fine grained, MG = medium grained, and MCG = medium-coarse grained.

$$4.57\left(\frac{\phi}{1-\phi}\frac{\lambda_l}{\lambda_{R,d}}\right)^{0.48m}\left(\frac{\rho_{R,l}}{\rho_{R,d}}\right)^{-4.30} \quad \ldots\ldots (B.49)$$

for the thermal conductivities of sandstone rocks at 68°F fully saturated with *one* liquid. In this equation, the subscript R,l refers to the reservoir rock at full liquid saturation, R,d refers to dry reservoir rock, l refers to liquid, a refers to air, and m is Archie's cementation factor.

The thermal conductivity of the dry consolidated sandstone at 68°F may be estimated from

$$\lambda_{R,d}=0.340\rho_{R,d}-3.20\phi+0.530k^{0.10}$$
$$+0.0130F-0.031. \quad \ldots\ldots\ldots\ldots (B.50)$$

The agreement between values of thermal conductivity reported in the literature and those calculated by Eq. B.49 is said to be within 15% for 85% of the values when the ratio $\lambda_{R,l}/\lambda_{R,d}$ is within the range of 1.20 to 2.30.

The effect of temperature on the thermal conductivity of liquid-saturated sandstones is given by

$$\lambda_{R,l}(T)=\lambda_{R,l}-0.71\times10^{-3}(T-528)$$
$$\cdot(\lambda_{R,l}-0.80)$$
$$\cdot[\lambda_{R,l}(T\times10^{-3})^{-0.55\lambda_{R,l}}+0.74],$$
$$\ldots\ldots\ldots\ldots\ldots\ldots (B.51)$$

where $\lambda_{R,l}$ is given by Eq. B.49, and T is in degrees Rankine. The effect of stress on the thermal conductivity of consolidated sandstones is to increase the thermal conductivity value by "1 to 2 percent per 1000 psi increase in effective stress."[49]

Fig. B.77 shows the measured thermal conductivity of Berea sandstone vs. temperature when the sample is saturated with four different fluids. Thermal conductivities of reservoir rocks containing several fluids are treated differently. Average values may be estimated from figures such as Fig. B.77 or from the values obtained with the equations given previously. Such averages must consider the relative volumes of the fluids present in the reservoir rock. According to Anand *et al.*,[49] the conductivity of the wetting phase has the dominant effect and the liquid conductivity used in Eq. B.49 should be biased accordingly.

TABLE B.13 – THERMAL CONDUCTIVITY OF ROCK-FORMING NATURAL MINERALS*

Mineral	Density (lbm/cu ft)	Thermal Conductivity (Btu/ft°F-D)
Calcite	169.9	49.8
Dolomite	178.4	76.35
Siderite	237.9	41.7
Aragonite	176.5	31.0
Anhydrite	185.9	65.97
Gypsum		17.4
Halite		84.7
Quartz	165.2	106.6
Chert	159.8	51.4
Pyrite	306.8	266.2
Hornblende	198.7	38.9
Albite	163.8	29.7
Microcline	159.7	34.2
Orthoclase	161.3	32.1
Kaolinite		38 †
Illite	181.3	30.49
Montmorillonite	176.6	112.8
Chlorite	171.8	71.36
Biotite	186.1	28.0
Muscovite	178.1	32.2

*Except for kaolinite, all data are from Horai,[51] obtained using naturally occurring minerals at room temperatures and pressures.
†Value used by Somerton *et al.*[48]

Table B.11 presents data on the thermal conductivity of several dry consolidated rock samples, calculated from thermal diffusivities and densities measured at 200°F. Repeat values were generally lower because of reactions that occurred in the samples as they were heated to temperatures of 1,800°F. Thus, the temperature history of a sample may have a significant effect on thermal (as well as other) properties. The effect of temperature on the thermal conductivities is shown in Fig. B.78. Initial sample properties are given in Tables B.11 and B.12.

In the absence of specific data, it appears that Eq. B.49 could be used, together with thermal conductivities of dry saturated rocks given in Tables B.11 and Fig. B.78, to estimate thermal conductivities of saturated consolidated rocks other than sandstones. Errors in this approach should be recognized as being potentially significant.

The following approximate averaging procedure is useful for estimating the value of the thermal conductivity of a porous medium containing water, oil, and gas when the value of the brine-saturated sample is available.

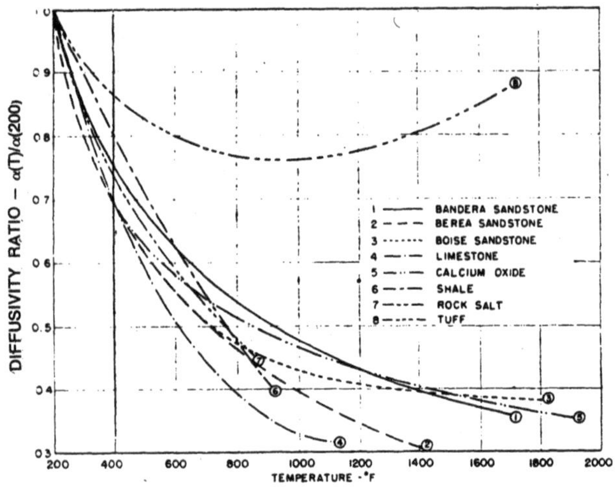

Fig. B.79 – Thermal diffusivity relative to that at 200°F – samples subjected to increasing temperatures.[50]

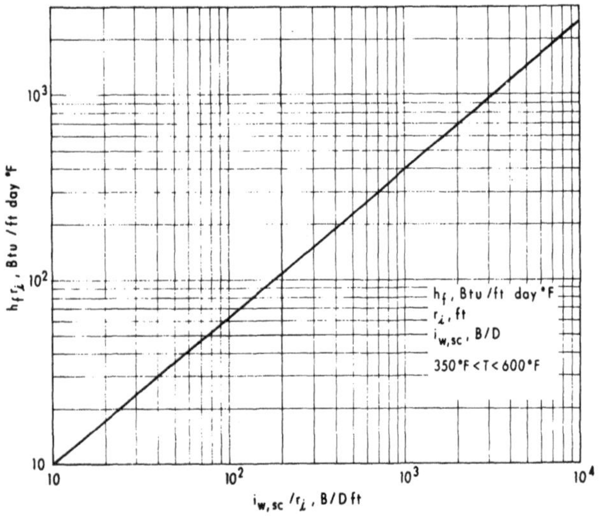

Fig. B.80 – Correlation for estimating film coefficient of heat transfer for water.

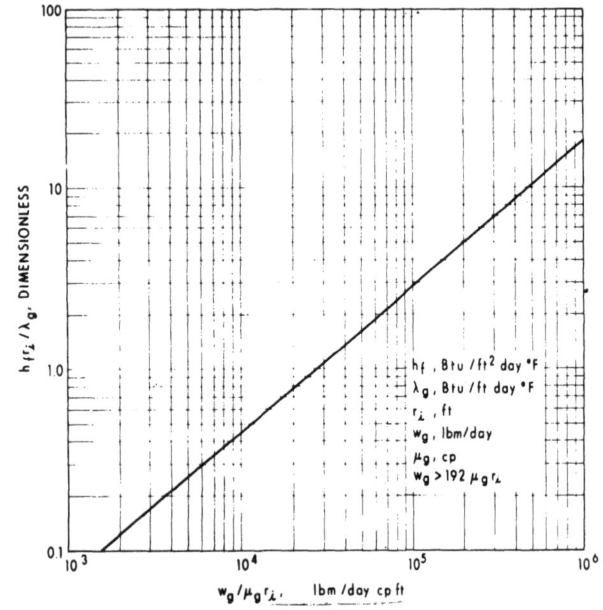

Fig. B.81 – Correlation for estimating film coefficient of heat transfer for gas flow at high velocities.

$$\lambda_R = \lambda_{R,w}\left(\frac{\lambda_o}{\lambda_{R,w}}\right)^{\phi S_o}\left(\frac{\lambda_g}{\lambda_{R,w}}\right)^{\phi S_g}, \quad \ldots\ldots\ldots (B.52)$$

where the value of $\lambda_{R,w}$ may be estimated from Eq. B.49 for consolidated samples and λ_o and λ_g are the thermal conductivities of the oil and gas phases. A similar volumetrically weighted geometric mean may be used to approximate any average value when no better procedure is available.

Table B.13 gives thermal conductivities of selected minerals found in reservoir rocks. Thermal conductivities of some metallic and nonmetallic materials are reported in Tables B.9 and B.10.

Fluid flow increases the apparent thermal conductivity of porous rocks. The effect has been shown in Fig. 3.1 for the apparent thermal conductivity in the direction of flow. This quantity also is known as the longitudinal thermal conductivity or longitudinal thermal dispersivity. Just as there is transverse fluid dispersivity, there is a corresponding transverse thermal conductivity or dispersivity. Generally, these effects are pronounced only near wells or in laboratory work, where the velocities are relatively high.[52,53]

B.7 Thermal Diffusivity

Thermal diffusivities are related to the thermal conductivity and volumetric heat capacity by the relation

$$\alpha = \lambda / M \ . \ \ldots\ldots\ldots\ldots\ldots\ldots (B.53)$$

Values calculated in this manner are quite adequate for engineering applications. Because of difficulties in the measurement techniques, calculated values actually may be better than directly measured values.

Table B.11 presents some information on thermal diffusivities of several dry consolidated rock samples at 200°F measured using two different techniques. Repeat measurements were always lower because of reactions that occurred in the samples as they were heated to temperatures of 1,800°F. The effect of temperature on the thermal diffusivity of dry consolidated rocks is shown in Fig. B.79. Initial sample properties are given in Tables B.11 and B.12.

B.8 Coefficients of Heat Transfer

For a discussion of the use of coefficients of heat transfer, refer to Chap. 10, Heat Losses from Surface and Subsurface Lines.

Values of h_f

For condensing steam, h_f, the film coefficient of heat transfer, is large, and it is generally adequate[54] to use

$$h_f = 48{,}000 \text{ Btu/sq ft-D-°F},$$

when

$$N_{Re} > 2{,}100,$$

where the Reynolds number N_{Re} is defined as

$$N_{Re} = \left(\frac{\pi}{2}(0.0616)\right)\frac{2 i_w \rho_{w,sc}}{\pi \mu_s r_i} \ . \quad \ldots\ldots\ldots (B.54)$$

Here, i_w is the steam equivalent injection rate as condensed water in barrels per day, $\rho_{w,sc}$ is the density of water at standard conditions (62.4 lbm/cu ft), μ_s is the viscosity of steam at the injection temperature in centipoise, and r_i is the inner radius of the conduit in feet.

Since the viscosity of saturated steam over typical temperature ranges is about 0.018 cp, turbulence will prevail ($N_{Re} > 2{,}100$) whenever i_w (in B/D) is greater than 9.85 r_i (in feet).

For hot water, the film coefficient of heat transfer is found[54] from

$$h_f = 0.0115 \frac{\lambda_w}{r_i} N_{Re}^{0.8} N_{Pr}^{0.4} \quad \text{. . . . . (B.55)}$$

when N_{Re} exceeds 2,100. N_{Re} also is given by Eq. B.54; λ_w is the thermal conductivity of water in Btu/ft-D-°F, the dimensionless Prandtl number is

$$N_{Pr} = \langle 58.1 \rangle (C\mu/\lambda)_w, \quad \text{. . . . . (B.56)}$$

C_w is the specific heat of water at constant pressure in Btu/lbm-°F, and μ_w is the viscosity of the hot water in centipoise. In the temperature range of 350 to 600°F, the value of factors representing the properties of water $[(\lambda_w/\mu_w^{0.8})(N_{Pr}^{0.4})]$ does not vary by more than ±10%. Accordingly, the approximate film coefficient is derived as follows.

$$h_f = 1.6\, i_w^{0.8} r_i^{-1.8}, \quad \text{. . . . . (B.57)}$$

when 350°F < T < 600°F, and $i_{w,sc}$ (in B/D) > r_i (in feet) and the subscript *sc* refers to standard conditions. Results calculated from Eq. B.57 are plotted in Fig. B.80.

For hot gases, the film coefficient is found by evaluating Eq. B.55 for gas properties. For gases it generally is found that

$$N_{Pr}^{0.4} \simeq 0.92 \pm 0.07,$$

and the Reynolds number, including the conversion factor associated with the oilfield units for gas, is given by

$$N_{Re} = \langle \pi(0.01096)/2 \rangle \frac{2}{\pi} \frac{w_g}{\mu_g r_i}. \quad \text{. . . . . (B.58)}$$

Accordingly, the film coefficient is given by

$$h_f \simeq 2.86 \times 10^{-4} \frac{\lambda_g}{r_i} \left[\frac{w_g}{\mu_g r_i} \right]^{0.8}, \quad \text{. . . . . (B.59)}$$

when

$$w_g > 192 \mu_g r_i,$$

and

w_g = $i\rho$, 10^3 lbm/D,
i = injection rate, Mscf/D,
ρ = density, lbm/cu ft,
μ_g = viscosity, cp,
λ_g = thermal conductivity, Btu/ft-D-°F, and
r_i = inner radius of the conduit, ft.

Values of h_f for hot-gas injection obtained from Eq. B.59 are shown in Fig. B.81.

Values of h_d

Values of the coefficient of heat transfer due to deposits of scale and dirt, h_d, seldom are known with any degree of confidence. Often values are specific to the project and seldom available unless measurements are made. Typical values of 48,000 Btu/sq ft-D-°F are used by McAdams,[54] and this value is recommended for use in the absence of better information. In injection and production lines, where low heat losses are desired, a significant scale deposit between pipe and insulation is an asset, since it corresponds to a relatively high resistance to heat flow. A scale deposit on the inside wall of the pipe also would be an asset if there were no danger of the deposit's being dislodged, thus possibly plugging the injection wellbore or other parts of the flow system.

In Chap. 10, the subscript d is replaced by one indicating the location of this coefficient of heat transfer.

Values of h_{fc}

For surface injection lines exposed to winds, the coefficient of heat transfer due to forced convection (air currents) at the outer surface of a conduit, h_{fc}, is given by McAdams[54] as

$$h_{fc} = \frac{0.12\lambda_a}{r_e} N_{Re}^{0.6}, \quad \text{. . . . . (B.60)}$$

for 1,000 < N_{Re} < 50,000, where

λ_a = thermal conductivity of air, Btu/ft-D-°F,
r_e = external radius of the conduit exposed to air, ft,
N_{Re} = $\left\langle \frac{1}{2}(4{,}364) \right\rangle \frac{2 r_e v_W \rho_{a,sc}}{\mu_a}$,
v_W = wind velocity normal to the pipe, mile/hr,
$\rho_{a,sc}$ = density of air, lbm/cu ft, and
μ_a = viscosity of air, cp.

For the surface temperatures at r_e of about 200°F, the film coefficient is given[54] approximately by

$$h_{fc} = \frac{18}{r_e} (r_e v_W)^{0.6}, \quad \text{. . . . . (B.61)}$$

when the given units are used. Although, strictly speaking, Eqs. B.60 and B.61 are valid only for wind velocities that are normal to the pipes, they generally are applied irrespective of wind direction, since there is no easy way to correct for differences in the wind direction. Radiation effects may be important for uninsulated exposed pipes carrying high-temperature fluids. For more on radiation effects, refer to the literature.[54-56]

For surface injection lines in buildings, where natural convection is applicable and radiation may be

TABLE B.14 – RADIATION-NATURAL CONVECTION COEFFICIENT OF HEAT TRANSFER*

Diameter of Pipe and Insulation (in.)	Surface Temperature (°F)														
	130	180	230	280	330	380	480	580	680	780	880	980	1,080	1,180	1,280
0.5	50.9	59.5	66.2	74.4	81.8	90.0	107	127	149	174	202	234	269	307	352
1	48.7	57.1	63.6	71.5	79.0	86.9	104	124	146	171	198	230	265	304	348
2	46.3	54.5	60.5	68.4	75.4	83.3	100	120	141	166	194	225	260	299	343
4	44.2	51.8	57.8	65.3	72.2	79.9	96.5	116	137	162	189	221	256	294	338
8	42.2	49.4	70.1	62.4	69.4	76.8	93.1	112	134	158	186	217	252	290	334
12	41.0	48.2	53.8	61.0	67.7	75.1	91.9	111	132	156	184	215	250	289	332
24	39.4	46.3	51.6	58.8	65.3	72.7	88.8	108	129	153	180	212	247	286	329

*Values of h_{rc} are for horizontal bare or insulated steel pipe with oxidized surfaces in a room at 80°F, adapted from Table 7.2 of Ref. 54, and are expressed in Btu/D-sq ft-°F.

important, h_{fc} is replaced by the coefficient of heat transfer by radiation and convection, h_{rc}. Values of h_{rc}, the sum of the coefficient of heat transfer by radiation and that by natural convection, h_c, are given in Table B.14, as adapted from McAdams.[54] These values are applicable to horizontal bare or insulated pipes in a room at 80°F. The coefficient of heat transfer by natural convection applicable under usual conditions[54] is

$$h_c = 0.13\left(\frac{T_s - T_A}{r_e}\right)^{1/4}, \quad \ldots\ldots\ldots\ldots\ldots \text{(B.62)}$$

where T_s and T_A are the surface and ambient temperatures (in °F), respectively, and r_e is the external radius of the conduit exposed to air (in ft).

Values of $h_{rc,an}$

For free convection in the annulus, the coefficient of heat transfer by radiation and convection in the annulus, $h_{rc,an}$ is given by Willhite[57] as

$$h_{rc,an} = 4.11\times 10^{-8}\left[\frac{1}{\epsilon_{ins}} + \frac{r_{ins}}{r_{ci}}\left(\frac{1}{\epsilon_{ci}} - 1\right)\right]^{-1}$$

$$\cdot F(T_{ins}, T_{ci}) + \frac{1}{r_e}\,\frac{\lambda_{a,a}}{\ln(r_{ci}/r_{ins})}, \quad \ldots \text{(B.63)}$$

where

$$F(T_{ins}, T_{ci}) = [(460 + T_{ins})^2 + (460 + T_{ci})^2]$$

$$\cdot (920 + T_{ins} + T_{ci}), \quad \ldots\ldots \text{(B.64)}$$

$$\lambda_{a,a} = 0.049\lambda_a N_{Gr}^{0.333} N_{Pr}^{0.407} \quad \ldots\ldots\ldots\ldots \text{(B.65)}$$

is the effective thermal conductivity of the air in the annulus,

$$N_{Gr} = \left\langle \frac{g_c}{g}(7.12\times 10^7)\right\rangle$$

$$\cdot \frac{(r_{ci} - r_{ins})^3\, g\, \rho_{an}^2 \beta_{an} (T_{ins} - T_{ci})}{g_c \mu_{an}^2}$$

$$\ldots\ldots\ldots\ldots\ldots\ldots\ldots\ldots \text{(B.66)}$$

is the Grashof number, and the Prandtl number N_{Pr} is given by Eq. B.56. Eq. B.63 is valid for $5\times 10^4 < N_{Gr}N_{Pr} < 2.5\times 10^8$.

In these equations, ϵ_{ins} and ϵ_{ci} are the emissivities of the inner and outer surfaces of the annulus (dimensionless), T_{ins} and T_{ci} are the temperatures at the inner and outer surfaces of the annulus (°F), β is the thermal volumetric expansion coefficient (°F^{-1}), and the fluid properties in the Grashof and Prandtl numbers are evaluated at the average temperature and pressure in the annulus. Emissivity is a measure of the radiant power emitted by a surface relative to that emitted by a perfect radiator. Emissivity values of some metallic surfaces are given in Table B.17. A more extensive list of values is given by McAdams.[54]

Examination of Eqs. B.63 and B.64 indicates that the value of $h_{rc,an}$ is strongly dependent on the temperatures on the surfaces of the annulus. To obtain these temperatures, use is made of the fact (already illustrated through Eq. 10.3) that at steady state the fractional temperature drop between two points is proportional to the fractional thermal resistance between the two points. Accordingly, in terms of the resistive components included in Eq. 10.6,

$$\frac{T_{ci} - T_A}{T_b - T_A} = \frac{1}{2\pi R_h}\left[\frac{\ln(r_{co}/r_{ci})}{\lambda_P} + \frac{\ln(r_w/r_{co})}{\lambda_{cem}}\right.$$

$$\left. + \frac{\ln(r_{Ea}/r_w)}{\lambda_{Ea}} + \frac{f(t_D)}{\lambda_E}\right], \quad \ldots \text{(B.67)}$$

or

$$T_{ci} = T_A + \frac{(T_b - T_A)}{2\pi R_h}\left[\frac{\ln(r_{co}/r_{ci})}{\lambda_P}\right.$$

$$\left. + \frac{\ln(r_w/r_{co})}{\lambda_{cem}} + \frac{\ln(r_{Ea}/r_w)}{\lambda_{Ea}} + \frac{f(t_D)}{\lambda_E}\right].$$

$$\ldots\ldots\ldots\ldots\ldots\ldots\ldots\ldots \text{(B.68)}$$

Similarly,

$$\frac{T_b - T_{ins}}{T_b - T_A} - \frac{1}{2\pi R_h}\left[\frac{1}{h_f r_i} + \frac{1}{h_{Pi} r_i}\right.$$

$$\left. + \frac{\ln(r_o/r_i)}{\lambda_P} + \frac{1}{h_{Po} r_o} + \frac{\ln(r_{ins}/r_o)}{\lambda_{ins}}\right],$$

$$\ldots\ldots\ldots\ldots\ldots\ldots\ldots\ldots \text{(B.69)}$$

or

TABLE B.15 – TUBING MINIMUM PERFORMANCE PROPERTIES[58]

Tubing Size Nom. (in.)	Tubing Size OD (in.)	Nominal Weight T & C Non-Upset (lb/ft)	Nominal Weight T & C Upset (lb/ft)	Nominal Weight Int. Jt. (lb/ft)	Grade	Wall Thickness (in.)	Inside Dia. (in.)	Threaded and Coupled Drift Dia. (in.)	Threaded and Coupled Coup. Outside Dia. Non-Upset (in.)	Threaded and Coupled Coup. Outside Dia. Upset Reg. (in.)	Threaded and Coupled Coup. Outside Dia. Upset Spec. (in.)	Integral Joint Drift Dia. (in.)	Integral Joint Box OD (in.)	Collapse Resistance (psi)	Internal Yield Pressure (psi)	Joint Yield Strength T & C Non-Upset (lb)	Joint Yield Strength T & C Upset (lb)	Joint Yield Strength Int. Jt. (lb)
3/4	1.050	1.14	1.20		H-40	.113	.824	.730	1.313	1.660				7,680	7,530	6,360	13,300	
	1.050	1.14	1.20		J-55	.113	.824	.730	1.313	1.660				10,560	10,360	8,740	18,290	
	1.050	1.14	1.20		C-75	.113	.824	.730	1.313	1.660				14,410	14,120	11,920	24,940	
	1.050	1.14	1.20		N-80	.113	.824	.730	1.313	1.660				15,370	15,070	12,710	26,610	
1	1.315	1.70	1.80	1.72	H-40	.133	1.049	.955	1.660	1.900		.955	1.550	7,270	7,080	10,960	19,760	15,970
	1.315	1.70	1.80	1.72	J-55	.133	1.049	.955	1.660	1.900		.955	1.550	10,000	9,730	15,060	27,160	21,960
	1.315	1.70	1.80	1.72	C-75	.133	1.049	.955	1.660	1.900		.955	1.550	13,640	13,270	20,540	37,040	29,940
	1.315	1.70	1.80	1.72	N-80	.133	1.049	.955	1.660	1.900		.955	1.550	14,550	14,160	21,910	39,510	31,940
1-1/4	1.660			2.10	H-40	.125	1.410					1.286	1.880	5,570	5,270			22,180
	1.660	2.30	2.40	2.33	H-40	.140	1.380	1.286	2.054	2.200		1.286	1.880	6,180	5,900	15,530	26,740	22,180
	1.660			2.10	J-55	.125	1.410					1.286	1.880	7,660	7,250			30,500
	1.660	2.30	2.40	2.33	J-55	.140	1.380	1.286	2.054	2.200		1.286	1.880	8,500	8,120	21,360	36,770	30,500
	1.660	2.30	2.40	2.33	C-75	.140	1.380	1.286	2.054	2.200		1.286	1.880	11,580	11,070	29,120	50,140	41,600
	1.660	2.30	2.40	2.33	N-80	.140	1.380	1.286	2.054	2.200		1.286	1.880	12,360	11,810	31,060	53,480	44,370
1-1/2	1.900			2.40	H-40	.125	1.650					1.516	2.110	4,920	4,610			26,890
	1.900	2.75	2.90	2.76	H-40	.145	1.610	1.516	2.200	2.500		1.516	2.110	5,640	5,340	19,090	31,980	26,890
	1.900			2.40	J-55	.125	1.650					1.516	2.110	6,640	6,330			36,970
	1.900	2.75	2.90	2.76	J-55	.145	1.610	1.516	2.200	2.500		1.516	2.110	7,750	7,350	26,250	43,970	36,970
	1.900	2.75	2.90	2.76	C-75	.145	1.610	1.516	2.200	2.500		1.516	2.110	10,570	10,020	35,800	59,960	50,420
	1.900	2.75	2.90	2.76	N-80	.145	1.610	1.516	2.200	2.500		1.516	2.110	11,280	10,680	38,180	63,960	53,780
2-1/16	2.063			3.25	H-40	1.56	1.751					1.657	2.325	5,590	5,290			35,690
	2.063			3.25	J-55	.156	1.751					1.657	2.325	7,690	7,280			49,070
	2.063			3.25	C-75	.156	1.751					1.657	2.325	10,480	9,920			66,910
	2.063			3.25	N-80	.156	1.751					1.657	2.325	11,180	10,590			71,370
2-3/8	2.375	4.00			H-40	.167	2.041	1.947	2.875					5,230	4,920	30,130		
	2.375	4.60	4.70		H-40	.190	1.995	1.901	2.875	3.063	2.910			5,890	5,600	35,960	52,170	
	2.375	4.00			J-55	.167	2.041	1.947	2.873					7,190	6,770	41,430		
	2.375	4.60	4.70		J-55	.190	1.995	1.901	2.875	3.063	2.910			8,100	7,700	49,450	71,730	
	2.375	4.00			C-75	.167	2.041	1.947	2.875					9,520	9,230	56,500		
	2.375	4.60	4.70		C-75	.190	1.995	1.901	2.875	3.063	2.910			11,040	10,500	67,430	97,820	
	2.375	5.80	5.95		C-75	.254	1.867	1.773	2.875	3.063	2.910			14,330	14,040	96,560	126,940	
	2.375	4.00			N-80	.167	2.041	1.947	2.875					9,980	9,840	60,260		
	2.375	4.60	4.70		N-80	.190	1.995	1.901	2.875	3.063	2.910			11,780	11,200	71,930	104,340	
	2.375	5.80	5.95		N-80	.254	1.867	1.773	2.875	3.063	2.910			15,280	14,970	102,990	135,400	
	2.375	4.60	4.70		P-105	.190	1.995	1.901	2.875	3.063	2.910			15,460	14,700	94,410	136,940	
	2.375	5.80	5.95		P-105	.254	1.867	1.773	2.875	3.063	2.910			20,060	19,650	135,180	177,710	
2-7/8	2.875	6.40	6.50		H-40	.217	2.441	2.347	3.500	3.668	3.460			5,580	5,280	52,780	72,480	
	2.875	6.40	6.50		J-55	.217	2.441	2.347	3.500	3.668	3.460			7,680	7,260	72,580	99,660	
	2.875	6.40	6.50		C-75	.217	2.441	2.347	3.500	3.668	3.460			10,470	9,910	90,070	136,900	
	2.875	8.60	8.70		C-75	.308	2.259	2.165	3.500	3.668	3.460			14,350	14,060	149,360	186,290	
	2.875	6.40	6.50		N-80	.217	2.441	2.347	3.500	3.668	3.460			11,160	10,570	105,570	144,960	
	2.875	8.60	8.70		N-80	.308	2.259	2.165	3.500	3.668	3.460			15,300	15,000	159,310	198,710	
	2.875	6.40	6.50		P-105	.217	2.441	2.347	3.500	3.000	3.460			14,010	13,870	138,560	190,260	
	2.875	8.60	8.70		P-105	.308	2.259	2.165	3.500	3.668	3.460			20,090	19,690	209,100	260,810	
3-1/2	3.500	7.70			H-40	.216	3.068	2.943	4.250					4,630	4,320	65,070		
	3.500	9.20	9.30		H-40	.254	2.992	2.867	4.250	4.500	4.180			5,380	5,080	79,540	103,610	
	3.500	10.20			H-40	.289	2.922	2.797	4.250					6,060	5,780	92,550		
	3.500	7.70			J-55	.216	3.068	2.943	4.250					5,970	5,940	89,470		
	3.500	9.20	9.30		J-55	.254	2.992	2.867	4.250	4.500	4.180			7,400	6,980	109,370	142,460	
	3.500	10.20			J-55	.289	2.922	2.797	4.250					8,330	7,950	127,250		
	3.500	7.70			C-75	.216	3.068	2.943	4.250					7,540	8,100	122,010		
	3.500	9.20	9.30		C-75	.254	2.992	2.867	4.250	4.500	4.180			10,040	9,520	149,140	194,260	
	3.500	10.20			C-75	.289	2.922	2.797	4.250					11,360	10,840	173,530		
	3.500	12.70	12.95		C-75	.375	2.750	2.625	4.250	4.500	4.180			14,350	14,060	230,990	276,120	
	3.500	7.70			N-80	.216	3.068	2.943	4.250					7,870	8,640	130,140		
	3.500	9.20	9.30		N-80	.254	2.992	2.867	4.250	4.500	4.180			10,530	10,160	159,090	207,220	
	3.500	10.20			N-80	.289	2.922	2.797	4.250					12,120	11,560	185,100		
	3.500	12.70	12.95		N-80	.375	2.750	2.625	4.250	4.500	4.180			15,310	15,000	246,390	294,530	
	3.500	9.20	9.30		P-105	.254	2.992	2.867	4.250	4.500	4.180			13,050	13,330	208,800	271,970	
	3.500	12.70	12.95		P-105	.375	2.750	2.625	4.250	4.500	4.180			20,090	19,690	323,390	386,570	
4	4.000	9.50			H-40	.226	3.548	3.423	4.750					4,060	3,960	72,000		
	4.000		11.00		H-40	.262	3.476	3.351		5.000				4,900	4,580		123,070	
	4.000	9.50			J-55	.226	3.548	3.423	4.750					5,110	5,440	99,010		
	4.000		11.00		J-55	.262	3.476	3.351		5.000				6,590	6,300		169,220	
	4.000	9.50			C-75	.226	3.548	3.423	4.750					6,350	7,420	135,010		
	4.000		11.00		C-75	.262	3.476	3.351		5.000				8,410	8,600		230,750	
	4.000	9.50			N-80	.226	3.548	3.423	4.750					6,590	7,910	144,010		
	4.000		11.00		N-80	.262	3.376	3.351		5.000				8,800	9,170		246,140	
4-1/2	4.500	12.60	12.75		H-40	.271	3.958	3.833	5.200	5.563				4,500	4,220	104,360	144,020	
	4.500	12.60	12.75		J-55	.271	3.958	3.833	5.200	5.563				5,720	5,800	143,500	198,030	
	4.500	12.60	12.75		C-75	.271	3.958	3.833	5.200	5.563				7,200	7,900	195,680	270,040	
	4.500	12.60	12.75		N-80	.271	3.958	3.833	5.200	5.563				7,500	8,430	208,730	288,040	

TABLE B.16 – CASING MINIMUM PERFORMANCE PROPERTIES[58]

OD Size (in.)	Dimensions: Weight Per Foot Nom. (Lbs.)	Wall (in.)	ID (in.)	Drift (in.)	Coupling OD (in.)	Minimum Collapse Pressure (psi)				Minimum Internal Yield Pressure (psi)				Joint Yield Strength (1000 Lbs.): Short Thread				Long Thread		
						H40	J55	N80	P110	H40	J55	N80	P110	H40	J55	N80	P110	J55	N80	P110
4-1/2	9.50	.205	4.090	3.965	5.000	2770	3310	...	...	3190	4380	...	...	77	101	...	...	...	...	...
	11.60	.250	4.000	3.875	5.000	...	4960	6350	7560	...	5350	7780	10690	...	154	...	...	162	223	279
	13.50	.290	3.920	3.795	5.000	...	...	8540	10670	...	...	9020	12410	...	...	...	...	...	270	338
	15.10	.337	3.826	3.701	5.000	...	...	...	14320	...	...	...	14420	...	...	...	...	...	...	406
5	11.50	.220	4.560	4.435	5.563	...	3060	...	...	...	4240	...	...	...	133	...	...	...	...	...
	13.00	.253	4.494	4.369	5.563	...	4140	...	...	...	4870	...	...	...	169	...	...	182	...	...
	15.00	.296	4.408	4.283	5.563	...	5550	7250	8830	...	5700	8290	11400	...	207	...	...	223	311	388
	18.00	.362	4.276	4.151	5.563	...	...	10490	13450	...	...	10140	13940	...	...	...	...	...	396	495
5-1/2	14.00	.244	5.012	4.887	6.050	2630	3120	...	...	3110	4270	...	...	130	172	...	...	...	...	...
	15.50	.275	4.950	4.825	6.050	...	4040	...	...	...	4810	...	...	...	202	...	...	217	...	...
	17.00	.304	4.892	4.767	6.050	...	4910	6280	7460	...	5320	7740	10640	...	229	...	...	247	348	445
	20.00	.361	4.778	4.653	6.050	...	...	8830	11080	...	...	9190	12640	...	...	...	...	...	428	548
	23.00	.415	4.670	4.545	6.050	...	...	11160	14520	...	...	9880	13580	...	...	...	...	...	502	643
6-5/8	20.00	.288	6.049	5.924	7.390	2520	2970	...	...	3040	4180	...	...	184	245	...	...	266	...	...
	24.00	.352	5.921	5.796	7.390	...	4560	5760	6710	...	5110	7440	10230	...	314	...	...	340	481	641
	28.00	.417	5.791	5.666	7.390	...	...	8170	10140	...	...	8810	12120	...	...	...	...	...	586	781
	32.00	.475	5.675	5.550	7.390	...	...	10320	13200	...	...	10040	13800	...	...	...	...	...	677	904
7	17.00	.231	6.538	6.413	7.656	1450	...	...	...	2310	...	...	...	122	...	...	...	...	...	...
	20.00	.272	6.456	6.331	7.656	1980	2270	...	...	2720	3740	...	...	176	234	...	...	...	...	...
	23.00	.317	6.366	6.241	7.656	...	3270	3830	...	...	4360	6340	...	...	284	...	...	313	442	...
	26.00	.362	6.276	6.151	7.656	...	4320	5410	6210	...	4980	7240	9960	...	334	...	...	367	519	693
	29.00	.408	6.184	6.059	7.656	...	...	7020	8510	...	...	8160	11220	...	...	...	...	...	597	797
	32.00	.453	6.094	5.969	7.656	...	...	8600	10760	...	...	9060	12460	...	...	...	...	...	672	897
	35.00	.498	6.004	5.879	7.656	...	...	10180	13010	...	...	9240	12700	...	...	...	...	...	746	996
	38.00	.540	5.920	5.795	7.656	...	...	11390	15110	...	...	9240	12700	...	...	...	...	...	814	1087
7-5/8	24.00	.300	7.025	6.900	8.500	2040	...	...	...	2750	...	...	...	212	...	...	...	...	...	...
	26.40	.328	6.969	6.844	8.500	...	2890	3400	...	...	4140	6020	...	...	315	...	...	346	490	...
	29.70	.375	6.875	6.750	8.500	...	...	4790	5340	...	...	6890	9470	...	...	...	...	...	575	769
	33.70	.430	6.765	6.640	8.500	...	...	6560	7850	...	...	7900	10860	...	...	...	...	...	674	901
	39.00	.500	6.625	6.500	8.500	...	...	8810	11060	...	...	9180	12620	...	...	...	...	...	798	1066
8-5/8	24.00	.264	8.097	7.972	9.625	...	1370	...	...	...	2950	...	...	...	244	...	...	...	...	...
	28.00	.304	8.017	7.892	9.625	1640	...	...	...	2470	...	...	...	233	...	...	...	...	...	...
	32.00	.352	7.921	7.796	9.625	2210	2530	...	...	2860	3930	...	...	279	372	...	...	417	...	...
	36.00	.400	7.825	7.700	9.625	...	3450	4100	...	...	4460	6490	...	...	434	...	...	486	688	...
	40.00	.450	7.725	7.600	9.625	...	...	5520	6380	...	...	7300	10040	...	...	...	...	...	788	1055
	44.00	.500	7.625	7.500	9.625	...	...	6950	8400	...	...	8120	11160	...	...	...	...	...	887	1186
	49.00	.557	7.511	7.386	9.625	...	...	8570	10720	...	...	9040	12430	...	...	...	...	...	997	1335
9-5/8	32.30	.312	9.001	8.845	10.625	1400	...	...	...	2270	...	...	...	254	...	...	...	...	...	...
	36.00	.352	8.921	8.765	10.625	1740	2020	...	...	2560	3520	...	...	294	394	...	...	453	...	...
	40.00	.395	8.835	8.679	10.625	...	2570	3090	...	...	3950	5750	...	...	452	...	...	520	737	...
	43.50	.435	8.755	8.599	10.625	...	...	3810	4430	...	...	6330	8700	...	...	...	...	...	825	1106
	47.00	.472	8.681	8.525	10.625	...	...	4750	5310	...	...	6870	9440	...	...	...	...	...	905	1213
	53.50	.545	8.535	8.379	10.625	...	...	6620	7930	...	...	7930	10900	...	...	...	...	...	1062	1422
10-3/4	32.75	.279	10.192	10.036	11.750	870	...	...	...	1820	...	...	...	205	...	...	...	...	...	...
	40.50	.350	10.050	9.894	11.750	1420	1580	...	...	2280	3130	...	...	314	420	...	...	...	...	...
	45.50	.400	9.950	9.794	11.750	...	2090	...	...	...	3580	...	...	...	493	...	...	...	...	...
	51.00	.450	9.850	9.694	11.750	...	2700	3220	3670	...	4030	5860	8060	...	565	804	1080	...	...	...
	55.50	.495	9.760	9.604	11.750	...	...	4020	4630	...	...	6450	8860	...	...	895	1203	...	...	...
	60.70	.545	9.660	9.504	11.750	...	...	...	5860	...	...	...	9760	...	...	...	1338	...	...	...
	65.70	.595	9.560	9.404	11.750	...	...	...	7490	...	...	...	10650	...	...	...	1472	...	...	...
11-3/4	42.00	.333	11.084	10.928	12.750	1070	...	...	...	1980	...	...	...	307	...	...	...	...	...	...
	47.00	.375	11.000	10.844	12.750	...	1510	...	...	...	3070	...	...	...	477	...	...	...	...	...
	54.00	.435	10.880	10.724	12.750	...	2070	...	...	...	3560	...	...	...	568	...	...	...	...	...
	60.00	.489	10.772	10.616	12.750	...	2660	3180	...	...	4010	5830	...	...	649	924	...	...	...	...
13-3/8	48.00	.330	12.715	12.559	14.375	740	...	...	...	1730	...	...	...	322	...	...	...	...	...	...
	54.50	.380	12.615	12.459	14.375	...	1130	...	...	...	2730	...	...	...	514	...	...	...	...	...
	61.00	.430	12.515	12.359	14.375	...	1540	...	...	...	3090	...	...	...	595	...	...	...	...	...
	68.00	.480	12.415	12.259	14.375	...	1950	...	...	...	3450	...	...	...	675	...	...	...	...	...
	72.00	.514	12.347	12.191	14.375	...	...	2670	...	...	...	5380	...	...	...	1040	...	...	...	...
16	65.00	.375	15.250	15.062	17.000	630	...	...	...	1640	...	...	...	439	...	...	...	...	...	...
	75.00	.438	15.124	14.936	17.000	...	1020	...	...	...	2630	...	...	...	710	...	...	...	...	...
	84.00	.495	15.010	14.822	17.000	...	1410	...	...	...	2980	...	...	...	817	...	...	...	...	...
18-5/8	87.50	.435	17.755	17.567	19.625	630	630	...	...	1630	2250	...	...	559	754	...	...	...	...	...
20	94.00	.438	19.124	18.936	21.000	520	520	...	...	1530	2110	...	...	581	784	...	...	907	...	...
	106.50	.500	19.000	18.812	21.000	...	770	...	...	...	2410	...	...	...	913	...	...	1057	...	...
	133.00	.635	18.730	18.542	21.000	...	1500	...	...	...	3060	...	...	...	1192	...	...	1380	...	...

$$T_{ins} = T_b - \frac{(T_b - T_A)}{2\pi R_h}\left[\frac{1}{h_f r_i} + \frac{1}{h_{Pi} r_i} + \frac{\ln(r_o/r_i)}{\lambda_P} + \frac{1}{h_{Po} r_o} + \frac{\ln(r_{ins}/r_o)}{\lambda_{ins}}\right]. \quad \text{(B.70)}$$

Since the four quantities R_h, T_{ins}, T_{ci} and $f(t_D)$ are interrelated by nonlinear expressions, the corresponding values of these quantities are determined by a successive trial-and-error procedure. An example calculation of such a procedure, patterned after Willhite,[57] is given as Example 10.2 in Chap. 10.

Values of $h_{rfc,an}$

The coefficient of heat transfer by radiant and forced convection of a gas in the annular space is $h_{rfc,an}$, approximated by

$$h_{rfc,an} = 4.11 \times 10^{-8}\left[\frac{1}{\epsilon_{ins}} + \frac{r_{ins}}{r_{ci}}\left(\frac{1}{\epsilon_{ci}} - 1\right)\right]^{-1} \cdot F(T_{ins}, T_{ci}) + \frac{1}{2} h_{f,an}. \quad \text{(B.71)}$$

The last term represents the contribution due to thermal resistance across the two films in the annulus: one at r_{ins} and one at r_{ci}. The film resistances are considered to have the same value and are calculated at the average conditions in the annulus, just like the last term in Eq. B.63. In an annular space, the film coefficient of heat transfer, $h_{f,an}$, for gas is calculated from Eq. B.59, but using the hydraulic (also known as equivalent) radius of the annulus instead of the pipe radius. Since the hydraulic radius is defined as twice the cross-sectional area divided by the total perimeter contacted by the flow, the hydraulic radius of the annulus is

$$r_H = 2\pi(r_{ci}^2 - r_{ins}^2)/2\pi(r_{ci} + r_{ins}) = r_{ci} - r_{ins}. \quad \text{(B.72)}$$

Accordingly, Eq. B.59 can be used to calculate $h_{f,an}$ when $i_{sc}\rho_{sc} > 192\mu_{g,an} r_H$.

B.9 Tubing and Casing Properties

Standard data on the weights and diameters of tubing and casing likely to be used in thermal projects are given in Tables B.15 and B.16, respectively.

B.10 Miscellaneous

Although not required for the examples given in the text, *K*-values (Chap. 3) are required for studies involving components present in more than one phase. Extensive information on measurements and correlations of *K*-values is given in Refs. 5, 6, 9, and 59. An example of the K_o values for hexane in Oklahoma City field crude is given in Fig. 3.6.

Recently, a significant contribution was made by Lee *et al.*,[60] who developed a procedure for determining the number of components that should be used to represent a crude in thermal numerical simulators.

TABLE B.17 – APPROXIMATE EMISSIVITIES OF SOME METALS

Metal	Temperature (°F)	Emissivity
Aluminum		
Oxidized or unpolished	400 to 1,100	0.11 to 0.19
Unoxidized or polished	400 to 1,100	0.04 to 0.08
Iron, steel		
Unoxidized or polished	350 to 1,200	0.05 to 0.30
Oxidized or unpolished	400 to 1,100	0.75 to 0.95
Stainless steel	75 to 210	0.07 to 0.30
Polished	450 to 1,600	0.5 to 0.6

Emissivities are reported in Table B.17. These are used in calculating radiation heat transfer, which is discussed in Chap. 10 and in Section B.8.

A correlation for the longitudinal dispersion coefficient in terms of rock and fluid properties and fluid fluxes is given by Eq. 3.35 and shown in Fig. 3.4. Perkins and Johnson[61] also provide correlations from transverse dispersion.

References

1. Matthews, C.S. and Russell, D.G.: *Pressure Buildup and Flow Tests in Wells*, Monograph Series, SPE, Dallas (1967) **1**.
2. Craig, F.F. Jr.: *The Reservoir Engineering Aspects of Waterflooding*, Monograph Series, SPE, Dallas (1971) **3**.
3. Earlougher, Robert C. Jr.: *Advances in Well Test Analysis*, Monograph Series, SPE, Dallas (1977) **5**.
4. Farouq Ali, S.M.: *Oil Recovery by Steam Injection*, Producers Publishing Co. Inc., Bradford, PA (1970).
5. *Engineering Data Book*, ninth edition, Gas Processors Suppliers Assn., Tulsa (1972) Sec. 16.
6. Edminster, W.C.: *Applied Hydrocarbon Thermodynamics*, Gulf Publishing Co., Houston (1961).
7. Trube, A.S.: "Compressibility of Undersaturated Hydrocarbon Reservoir Fluids," *Trans.*, AIME (1956) **210**, 341-344.
8. "Measuring, Sampling, and Testing Crude Oil," API Standard 2500, API, Dallas; reproduced in *Petroleum Production Handbook*, McGraw-Hill Book Co. Inc., New York City (1962) **1**, Chap. 16.
9. Brown, G., Katz, D.L., Oberfell, G.G., and Alden, R.C.: "Natural Gasoline and the Volatile Hydrocarbons," Natural Gasoline Assn. of America, Tulsa (1948).
10. Standing, M.B.: *Volumetric and Phase Behavior of Oil Field Hydrocarbon Systems*, first edition, Reinhold Publishing Corp., New York City (1952).
11. Cronquist, C.: "Dimensionless PVT Behavior of Gulf Coast Reservoir Oils," *J. Pet. Tech.* (May 1973) 538-542.
12. Standing, M.B.: *Volumetric and Phase Behavior of Oil Field Hydrocarbon Systems*, second edition, SPE, Dallas (1977).
13. Standing, M.B. and Katz, D.L.: "Density of Natural Gases," *Trans.*, AIME (1942) **146**, 140-149.
14. Dodson, C.R. and Standing, M.B.: "Pressure-Volume-Temperature and Solubility Relations for Natural-Gas-Water Mixtures," *Drill. and Prod. Prac.*, API (1944) 173-179.
15. *Steam Tables, Properties of Saturated and Superheated Steam*, Combustion Engineering Inc., third edition, 23rd Printing, Windsor, CT (1940).
16. Keenan, J.H. and Keyes, F.G.: *Thermodynamic Properties of Steam*, John Wiley and Sons Inc., New York City (1936).
17. Helgeson, H.C.: "Thermodynamics of Hydrothermal Systems at Elevated Temperatures and Pressures," *American J. Science* (1967) **267**, 729-804.
18. *Handbook of Physical Constants*, S.P. Clark Jr. (ed.), revised edition, Geological Soc. of America (1966) Memoir 97.
19. Grim, R.F.: *Clay Mineralogy*, McGraw-Hill Book Co., Inc., New York City (1968).
20. Hall, Howard N.: "Compressibility of Reservoir Rocks," *Trans.*, AIME (1953) **198**, 309-311.
21. van der Knaap, W.: "Nonlinear Behavior of Elastic Porous

Media," *Trans.*, AIME (1959) **216**, 179-187.
22. Newman, G.H.: "Pore-Volume Compressibility of Consolidated, Friable, and Unconsolidated Reservoir Rocks Under Hydrostatic Loading," *J. Pet. Tech.* (Feb. 1973) 129-134.
23. Ramey, H.J. Jr.: "Rapid Method of Estimating Reservoir Compressibility," *J. Pet. Tech.* (April 1964) 447-454; *Trans.*, AIME, **231**.
24. Long, G. and Chierici, G.: "Salt Content Changes Compressibility of Reservoir Brines," *Pet. Eng.* (July 1961) B-25 through B-31.
25. Trube, Albert S.: "Compressibility of Natural Gases," *Trans.*, AIME (1957) **210**, 355-357.
26. Braden, W.B.: "A Viscosity-Temperature Correlation at Atmospheric Pressure for Gas-Free Oils," *J. Pet. Tech.* (Nov. 1966) 1487-1490; *Trans.*, AIME, **237**.
27. Chew, Ju-Nam and Connally, Carl A. Jr.: "A Viscosity Correlation for Gas-Saturated Crude Oils," *Trans.*, AIME (1959) **216**, 264-272.
28. Davidson, L.B.: "The Effect of Temperature on the Permeability Ratio of Different Fluid Pairs in Two-Phase Systems," *J. Pet. Tech.* (Aug. 1969) 1037-1046.
29. *Chemical Engineers' Handbook*, R.H. Perry and C.H. Chilton (eds.), fifth edition, McGraw-Hill Book Co. Inc., New York City (1973).
30. Beal, C.: "The Viscosity of Air, Water, Natural Gas, Crude Oil, and Its Associated Gases at Oil-Field Temperatures and Pressures," *Trans.*, AIME (1946) **165**, 94-115.
31. Rojas, G.G., Barrios, T., Scudiero, B., and Ruiz, J.: "Rheological Behavior of Extra-Heavy Crude Oils from the Orinoco Oil Belt," *Proc.*, Oil Sands of Canada-Venezuela, CIM Special Vol. 17 (1977) 284-302.
32. Scudiero, B. and Rojas, G.G.: "Efecto de la temperatura sobre las propiedades reológicas de varios crudos de la Faja Petrolífera del Orinoco," *Proc.*, Simposio Sobre Crudos Pesados, U. del Zulia, Inst. de Investigaciones Petroleras, Maracaibo, Venezuela (1974) 524-550.
33. Carr, N.L., Kobayashi, R., and Burrows, D.B.: "Viscosity of Hydrocarbon Gases Under Pressure," *Trans.*, AIME (1954) **201**, 264-272.
34. Touloukian, Y.S., Kirby, R.K., Taylor, R.E., and Lee, T.Y.R.: "Viscosity," *Thermophysical Properties of Matter*, IFI/Plenum, New York City (1970) **11**.
35. Eckert, E.R.G. and Drake, R.M. Jr.: *Heat and Mass Transfer*, McGraw-Hill Book Co. Inc., New York City (1959).
36. Kesler, M.G. and Lee, B.I.: "Improve Prediction of Enthalpy of Fractions," *Hydrocarbon Processing* (March 1976) 153-158.
37. *Technical Data Book*, API, New York City (1964).
38. Tarakad, R.R. and Danner, R.P.: "A Comparison of Enthalpy Prediction Methods," *AIChE J.* (March 1976) **22**, No. 2, 409-411.
39. Gambill, W.R.: "You Can Predict Heat Capacities," *Chemical Engineering* (June 1957).
40. Touloukian, Y.S., Kirby, R.K., Taylor, R.E., and Lee, T.Y.R.: "Specific Heat – Nonmetallic Liquids and Gases," *Thermophysical Properties of Matter*, IFI/Plenum, New York City (1970) **6**.
41. Somerton, W.H.: "Some Thermal Characteristics of Porous Rocks," *Trans.*, AIME (1958) **213**, 375-378.
42. Touloukian, Y.S., Kirby, R.K., Taylor, R.E., and Lee, T.Y.R.: "Specific Heat – Nonmetallic Solids," *Thermophysical Properties of Matter*, IFI/Plenum, New York City (1970) **5**.
43. Rohsenow, W.M. and Choi, H.Y.: *Heat, Mass and Momentum Transfer*, Prentice-Hall Inc., Englewood Cliffs, NJ (1961).
44. Touloukian, Y.S., Kirby, R.K., Taylor, R.E., and Lee, T.Y.R.: "Thermal Conductivity – Nonmetallic Liquids and Gases," *Thermophysical Properties of Matter*, IFI/Plenum, New York City (1970) **3**.
45. Cragoe, C.S.: Miscellaneous Publication No. 97, U.S. Bureau of Standards (1929).
46. Lenoir, J.M.: "Effect of Pressure on Thermal Conductivity of Liquids," *Petroleum Refiner* (1956) **36**, No. 8, 162-164.
47. Lenoir, J.M. and Comings, E.W.: "Thermal Conductivity of Gases, Measurements at High Pressure," *Chem. Eng. Prog.* (1951) **47**, 223-231.
48. Somerton, W.H., Keese, J.A., Chu, S.L.: "Thermal Behavior of Unconsolidated Oil Sands," *Soc. Pet. Eng. J.* (Oct. 1974) 513-521.
49. Anand, J., Somerton, W.H., Gomaa, E.: "Predicting Thermal Conductivities of Formations From Other Known Properties," *Soc. Pet. Eng. J.* (Oct. 1973) 267-273.
50. Somerton, W.H. and Boozer, G.D.: "Thermal Characteristics of Porous Rocks at Elevated Temperatures," *J. Pet. Tech.* (June 1960) 77-81.
51. Horai, Ki-iti: "Thermal Conductivity of Rock-Forming Minerals," *J. Geophys. Research* (1971) **76**, No. 5, 1278-1308.
52. Green, D.W.: "Heat Transfer with Flowing Fluid Through Porous Media," PhD dissertation, U. of Oklahoma, Norman (1963).
53. Bear, J.: *Dynamics of Fluids in Porous Media*, American Elsevier Publishing Co., New York City (1972).
54. McAdams, W.H.: *Heat Transmission*, third edition, McGraw-Hill Book Co. Inc., New York City (1954).
55. Nelson, W.L.: *Petroleum Refinery Engineering*, fourth edition, McGraw-Hill Book Co. Inc., New York City (1958).
56. Jakob, M.: *Heat Transfer*, John Wiley and Sons Inc., New York City (1950) **1**.
57. Willhite, G.P.: "Overall Heat Transfer Coefficients in Steam and Hot Water Injection Wells," *J. Pet. Tech.* (May 1967) 607-615.
58. Allen, T.O. and Roberts, A.P.: *Production Operations*, Oil and Gas Consultants Intl. Inc., Tulsa (1978) **1**.
59. Katz, D.L., Cornell, D., Kobayashi, R., Poettmann, F.H., Elenbaas, J.R., and Weinaug, C.F.: *Handbook of Natural Gas Engineering*, McGraw-Hill Book Co. Inc., New York City (1959).
60. Lee, S.T. *et al.*: "Experimental and Theoretical Studies on the Fluid Properties Required for Stimulation of Thermal Processes," *Soc. Pet. Eng. J.* (Oct. 1981) 535–50.
61. Perkins, T.K. and Johnston, O.C.: "A Review of Diffusion and Dispersion in Porous Media," *Soc. Pet. Eng. J.* (1963) 70-84; *Trans.*, AIME, **228**.

Nomenclature

Symbol	Description	Dimensions	Customary Units	Metric Units	Customary to Metric Conversion Factors
a_R	Air required to burn through formation	–	Mscf/cu ft	std m^3/m^3	1.011 285 E+00[a]
a_R^*	Air injected to burn through formation	–	Mscf/cu ft	std m^3/m^3	1.011 285 E+00[a]
a_1, a_2	Constants defined by Eq. 5.16	–	–	–	–
A	Area	L^2	sq ft	m^2	9.290 304* E−02
A_c	Pre-exponential constant appearing in Eq. 8.13	$m^n/L^n t^{2n+1}$	sec^{-1}atm^{-n}	Same	1.0* E+00
A_P	Area of well pattern	L^2	acre	ha	4.046 856 E−01
A_ψ	Cumulative area long a streamtube	L^2	acre	ha	4.046 856 E−01
b	Parameter defined by Eq. 9.10	–	–	–	–
B_g	Gas formation volume factor	–	RB/scf	res m^3/std m^3	5.551 931 E+00[a]
B_o	Oil formation volume factor	–	RB/STB	res m^3/stock-tank m^3	1.0* E+00
B_w	Water formation volume factor	–	RB/STB	res m^3/stock-tank m^3	1.0* E+00
c	Compressibility (Chaps. 4 and 13))	Lt^2/m	psi^{-1}	kPa^{-1}	1.450 377 E−01
c	Concentration (Chaps. 3 and 8)	m/L^3	lbm/cu ft	kg/m^3	1.601 846 E+01
		–	wt%	kg/kg	1.0* E−02
		–	vol%	m^3/m^3	1.0* E−02
		N/L^3	lbm mol/cu ft	kmol/m^3	1.601 846 E+01
		m/N	lbm/lbm mol	kg/kmol	1.0* E+00
c_{CO_2}	Carbon dioxide concentration	–	lbm mol/lbm mol	mol/mol	1.0* E+00
c_g	Gas compressibility	Lt^2/m	psi^{-1}	kPa^{-1}	1.450 377 E−01
c_{N_2}	Nitrogen concentration	–	lbm mol/lbm mol	mol/mol	1.0* E+00
c_o	Oil compressibility	Lt^2/m	psi^{-1}	kPa^{-1}	1.450 377 E−01
c_{oa}	Apparent oil compressibility	Lt^2/m	psi^{-1}	kPa^{-1}	1.450 377 E−01
c_{O_2}	Oxygen concentration	–	lbm mol/lbm mol	mol/mol	1.0* E+00
c_{pr}	Pseudoreduced compressibility	–	–	–	–
c_w	Water compressibility or compressibility of aqueous phase	Lt^2/m	psi^{-1}	kPa^{-1}	1.450 377 E−01
c_{wa}	Apparent compressibility of water	Lt^2/m	psi^{-1}	kPa^{-1}	1.450 377 E−01
C	Isobaric specific heat	m/Lt^2T	Btu/lbm-°F	kJ/kg·K	4.186 8* E+00

[a]The standard cubic foot (scf) and barrel (bbl) are measured at 60°F and 14.696 psia; the cubic meter is measured at 15°C and 100 kPa (1 bar).
[b]Money.
[c]Not applicable.
[d]This is not a factor. Perform the operation indicated.

Symbol	Description	Dimensions	Customary Units	Metric Units	Customary to Metric Conversion Factors
C_j	Capital investment in *j*th year (Chaps. 12 and 14)	M[b]	Various	NA[c]	NA[c]
C_o	Isobaric specific heat of oil	m/Lt^2T	Btu/lbm-°F	kJ/kg·K	4.186 8* E+00
C_w	Isobaric specific heat of water	m/Lt^2T	Btu/lbm-°F	kJ/kg·K	4.186 8* E+00
C_V	Heat capacity at constant volume	m/Lt^2T	Btu/lbm-°F	kJ/kg·K	4.186 8* E+00
C_σ	Isobaric specific heat of solids in reservoir matrix	m/Lt^2T	Btu/lbm-°F	kJ/kg·K	4.186 8* E+00
d	Grain diameter*	L	cm	cm	1
D	Dispersion coefficient (Chap. 3)	L^2/t	sq ft/D	m^2/D	9.290 304* E−02
D	Depth below surface (Chap. 10)	L	ft	m	3.048* E−01
D_c	Depth at which steam quality becomes zero in an injection well (Chaps. 10 and 11)	L	ft	m	3.048* E−01
D_M	Molecular diffusivity (Chap. 3)	L^2/t	sq ft/D	m^2/d	9.290 304* E−02
D_1	Parameter defined by Eq. 10.16	L	ft	m	3.048* E−01
e	Internal energy per unit mass (Chap. 3)	L^2/t^2	Btu/lbm	kJ/kg	2.326 000 E+00
e_t	Total energy per unit mass	L^2/t^2	Btu/lbm	kJ/kg	2.326 000 E+00
erfc(x)	Complementary error function, defined by Eq. 5.7	–	–	–	–
E	Activation energy (Chap. 8)	L^2/t^2	Btu/lbm mol	kJ/kmol	2.326 000 E+00
E	Efficiency	–	–	–	–
E_{cb}	Fraction of oil displaced from burned volume produced during wet combustion	–	–	–	–
E_c	Fraction of oil displaced that is produced	–	–	–	–
E_{comp}	Volumetric compression efficiency, fraction	–	–	–	–
E_h	Heat efficiency, the fraction of the injected heat present in the reservoir	–	–	–	–
$E_{h,s}$	Steam zone heat efficiency, the fraction of the injected heat present within the steam zone	–	–	–	–
$E_{s,V}$	Vertical sweep efficiency of steam zone (Fig. 7.10)	–	–	–	–
E_R	Oil recovery efficiency, fraction	–	–	–	–
E_{Rhw}	Potential fractional recovery of the initial oil in place due to a hot waterflood	–	–	–	–
E_{O_2}	Oxygen consumption efficiency, fraction	–	–	–	–
$\bar{E}_{O_2}$	Average oxygen consumption efficiency, fraction	–	–	–	–

f	Volume fraction of noncondensable gas in gas phase	–	–	–	–
f_{cy}	Fractional cylinder clearance	–	–	–	–
f_{hv}	Fraction of heat injected in vapor form	–	–	–	–
f_{HD}	Function, given in Fig. 9.4, representing heat lost horizontally from steam zone during cyclic steam injection	–	–	–	–
f_o	Fractional oil production or fractional oil cut	–	–	–	–
f_{pD}	Function, defined by Eq. 9.3, representing heat loss via produced fluids	–	–	–	–
f_q	Fraction volume content of quartz in sand	–	–	–	–
f_s	Steam quality or the weight fraction of dry steam in a mixture of steam and water	–	–	–	–
f_{sdh}	Quality of injected steam at downhole conditions	–	–	–	–
f_{sb}	Quality of steam at boiler outlet	–	–	–	–
$f(t_D)$	Time function reflecting the effective changing thermal resistance of the earth	–	–	–	–
$f_{\dot{Q}}$	Fraction of heat generated in situ transferred downstream of combustion front	–	–	–	–
f_{VD}	Function, given in Fig. 9.4, representing heat lost vertically from steam zone during cyclic steam injection	–	–	–	–
f_w	Fractional flow of water or fractional water cut	–	–	–	–
f'_w	Derivative of the f_w vs. S_w curve	–	–	–	–
f_3, f_4	Function, defined in Eq. 5.14	–	–	–	–
F_{ao}	Air injected per unit oil produced, air/oil ratio	–	Mscf/bbl	std m^3/m^3	1.801 175 E+02[a]
F_{comp}	Total or overall compression ratio	–	–	–	–
F_G	Geometric factor (Eq. 4.6)	–	–	–	–
F_I	Inhomogeneity factor (Fig. 3.4)	–	–	–	–
$F(n_{CS})$	Function discussed below Eq. 11.7	–	–	–	–
F_o	Fraction of bulk volume occupied by displaceable oil, defined by Eq. 12.7	–	–	–	–
F_{os}	Oil produced per unit steam injected or oil/steam ratio	–	bbl/bbl	m^3/m^3	1.0* E+00
F_p	Function defined by Eq. 13.2	–	–	–	–
F_R	Formation resistivity factor	–	–	–	–
F_{SC}	Stage compression ratio	–	–	–	–

Symbol	Description	Dimensions	Customary Units	Metric Units	Customary to Metric Conversion Factors
F_{so}	Steam injected per unit of oil produced or steam/oil ratio	–	bbl/bbl	m^3/m^3	1.0* E+00
F_v	Ratio of convection-front velocity to combustion-front velocity	–	–	–	
F_{wa}	Water/air ratio	–	bbl/Mscf	m^3/std m^3	5.551 931 E−03[a]
$F_{\Delta h}$	Steam/reservoir enthalpy ratio, defined by Eq. 12.8	–	–	–	–
F_μ	Ratio of viscosities defined by Eq. 6.8	–	–	–	–
F_η	Ratio of vertical dispersive oxygen flux to horizonal convective heat flux (Fig. 5.16)	–	–	–	–
F_1, F_2	Constants defined in Table 9.2	–	–	–	–
g	Acceleration constant due to gravity	L/t^2	32.17405 ft/s^2	9.806650 m/s^2	3.048* E−01
g_c	Conversion factor in Newton's second law of motion	–	32.17405 lbm-ft/lbf-s^2	1.0 kg·m/N·s^2	3.108 095 E−02
g_G	Geothermal temperature gradient	T/m	°F/ft	K/m	1.822 689 E+00
G	Function defined by Eq. 5.13	–	–	–	–
G_a	Total air injected	L^3	MMscf	std m^3	2.863 640 E+04[a]
ΔG	Change in G	–	–	–	–
h	Enthalpy per unit mass (Chap. 3)	L^2/t^2	Btu/lbm	kJ/kg	2.326 000 E+00
h	Coefficient of heat transfer (always with subscript identifying type, Chap. 10)	m/t^3T	Btu/sq ft-D-°F	kJ/m^2·d·K kW/m^2·K	2.044 175 E+01 1.362 783 E−01
h	Reservoir thickness (usually with *n* or *t* subscript)	L	ft	m	3.048* E−01
h_d	Coefficient of heat transfer due to dirt and scale	m/t^3T	Btu/sq ft-D-°F	kJ/m^2·d·K kW/m^2·K	2.044 175 E+01 1.362 783 E−01
h_f	Fluid enthalpy (Chap. 3)	L^2/t^2	Btu/lbm	kJ/kg	2.326 000 E+00
h_f	Film coefficient of heat transfer	m/t^3T	Btu/sq ft-D-°F	kJ/m^2·d·K	2.044 175 E+01
h_{fc}	Coefficient of heat transfer due to forced convection	m/t^3T	Btu/sq ft-D-°F	kJ/m^2·d·K	2.044 175 E+01
h_i	Enthalpy of phase *i*	L^2/t^2	Btu/lbm	kJ/kg	2.326 000 E+00
h_n	Net reservoir thickness	L	ft	m	3.048* E−01
h_{Pi}	Film coefficient of heat transfer at inner radius of pipe	m/t^3T	Btu/sq ft-D-°F	kJ/m^2·d·K	2.044 175 E+01
h_{Po}	Film coefficient of heat transfer at outer radius of pipe	m/t^3T	Btu/sq ft-D-°F	kJ/m^2·d·K	2.044 175 E+01
h_{rc}	Coefficient of heat transfer due to radiation and natural convection	m/t^3T	Btu/sq ft-D-°F	kJ/m^2·d·K	2.044 175 E+01

$h_{rcf,an}$	Coefficient of annular heat transfer due to radiation and forced convection	m/t^3T	Btu/sq ft-D-°F	$kJ/m^2 \cdot d \cdot K$	2.044 175 E+01
$h_{rc,an}$	Coefficient of annular heat transfer due to radiation and convection	m/t^3T	Btu/sq ft-D-°F	$kJ/m^2 \cdot d \cdot K$	2.044 175 E+01
h_s	Enthalpy of dry steam	L^2/t^2	Btu/lbm	kJ/kg	2.326 000 E+00
h_t	Gross reservoir thickness	L	ft	m	3.048* E−01
h_w	Enthalpy of liquid water	L^2/t^2	Btu/lbm	kJ/kg	2.326 000 E+00
i	Discount rate (Chap. 12)	–	–	–	–
i	Injection rate	L^3/t	B/D	m^3/d	1.589 873 E−01
i_a	Air injection rate	L^3/t	Mscf/D	std m^3/d	2.863 640 E−02[a]
i_D	Dimensionless injection rate	–	–	–	–
i_J	Injection rate at Well J	L^3/t	B/D	m^3/d	1.589 873 E−01
i_r	Rate of return (Chap. 12)	–	–	–	–
i_s	Steam injection rate	L^3/t	B/D	m^3/d	1.589 873 E−01
i_t	Total injection rate	L^3/t	B/D	m^3/d	1.589 873 E−01
i_w	Water injection rate	L^3/t	B/D	m^3/d	1.589 873 E−01
I	Electric current (Chap. 14)	q/t	A	A	1
I	Injectivity or flow capacity between wells	L^4t/m	bbl/psi-D	$m^3/kPa \cdot d$	2.305 916 E−02
I_D	Dimensionless injectivity	–	–	–	–
I_j	Operating cash income in *j*th year (Chap. 12)	M[b]	Various	NA[c]	NA[c]
J	Mechanical equivalent of heat (Chaps. 3 and 10)	–	778 ft-lbf/Btu	1 kJ/kJ	1.285 E−03
J	Productivity or flow capacity between wells	L^4t/m	bbl/psi-D	$m^3/kPa \cdot d$	2.305 916 E−02
J_h	Productivity after thermal stimulation	L^4t/m	bbl/psi-D	$m^3/kPa \cdot d$	2.305 916 E−02
$J_{J,K}$	Productivity between Wells J and K	L^4t/m	bbl/psi-D	$m^3/kPa \cdot d$	2.305 916 E−02
k	Permeability	L^2	md	μm^2	9.869 233 E−04
$\bar{k}$	Average permeability	L^2	md	μm^2	9.869 233 E−04
k_a	Absolute permeability	L^2	md	μm^2	9.869 233 E−04
$k_{a,d}$	Absolute permeability on downstream side of displacement front	L^2	md	μm^2	9.869 233 E−04
$k_{a,u}$	Absolute permeability on upstream side of displacement front	L^2	md	μm^2	9.869 233 E−04
k_e	Effective permeability	L^2	md	μm^2	9.869 233 E−04
k_g	Permeability to gas phase	L^2	md	μm^2	9.869 233 E−04
k_i	Permeability to phase *i* (Chap. 3)	L^2	md	μm^2	9.869 233 E−04
k_r	Relative permeability	–	–	–	–
k_{rg}	Relative permeability to gas	–	–	–	–
k_{ro}	Relative permeability to oil	–	–	–	–

Symbol	Description	Dimensions	Customary Units	Metric Units	Customary to Metric Conversion Factors
k_{rs}	Relative permeability to steam	–	–	–	–
k_{rw}	Relative permeability to water	–	–	–	–
$k_{rw,d}$	Relative permeability to water on downstream side of displacement front	–	–	–	–
k_s	Permeability to steam	L^2	md	μm^2	9.869 233 E−04
k_H	Horizontal or bedding-plane permeability	L^2	md	μm^2	9.869 233 E−04
k_V	Vertical permeability or that normal to bedding plane	L^2	md	μm^2	9.869 233 E−04
k_σ	Permeability at one standard deviation	L^2	md	μm^2	9.869 233 E−04
K	Rate of oxygen reacted per unit of fuel	1/t	lbm/D-lbm	kg/d·kg	1.0* E+00
K	Secenov's coefficient	–	–	–	–
K_{oj}, K_{wj}	Equilibrium ratios defined by Eqs. 3.37 and 3.38	–	lbm mol/lbm mol	mol/mol	1.0* E+00
$K(m)$	Complete elliptic integral of first kind	–	–	–	–
ln	Natural logarithm, base e	–	–	–	–
L	Distance between wells	L	ft	m	3.048* E−01
L	Length	L	ft	m	3.048* E−01
			Å	nm	1.0* E−01
			in.	mm	2.54* E+01
			mile	km	1.609 344* E+00
L', L''	Characteristic lengths of a repeated well pattern (Table 4.2)	L	ft	m	3.048* E−01
$L_{J,K}$	Distance between Wells J and K	L	ft	m	3.048* E−01
L_v	Latent heat of vaporization	L^2/t^2	Btu/lbm	kJ/kg	2.326 000 E+00
L_{vb}	Latent heat of vaporization at boiler outlet	L^2/t^2	Btu/lbm	kJ/kg	2.326 000 E+00
L_{vdh}	Latent heat of vaporization at downhole conditions	L^2/t^2	Btu/lbm	kJ/kg	2.326 000 E+00
m	Archie's cementation factor, Eq. B.49	–	–	–	–
m	Mass	m	lbm	kg	4.535 924 E−01
m	Mole ratio of CO_2/CO	–	–	–	–
m	Parameter defined in Table 4.2	–	–	–	–
m'	Mole ratio of $CO/(CO+CO_2)$	–	–	–	–
m_E	Mass of fuel burned per unit volume of laboratory pack	m/L^3	lbm/cu ft	kg/m^3	1.601 846 E+01
m_{O_2}	Mass of oxygen	m	lbm	kg	4.535 924 E−01

m_R	Mass of fuel burned per unit reservoir volume	m/L^3	lbm/cu ft	kg/m^3	1.601 846 E+01
M	Mobility ratio of coflowing fluids, Eq. 6.5	–	–	–	–
M	Molecular weight (Chap. 3)	–	–	–	–
M	Volumetric isobaric heat capacity (ρC)	$\dot{m}/Lt^2T$	Btu/cu ft-°F	$kJ/m^3 \cdot K$	6.706 611 E+01
M_a	Volumetric heat capacity of air	m/Lt^2T	Btu/cu ft-°F	$kJ/m^3 \cdot K$	6.706 611 E+01
$M_{calcite}$	Volumetric heat capacity of calcite	m/Lt^2T	Btu/cu ft-°F	$kJ/m^3 \cdot K$	6.706 611 E+01
M_{eq}	Equivalent mobility ratio for condensing fluid drives	–	–	–	–
M_f	Volumetric heat capacity of fluid	m/Lt^2T	Btu/cu ft-°F	$kJ/m^3 \cdot K$	6.706 611 E+01
M_g	Molecular weight of gas (Chap. 3)	–	–	–	–
M_g	Volumetric heat capacity of gas	m/Lt^2T	Btu/cu ft-°F	$kJ/m^3 \cdot K$	6.706 611 E+01
M_j	Molecular weight of component j (Chap. 3)	–	–	–	–
M_{ls}	Volumetric heat capacity of limestone	m/Lt^2T	Btu/cu ft-°F	$kJ/m^3 \cdot K$	6.706 611 E+01
M_o	Molecular weight of oil (Chap. 3)	–	–	–	–
M_o	Volumetric heat capacity of oil	m/Lt^2T	Btu/cu ft-°F	$kJ/m^3 \cdot K$	6.706 611 E+01
M_R	Volumetric heat capacity of reservoir	m/Lt^2T	Btu/cu ft-°F	$kJ/m^3 \cdot K$	6.706 611 E+01
M_{Rse}	Effective volumetric heat capacity of the steam zone neglecting contributions due to steam; used in Eq. 7.19	m/Lt^2T	Btu/cu ft-°F	$kJ/m^3 \cdot K$	6.706 611 E+01
M_s	Volumetric heat capacity of steam	m/Lt^2T	Btu/cu ft-°F	$kJ/m^3 \cdot K$	6.706 611 E+01
M_S	Volumetric heat capacity of formations adjacent to surrounding heated zone	m/Lt^2T	Btu/cu ft-°F	$kJ/m^3 \cdot K$	6.706 611 E+01
M_{sh}	Volumetric heat capacity of shale	m/Lt^2T	Btu/cu ft-°F	$kJ/m^3 \cdot K$	6.706 611 E+01
M_w	Volumetric heat capacity of water	m/Lt^2T	Btu/cu ft-°F	$kJ/m^3 \cdot K$	6.706 611 E+01
$M_{w,o}$	Water/oil mobility ratio	–	–	–	–
M_σ	Volumetric heat capacity of reservoir solids	m/Lt^2T	Btu/cu ft-°F	$kJ/m^3 \cdot K$	6.706 611 E+01
$(M\sqrt{\alpha})_S$	Average value of $M\sqrt{\alpha}$ in the formations surrounding the steam zone	$m/t^{5/2}T$	Btu/sq ft-$D^{1/2}$-°F	$kJ/m^2 \cdot d^{1/2} \cdot K$	2.044 175 E+01
n	Concentration of dissolved solids, Eq. B.17	–	–	–	–
n	Index of summation in Eq. 5.14	–	–	–	–
n	Order of reaction (Chap. 8)	–	–	–	–
n_C	Number of components	–	–	–	–
n_{CS}	Number of compression stages	–	–	–	–
n_i	Number of injectors	–	–	–	–
n_p	Number of producers	–	–	–	–
n_P	Number of phases	–	–	–	–
n_{str}	Number of strokes per minute	$1/t$	1/min	1/min	1
n_T	Number of temperature zones	–	–	–	–

Symbol	Description	Dimensions	Customary Units	Metric Units	Customary to Metric Conversion Factors
n_y	Number of years for which evaluation is made	–	–	–	–
n_Δ	Number of step changes	–	–	–	–
$\hat{n}$	Mole ratio of oxygen reacted to CO_2 generated	–	–	–	–
N_{Gr}	Grashof number (Chap. 10)	–	–	–	–
N_p	Cumulative oil production	L^3	bbl	m^3	1.589 873 E−01
N_{Pr}	Prandtl number (Chap. 10)	–	–	–	–
N_{Re}	Reynolds number (Chap. 10)	–	–	–	–
O_j	Operating cost in *j*th year (Chap. 12)	M[b]	Various	NA[c]	NA[c]
p	Pressure	m/Lt^2	psi	kPa	6.894 757 E+00
p_{ab}	Absolute pressure	m/Lt^2	psia	kPa	6.894 757 E+00
$p_{ab,dis}$	Compression stage discharge pressure	m/Lt^2	psia	kPa	6.894 757 E+00
$p_{ab,fin}$	Final compressor pressure	m/Lt^2	psia	kPa	6.894 757 E+00
$p_{ab,int}$	Compression stage intake pressure	m/Lt^2	psia	kPa	6.894 757 E+00
$p_{ab,suc}$	Intake pressure at first compression stage	m/Lt^2	psia	kPa	6.894 757 E+00
p_b	Bubble-point pressure	m/Lt^2	psia	kPa	6.894 757 E+00
p_{cj}	Critical pressure of component *j*	m/Lt^2	psia	kPa	6.894 757 E+00
p_g	Pressure of the gas phase	m/Lt^2	psi	kPa	6.894 757 E+00
p_i	Initial pressure; pressure of phase *i* (Chap. 3)	m/Lt^2	psi	kPa	6.894 757 E+00
p_{idh}	Bottomhole injection pressure	m/Lt^2	psi	kPa	6.894 757 E+00
p_{inj}	Injection pressure	m/Lt^2	psi	kPa	6.894 757 E+00
p_p	Production pressure	m/Lt^2	psi	kPa	6.894 757 E+00
p_{pc}	Pseudocritical pressure	m/Lt^2	psia	kPa	6.894 757 E+00
p_{pdh}	Bottomhole production pressure	m/Lt^2	psi	kPa	6.894 757 E+00
p_{pr}	Pseudoreduced reference pressure	–	psia	kPa	6.894 757 E+00
p_{Rs}	Static reservoir pressure	m/Lt^2	psi	kPa	6.894 757 E+00
p_s	Partial steam pressure, pressure of saturated steam	m/Lt^2	psia	kPa	6.894 757 E+00
p_{sc}	Pressure at standard conditions	m/Lt^2	psia	kPa	6.894 757 E+00
p_{O_2}	Partial pressure of oxygen	m/Lt^2	psia	kPa	6.894 757 E+00
p_r	Reference pressure	m/Lt^2	psia	kPa	6.894 757 E+00
$\partial p/\partial n$	Pressure gradient normal to displacement front	m/L^2t^2	psi/ft	kPa/m	2.262 059 E+01
P	Power	mL^2/t^3	kW	kW	1
			hp(550 ft·lbf/s)	kW	7.456 999 E−01

$P_{c,og}$	Capillary pressure between oil and gas phases (Chap. 3)	m/Lt^2	psi	kPa	6.894 757 E+00
$P_{c,wo}$	Capillary pressure between water and oil (Chap. 3)	m/Lt^2	psi	kPa	6.894 757 E+00
P_j	Undiscounted cash flow in *j*th year	M[b]	Various	NA[c]	NA[c]
P_{PV}	Present value cash flow	M[b]	Various	NA[c]	NA[c]
P_r	Power at reference location	mL^2/t^3	kW	kW	1
q	Production rate, flow rate	L^3/t	B/D	m^3/d	1.589 873 E−01
q_{gh}	Gas production rate after thermal stimulation	L^3/t	B/D	m^3/d	1.589 873 E−01
q_K	Production rate from Well K	L^3/t	B/D	m^3/d	1.589 873 E−01
q_o	Oil production rate	L^3/t	B/D	m^3/d	1.589 873 E−01
q_{oh}	Oil production rate after thermal stimulation	L^3/t	B/D	m^3/d	1.589 873 E−01
$(q_{oh})_{max}$	Maximum oil production rate after thermal stimulation	L^3/t	B/D	m^3/d	1.589 873 E−01
q_s	Steam flow rate	L^3/t	B/D	m^3/d	1.589 873 E−01
q_{sc}	Flowrate at standard conditions	L^3/t	B/D	m^3/d	1.589 873 E−01
q_t	Total flow rate	L^3/t	B/D	m^3/d	1.589 873 E−01
q_w	Water production rate	L^3/t	B/D	m^3/d	1.589 873 E−01
q_{wh}	Water production rate after thermal stimulation	L^3/t	B/D	m^3/d	1.589 873 E−01
q_ψ	Flow rate within a streamtube	L^3/t	B/D	m^3/d	1.589 873 E−01
Q	Amount of heat in reservoir	mL^2/t^2	Btu	kJ	1.055 056 E+00
Q_i	Cumulative heat injected	mL^2/t^2	Btu	kJ	1.055 056 E+00
Q_{iD}	Dimensionless cumulative heat injected	–	–	–	–
Q_l	Amount of heat lost	mL^2/t^2	Btu	kJ	1.055 056 E+00
Q_{ls}	Heat loss per unit length	mL/t^2	Btu/ft	kJ/m	3.461 147 E+00
Q_{max}	Heat in reservoir at end of steam injection cycle	L^3/t	B/D	m^3/d	1.589 873 E−01
$\dot{Q}$	Rate of energy input from sources	mL^2/t^3	Btu/D	kJ/d	1.055 056 E+00
$\dot{Q}_i$	Heat injection rate	mL^2/t^3	Btu/D	kJ/d	1.055 056 E+00
$\dot{Q}_{iD}$	Dimensionless heat injection rate	–	–	–	–
$\dot{Q}_{i,N}$	Heat injection rate after *N*th step change	mL^2/t^3	Btu/D	kJ/d	1.055 056 E+00
$\dot{Q}_l$	Rate of heat loss	mL^2/t^3	Btu/D	kJ/d	1.055 056 E+00
$\dot{Q}_{ls}$	Rate of heat loss per unit length	mL/t^2	Btu/ft	kJ/m	3.461 147 E+00
$\dot{Q}_{p,dh}$	Rate of heat withdrawal from reservoir via produced fluids	mL^2/t^3	Btu/D	kJ/d	1.055 056 E+00
$\dot{Q}_v$	Rate of injection of latent heat	mL^2/t^3	Btu/D	kJ/d	1.055 056 E+00
				W	1.221 130 E−02
r	Radius	L	ft	m	3.048* E−01
r_B	Radial distance to bottom of steam zone	L	ft	m	3.048* E−01

Symbol	Description	Dimensions	Customary Units	Metric Units	Customary to Metric Conversion Factors
r_{ci}	Inner casing radius	L	ft	m	3.048* E−01
r_{co}	Outer casing radius	L	ft	m	3.048* E−01
r_e	External radius	L	ft	m	3.048* E−01
r_{eD}	r_e/r_w, dimensionless exterior radius	–	–	–	–
r_{Ea}	Altered radius in the earth around wellbore	L	ft	m	3.048* E−01
r_{gr}	Grain radius	L	ft	m	3.048* E−01
r_h	Radius of heated or steam zone	L	ft	m	3.048* E−01
r_H	Hydraulic equivalent radius defined as twice the cross-sectional area to flow divided by the total perimeter contacted by the flow	L	ft	m	3.048* E−01
r_{ins}	External radius of insulation	L	ft	m	3.048* E−01
r_i	Inner radius	L	ft	m	3.048* E−01
r_o	Outer radius	L	ft	m	3.048* E−01
r_T	Radial distance to top of steam zone	L	ft	m	3.048* E−01
r_w	Well radius	L	ft	m	3.048* E−01
r_{wa}	Apparent well radius	L	ft	m	3.048* E−01
r_1	Specific radius	L	ft	m	3.048* E−01
R	Universal gas constant	L^2/t^2T	1.9858 Btu/°R-lbm mol	8.3143 kJ/kmol·K	4.1868* E+00
R_e	Electric resistance	mL^3/tq	Ω	Ω	1
R_h	Overall specific thermal resistance	Tt^3/mL	°F-ft-D/Btu	K·m·d/kJ	1.604 970 E−01
R'_h	Overall specific thermal resistance from the wellbore	Tt^3/mL	°F-ft-D/Btu	K·m·d/MJ	1.604 970 E−02
				K·m·d/kJ	1.604 970 E−01
				K·m·d/kW	2.407 455 E+01
R_s	Solution GOR, amount of gas dissolved per unit oil	–	Mscf/bbl	std m^3/m^3	1.801 175 E+02[a]
R_{sw}	Amount of gas dissolved in water	–	Mscf/bbl	std m^3/m^3	1.801 175 E+02[a]
s	Skin	–	–	–	–
s_h	Skin after thermal stimulation	–	–	–	–
S	Saturation	–	–	–	–
S_g	Gas saturation	–	–	–	–
S_{gi}	Initial gas saturation	–	–	–	–
S_i	Saturation of phase *i* (Chap. 3)	–	–	–	–
S_{iw}	Irreducible water saturation	–	–	–	–

S_o	Oil saturation	–	–	–	–
S_{oF}	Oil saturation burned	–	–	–	–
S_{oi}	Initial oil saturation	–	–	–	–
$S_{o,ob}$	Oil saturation in oil bank	–	–	–	–
S_{or}	Residual oil saturation	–	–	–	–
S_r	Reduced saturation (Table 6.2)	–	–	–	–
S_s	Steam saturation	–	–	–	–
S_w	Water saturation	–	–	–	–
$S_{w,d}$	Water saturation at downstream side of displacement front	–	–	–	–
S_{wF}	Water saturation resulting from combustion	–	–	–	–
S_{wi}	Initial water saturation	–	–	–	–
$\bar{S}$	Average saturation	–	–	–	–
$\bar{S}_{o,wf}$	Average oil saturation in the swept zone in a cold waterflood	–	–	–	–
$\bar{S}_o(T_i)$	Average oil saturation in the swept zone in a hot waterflood	–	–	–	–
$\bar{S}_w$	Average water saturation	–	–	–	–
t	Time	t	D	d	1
t_{cD}	Dimensionless critical time	–	–	–	–
t_D	Dimensionless time	–	–	–	–
t_{D1}	Dimensionless time at 1 year, or at rate step change 1	–	–	–	–
t_{ig}	Ignition time	t	D	d	1
t_j	Time at which *j*th step change occurs	t	D	d	1
t_{min}	Minimum time required to approach steady state (Eq. 9.12)	t	D	d	1
t_p	Production time	t	D	d	1
t_{1D}	Dimensionless time defined by Eq. 5.15	–	–	–	–
T	Temperature	T	°F	°C	(°F − 32)/1.8[d]
$\mathfrak{I}$	Transmissivity (kh/μ)	L^4t/m	md-ft/cp	$\mu m^2 \cdot m/Pa \cdot s$	3.008 142 E − 01
$\mathfrak{I}_{ai}$	Effective transmissivity to phase *i*	L^4t/m	md-ft/cp	$\mu m^2 \cdot m/Pa \cdot s$	3.008 142 E − 01
$\mathfrak{I}_{as}$	Effective transmissivity to steam	L^4t/m	md-ft/cp	$\mu m^2 \cdot m/Pa \cdot s$	3.008 142 E − 01
$\mathfrak{I}_i$	Transmissivity to phase *i*	L^4t/m	md-ft/cp	$\mu m^2 \cdot m/Pa \cdot s$	3.008 142 E − 01
T_A	Ambient temperature	T	°F	°C	(°F − 32)/1.8[d]
T_{ab}	Absolute temperature	T	°R	K	5/9
$T_{ab,dis}$	Discharge temperature	T	°R	K	5/9
$T_{ab,int}$	Intake temperature	T	°R	K	5/9

Symbol	Description	Dimensions	Customary Units	Metric Units	Customary to Metric Conversion Factors
T_{abi}	Initial absolute temperature	T	°R	K	5/9
T_b	Bulk temperature of fluid	T	°F	°C	(°F − 32)/1.8[d]
T_{ci}	Temperature at inner radius of casing	T	°F	°C	(°F − 32)/1.8[d]
T_{cj}	Critical temperature of component *j*	T	°R	K	5/9
T_S	Surface temperature (Chap. 14)	T	°F	°C	(°F − 32)/1.8[d]
$T_E(z)$	Undisturbed earth temperature at depth *z*	T	°F	°C	(°F − 32)/1.8[d]
T_i	Initial temperature	T	°F	°C	(°F − 32)/1.8[d]
T_{idh}	Downhole injection steam temperature	T	°F	°C	(°F − 32)/1.8[d]
T_{inj}	Injection temperature	T	°F	°C	(°F − 32)/1.8[d]
T_{ins}	Temperature at exposed face of insulation	T	°F	°C	(°F − 32)/1.8[d]
T_j	Temperature in *j*th zone	T	°F	°C	(°F − 32)/1.8[d]
T_{pc}	Pseudocritical temperature	T	°R	K	5/9
T_{pdh}	Downhole producing temperature	T	°F	°C	(°F − 32)/1.8[d]
T_{pr}	Pseudoreduced temperature	–	–	–	–
T_r	Reference temperature	T	°F	°C	(°F − 32)/1.8[d]
T_s	Temperature of steam	T	°F	°C	(°F − 32)/1.8[d]
T_{sb}	Steam temperature at boiler outlet	T	°F	°C	(°F − 32)/1.8[d]
T_{sc}	Temperature at standard conditions	T	°R	K	5/9
T_S	Surface temperature (Chap. 14)	T	°F	°C	(°F − 32)/1.8[d]
$\bar{T}$	Average temperature	T	°F	°C	(°F − 32)/1.8[d]
$\bar{T}_{an}$	Average temperature in annulus	T	°F	°C	(°F − 32)/1.8[d]
u	Volumetric flux	L/t	cu ft/sq ft-D	m/d	3.048* E − 01
u_a	Air flux	L/t	scf/sq ft-D	std $m^3/m^2 \cdot d$	3.082 396 E − 01[a]
u_d	Volumetric flux at downstream side of displacement front	L/t	cu ft/sq ft-D	m/d	3.048* E − 01
u_D	Dimensionless flux	–	–	–	–
u_e	Electric charge flux or current density (Chap. 14)	q/m^2t	A/sq ft	A/m^2	1.076 074 E + 01
u_e	Total energy flux rate	m/t^3	Btu/sq ft-D	$kJ/m^2 \cdot d$	1.135 653 E + 01
u_h	Heat flux	m/t^3	Btu/sq ft-D	$kJ/m^2 \cdot d$	1.135 653 E + 01
u_i	Volumetric flux of phase *i*	L/t	cu ft/sq ft-D	m/d	3.048* E − 01
u_m	Total mass flux	m/L^2t	lbm/sq ft-D	$kg/m^2 \cdot d$	4.882 428 E + 00
u_{min}	Minimum air flux	L/t	scf/sq ft-D	std $m^3/m^2 \cdot d$	3.082 396 E − 01[a]
u_r	Radiative heat flux	m/t^3	Btu/sq ft-D	$kJ/m^2 \cdot d$	1.135 653 E + 01

u_s	Steam flux	L/t	cu ft/sq ft-D	m/d	3.048* E−01
u_T	Convective heat flux	m/t^3	Btu/sq ft-D	kJ/m^2·d	1.135 653 E+01
u_u	Volumetric flux at upstream side of displacement front	L/t	cu ft/sq ft-D	m/d	3.048* E−01
u_w	Water flux	L/t	cu ft/sq ft-D	m/d	3.048* E−01
u_η	Diffusive mass flux	m/L^{2}t	lbm/sq ft-D	kg/m^2·d	4.882 428 E+00
u_λ	Conductive heat flux	m/t^3	Btu/sq ft-D-°F	kJ/m^2·d	1.135 653 E+01
U	Overall coefficient of heat transfer	m/t^{3}T	Btu/sq ft-D-°F	kJ/m^2·d·K	2.044 175 E+01
				kW/m^2·d	1.362 783 E−01
$U(x)$	Unit function, equals 0 for $x<0$, equals 1 for $x>0$	–	–	–	–
v	Fluid velocity (u/ϕ)	L/t	ft/D	m/d	3.048* E−01
v	Specific volume (reciprocal density), (Appendix B)	L^3/m	cu ft/lbm	m^3/kg	6.242 796 E−02
v_b	Rate of advance of the combustion front	L/t	ft/D	m/d	3.048* E−01
v_T	Frontal rate of advance of convective heat front	L/t	ft/D	m/d	3.048* E−01
v_w	Specific volume of water (Appendix B)	L^3/m	cu ft/lbm	m^3/kg	6.242 796 E−02
v_W	Wind speed	L/t	miles/hr	km/h	1.609 344 E+00
V	Permeability variation	–	–	–	–
V	Volume	L^3	cu ft	m^3	2.831 685 E−02
			Mscf	std m^3	2.863 640 E+01[a]
			scf	std m^3	2.863 640 E−02[a]
V_d	Bulk volume displaced	L^3	acre-ft	m^3	1.233 489 E+03
				ha·m	1.233 489 E−01
V_e	Voltage (Chap. 14)	mL^2/t^{2}q	V	V	1
$V_{e,w}$	Voltage at wellbore	mL^2/t^{2}q	V	V	1
V_j	Gross revenue in *j*th year (Chap. 12)	M[b]	Various	NA[c]	NA[c]
V_o	Oil volume	L^3	bbl	m^3	1.589 873 E−01
V_{ob}	Oil bank volume	L^3	acre-ft	m^3	1.233 489 E+03
				ha·m	1.233 489 E−01
V_{oi}	Initial oil volume	L^3	bbl	m^3	1.589 873 E−01
V_p	Pore volume	L^3	cu ft	m^3	2.831 685 E−02
V_P	Pattern volume	L^3	acre-ft	m^3	1.233 489 E+03
				ha·m	1.333 489 E−01
V_R	Reservoir volume	L^3	acre-ft	m^3	1.233 489 E+03
V_{Rb}	Reservoir volume burned	L^3	acre-ft	m^3	1.233 489 E+03
$(V_{Rb})_{max}$	Maximum reservoir volume burned	L^3	acre-ft	m^3	1.233 489 E+03
V_s	Steam zone volume	L^3	acre-ft	m^3	1.233 489 E+03
V_{sD}	Dimensionless steam zone volume	–	–	–	–

Symbol	Description	Dimensions	Customary Units	Metric Units	Customary to Metric Conversion Factors
$V_{s,ss}$	Steady-state steam zone volume	L^3	acre-ft	m^3	1.233 489 E+03
$(V_s)_{max}$	Steam zone volume at end of steam injection	L^3	acre-ft	m^3	1.233 489 E+03
V_{str}	Maximum volume displaced by a piston	L^3	cu in.	cm^3	1.638 706 E+01
$V_{u,j}$	Unit crude price	M^b/L^3	Various	NA[c]	NA[c]
w	Mass flow rate	m/t	lbm/D	kg/d	4.535 924 E−01
w_i	Mass rate of injection	m/t	lbm/D	kg/d	4.535 924 E−01
w_{is}	Mass injection rate of steam	m/t	lbm/D	kg/d	4.535 924 E−01
w_j	Rate of mass input of component j from sources	m/t	lbm/D	kg/d	4.535 924 E−01
w_s	Mass flow rate of dry steam	m/t	lbm/D	kg/d	4.535 924 E−01
w_t	Total mass flow rate	m/t	lbm/D	kg/d	4.535 924 E−01
w_w	Mass flow rate of water	m/t	lbm/D	kg/d	4.535 924 E−01
W_i	Cumulative water injected	L^3	bbl	m^3	1.589 873 E−01
W_{iD}	Dimensionless cumulative water injected	–	–	–	–
W_p	Cumulative water production	L^3	bbl	m^3	1.589 873 E−01
W_{sD}	Dimensionless volume of steam injected	–	–	–	–
$W_{s,eq}$	Equivalent volume of steam injected	L^3	bbl	m^3	1.589 873 E−01
x	Atoms of hydrogen per atom of carbon in burned fuel (Chap. 8)	–	–	–	–
x	Distance along the x ordinate	L	ft	m	3.048* E−01
x'	Variable of integration	–	–	–	–
$\|x\|$	Magnitude of x	L	ft	m	3.048* E−01
x,y,z	Distances from the origin along x,y,z ordinates of a Cartesian coordinate system	L	ft	m	3.048* E−01
x_D	Dimensionless distance	–	–	–	–
x_{oj}	Mole fraction of the oleic phase made up of component j	–	lbm mol/lbm mol	mol/mol	1.0* E+00
x_{wj}	Mole fraction of the aqueous phase made up of component j	–	lbm mol/lbm mol	mol/mol	1.0* E+00
y	Distance along the y coordinate	L	ft	m	3.048* E−01
y_j	Mole fraction of the gaseous phase made up of component j	–	–	–	–
z	Distance along vertical ordinate, positive upward unless otherwise indicated	L	ft	m	3.048* E−01

z	Supercompressibility factor (Chaps. 4 and 11)	–	–	–	–
z_D	Dimensionless vertical distance	–	–	–	–
z_{fin}	Supercompressibility at delivery side of last compression stage	–	–	–	–
Δz_s	Thickness of steam zone	L	ft	m	3.048* E−01
z_{suc}	Supercompressibility at intake side of first compression stage	–	–	–	–
Z	Real quantity equal to $x_D/2\sqrt{t_D - x_D}$	–	–	–	–
α	Thermal diffusivity	L^2/t	sq ft/D	m^2/d	9.290 304* E−02
α_e	Electromagnetic power absorption coefficient (Chap. 14)	1/L	1/ft	1/m	3.280 840 E+00
α_E	Thermal diffusivity of earth	L^2/t	sq ft/D	m^2/d	9.290 304* E−02
α_{ls}	Thermal diffusivity of limestone	L^2/t	sq ft/D	m^2/d	9.290 304* E−02
α_R	Thermal diffusivity of reservoir	L^2/t	sq ft/D	m^2/d	9.290 304* E−02
α_S	Thermal diffusivity of formations adjacent to the heated zone	L^2/t	sq ft/D	m^2/d	9.290 304* E−02
α_{sh}	Thermal diffusivity of shale	L^2/t	sq ft/D	m^2/d	9.290 304* E−02
β	Mole ratio of CO/CO_2 (Chap. 8)	–	lbm mol/lbm mol	mol/mol	1.0* E+00
β_g	Volumetric thermal expansion coefficient of gas	1/T	1/°F	1/K	1.8* E+00
β_o	Volumetric thermal expansion coefficient of oil	1/T	1/°F	1/K	1.8* E+00
γ	Specific gravity	–	–	–	–
γ_g	Specific gravity of gas	–	–	–	–
γ_o	Specific gravity of oil	–	–	–	–
δ	Coefficient defined in Eq. 10.11	–	–	–	–
$\Delta(\ )$	Increment or decrement in ( )		Same as for quantity within ( )		
Δh_a	Heat of combustion per unit volume of air reacted	m/Lt^2	Btu/scf	kJ/m^3	3.684 318 E+01[a]
Δh_F	Heat of combustion per mass of fuel	L^2/t^2	Btu/lbm	kJ/kg	2.326 000 E+02
Δh_{O_2}	Heat of reaction per unit oxygen	L^2/t^2	Btu/lbm	kJ/kg	2.326 000 E+02
Δh_r	Heat of reaction	L^2/t^2	Btu/lbm	kJ/kg	2.326 000 E+02
		L^2/t^2N	Btu/lbm mol	kJ/kmol	2.326 000 E+02
$\Delta N_{p,j}$	Annual oil production in *j*th year	L^3	bbl	m^3	1.589 873 E−01
Δp	Pressure drop	m/Lt^2	psi	kPa	6.894 757 E+00
Δp^2	Difference in pressure squared	m^2/L^2t^4	psi^2	kPa^2	4.753 767 E+01
Δp_D	Dimensionless pressure drop	–	–	–	–
$\Delta p_{J,K}$	Pressure drop between Wells J and K	m/Lt^2	psi	kPa	6.894 757 E+00
$\Delta \dot{Q}_j$	*j*th step change in heat injection rate	mL^2/t^3	Btu/D	kJ/d	1.055 056 E+00
Δr	Radial distance	L	ft	m	3.048* E−01

Symbol	Description	Dimensions	Customary Units	Metric Units	Customary to Metric Conversion Factors
ΔS_o	Displaceable oil saturation	–	–	–	–
Δt	Time elapsed since steam first arrived where Δz_s is to be calculated	t	D	d	1
ΔT	Temperature rise	T	°F	°C	5/9
ΔT_D	Dimensionless temperature rise (Eq. 9.9)	–	–	–	–
ΔT_i	Injection temperature above the initial reservoir temperature	T	°F	°C	5/9
ΔT_{max}	Maximum temperature rise	T	°F	°C	5/9
ΔT_{pdh}	Temperature of produced fluids above the initial reservoir temperature	T	°F	°C	5/9
T_{te}	Temperature rise at which casing is in tension (after it has yielded)	T	°F	°C	5/9
ΔT_{yp}	Temperature rise at which casing under compression reaches its yield point	T	°F	°C	5/9
ΔV_e	Voltage drop	mL^2/t^2q	V	V	1
Δz	Increment in z	L	ft	m	3.048* E−01
Δz_D	Change in dimensionless vertical distance	–	ft	m	3.048* E−01
$\Delta \rho$	Density difference	m/L^3	lbm/cu ft	kg/m^3	1.601 846 E-01
$\Delta \psi$	Increment in stream function	Various	–	–	–
ϵ	Emissivity	–	–	–	–
ϵ_i	Emissivity at inner casing radius	–	–	–	–
ϵ_o	Emissivity at exposed surface of insulation	–	–	–	–
η_h	Fraction of heat generated in borehole heaters which enters the formation	–	–	–	–
θ	Parameter defined by Eq. 5.15	–	–	–	–
θ_c	Contact angle	–	deg(°)	rad	1.745 329 E−02
Θ	Angle of dip	–	deg(°)	rad	1.745 329 E−02
λ	Fluid mobility (Chap. 7)	L^3t/m	md/cp	md/mPa·s	9.869 233 E−04
λ	Thermal conductivity	mL/t^3T	Btu/ft-D-°F	kJ/m·d·K	6.230 646 E+00
				W/m·K	4.153 764 E+01
λ_a	Thermal conductivity of air	mL/t^3T	Btu/ft-D-°F	kJ/m·d·K	6.230 646 E+00
$\lambda_{a,an}$	Thermal conductivity of air in annulus	mL/t^3T	Btu/ft-D-°F	kJ/m·d·K	6.230 646 E+00
λ_{cem}	Thermal conductivity of cement	mL/t^3T	Btu/ft-D-°F	kJ/m·d·K	6.230 646 E+00

λ_E	Thermal conductivity of the earth	mL/t^3T	Btu/ft-D-°F	kJ/m·d·K	6.230 646 E+00
λ_{Ea}	Thermal conductivity of earth in altered zone around wellbore	mL/t^3T	Btu/ft-D-°F	kJ/m·d·K	6.230 646 E+00
λ_{ins}	Thermal conductivity of insulation	mL/t^3T	Btu/ft-D-°F	kJ/m·d·K	6.230 646 E+00
λ_l	Thermal conductivity of liquid	mL/t^3T	Btu/ft-D-°F	kJ/m·d·K	6.230 646 E+00
λ_P	Thermal conductivity of pipe	mL/t^3T	Btu/ft-D-°F	kJ/m·d·K	6.230 646 E+00
λ_R	Thermal conductivity of reservoir	mL/t^3T	Btu/ft-D-°F	kJ/m·d·K	6.230 646 E+00
$\lambda_{R,d}$	Thermal conductivity of dry formation	mL/t^3T	Btu/ft-D-°F	kJ/m·d·K	6.230 646 E+00
$\lambda_{R,l}$	Thermal conductivity of saturated formation	mL/t^3T	Btu/ft-D-°F	kJ/m·d·K	6.230 646 E+00
$\lambda_{R,o}$	Thermal conductivity of reservoir in the absence of fluid flow	mL/t^3T	Btu/ft-D-°F	kJ/m·d·K	6.230 646 E+00
λ_S	Thermal conductivity of formations adjacent to heated zone	mL/t^3T	Btu/ft-D-°F	kJ/m·d·K	6.230 646 E+00
λ_{sh}	Thermal conductivity of shale	mL/t^3T	Btu/ft-D-°F	kJ/m·d·K	6.230 646 E+00
λ_w	Thermal conductivity of water	mL/t^3T	Btu/ft-D-°F	kJ/m·d·K	6.230 646 E+00
μ	Viscosity	m/Lt	cp	Pa·s	1.0* E−03
μ_a	Air viscosity	m/Lt	cp	Pa·s	1.0* E−03
μ_g	Gas viscosity	m/Lt	cp	Pa·s	1.0* E−03
μ_{ga}	Gas viscosity at 1 atm	m/Lt	cp	Pa·s	1.0* E−03
μ_i	Viscosity of phase *i* (Chap. 3)	m/Lt	cp	Pa·s	1.0* E−03
μ_o	Oil viscosity	m/Lt	cp	Pa·s	1.0* E−03
μ_{oh}	Oil viscosity after thermal stimulation	m/Lt	cp	Pa·s	1.0* E−03
μ_s	Steam viscosity	m/Lt	cp	Pa·s	1.0* E−03
μ_w	Water viscosity	m/Lt	cp	Pa·s	1.0* E−03
ν	Kinematic viscosity	L^2/t	cSt	m^2/s	1.0* E−06
ν_s	Kinematic viscosity of steam	L^2/t	cSt	m^2/s	1.0* E−06
ν_w	Kinematic viscosity of water	L^2/t	cSt	m^2/s	1.0* E−06
$\nu_{w,d}$	Kinematic viscosity of water on downstream side of displacement front	L^2/t	cSt	m^2/s	1.0* E−06
ρ	Density	m/L^3	lbm/cu ft	kg/m^3	1.601 846 E+01
ρ_f	Fluid density	m/L^3	lbm/cu ft	kg/m^3	1.601 846 E+01
ρ_g	Density of gas	m/L^3	lbm/cu ft	kg/m^3	1.601 846 E+01
ρ_i	Density of phase *i* (Chap. 3)	m/L^3	lbm/cu ft	kg/m^3	1.601 846 E+01
ρ_{wdh}	Density of water at downhole injection conditions	m/L^3	lbm/cu ft	kg/m^3	1.601 846 E+01
ρ_o	Density of oil	m/L^3	lbm/cu ft	kg/m^3	1.601 846 E+01
ρ_s	Density of dry steam	m/L^3	lbm/cu ft	kg/m^3	1.601 846 E+01
ρ_w	Density of aqueous phase, or water	m/L^3	lbm/cu ft	kg/m^3	1.601 846 E+01

Symbol	Description	Dimensions	Customary Units	Metric Units	Customary to Metric Conversion Factors
ρ_σ	Density of solids in formation	m/L^3	lbm/cu ft	kg/m^3	1.601 846 E+01
σ	Electric conductivity (Chap. 14)	tq^2/mL	mho/m	S/m	1.0* E+00
σ	Interfacial tension (Chap. 2)	m/t^2	dyne/cm	mN/m	1.0* E+00
σ	Stefan-Boltzmann constant (Chap. 3)	mL^2/t^2T^4	4.13×10^{-8} Btu/ft-D-°R^4	1.5×10^{-6} kJ/m·d·K^4	3.633 713 E+01
ϕ	Porosity	—	—	—	—
ϕ_E	Porosity in laboratory experiments	—	—	—	—
Φ	Fluid potential	Various			
ψ	Stream function	Various			

[a]The standard cubic foot (scf) and barrel (bbl) are measured at 60°F and 14.696 psia; the cubic meter is measured at 15°C and 100 kPa (1 bar).
[b]Money.
[c]Not applicable.
[d]This is not a factor. Perform the operation indicated.

Subscripts

a = air; altered (with radius); absolute (with permeability); apparent or effective
ab = absolute
abi = initial absolute value
an = annulus
ao = air/oil (with ratio)
A = ambient conditions
b = bulk (with fluid temperature); burned, at boiler outlet
B = bottom (Fig. 7.10)
c = contact (with angle); conversion (with conversion factor in Newton's second law of motion, g_c); critical; capture (with efficiency); capillary (with pressure)
calcite = calcite
cem = cement
$c\iota$ = inner radius of casing
co = outer radius of casing
comp = compression
cy = cylinder
C = component
CO_2 = carbon dioxide
CS = compression stage
d = downstream side of displacement front; dirt or scale (with coefficient of heat transfer); dry
dh = downhole
dis = discharge side of compression stage
D = dimensionless
e = energy; exterior (with radius); electric (Chap. 14)
eq = equivalent
E = earth (in Chap. 10); experimental (in Chap. 8)
f = film (with heat-transfer coefficient); fluid; front (with radius)
fc = forced convection
fin = discharge side of last compression stage
F = fuel
g = gas, gas phase, gaseous phase
gr = grain
G = geometric; geothermal (with gradient)
Gr = Grashof (with N)
h = thermal; borehole (with radius)
hw = hot-water flood
H = horizontal
i = initial; phase (in Chap. 3, stands for g, o, or w); injection
ι = inner (with radius)
idh = injection well downhole conditions
ig = ignition
inj = injection
int = intake side of compression stage
ins = insulation
iw = irreducible water
I = inhomogeneity
j = index (j=1,2,3, ...)
J = index (J=1,2,3, ...)
k = index (k=1,2,3, ...)
K = index (K=1,2,3, ...)
l = loss; liquid in Appendix B
ls = limestone
m = mass; mineral

max = maximum
min = minimum
M = molecular
n = net
N_2 = nitrogen
o = oil; outer (with radius)
oa = oil, apparent
ob = oil bank
oF = oil burned
oi = initial oil
or = residual oil
O_2 = oxygen
p = production well, production, or producing
pay = pay interval
pc = pseudocritical
pdh = production well downhole conditions
P = phase (in Chap. 3); pipe; pattern (with reservoir volume)
Pr = Prandtl (with N)
PV = present value
P_i = inner radius of pipe
P_o = outer radius of pipe
pr = pseudoreduced
p_r = pseudoreduced
q = quartz
$\dot{Q}$ = heat rate
r = reaction (with quantity of heat); reference conditions; return (with discount rate)
rg = relative, gas (with permeability)
ro = relative, oil (with permeability)
$\imath$ = radiation
$\imath c$ = radiation-convection
Rb = reservoir burned
Re = Reynolds (with N)
$\imath fc$ = radiation-forced convection
Rhw = reservoir hot-water flood
Rs = static, reservoir (with pressure)
s = specific, per unit length (with rate of heat loss); steam; solution (with ratio)
sc = standard conditions
sh = shale
ss = steady state
str = stroke
suc = suction
S = surface; formations surrounding the heated interval
SC = stage compression
SV = vertical sweep
t = total
te = tension
T = temperature; top (Fig. 7.10); convective
u = unit of production (with gross revenue); upstream side of displacement front
v = vaporization (with latent heat); velocity
V = vertical; volume
w = water, aqueous phase; wellbore (with radius)
wa = wellbore, apparent (with radius); water/air (with ratio); water, apparent
wF = water from combustion
wi = initial water
W = wind
x,y,z = components in $x,y,$ and z directions, respectively
y = years
yp = yield point
Δ = intervals; change
Δh = enthalpy
η = dispersive
λ = conductive
μ = viscosity
σ = at one standard deviation (with permeability); total solids in reservoir
Φ = potential energy
ψ = streamline

Bibliography

A

Abernethy, E.R.: "Production Increase of Heavy Oil by Electromagnetic Heating," *J. Cdn. Pet. Tech.* (July-Sept. 1976) 91-97.

Abstracts of Third Annual Oil Shale Conversion Symposium, *Cameron Synthetic Fuels Report,* Cameron Engineers Inc., Denver (March 1980) **17,** No. 1.

"AC Current Heats Heavy Oil for Extra Recovery," *World Oil* (May 1970) 83-86.

Acharya, U.K. and Somerton, W.H.: "Theoretical Study of In-Situ Combustion in Thick Inclined Oil Reservoirs," paper SPE 7967 presented at the 1979 SPE California Regional Meeting, Ventura, April 18-20.

Acton, J.F.: "Pump-Off Controller Application for Midway-Sunset Cyclic Steam Operations," paper SPE 9915 presented at the 1981 SPE California Regional Meeting, Bakersfield, March 25-26.

Adams, R.H. and Khan, A.M.: "Cyclic Steam Injection Project Performance Analysis and Some Results of a Continuous Steam Displacement Pilot," *J. Pet. Tech.* (Jan. 1969) 95-100; *Trans.,* AIME, **246.**

Adivarahan, P., Kunii, D., and Smith, J.M.: "Heat Transfer in Porous Rocks Through Which Single-Phase Fluids are Flowing," *Trans.,* AIME (1962) **225**, 290.

Afoeju, B.I.: "Conversion of Steam Injection to Waterflood, East Coalinga Field," *J. Pet. Tech.* (Nov. 1974) 1227-1232.

Alexander, J.D., Martin, W.L., and Dew, J.N.: "Factors Affecting Fuel Availability and Composition During In-Situ Combustion," *J. Pet. Tech.* (Oct. 1962) 1154-1164; *Trans.,* AIME, **225.**

Allen, H.E. and Davis, R.K.: "Electric Formation Heaters Boost Cut Bank Production," *Oil and Gas J.* (June 14, 1954) 125-137.

Allen, T.O. and Roberts, A.P.: *Production Operations,* Oil and Gas Consultants Intl. Inc., Tulsa (1978) **1.**

Anand, J., Somerton, W.H., and Gomaa, E.: "Predicting Thermal Conductivities of Formations from Other Known Properties," *Soc. Pet. Eng. J.* (Oct. 1973) 267-273.

Anderson, G.W.: "New Methods Make Downhole Liquid Plugging Practical," *World Oil* (Feb. 1, 1978) 37-43.

Annual Review of California Oil and Gas Production, Conservation Committee of California Oil Producers, Los Angeles (yearly through 1979).

Annual Review, Pacific Oil World, January issue of each year, 1970-1978.

Anthony, M.J., Taylor, T.D., and Gallagher, B.: "Fireflooding a High-Gravity Crude in a Watered-Out West Texas Sandstone," *J. Pet. Tech.* (Oct. 1982) 2244-2250.

Araujo, J.: "Estimulación Cíclica con Aqua Caliente en las Arenas del Grupo I—Campo Morichal," Petróleos de Venezuela—I Simposio de Crudos Extra-Pesados, Maracay, Oct. 13-15, 1976.

Araujo, J.: "Proyecto Piloto de Combustión Inversa (Grupo II) en el Campo Morichal," Petróleos de Venezuela—I Simposio de Crudos Extra-Pesados, Maracay, Oct. 13-15, 1976.

Araujo, J.: "Prueba Piloto de Combustión In-Situ Húmeda, Complejo Melones—14," Petróleos de Venezuela—I Simposio de Crudos Extra-Pesados, Maracay, Oct. 13-15, 1976.

Aziz, K., Bories, S.A., and Combarnous, M.A.: "The Influence of Natural Convection in Gas, Oil and Water Reservoirs," *J. Cdn. Pet. Tech.* (April-June 1973) 41-47.

B

Bader, B.E., Fox, R.L., Johnson, D.R., and Donaldson, A.B.: *Deep Steam Project, Quarterly Report for April 1-June 30, 1979,* Sand 79-2091, Sandia Laboratories, Albuquerque, NM (1979).

Bailey, H.R.: "Heat Conduction from a Cylindrical Source With Increasing Radius," *Quarterly of Applied Mathematics* (1959) **17,** 255.

Bailey, H.R. and Larkin, B.K.: "Conduction-Convection in Underground Combustion," *Trans.,* AIME (1960) **219,** 320-331.

Baker, P.E.: "Effect of Pressure and Rate of Steam Zone Development in Steamflooding," *Soc. Pet. Eng. J.* (Oct. 1973) 274-284; *Trans.,* AIME, **255.**

Baker, P.E.; "An Experimental Study of Heat Flow in Steam Flooding," *Soc. Pet. Eng. J.* (March 1969) 89-99; *Trans.,* AIME, **246.**

Baker, P.E.; "Heat Wave Propagation and Losses in Thermal Oil Recovery Processes," *Proc.,* Seventh World Petroleum Cong., Mexico City (1967) **3,** 459-470.

Baker, P.E.: "Temperature Profiles in Underground Combustion," *Soc. Pet. Eng. J.* (March 1962) 21-27.

Ballard, J.R., Lanfranchi, E.E., and Vanags, P.A.: "Thermal Recovery in the Venezuelan Heavy Oil Belt," *J. Cdn. Pet. Tech.* (April-June 1977) 22-27.

Banderob, L.D. and Fazio, P.J.: "Thermal Well Problems and Completion Techniques in the Midway Sunset Field," *Proc.,* Pet. Ind. Conf. on Thermal Recovery, Los Angeles (1966) 56-60.

Bansbach, P.L.: "Treating Emulsions Produced by Thermal Recovery Operations," paper SPE 1328 presented at the 1965 SPE California Regional Meeting, Bakersfield, Nov. 4-5.

Bansbach, P.L.: "Treating Emulsions Produced by Thermal Recovery Operations," paper SPE 1328 presented at the SPE 36th Annual California Regional Meeting, Bakersfield, Nov. 4-5, 1965.

Barber, R.M.: "Trix Liz Fireflood Looks Good," *Oil and Gas J.* (Jan. 1, 1973) 36-39.

Barnes, A.L.: "Results from a Thermal Recovery Test in a Watered-Out Reservoir," *J. Pet. Tech.* (Nov. 1965) 1343-1353.

Bartley, L.R. and Bullen, R.S.: "Calculate Your Steam Generation Costs," *Canadian Petroleum* (August 1967) 28-31.

Beal, C.: "The Viscosity of Air, Water, Natural Gas, Crude Oil, and Its Associated Gases at Oil-Field Temperatures and Pressures," *Trans.,* AIME (1946) **165,** 94-115.

Bear, J.: *Dynamics of Fluids in Porous Media*, American Elsevier Publishing Co., New York City (1972).

Beckers, H.L. and Harmsen, G.J.: "The Effect of Water Injection on Sustained Combustion in a Porous Medium," *Soc. Pet. Eng. J.* (June 1970) 145-163; *Trans.,* AIME, **249.**

Bell, C.W. and Titus, C.H.: "Electro-Thermal Process for Promoting Oil Recovery," Canada Patent 932,657 (Aug. 28, 1973).

Benham, A.L. and Poettmann, F.H.: "The Thermal Recovery Process—An Analysis of Laboratory Combustion Data," *Trans.,* AIME (1958) **213,** 406-408.

Benton, J.P. *et al.*: "Pressure Maintenance by In-Situ Combustion, West Heidelberg Unit, Jasper County, Mississippi," *J. Pet. Tech.* (Oct. 1983) 1877-1883.

Bentsen, R.G. and Donohue, D.A.T.: "A Dynamic Programming Model of the Cyclic Steam Injection Process," *J. Pet. Tech.* (Nov. 1969) 1582-1596; *Trans.,* AIME, **246.**

Berry, K.L.: "Conceptual Design of Combined In-Situ and Surface Retorting of Oil Shale," *Proc.*, Eleventh Oil Shale Symposium, Colorado School of Mines, Golden (1978) **73,** 176–183.

Berry, V.J. Jr. and Parrish, D.R.: "A Theoretical Analysis of Heat Flow in Reverse Combustion," *Trans.*, AIME (1960) **219,** 124–131.

Bertness, T.A.: "Thermal Recovery: Principles and Practices of Oil Treatment," paper SPE 1266 presented at the 1965 SPE Annual Meeting, Denver, Oct. 3–6.

Betz Handbook of Industrial Water Conditioning, Betz Laboratories Inc., Trevose, PA (1962)

Bheda, M. and Wilson, D.B.: "A Foam Process for Treatment of Sour Water," *Water—1969,* Chem. Eng. Prog. Symposium Series (1969) 274.

Bia, P. and Combarnous, M.: "Transfert de Chaleur et de Masse," Revue de l'Institute du Pétrole (June 1975) 361–376.

Binder, G.G. Jr., Elzinga, E.R., Tarmy, B.L., and Willman, B.T.: "Scaled Model Tests of In-Situ Combustion in Massive Unconsolidated Sands," *Proc.*, Seventh World Pet. Cong., Mexico City (1977) **3,** Paper PD-12(4), 477–485.

Birch, F. and Clark, H.: "The Thermal Conductivity of Rocks and Its Dependence Upon Temperature and Composition: Part I," *Am. J. Sci.* (Aug. 1940) 529. "Part II," *Am. J. Sci.* (Sept. 1940) 613.

Bird, R.B., Stewart, W.E., and Lightfood, E.N.: *Transport Phenomena,* John Wiley and Sons Inc., New York City (1960).

Bissoondatt, J.C.: "Casing Failure in Steam Stimulated Wells—A Case History of the Guapo Field," paper SPE 5951 presented at the 1976 SPE Trinidad and Tobago Section Conference, Trinidad, April 2–3.

Bleakley, W.B.: "Journal Survey Shows Recovery Projects Up," *Oil and Gas J.* (March 25, 1974) 69–76.

Bleakley, W.B.: "Steamed Wells Need Good Completions," *Oil and Gas J.* (April 4, 1966) 136–138.

Blevins, T.R., Aseltine, R.J., and Kirk, R.S.: "Analysis of a Steam Drive Project, Inglewood Field, California," *J. Pet. Tech.* (Sept. 1969) 1141–1150.

Blevins, T.R. and Billingsley, R.H.: "The Ten-Pattern Steamflood, Kern River Field, California," *J. Pet. Tech.* (Dec. 1975) 1505–1514; *Trans.*, AIME, **259.**

Boberg, T.C. and Lantz, R.B.: "Calculation of the Production Rate of a Thermally Stimulated Well," *J. Pet. Tech.* (Dec. 1966) 1613–1623; *Trans.*, AIME, **237.**

Boberg, T.C., Penberthy, W.L. Jr., and Hagedorn, A.R.: "Calculating the Steam-Stimulated Performance of Gas-Lifted and Flowing Heavy-Oil Wells," *J. Pet. Tech.* (Oct. 1973) 1207–1215; *Trans.*, AIME, **255.**

Boberg, T.C. and West, R.C.: "Correlation of Steam Stimulation Performance," *J. Pet. Tech.* (Nov. 1972) 1367–1368.

Bookout, D.E., Glenn, J.J. Jr., and Schaller, H.E.: "Injection Profiles During Steam Injection," paper 801-43C presented at the API Prod. Div. Pacific Coast Dist. Meeting, Los Angeles, May 2–4, 1967.

Borregales, C.: "Inyección Alternada de Vapor en la Costa Bolívar," I Simposio de Crudos Extra-Pesados, Petróleos de Venezuela, Maracay, Oct. 13–15, 1976.

Borregales, C., Kadi, N., Aymard, R., and Lanfranchi, E.: "Espaciamiento de Pozos," Petróleos de Venezuela—I Simposio de Crudos Extra-Pesados, Maracay, Oct. 13–15, 1976.

Bott, R.C.: "Cyclic Steam Project in a Virgin Tar Reservoir," *J. Pet. Tech.* (May 1967) 585–591.

Bousaid, I.S. and Ramey, H.J. Jr.: "Oxidation of Crude Oil in Porous Media," *Soc. Pet. Eng. J.* (June 1968) 137–148.

Bowman, C.H. and Gilbert, S.: "A Successful Cyclic Steam Injection Project in Santa Barbara Field, Eastern Venezuela," *J. Pet. Tech* (Dec. 1969) 1531–1539.

Bowman, C.H.: "A Two-Spot Combustion Recovery Project," *J. Pet. Tech.* (Sept. 1965) 994–998.

Braden, W.B.: "A Viscosity-Temperature Correlation at Atmospheric Pressure for Gas-Free Oils," *J. Pet. Tech.* (Nov. 1966) 1487–1490; *Trans.*, AIME, **237.**

Bradley, B.W. and Gatzke, L.K.: "Steam Flood Heater Scale and Corrosion," *J. Pet. Tech.* (Feb. 1975) 171–178.

Brandt, H., Poynter, W.G., Hummell, J.D.: "Stimulating Heavy Oil Reservoirs With Downhole Air Gas Burners," *World Oil* (Sept. 1965) 91–95.

Bridges, J.E., Taflone, A., and Snow, R.H.: "Net Energy Recoveries for the In-Situ Dielectric Heating of Oil Shale," *Proc.*, 11th Oil Shale Symposium, Colorado School of Mines, Golden (1978) 311–330.

Brigham, W.E., Satman, A., and Soliman, M.Y.: "Recovery Correlations for In-Situ Combustion Field Projects and Application to Combustion Pilots," *J. Pet. Tech.* (Dec. 1980) 2132–2138.

Brown, G., Katz, D.L., Oberfell, G.G., and Alden, R.C.. "Natural Gasoline and the Volatile Hydrocarbons," Natural Gasoline Assn. of America, Tulsa (1948).

Bruges, E.A., Latto, B., and Ray, A.K.: New Correlations and Tables of the Coefficient of Viscosity of Water and Steam up to 1000 Bar and 1000°C," *Intl. J. Heat and Mass Transfer* (May 1966) 465–480.

Brusset, M.J. and Edgington, A.N.: "Equipment Performance in an Alberta High-Pressure Steam Injection Project," paper presented at the Pet. Soc. of CIM Tech. Meeting, Banff, May 24–26, 1967.

Buchwald, R.W. Jr., Hardy, W.C., and Neinast, G.S.: "Case Histories of Three In-Situ Combustion Projects," *J. Pet. Tech.* (July 1973) 784–792.

Buckles, R.S.: "Steam Stimulation Heavy Oil Recovery at Cold Lake, Alberta," paper SPE 7994 presented at the 1979 SPE California Regional Meeting, Ventura, April 18–20.

Buckley, S.E. and Leverett, M.C.: "Mechanism of Fluid Displacement in Sands," *Trans.*, AIME (1942) **146,** 107–116.

Bueno, N., Franco, L., Gonzalez, A., and Mayer, A.: "Plantas y Equipos en Proyectos Térmicos," paper presented at Petróleos de Venezuela—I Simposio de Crudos Extra-Pesados, Maracay, Oct. 13–15, 1976.

Buesse, H.: "Experimental Investigations of Fuel Recovery by Underground Partial Combustion of Petroleum Reservoirs," *Erdoel, Erdgas A.* (1971) **87,** No. 12, 414–427 (English translation).

Burger, J.G.: "Spontaneous Ignition in Oil Reservoirs," *Soc. Pet. Eng. J.* (April 1976) 73–81.

Burger, J., Aldea, Gh., Carcoana, A., Petcovici, V., Sahuquet, B., and Delye, M.: "Recherches de Base Sur La Combustion In-Situ et Resultats Récent sur Champ," *Proc.*, Ninth World Pet. Cong., Tokyo (1975) **3,** 279–289.

Burger, J.G. and Sahuquet, B.C.: "Chemical Aspects of In-Situ Combustion—Heat of Combustion and Kinetics," *Soc. Pet. Eng. J.* (Oct. 1972) 410–422; *Trans.*, AIME, **253.**

Burger, J.G. and Sahuquet, B.C.: "Laboratory Research on Wet Combustion," *J. Pet. Tech.* (Oct. 1973) 1137–1146.

Burger, J. and Sahuquet, B.: "Les Méthodes Thermiques de Production des Hydrocarbures," *Revue IFP* (March–April 1977) Chap. 5, 141–188.

Burger, M.: "Laboratory Research on Wet Combustion," *Proc.*, Simposio Sobre Crudos Pesados, U. del Zulia, Inst. de Investigaciones Petroleras, Maracaibo, July 1–3, 1974.

Burke, R.G.: "Combustion Project is Making a Profit," *Oil and Gas J.*, (Jan. 18, 1965) 44–46.

Burkill, G.C.C. and Leal, A.J.: "Aspectos Operacionales de la Inyección Selectiva de Vapor en Pozos de la C.S.V. de la Costa Bolívar, Estado Zulia," *Proc.*, Simposio Sobre Crudos Pesados, Inst. de Investigaciones Petroleras, U. del Zulia, Maracaibo, July 1–3, 1974.

Burns, J.: "A Review of Steam Soak Operation in California," *J. Pet. Tech.* (Jan. 1969) 25–34.

Burns, W.C.: "Water Treatment for Once-Through Steam Generators," *J. Pet. Tech.* (April 1965) 417–421.

Burris, D.R.: "Dual Polarity Oil Dehydration," *Pet. Eng.* (Aug. 1977) 30–41.

Bursell, C.G.: "Steam Displacement—Kern River Field," *J. Pet. Tech.* (Oct. 1970) 1225–1231.

Bursell, C.G., Taggert, H.J., and DeMirjian, H.A.: "Thermal Displacement Tests and Results, Kern River Field, California," *Prod. Monthly*, (Sept. 1966) **30,** No. 9, 18–21.

Burwell, E.L., Sterner, T.E., and Carpenter, H.C.: "In-Situ Retorting of Oil Shale," RI 7783, USBM (1973).

Bussey, L.E.: *The Economic Analysis of Industrial Projects*, Prentice-Hall Inc., Englewood Cliffs, NJ (1978).

Buxton, T.S. and Craig, F.F. Jr.: "Effect of Injected Air-Water Ratio and Reservoir Oil Saturation on the Performance of a Combination of Forward Combustion and Waterflooding," AIChE Symposium Series (1973) **69,** No. 127, 27–30.

Buxton, T.S. and Pollock, C.B.: "The Sloss COFCAW Project—Further Evaluation of Performance During and After Air Injection," *J. Pet. Tech.* (Dec. 1974) 1439–1448; *Trans.*, AIME, **257.**

C

Cady, G.V. and Moss, J.T.: "The Design and Installation of an Oxygen Supported Combustion Project in the Lindbergh Field, Alberta, Canada," paper presented at the 1981 World Oil and Gas Show, Dallas, Dec. 14–17.

Cameron Synthetic Fuels Report, Cameron Engineers Inc., Denver (June 1980) **17,** No. 2.

Cameron Synthetic Fuels Report, Cameron Engineers Inc., Denver (Sept. 1980) **17,** No. 3.

Campbell, G.G., Burwell, E.L., Sterner, T.E., and Core, L.L.: "Underground Combustion Oil-Recovery Experiments in the Venango Second Sand, Reno Pool, Venango County, PA," RI 6942, USBM (1967).

Campbell, G.G., Burwell, E.L., Sterner, T.E., and Core, L.L.: "Why a Fire Flood Project Failed," *World Oil* (Feb. 1966) 46–50.

Campbell, G.G., Scott, W.G., and Miller, J.S.: "Evaluation of Oil Shale Fracturing Tests Near Rock Spring, Wyoming," RI 7397, USBM (1970).

Carr, N.L., Kobayashi, R., and Burrows, D.B.: "Viscosity of Hydrocarbon Gases Under Pressure," *Trans.*, AIME (1954) **201,** 264–272.

Carslaw, H.S. and Jaeger, J.C.: *Conduction of Heat in Solids*, Oxford U. Press, Amen House, London (1950).

Casey, T.J.: "A Field Test of the In-Situ Combustion Process in a Near-Depleted Water Drive Reservoir," *J. Pet. Tech.* (Feb. 1971) 153–160.

Cato, B.W. and Frnka, W.A.: "Getty Oil Reports Fireflood Pilot is Successful Project," *Oil and Gas J.* (Feb. 12, 1968) 93–97.

Cato, B.W. and Frnka, W.A.: "Results of an In-Situ Combustion Pilot Project," *Prod. Monthly* (June 1968) 20–24.

Chappelear, J.E. and Volek, C.W.: "The Injection of a Hot Liquid Into a Porous Medium," *Soc. Pet. Eng. J.* (March 1969) 100–114; *Trans.*, AIME, **246.**

Chemical Engineers' Handbook, R.H. Perry and C.H. Chilton (eds.), fifth edition, McGraw-Hill Book Co. Inc., New York City (1973).

Chemistry of Coal Utilization—Supplementary Volume, H.H. Lowry (ed.), John Wiley & Sons Inc., New York City (1963).

Chew, Ju-Nam and Connally, Carl A. Jr.: "A Viscosity Correlation for Gas-Saturated Crude Oils," *Trans.*, AIME (1959) **216** 23–25.

Chirinos, F., Fuenmayor, I., Cerrada, R., Garcia, E., Zerpa, C., and Chacín, E.: "Levantamiento Artificial," Petróleos de Venezuela—I Simposio de Crudos Extra-Pesados, Maracay, Oct. 13–15, 1976.

Chu, C.: "Kinetics of Methane Oxidation Inside Sandstone Matrices," *Soc. Pet. Eng. J.* (June 1971) 145–151.

Chu, C.: "State-of-the-Art Review of Fireflood Field Projects," *J. Pet. Tech.* (Jan. 1982) 19–36.

Chu, C. and Trimble, A.E.: "Numerical Simulation of Steam Displacement—Field Applications," *J. Pet. Tech.* (June 1975) 765–776.

Chuoke, R.L., van Meurs, P., and van der Poel, C.: "The Instability of Slow, Immiscible, Viscous Liquid-Liquid Displacements in Porous Media," *Trans.*, AIME (1959) **216,** 188–194.

Clark, C.E. and Varisco, D.C.: "Net Energy and Oil Shale," *Proc.*, Eighth Oil Shale Symposium, Colorado School of Mines, Golden (July 1975) **70,** No. 3, 3–19.

Clark, G.A., Jones, R.G., Kinney, W.L., Schilson, R.E., Surkalo, H., and Wilson, R.S.: "The Fry In-Situ Combustion Test—Field Operations," *J. Pet. Tech.* (March 1965) 343–347.

Clark, G.A., Jones, R.G., Kinney, W.L., Schilson, R.E., Surkalo, H., and Wilson, R.S.: "The Fry In-Situ Combustion Test—Performance," *J. Pet. Tech.* (March 1965) 348–353.

Closmann, P.J., Jones, E.R., and Vogel, E.A.: "Effect of a Heat-Conducting Well Casing on the Temperature Distribution in an Observation Well," *Trans.*, ASME, **101,** 20–27.

Closmann, P.J.: "Steam Zone Growth During Multiple-Layer Steam Injection," *Soc. Pet. Eng. J.* (March 1967) 1–10.

Closmann, P.J.: "Steam Zone Growth in a Preheated Reservoir," *Soc. Pet. Eng. J.* (Sept. 1968) 313–320.

Closmann, P.J., Ratliff, N.W., and Truitt, N.E.: "A Steam-Soak Model for Depletion-Type Reservoirs," *J. Pet. Tech.* (June 1970) 757–770; *Trans.*, AIME, **249.**

Closmann, P.J. and Smith, R.A.: "Temperature Observations and Steam-Zone Rise in the Vicinity of a Steam-Heated Fracture," *Soc. Pet. Eng. J.* (Aug. 1983) 575–586.

Coats, K.H.: "In-Situ Combustion Model," *Soc. Pet. Eng. J.* (Dec. 1980) 533–554.

Coats, K.H.: "Simulation of Steamflooding with Distillation and Solution Gas," *Soc. Pet. Eng. J.* (Oct. 1976) 235–247.

Coats, K.H., George, W.D., Chu, C., and Marcum, B.E.: "Three-Dimensional Simulation of Steamflooding," *Soc. Pet. Eng. J.* (Dec. 1974) 573–592; *Trans.*, AIME, **257.**

Coats, K.H., Ramesh, A.B., Todd, M.R., and Winestock, A.G.: "Numerical Modeling of Thermal Reservoir Behavior," *Proc.*, CIM Canada/Venezuela Oil Sands Symposium, Edmonton, Alta. (1977) Special Vol. 17, 399–410.

Coberly, C.J. and Wagner, E.M.: "Some Considerations in Selection and Installation of Gravel Packs for Oil Wells," *Pet. Tech.* (Aug. 1938) 1–20.

Coleman, D.M. and Walker, E.L.: "Battrum Fire Flood Encourages Mobil," *Cdn. Pet.* (Dec. 1967) 19-22, 24.

Collins, R.E.: *Flow of Fluids Through Porous Materials*, Reinhold Publishing Corp., New York City (1961); reprinted by Petroleum Publishing Co., Tulsa, OK (1976).

Combarnous, M., and Bia, P.: "Combined Free and Forced Convection in Porous Media," *Soc. Pet. Eng. J.* (Dec. 1971) 399-405.

Combarnous, M. and Pavan, J.: "Déplacement Par L'Eau Chaude D'Huiles en Place Dans Un Milieu Poreux," *Comptes Rendus du Troisième Colloque, Assn. de Recherche sur les Techniques de Forage et de Production*, Editions Technip, Paris (1969) 737-757.

Combarnous, M. and Sorieau, P.: "Les Méthodes Thermiques de Production des Hydrocarbures," *Revue*, Inst. Français du Pétrole (July-Aug. 1976) Chap. 3, 543-577.

Compressed Air Handbook, second edition, Compressed Air and Gas Inst., McGraw-Hill Book Co. Inc., New York City (1954).

"Conditioning Water for Boilers," Nalco Chemical Co., Chicago (1962).

Connally, C.A. Jr. and Marberry, J.E.: "Steam Flood Pilot, Nacatoch Sand—Troy Field, Arkansas," paper SPE 4755 presented at the 1974 SPE Improved Oil Recovery Symposium, Tulsa, April 22-24.

"Continuous Steam Injection, Heat Scavenging Spark Kern River," *Oil and Gas J.* (Jan. 27, 1975) 127-141.

Cooperman, P.: "Some Criteria for the In-Situ Combustion of Crude Oil," *J. Appl. Phys.* (1959) **30,** 1376-1380.

Couch, E.J. and Selig, F.: "Further Discussion of a Study of Forward Combustion in a Radial System Bounded by Permeable Media," *J. Pet. Tech.* (Dec. 1963) 1370-1371.

Counihan, T.M.: "A Successful In-Situ Combustion Pilot in the Midway-Sunset Field, California," paper SPE 6525 presented at the 1977 SPE California Regional Meeting, Bakersfield, April 13-15.

Coutret, H.C. Jr.: "A Comparison of the Thermal Displacement Processes—Fire and Steam," *Drill Bit* (April 1968) 1, 4, 6, 8, 9.

Cragoe, C.S.: Miscellaneous Publication No. 97, U.S. Bureau of Standards (1929).

Craig, F.F. Jr.: *The Reservoir Engineering Aspects of Waterflooding*, Monograph Series, Society of Petroleum Engineers of AIME, Dallas (1971) **3.**

Craig, F.F. Jr. and Parrish, D.R.: "A Multipilot Evaluation of the COFCAW Process," *J. Pet. Tech.* (June 1974) 659-666.

Crawford, P.B.: "Combustion Calculations for Steam and Hot Water Generators," *Prod. Monthly* (Jan. 1967) 16-17.

Crawford, P.B.: "Thermal Recovery—A Summary," *Interstate Oil Compact Comm.* (Dec. 1966) 35-36.

Croes, G.A. and Schwarz, N.: "Dimensionally Scaled Experiments and Theories and the Water-Drive Process," *Trans.*, AIME (1955) **240,** 35-42.

Crolla, A., Herrera, A., and Menon, K.C.K.: "An Empirical Method for Prediction and Optimizing Steam Soak Performance, Bolivar Coast Field," *Proc.*, Simposio Sobre Crudos Pesados, U. del Zulia, Inst. de Investigaciones Petroleras, Maracaibo, July 1-3, 1974.

Cronquist, C.: "Dimensionless PVT Behavior of Gulf Coast Reservoir Oils," *J. Pet. Tech* (May 1973) 538-542.

Crookston, R.B., Culham, W.E., and Chen,W.H.: "A Numerical Simulation Model for Thermal Recovery Processes," *Soc. Pet. Eng. J.* (Feb. 1979) 37-58; *Trans.*, AIME, **267.**

Crude Petroleum and Petroleum Products Annual Petroleum Statements, published annually by the USBM, Washington, DC.

Cummins, J.J. and Robinson, W.E.: "Thermal Degradation of Green River Kerogen at 150°C to 350°C," RI 7620, LERC, USBM (1972).

D

Dabbous, M.K.: "In-Situ Oxidation of Crude Oils in Porous Media," PhD dissertation, U. of Pittsburgh, (1971).

Dabbous, M.K. and Fulton, P.F.: "Low-Temperature Oxidation Reaction Kinetics and Effects on the In-Situ Combustion Process," *Soc. Pet. Eng. J.* (June 1974) 253-262.

Dabbous, M.K., Reznik, A.A., Taber, J.J., and Fulton, P.F.: "The Permeability of Coal to Gas and Water," *Soc. Pet. Eng. J.* (Dec. 1974) 563-572; *Trans.*, AIME, **257.**

Dafter, R.: "Scraping the Barrel, The Worldwide Potential for Enhanced Oil Recovery," Management Reports, *Financial Times*, London (1980) 41.

Dafter, R.: "Winning More Oil," *Financial Times*, London (1981).

Dalton, R.L. Jr., Rapoport, L.A., and Carpenter, C.W. Jr.: "Laboratory Studies of Pilot Waterfloods," *Trans.*, AIME (1960) **219,** 24-30.

Datta, R.: "An Ammonia Injection Scheme for Fracturing and Permeability Enhancement of Coal Beds," *Proc.*, Second Annual Underground Coal Gasification Symposium, Morgantown Energy Research Center, U.S. DOE, Morgantown, WV (1976) MERC/SP-76/3.

Davidson, L.B.: "The Effect of Temperature on the Permeability Ratio of Different Fluid Pairs in Two-Phase Systems," *J. Pet. Tech.* (Aug. 1969) 1037-1046.

Davies, L.G., Silberberg, I.H., and Caudle, B.H.: "A Method of Predicting Oil Recovery in a Five-Spot Steamflood," *J. Pet. Tech.* (Sept. 1968) 1050-1058; *Trans.*, AIME, **243.**

de Haan, H.J. and Schenk, L.: "Performance and Analysis of a Major Steam Drive Project in the Tia Juana Field, Western Venezuela," *J. Pet. Tech.* (Jan. 1969) 111-119; *Trans.*, AIME **246.**

de Haan, H.J. and van Lookeren, J.: "Early Results of the First Large-Scale Steam Soak Project in the Tia Juana Field, Western Venezuela," *J. Pet. Tech.* (Jan. 1969) 101-110; *Trans.*, AIME, **246.**

Deppe, J.C.: "Injection Rates—The Effect of Mobility Ratio, Area Swept, and Pattern," *Soc. Pet. Eng. J.* (June 1961) 81-91; *Trans.*, AIME, **222.**

DePriester, C.L. and Pantaleo, A.J.: "Well Stimulation by Downhole Gas-Air Burner," *J. Pet. Tech.* (Dec. 1963) 1297-1302.

Dew, J.N. and Martin, W.L.: "Air Requirements for Forward Combustion," *Pet. Eng.* (Dec. 1964) 82-86.

Dewan, J.T. and Elfarr, J.: "Lifting of Heavy Oil with Inert-Gas-Operated Chamber Pumps," paper SPE 9913 presented at the 1981 SPE California Regional Meeting, Bakersfield, March 25-26.

Diaz-Muñoz, J. and Farouq Ali, S.M.: "Effectiveness of Hot-Water Stimulation of Heavy-Oil Formations," *J. Cdn. Pet. Tech.* (July-Sept. 1975) 66-76.

Dietrich, W.K. and Willhite, G.P.: "Steam Soak Results, Sisquoc Pool—Cat Canyon Field, Santa Barbara County," paper presented at Petroleum Industry Conf. on Thermal Oil Recovery, Los Angeles (June 1966).

Dietz, D.N.: "Hot Water Drive," *Proc.*, Seventh World Pet. Cong., Mexico City (1967) 3, 451-457.

Dietz, D.N.: "A Theoretical Approach to the Problem of Encroaching and Bypassing Edge Water," *Proc.*, Koninkl. Ned. Akad, Wetenschap (1953) B56, 83.

Dietz, D.N.: "Wet Underground Combustion, State of the Art," *J. Pet. Tech.* (May 1970) 605-617; *Trans.*, AIME, **249.**

Dietz, D.N. and Weijdema, J.: "Reverse Combustion Seldom Feasible," *Prod. Monthly* (May 1968) 10.

Dietz, D.N. and Weijdema, J.: "Wet and Partially Quenched Combustion," *J. Pet. Tech.* (April 1963) 411–415; *Trans.*, AIME, **243.**

Dillabough, J.A. and Prats, M.: "Proposed Pilot Test for the Recovery of Crude Bitumen from the Peace River Tar Sands Deposit—Alberta, Canada," *Proc.*, Simposio Sobre Crudos Pesados, Inst. de Investigaciones Petroleras, U. del Zulia, Maracaibo, Venezuela, July 1–3, 1974.

Dirksen, C.: "Natural Convection in Porous Media and Its Effect on Segregated Forward Combustion," *Soc. Pet. Eng. J.* (Sept. 1966) 267–280.

Dodson, C.R. and Standing, M.B.: "Pressure-Volume-Temperature and Solubility Relations for Natural-Gas-Water Mixtures," *Drill. and Prod. Prac.*, API (1944) 173–179.

Doscher, T.M. and Ershaghi, I.: "Current Economic Appraisal of Steam and Combustion Drives," paper SPE 7073 presented at the 1978 SPE Improved Oil Recovery Symposium, Tulsa, April 17–19.

Doscher, T.M. and Ghassemi, F.: "The Influence of Oil Viscosity and Thickness on the Steam Drive," *J. Pet. Tech.* (Feb. 1983) 291–298.

Doscher, T.M., Labelle, R.W., Sawatsky, L.H., and Zwicky, R.W.: "Steam Drive Successful in Canada's Oil Sands," *Pet. Eng.* (Jan. 1964) 71–78.

Doscher, T.M., Omoregie, O.S., and Ghassemi, F.: "Steam Drive Definition and Enhancement," *J. Pet. Tech.* (July 1982) 1543–1545.

Dougan, J.L.: "Recovery of Petroleum Products from Oil Shale," U.S. Patent No. 3, 241, 611 (March 22, 1966).

Dougan, P.M., Reynolds, F.S., and Root, P.J.: "The Potential for In-Situ Retorting of Oil Shale in the Piceance Creek Basin of Northwestern Colorado," *Quarterly*, Colorado School of Mines, Golden (Oct. 1971) 57–72.

Dowd, J.J., Elder, J.L., Capp, J.P., and Cohen, P.: "Field Scale Experiments in Underground Gasification of Coal at Gorgas, Alabama. Use of Electrolinking-Carbonization as Means of In-Situ Preparation," RI 5357, USBM (1957).

Drinkard, G., Prats, M., and O'Brien, S.M.: "Coal Disaggregation by Basic Aqueous Solution for Slurry Recovery," U.S. Patent No. 4,023,193 (June 1977).

Dubrovai, K.K., Sheinman, A.B., Sorokin, N.A., Sacks, C.L., Pronin, V.I., and Charuigin, M.M.: "Experiments on Thermal Recovery in the Chusovsk Town," *Pet. Economy* (Nov. 5, 1936).

Duerksen, J.H. and Gomaa, E.E.: "Status of the Section 26C Steamflood, Midway-Sunset Field, California," paper SPE 6748 presented at the 1977 SPE Annual Technical Conference and Exhibition, Denver, Oct. 9–12.

Duerksen, J.H. and Hsueh, L.M.: "Steam Distillation of Crude Oils," *Soc. Pet. Eng. J.* (April 1983) 265–271.

Duncan, D.C. and Swanson, V.E.: "Organic-Rich Shale of the U.S. and World Land Areas," USGS Circular 523 (1965).

Dunn, K.: "Use of Incremental Analysis in the Operations of Cyclic Steam Projects," paper SPE 2258 presented at the 1968 SPE Annual Meeting, Houston, Sept. 29–Oct. 2.

Durie, R.W.: "Temperature Distribution Ahead of an Advancing Steam Zone," *J. Cdn. Pet. Tech.* (Jan.–March 1969) 30–34.

Dyes, A.B.: "Production of Water-Driven Reservoirs Below Their Bubble Point," *Trans.*, AIME (1954) **201,** 240–244.

Dyes, A.B., Caudle, B.H., and Erickson, R.A.: "Oil Production After Breakthrough–As Influenced by Mobility Ratio," *Trans.*, AIME, (1954) **201,** 81–86.

Dykstra, H. and Parsons, H.L.: "The Prediction of Oil Recovery by Waterflooding," *Secondary Recovery of Oil in the United States*, second edition, API, New York City (1950) 160–174.

E

Earlougher, R.C. Jr.: *Advances in Well Test Analysis*, Monograph Series, Society of Petroleum Engineers, Dallas (1977) **5.**

Earlougher, R.C. Jr.: "Some Practical Considerations in the Design of Steam Injection Wells," *J. Pet. Tech.* (Jan. 1969) 79–86; *Trans.*, AIME, **246.**

Earlougher, R.C. Jr., Galloway, J.R., and Parsons, R.W.: "Performance of the Fry In-Situ Combustion Project," *J. Pet. Tech.* (May 1970) 551–557.

Eckert, E.R.G. and Drake, R.M. Jr.: *Heat and Mass Transfer*, McGraw-Hill Book Co. Inc., New York City (1959).

Eckert, E.R.G., Sparrow, E.M., Goldstein, R.J., Scott, C.J., and Pfender, E.: "Heat Transfer—A Review of 1971 Literature," *Intl. J. Heat Mass Transfer* (Nov. 1972) 1969–2010.

Edminster, W.C.: *Applied Hydrocarbon Thermodynamics*, Gulf Publishing Co., Houston (1961).

Edmondson, T.A.: "Effect of Temperature on Waterflooding," *J. Cdn. Pet. Tech.* (Oct.–Dec. 1965) 236–242.

Effinger, A.W. and Wasson, J.A.: "Applying Marx and Langenheim Calculations to the Prediction of Oil Recovery by Steamflooding in Venango Sands," USBM Circ. 8432 (Nov. 1969).

Eickmeier, J.R., Ersoy, D., and Ramey, H.J. Jr.: "Wellbore Temperatures and Heat Losses During Production or Injection Operations," *J. Cdn. Pet. Tech.* (April-June 1970) 115–121.

Eisenhawer, S.W., Johnson, D.R., Vigil, W.J., Medlau, R.F., and Gilby, D.: "A System to Evaluate the Performance of Insulated Tubulars in Steam Injection Wells," paper SPE 9911 presented at the 1981 SPE California Regional Meeting, Bakersfield, March 25–26.

Ejiogu, G.J., Bennion, D.W., Moore, R.G. and Donnelly, J.K.: "Wet Combustion—A Tertiary Recovery Process for the Pembina Cardium Reservoir," *J. Cdn. Pet. Tech.* (July–Sept. 1979) 58–65.

Elder, J.L.: "The Underground Gasification of Coal," *Chemistry of Coal Utilization*, Supplementary Volume, H.H. Lowry (ed.), John Wiley and Sons Inc., New York City (1963) 1023–1040.

Elder, R.B. and Wu, C.H.: "Correlation of Crude Oil Steam Distillation Yields with Basic Crude Oil Properties," *Soc. Pet. Eng. J.* (Dec. 1983) 937–945.

Elias, R. Jr., Johnstone, J.R., Krause, J.D., Scanlan, J.C., and Young, W.W.: "Steam Generation with High TDS Feedwater," paper SPE 8819 presented at the 1980 SPE California Regional Meeting, Los Angeles, April 9–11.

Elkins, L.F., Skov, A.M., Martin, P.J., and Lutton, D.R.: "Experimental Fireflood—Carlyle Field, Kansas," paper SPE 5014 presented at the 1974 SPE Annual Meeting, Houston, Oct. 6–9.

Elson, T.D. and Marsden, S.S.: "The Effectiveness of Foaming Agents at Elevated Temperatures over Extended Periods of Time," paper SPE 7116 presented at the 1978 SPE California Regional Meeting, San Francisco, April 12–14.

Elson, T.D.: "Phase Separation of Two-Phase Fluid in an Injection Wellbore," paper SPE 9916 presented at the 1981 SPE California Regional Meeting, Bakersfield, March 25-26.

Emery, L.W.: "Results from a Multi-Well Thermal-Recovery Test in Southeastern Kansas," *J. Pet. Tech.* (June 1962) 671-678; *Trans.*, AIME, **225.**

Energy Alternatives: A Comparative Analysis, The Science and Public Policy Program, U. of Oklahoma, Norman (1975) Chap. 15; U.S. Govt. Printing Office No. 041-011-00025-4.

Energy Study, Oregon Office of Energy Research and Planning, Office of the Governor, State of Oregon, Salem (1974) 150-180.

Engineering Data Book, ninth edition, Gas Processors Suppliers Assn., Tulsa (1979).

Enhanced Oil Recovery Field Reports, Society of Petroleum Engineers, Dallas, TX (issued biannually).

"Enhanced Recovery Report—Canadians, Venezuelans Exchange Views at Edmonton Oil Symposium," *Oilweek* (June 20, 1977) **9,** 12-32.

"Ensayan Varios Métodos de Producción," *Petróleo y Petroquímica Internacional* (Nov. 1973) 16-19.

Eson, R.L., Fitch, J.P., and Shannon, A.M.: "North Kern Front Field Steam Drive with Ancillary Materials," paper SPE 9778 presented at the 1981 SPE California Regional Meeting, Bakersfield, March 25-26.

Evers, R.H.: "Mixed Media Filtration of Oily Waste Waters," *J. Pet. Tech.* (Feb. 1975) 157-163.

F

Fanaritis, J.P. and Kimmel, J.D.: "Review of Once-Through Steam Generators," *J. Pet. Tech.* (April 1965) 409-416.

Farouq Ali, S.M.: "A Current Appraisal of In-Situ Combustion Field Tests," *J. Pet. Tech.* (April 1972) 477-486.

Farouq Ali, S.M.: "Current Status of Steam Injection as a Heavy Oil Recovery Method," *J. Cdn. Pet. Tech.* (Jan.-March 1974) 54-68.

Farouq Ali, S.M.: "Effect of Temperature and Pressure on the Viscosity of Steam and Other Gases," *Prod. Monthly* (March 1967).

Farouq Ali, S.M.: "Estimation, Correlation and Measurement of Oil Viscosities at High Temperatures," *Prod. Monthly* (March 1967).

Farouq Ali, S.M.: "Fluid Viscosity, Its Estimation, Measurement and Role in Thermal Recovery," *Prod. Monthly* (Jan. 1967) 20.

Farouq Ali, S.M.: *Oil Recovery by Steam Injection,* Producers Publishing Co. Inc., Bradford, PA (1970).

Farouq Ali, S.M.: "Well Stimulation by Downhole Thermal Methods," *Pet. Eng.* (Oct. 1973) 25-35.

Farouq Ali, S.M.: "Well Stimulation by Thermal Methods," *Prod. Monthly* (April 1968) 23-27.

Farouq Ali, S.M. and Meldau, R.F.: "Current Steamflood Technology," *J. Pet. Tech.* (Oct. 1979) 1332-1342.

Fassihi, M.R., Brigham, W.E., and Ramey, H.J. Jr.: "Reaction Kinetics of In-Situ Combustion: Part 1—Observations, Part 2—Modeling," *Soc. Pet. Eng. J.* (Aug. 1984) 399-416.

Fassihi, M.R., Gobran, B.D., and Ramey, H.J. Jr.: "Algorithm Calculates Performance of In-Situ Combustion," *Oil and Gas J.* (Nov. 16, 1981) 90-98.

Faulkner, B.L.: "Reservoir Engineering Design of Tertiary Miscible Gas Drive Pilot Project," paper SPE 5539 presented at the 1975 SPE Annual Technical Conference and Exhibition, Dallas, Sept. 28-Oct. 1.

Fazio, P.J. and Banderob, L.D.: "Thermal Well Problems and Completion Techniques in the Midway-Sunset Field," *Prod. Monthly* (Nov. 1966) 10-15.

Felber, B.J., Dauben, D.L., and Marrs, R.E.: "Lignosulfonates for High-Temperature Plugging," Canadian Patent No. 1,041,900 (Nov. 7, 1978).

Fiorillo, G.: "Exploration of the Orinoco Oil Belt—Review and General Strategy," paper UNITAR/CF10/V/3 presented at the 11th Intl. Conference on Heavy Crude and Tar Sands, Caracas Feb. 7-17, 1982.

Fischer, D.D., Brandenburg, C.F., King, S.B., Boyd, R.M., and Hutchinson, H.L.: "Status of the Linked Vertical Well in Process in Underground Coal Gasification," *Proc.*, Second Annual Underground Coal Gasification Symposium, Morgantown Energy Research Center, U.S. DOE, Morgantown, WV (1976) MERC/SP-76/3.

Fitch, J.P. and Minter, R.B.: "Chemical Diversion of Heat Will Improve Thermal Recovery," paper SPE 6172 presented at the 1976 SPE Annual Technical Conference and Exhibition, New Orleans, Oct. 3-6.

Fouda, A.E., Gregory, G.A., Rhodes, E., and Scott, D.S.: "Problems of Two-Phase Flow in Networks and Metering of Quality Mass Flow and Void Fraction," *The Oil Sands of Canada-Venezuela,* CIM (1977) Special Vol. 17, 663-671.

Fournier, K.P.: "A Numerical Method for Computing Recovery of Oil by Hot Water Injection in a Radial System," *J. Pet. Tech.* (June 1965) 131-140.

Fox, R.L., Donaldson, A.B., and Mulac, A.J.: "Development of Technology for Downhole Steam Production," paper SPE 9776 presented at the 1981 SPE California Regional Meeting, Bakersfield, March 25-26.

Franco, A.: "How MGO Handles Heavy Crude," *Oil and Gas J.*, (June 24, 1974) 105-109.

French, M.S. and Howard, R.L.: "Thermal Recovery. Part 1—The Steamflood Job: Hefner Sho-Vel-Tum," *Oil and Gas J.* (July 17, 1967) 64-66.

Friske, E.W.: "Costs of Steam Injection in the San Joaquin Valley," API paper 801-43i, presented at Spring Meeting of the Pacific Coast Dist., API Div. of Production, Los Angeles, May 2-4, 1967.

G

Gadelle, C.P., Burger, T.G., and Bardon, C.: "Heavy-Oil Recovery by In-Situ Combustion—Two Field Cases in Rumania," *J. Pet. Tech.* (Nov. 1981) 2057-2066.

Gambill, W.R.: "You Can Predict Heat Capacities," *Chemical Engineering* (June 1957).

Garon, A.M., Geisbrecht, R.A., and Lowry, W.E. Jr.: "Scaled Model Experiments of Fireflooding in Tar Sands," paper SPE 9949 presented at the 1980 SPE Annual Technical Conference and Exhibition, Dallas, Sept. 21-24.

Garon, A.M. and Wygal, R.J. Jr.: "A Laboratory Investigation of Fire-Water Flooding," *Soc. Pet. Eng. J.* (Dec. 1974) 537-544.

Garon, A.M. and Wygal, R.J. Jr.: "A Laboratory Investigation of Wet Combustion Parameters," *Soc. Pet. Eng. J.* (Dec. 1974) 537-544.

Gates, C.F., and Brewer, S.W.: "Steam Injection Into the D and E Zones, Tulare Formation, South Belridge Field, Kern County, California," *J. Pet. Tech.* (March 1975) 343-348.

Gates, C.F. and Holmes, B.G.: "Thermal Well Completion and Operation," *Proc.*, Seventh World Pet. Cong., Mexico City (1967) 419-429.

Gates, C.F., Jung, K.D., and Surface, R.A.: "In-Situ Combustion in the Tulare Formation, South Belridge Field, Kern County, California," *J. Pet. Tech.* (March 1978) 798–806; *Trans.*, AIME, **265.**

Gates, C.F. and Ramey, H.J. Jr.: "Field Results of South Belridge Thermal Recovery Experiment," *Trans.*, AIME (1958) **213,** 236–244.

Gates, C.F. and Ramey, H.J. Jr.: "A Method for Engineering In-Situ Combustion Oil Recovery Projects," *J. Pet. Tech.* (Feb. 1980) 285–294.

Gates, C.F. and Sklar, I.: "Combustion as a Primary Recovery Process—Midway Sunset Field," *J. Pet. Tech.* (Aug. 1971) 981–986; *Trans.*, AIME, **251.**

Gaucher, D.H. and Lindley, D.C.: "Waterflood Performance in a Stratified Five-Spot Reservoir—A Scaled-Model Study," *Trans.*, AIME (1960) **218,** 208–215.

Geertsma, J., Croes, G.A., and Schwartz, N.: "Theory of Dimensionally Scaled Models of Petroleum Reservoirs," *Trans.*, AIME (1956) **207,** 118–123.

Geffen, T.M.: "Oil Production to Expect from Known Technology," *Oil and Gas J.* (May 7, 1973) 66–68, 73–76.

"Getty Expands Bellevue Fire Flood," *Oil and Gas J.* (Jan. 13, 1975) 45–49.

Gibbon, A.: "Thermal Principle Applied to Secondary Oil Recovery," *Oil Weekly* (Nov. 6, 1944) 170–172.

Giguere, R.J.: "An In-Situ Recovery Process for the Oil Sands of Alberta," paper presented at the 26th Canadian Chemical Engineering Conf., Toronto, Oct. 3–6, 1976.

Giguere, R.J.: "An In-Situ Recovery Process for the Oil Sands of Alberta," *Energy Process Cdn.* (Sept.-Oct. 1977) **70,** No. 1, 36–40.

Giusti, L.E.: "CSV Makes Steam Soak Work in Venezuela Field," *Oil and Gas J.* (Nov. 4, 1974) 88–93.

Giusti, L.E.: "Experiencias de la C.S.V. con la Inyección Alternada de Vapor en la Costa Bolívar, Estado Zulia," *Proc.*, Simposio Sobre Crudos Pesados, U. del Zulia, Inst. de Investigaciones Petroleras, Maracaibo, Venezuela, July 1–3, 1974.

Glassett, J.M. and Glassett, J.A.: "The Production of Oil from Intermountain West Tar Sands Deposits," BUMines-OFR-92-76, USBM (1976).

Gobran, B.D., Brigham, W.E., and Ramey, H.J. Jr.: "Absolute Permeability as a Function of Confining Pressure, Pore Pressure, and Temperature," paper SPE 10156 presented at the 1981 SPE Annual Technical Conference and Exhibition, San Antonio, Oct. 5–7.

Gomaa, E.E., Duerksen, J.H., and Woo, P.T.: "Designing a Steamflood Pilot in the Thick Monarch Sand of the Midway-Sunset Field," *J. Pet. Tech.* (Dec. 1977) 1559–1568.

Gomaa, E.E. and Somerton, W.H.: "Thermal Behavior of Multifluid-Saturated Formations, Part I: Effect of Wettability, Saturation and Grain Structure," paper SPE 4896A presented at the 1971 SPE California Regional Meeting, San Francisco, April 4–5.

Gomaa, E.E. and Somerton, W.H.: "Thermal Behavior of Multifluid-Saturated Formations, Part II: Effect of Vapor Saturation—Heat Concept and Apparent Thermal Conductivity," paper SPE 4896B presented at the 1971 SPE California Regional Meeting, San Francisco, April 4–5.

Gottfried, B.S.: "A Mathematical Model of Thermal Oil Recovery in Linear Systems," *J. Pet. Tech.* (Sept. 1965) 196–210; *Trans.*, AIME, **234.**

Gottfried, B.S.: "Some Theoretical Aspects of Underground Combustion in Segregated Oil Reservoirs," *Soc. Pet. Eng. J.* (Sept. 1966) 281–291.

Grant, B.F.: "Retorting Oil Shale Underground—Problems and Possibilities," *Quarterly*, Colorado School of Mines, Golden (July 1974) 39–46.

Grant, B.R. and Szasz, S.E.: "Development of Underground Heat Wave for Oil Recovery," *Trans.*, AIME (1954) **201,** 108–118.

Gray, J.E.: *Making Thermal Recovery Projects Pay—Exploration and Economics of the Petroleum Industry*, Gulf Publishing Co. (1967) **5,** 239.

Green, D.W.: "Heat Transfer with Flowing Fluid through Porous Media," PhD dissertation, U. of Oklahoma, Norman (1963).

Green, D.W. and Perry, R.H.: "Heat Transfer With a Flowing Fluid Through Porous Media," paper 9 presented at the Fourth Natl. Heat Transfer Conference, Buffalo, NY (1960).

Green, K.B.: "Thermal Recovery. Part 2—The Fireflood: Cox Penn Sand," *Oil and Gas J.* (July 17, 1967) 66–69.

Gregg, D.W., Hill, R.W., and Olness, D.U.: "An Overview of the Soviet Effort in Underground Coal Gasification," UCRL-52004, Lawrence Livermore Natl. Laboratory, Livermore, CA (Jan. 29, 1976).

Grim, R.F.: *Clay Mineralogy*, McGraw-Hill Book Co. Inc., New York City (1968).

Gringarten, A.C. and Sauty, J.P.: "A Theoretical Study of Heat Extraction from Aquifers with Uniform Regional Flow," *J. Geophys. Research* (Dec. 1975) **80,** No. 35, 4956–4962.

Gunn, R.D., Gregg, D.W., and Whitman, D.L.: "A Theoretical Analysis of Soviet In-Situ Coal Gasification Field Tests," *Proc.*, Second Annual Underground Coal Gasification Symposium, Morgantown Energy Research Center, U.S. DOE, Morgantown, WV (1976) MERC/SP-76/3.

Gunn, R.D. and Krantz, W.B.: "Reverse Combustion Instabilities in Tar Sands and Coal," *Soc. Pet. Eng. J.* (Aug. 1980) 267–277.

H

Hagoort, H., Leijnse, A., and van Poelgeest, F.: "Steam-Strip Drive: A Potential Tertiary Recovery Process," *J. Pet. Tech.* (Dec. 1976) 1409–1419.

Hall, A.L. and Bowman, R.W.: "Operation and Performance of the Slocum Thermal Recovery Project," *J. Pet. Tech.* (April 1973) 402–408.

Hall, Howard N.: "Compressibility of Reservoir Rocks," *Trans.*, AIME (1953) **198,** 309–311.

Hampton, L.A.: "How Various Fuels Affect the Design and Operating Cost of Steam Generators," paper 6812 presented at the 19th Annual Petroleum Society of CIM Meeting, Calgary, May 7–10, 1968.

Handbook of Physical Constants, S.P. Clark Jr. (ed.), revised edition, Geological Soc. of America (1966) Memoir 97.

Handy, L.L., Ersaghi, I., Amaefule, J., and Hsu, J.: "The Use of Chemical Additives with Steam Injection to Increase Oil Recovery," *Proc.*, Fourth Annual DOE Symposium, Tulsa, OK (Aug. 1978) **1B**-Oil, D-9/1 through D-9/14.

Hanzlik, E.J.: "Steamflooding as an Alternative EOR Process for Light Oil Reservoirs," paper SPE 10319 presented at the 1981 SPE Annual Technical Conference and Exhibition, San Antonio, Oct. 5–7.

Hardy, W.C.: "Deep-Reservoir Fireflooding Economics for Independents," *Oil and Gas J.* (Jan. 18, 1971) 60–64.

Hardy, W.C.: "Sun Oil In-Situ Combustion Operations in U.S.A.," paper presented at the Symposium on Heavy Oil Recovery, Maracaibo, July 1-3, 1974.

Hardy, W.C., Fletcher, P.B., Shepard, J.C., Dittman, E.W., and Zadow, D.W.: "In-Situ Combustion Performance in a Thin Reservoir Containing High-Gravity Oil," *J. Pet. Tech.* (Feb. 1972) 199-208; *Trans.*, AIME, **253.**

Hardy, W.C. and Raiford, J.D.: "In-Situ Combustion in a Bartlesville Sand—Allen County, Kansas," *Proc.*, Tertiary Oil Recovery Conf., Wichita, KS, Oct. 23-24, 1975; Inst. of Mineral Resources Research, U. of Kansas, Lawrence.

Harmsen, G.J.: "A Note on COFCAW," *J. Pet. Tech.* (July 1969) 801-802.

Harmsen, G.J.: "Oil Recovery by Hot-Water and Steam Injection," *Proc.*, Seventh World Pet. Cong., Mexico City (1967) **3**, 243-251.

Hearn, C.L.: "Effect of Latent Heat Content of Injected Steam in a Steam Drive," *J. Pet.Tech.* (April 1969) 374-375.

Hearn, C.L.: "The El-Dorado Steam Drive—A Pilot Tertiary Recovery Test," *J. Pet. Tech.* (Nov. 1972) 1377-1384.

Heflin, J.D., Lawrence, T.D., Oliver, D., and Koenn, L.: "California Applications of the Continuous Carbon/Oxygen Log," paper presented at 1977 API Meeting, Bakersfield, CA, Oct. 25-27.

Helgeson, H.C.: "Thermodynamics of Hydrothermal Systems at Elevated Temperatures and Pressures," *American J. Science* (1967) **267,** 729-804.

Herrera, A.J.: "M-6 Steam Drive Project Design and Implementation," *J. Cdn. Pet. Tech.* (July-Sept. 1977) 62-71.

Herrera, A.J.: "The M6 Steam Drive Project Design and Implementation," *The Oil Sands of Canada-Venezuela, 1977* (1977) CIM Special Volume 17, 551-560.

Hess, P.H.: "Chemical Method for Formation Plugging," *J. Pet. Tech.* (May 1971) 559-564.

Hewitt, C.H. and Morgan, J.T.: "The Fry In-Situ Combustion Test—Reservoir Characteristics," *J. Pet. Tech.* (March 1965) 337-342.

Heymer, D.: "Sekundärverfahren unter Anwendung von Wärme im Feld Emlichheim," *Z. deutsche geolog. Ges.*, (Aug. 1967) **119,** 570-573.

Higgins, R.V. and Leighton, A.J.: "A Computer Method to Calculate Two-Phase Flow in Any Irregularly Bounded Porous Medium," *J. Pet. Tech.* (June 1962) 679-683; *Trans.*, AIME, **225.**

Higgins, R.V. and Leighton, A.J.: "Computer Prediction of Water Drive of Oil and Gas Mixtures through Irregularly Bounded Porous Media—Three-Phase Flow," *J. Pet. Tech.* (Sept. 1962), 1048-1054; *Trans.*, AIME, **228.**

Hill, F.L. and Land, P.E.: " 'MOCO' Combustion Projects, Midway-Sunset Oil Field," *Sum. Oper. Calif. Oil Fields* (Jan. 1971) 51-59.

History of Petroleum Engineering, API Div. of Production, Dallas (1961).

Hodgkinson, R.J. and Hugli, A.D.: "Determination of Steam Quality Anywhere in the System," *The Oil Sands of Canada-Venezuela,* CIM (1977) Special Vol. 17, 672-675.

Holke, D.C. and Huebner, W.B.: "Thermal Stimulation and Mechanical Techniques Permit Increased Recovery from Unconsolidated Viscous Oil Reservoir," paper SPE 3671 presented at the 1971 SPE California Regional Meeting, Los Angeles, Nov. 4-5.

Holliday, G.H.: "Calculation of Allowable Maximum Casing Temperature to Prevent Tension Failures in Thermal Wells," paper 69-PET-10 presented at the ASME Meeting, Sept. 21-25, 1969.

Holst, P.H. and Flock, D.L.: "Wellbore Behaviour During Saturated Steam Injection," *J. Cdn. Pet. Tech.* (1966).

Holst, P.H. and Karra, P.S.: "The Size of the Steam Zone in Wet Combustion," *Soc. Pet. Eng. J.* (Feb. 1975) 13-18.

Hong, K.C.: "Two-Phase Splitting at a Pipe Tee," *J. Pet. Tech.* (Feb. 1978) 290-296; *Trans.*, AIME, **265.**

Hong, K.C. and Jensen, R.B.: "Optimization of Multicycle Steam Stimulation," *Soc. Pet. J.* (Sept. 1969) 357-367; *Trans.*, AIME, **246.**

Horai, Ki-iti: "Thermal conductivity of Rock-Forming Minerals," *J. Geophys. Research* (1971) **76,** No. 5, 1278-1308.

Horne, J.S., Bousaid, J.S., Dore, T.L., and Smith, L.B.: "Initiation of an In-Situ Combustion Project in a Thin Oil Column Underlain by Water," *J. Pet. Tech.* (Oct. 1982) 2233-2243.

Howard, C.E., Widmeyer, R.H., and Haynes, S. Jr.: "The Charco Redondo Thermal Recovery Pilot," *J. Pet. Tech.* (Dec. 1977) 1522-1532.

Howard, F.A.: "Method of Operating Oil Wells," U.S. Patent No. 1,473,348 (filed Aug. 9, 1920; issued Nov. 6, 1923).

Hull, R.J.: "The Thermosludge Water Treating and Steam Generation Process," *J. Pet. Tech.* (Dec. 1967) 1537-1540.

Hutchison, V.V.: *A Bibliography of Thermal Methods of Oil Recovery,* Independent Petroleum Assn. of America, Tulsa (July 1965).

Huygen, H.A. and Huitt, J.L.: "Well-Bore Heat Losses and Casing Temperatures During Steam Injection," *Prod. Monthly* (Aug. 1966) 2-8.

Huygen, H.H.A. and Huitt, J.L.: "Wellbore Heat Losses and Casing Temperatures During Steam Injection," *Drill. and Prod. Prac.*, API (1967) 25-32.

Hwang, M.K., Jines, W.R., and Odeh, A.S.: "An In-Situ Combustion Process Simulator With a Moving Front Representation," *Soc. Pet. Eng. J.* (April 1982) 271-279.

I

"Imperial Poised to Launch Project at Cold Lake Heavy Oil Region," *Oilweek* (Nov. 21, 1977) 16, 17.

J

Jakob, M.: *Heat Transfer,* John Wiley and Sons Inc., New York City (1949) **1.**

Jenkins, R. and Aronofsky, J.S.: "Analysis of Heat Transfer Processes in Porous Media—New Concepts in Reservoir Heat Engineering," *Proc.*, 18th Technical Conf. on Petroleum Production, Pennsylvania State U., University Park, PA (1954) 69.

Jenkins, R. and Aronofsky, J.S.: "Analysis of Heat Transfer Processes in Porous Media—New Concepts in Reservoir Heating Engineering," *Prod. Monthly* (May 1955) 37.

Johnson, C.E. Jr.: "Prediction of Oil Recovery by Water Flood—A Simplified Graphical Treatment of the Dykstra-Parsons Method," *Trans.*, AIME (1956) **207,** 345-356.

Johnson, D.R. and Fox, R.L.: "Examination of Techniques for Thermally Efficient Delivery of Steam to Deep Reservoirs," paper SPE 8820 presented at the 1980 SPE/DOE Enhanced Oil Recovery Symposium, Tulsa, April 20-23.

Johnson, F.S., Bayazeed, A.F., and Dutcher, H.: "An Independent's Struggle With Steam," *Indep. Pet. Mon.* (July 1969) 16-22.

Johnson, F.S., Walker, S.J., and Bayazeed, A.F.: "Oil Vaporization During Steamflooding," *J. Pet. Tech.* (June 1971) 731-742.

Johnson, L.A. Jr., Fahy, L.J., Romanowski, L.J. Jr., and Thomas, K.P.: "An Evaluation of a Steamflood Experiment in a Utah Tar Sand Deposit," *J. Pet. Tech.* (May 1982) 1119-1126.

Johnson, L.A. Jr., Fahy, L.J., Romanowski, L.J. Jr., and Thomas, K.P.: "An Evaluation of a Steamflood Experiment in a Utah Tar Sand Deposit," paper SPE 10228 presented at the SPE 56th Annual Technical Conference and Exhibition, San Antonio, Oct. 5-7, 1981.

Joint Association Survey of the U.S. Oil and Gas Producing Industry, sponsored by the API, IPAA, and the Mid-Continent Oil and Gas Assn., published annually by API, Dallas.

Jones, J.: "Steam Drive Model for Hand-Held Programmable Calculators," *J. Pet. Tech.* (Sept. 1981) 1583-1598.

Jordan, J.K., Rayne, J.R., and Marshall, S.W.III: "A Calculation Procedure for Estimating the Production History During Hot Water Injection in Linear Reservoirs," paper presented at the Twentieth Technical Conf. on Petroleum Production, Pennsylvania State U., University Park, May 9-10, 1957.

Jorden, J.R. Jr. and Campbell, F.L.: *Well Logging,* Monograph Series, Society of Petroleum Engineers of AIME, Dallas (in preparation).

Joseph, C. and Pusch, W.H.: "A Field Comparison of Wet and Dry Combustion," *J. Pet. Tech.* (Sept. 1980) 1523-1528.

K

Kahn, S.A. and Marsden, S.S.: "Foam Proves Effective Way To Solve Production Problems," *Oil and Gas J.* (June 5, 1967) 126-129.

Katz, D.L., Cornell, D., Kobayashi, R., Poettmann, F.H., Elenbaas, J.R., and Weinaug, C.F.: *Handbook of Natural Gas Engineering,* McGraw-Hill Book Co. Inc., New York City (1959).

Katz, D.L. and Hachmuth, K.H.: "Vaporization Equilibrium Constants in a Crude Oil—Natural Gas System," *Ind. Eng. Chem.* (1937) **29,** No. 9.

Kazemi, H.: "Locating a Burning Front by Pressure Transient Measurements," *J. Pet. Tech.* (Feb. 1966) 227-232; *Trans.*, AIME, **237.**

Keenan, J.H. and Keyes, F.G.: *Thermodynamic Properties of Steam,* John Wiley and Sons Inc., New York City (1936).

Keller, H.H. and Couch, E.J.: "Well Cooling by Downhole Circulation of Water," *Soc. Pet. Eng. J.* (Dec. 1968) 405-412; *Trans.*, AIME, **243.**

Kemp, P.B.: "Steam Stimulation Tidelands Operations, Wilmington Oil Field," paper 801-46G presented at the API Prod. Div. Pacific Coast Dist. Spring Meeting, Los Angeles, May 12-14, 1970.

Kendall, G.H.: "Importance of Reservoir Description in Evaluating In-Situ Recovery Methods for Cold Lake Heavy Oil—Part 1, Reservoir Description," paper 7620 presented at the CIM 27th Annual Technical Meeting, Calgary, Canada, June 7-11, 1976.

Kennedy, J.L.: "Study Analyzes Effect of Oil Price on Enhanced Recovery's Future Role," *Oil and Gas J.* (Dec. 27, 1976) 185-191.

Kesler, M.G. and Lee, B.I.: "Improve Prediction of Enthalpy of Fractions," *Hydrocarbon Processing* (March 1976) 153-158.

Koch, R.L.: "Practical Use of Combustion Drive at West Newport Field," *Pet. Eng.* (Jan. 1956) **72.**

Konopnicki, D.T., Traverse, E.F., Brown, A., and Deibert, A.D.: "Design and Evaluation of the Shiells Canyon Field Steam-Distillation Drive Pilot Projects," *J. Pet. Tech.* (May 1979) 546-560.

Kuhn, C.S. and Koch, R.L.: "In-Situ Combustion—Newest Method of Increasing Oil Recovery," *Oil and Gas J.* (Aug. 10, 1953) **52,** 92.

Kunii, D. and Smith, J.M.: "Heat Transfer Characteristics of Porous Rocks: II: Thermal Conductivity of Unconsolidated Particles With Flowing Fluids," *AIChE J.* (March 1961) 29.

Kunii, D. and Smith, J.M.: "Thermal Conductivities of Porous Rocks Filled With Stagnant Fluid," *Soc. Pet. Eng. J.* (March 1961) 37-42.

Kuo, C.H.: "A Heat Transfer Study for the In-Situ Combustion Process," paper SPE 2651 presented at the 1969 SPE Annual Meeting, Denver, Sept. 28-Oct. 1.

Kuo, C.H., Shain, S.A., and Phocas, D.M.: "A Gravity Drainage Model for the Steam-Soak Process," *Soc. Pet. Eng. J.* (June 1970) 119-126.

Kuuskraa, V.A., Hammershaimb, E.C., and Doscher, T.M.: "Improved Steam Flooding by Use of Additives," U.S. DOE, Oakland, CA (Dec. 1979).

Kyte, J.R., Stanclift, R.J. Jr., Stephan, S.C. Jr., and Rapoport, L.A.: "Mechanism of Water Flooding in the Presence of Free Gas," *Trans.*, AIME (1956) **207,** 215-221.

L

Land, C.S., Cupps, C.Q., Marchant, L.C., and Carlson, F.M.: "Field Test of Reverse-Combustion Oil Recovery from a Utah Tar Sand," *J. Cdn. Pet. Tech.* (April-June 1977) 34-38.

Landrum, B.L., Smith, J.E., and Crawford, P.B.: "Calculation of Crude Oil Recoveries by Steam Injection," *Trans.*, AIME (1960) **219,** 251.

Langnes, G.L. and Beeson, C.M.: "In-Situ Combustion Combined With Waterflooding, Part I," *Pet. Eng.* (July 1965) 92, 93, 96, 101. "Part II," *Pet. Eng.* (Aug. 1965) 8, 98.

Lauwerier, H.A.: "The Transport of Heat in an Oil Layer Caused by Injection of Hot Fluid," *Applied Science Research (1955)* **A5,** 145-150.

Lee, S.T., Jacoby, R.H., Chen, W.H., and Culham, W.E.: "Experimental and Theoretical Studies on the Fluid Properties Required for Simulation of Thermal Processes," *Soc. Pet. Eng. J.* (Oct. 1981) 535-550.

Lekas, M.A.: "Progress Report on the Geokinetics Horizontal In-Situ Retorting Process," *Proc.*, 12th Oil Shale Symposium, Colorado School of Mines, Golden (1979) 228-236.

Lenoir, J.M.: "Effect of Pressure on Thermal Conductivity of Liquids," *Petroleum Refiner* (1956) **36,** No. 8, 162-164.

Lenoir, J.M. and Comings, E.W.: "Thermal Conductivity of Gases, Measurements at High Pressure," *Chem. Eng. Prog.* (1951) **47,** 223-231.

"Letter Symbols for Petroleum Reservoir Engineering, Natural Gas Engineering, and Well Logging Quantities," *Trans.*, AIME (1965) **234,** 1463-1496.

Leutwyler, K.: "Casing Temperature Studies in Steam Injection Wells," *J. Pet. Tech.* (Sept. 1966) 1157-1162.

Leutwyler, K. and Bigelow, H.L.: "Temperature Effects on Subsurface Equipment in Steam Injection Systems," *J. Pet. Tech.* (Jan. 1965) 93-101; *Trans.*, AIME, **234.**

Lewis, J.O.: "Methods of Increasing the Recovery from Oil Sands," *Bull. 148, Petroleum Technology,* USBM (1971) **37.**

Licht, W. and Stechert, D.C.: "The Variation of the Viscosity of Gases and Vapours with Temperature," *J. Phys. Chem.* (1944) **48,** 23.

Lillie, W.H.E. and Springer, F.P.: "Status of the Steam Drive Pilot in the Georgsdorf Field, Federal Republic of Germany," *J. Pet. Tech.* (Jan. 1981) 173–180.

Lindsly, B.E.: "Recovery by Use of Heated Gas," *Oil and Gas J.* (Dec. 20, 1928) 27.

Lombard, D.B. and Carpenter, H.C.: "Recovering Oil by Retorting a Nuclear Chimney in Oil Shale," *J. Pet. Tech.* (June 1967) 727–734.

Long, G. and Chierici, G.: "Salt Content Changes Compressibility of Reservoir Brines," *Pet. Eng.* (July 1961) B-25 through B-31.

Long, R.J.: "Case History of Steam Soaking in the Kern River Field, California," *J. Pet. Tech.* (Sept. 1965) 989–993.

López C., F.F. and Rivera R., J.: "Simulación Matemática de los Mecanismos de Transferencia de Calor hacia las Formaciones que Atraviezan los Pozos Injectores de Vapor," Publication No. 72BH/089, Inst. Mexicano del Petróleo (Nov. 1971).

Lucas, R.N.: "Dehydration of Heavy Crudes by Electrical Means," paper SPE 1506 presented at the 1966 SPE Annual Meeting, Dallas, Oct. 2–5.

Lytle, W.S.: "How Steam Rates in Pennsylvania," *Oil and Gas J.* (Jan. 23, 1967) 5, 101, 104, 106.

M

Malofeev, G.E.: "Calculation of the Temperature Distribution in a Formation When Pumping Hot Fluid into a Well," *Neft i Gaz* (1960) **3,** No. 7, 59–64.

Mandl, G. and Volek, C.W.: "Heat and Mass Transport in Steam-Drive Processes," *Soc. Pet. Eng. J.* (March 1969) 59–79; Trans., AIME, **246.**

Marberry, J.E. and Bhatia, S.K.: "Fosterton Northwest—A Tertiary Combustion Case History," paper SPE 4764 presented at the 1974 SPE Improved Oil Recovery Symposium, Tulsa, April 22–24.

Martin, J.C.: "A Theoretical Analysis on Steam Stimulation," *J. Pet. Tech.* (March 1967) 411–418; *Trans.*, AIME, **240.**

Martin, J.D. and Levanas, L.D.: "Air Oxidation of Sulfide in Process Water," *Oil and Gas J.* (June 11, 1962) 184–187.

Martin, W.L., Alexander, J.D., and Dew, J.N.: "Process Variables of In-Situ Combustion," *Trans.*, AIME (1958) **213,** 28–35.

Martin, W.L., Alexander, J.D., Dew, J.N., and Tynan, J.W.: "Thermal Recovery at North Tisdale Field, Wyoming," *J. Pet. Tech.* (May 1972) 606–616.

Martin, W.L., Dew, J.N., Powers, M.L., and Steves, H.B.: "Results of a Tertiary Hot Waterflood in a Thin Sand Reservoir," *J. Pet. Tech.* (July 1968) 739–750; *Trans.*, AIME, **243.**

Marx, J.W. and Langenheim, R.H.: "Reservoir Heating by Hot Fluid Injection," *Trans.*, AIME (1959) **216,** 312–315.

Marx, J.W. and Langenheim, R.H.: "Authors' Reply to H.J. Ramey, Jr.," in response to "Further Discussion of Paper Published in Transactions Volume 216," *Trans.*, AIME (1959) **216,** 365.

Marx, J.W. and Trantham, J.C.: "Reverse Combustion Produces Altered Crude," *Oil and Gas J.* (May 17, 1965) 123, 124, 127.

Matheny, L.S. Jr.: "EOR Methods Help Ultimate Recovery," *Oil and Gas J.* (March 31, 1980) 79–124.

Matthews, C.S. and Russell, D.G.: *Pressure Buildup and Flow Tests in Wells,* Monograph Series, Society of Petroleum Engineers, Dallas (1967) **1.**

Matzick, A. and Dannenberg, R.O.: "USBM Gas Combustion Retort for Oil Shale—A Study of Process Variable," RI 5545, USBM (1960).

McAdams, W.H.: *Heat Transmission,* third edition, McGraw-Hill Book Co. Inc., New York City (1954).

McCarthy, H.E. and Cha, C.Y.: "OXY Modified In-Situ Process Development and Update," *Quarterly,* Colorado School of Mines, Golden (April 1976).

McCray, A.W.: *Petroleum Evaluations and Economic Decisions,* Prentice-Hall Inc., Englewood Cliffs, NJ (1975).

McKetta, J.J. Jr. and Katz, D.L.: "Phase Relationships of Hydrocarbon-Water Systems," *Trans.*, AIME (1974) **170,** 34–43.

McLennan, I.C.: "Raising of High Pressure Steam from Brackish Lake Maracaibo Water," *J. Inst. Petroleum* (Aug. 1968) 212–232.

McNamara, P.H., Peil, C.A., and Washington, L.J.: "Characterization, Fracturing, and True In-Situ Retorting in the Antrim Shale of Michigan," *Proc.*, 12th Oil Shale Symposium, Colorado School of Mines, Golden (1979) 353–365.

McNeil, J.S. Jr. and Moss, J.T.: "Recent Progress in Oil Recovery by In-Situ Combustion," *Pet. Eng.* (July 1958) B29–B32, B36, B41, B42.

McNeil, J.S. Jr. and Nelson, T.W.: "Thermal Methods Provide Three Ways to Improve Oil Recovery," *Oil and Gas J.* (Jan. 19, 1959) 86–98.

"Measuring, Sampling, and Testing Crude Oil," API Standard 2500, API, Dallas; reproduced in *Petroleum Production Handbook,* McGraw-Hill Book Co. Inc., New York City (1962) **1,** Chap. 16.

Melcher, A.G.: "Net Energy Analysis: An Energy Balance Study of Western Fossil Fuels," *Proc.*, Ninth Oil Shale Symposium, Colorado School of Mines, Golden (July 1976) **71,** No. 3, 11–31.

Meldau, R.F. and Lumpkin, W.B.: "Phillips Tests Methods to Improve Drawdown and Producing Rates in Venezuela Fire Flood," *Oil and Gas J.* (Aug. 12, 1974) 127–134.

Meldau, R.F. and Lumpkin, W.B.: "Producing Venezuelan Fire Floods," *Proc.*, Simposio Sobre Crudos Pesados, U. del Zulia, Inst. de Investigaciones Petroleras, Maracaibo, Venezuela, July 1–3, 1974.

Miller, C.A.: "Stability of Moving Surfaces in Fluid Systems With Heat and Mass Transport; III. Stability of Displacement Fronts in Porous Media," *AIChE J.* (May 1975) 474–479.

Morel-Seytoux, H.J.: "Analytical-Numerical Method in Waterflooding Predictions," *Soc. Pet. Eng. J.* (Sept. 1965) 247–258; *Trans.*, AIME, **234.**

Morel-Seytoux, H.J.: "Unit Mobility Ratio Displacement Calculations for Pattern Floods in a Homogeneous Medium," *Soc. Pet. Eng. J.* (Sept. 1966) 217–227; *Trans.*, AIME, **237.**

Moss, J.T.: "Laboratory Investigation of the Oxygen Combustion Process for Heavy Oil Recovery," paper SPE 10706 presented at the 1982 SPE/DOE Enhanced Oil Recovery Symposium, Tulsa, April 5–8.

Moss, J.T.: "Mini-Tests Evaluate Thermal-Drive Variables," *Oil and Gas J.* (April 1, 1974) 111–114.

Moss, J.T.: "Predicting Oil Displacement By Thermal Methods," paper SPE 9944 presented at the 1981 SPE California Regional Meeting, Bakersfield, March 25–26.

Moss, J.T. and White, P.D.: "How to Calculate Temperature Profiles in a Water-Injection Well," *Oil and Gas J.* (March 9, 1959) 174–178.

Moss, J.T., White, P.D., and McNeil, J.S. Jr.: "In-Situ Combustion Process—Results of a Five-Well Field Experiment in Southern Oklahoma," *J. Pet. Tech.* (April 1959) 55–64; Trans., AIME, **216.**

Muskat, M.: *The Flow of Homogeneous Fluids Through Porous Media,* J.W. Edwards Inc., Ann Arbor, MI (1946) 458–465.

Muskat, M.: *Physical Principles of Oil Production*, McGraw-Hill Book Co. Inc., New York City (1949) 302.

Myal, F.R., and Farouq Ali, S.M.: "Recovery of Penn-Grade Crude Oils by Steam," *J. Pet. Tech.* (June 1970) 705-710.

Myhill, N.A. and Stegemeier, G.L.: "Steam Drive Correlation and Prediction," *J. Pet. Tech.* (Feb. 1978) 173-182.

N

Natland, M.L.: "Project Oil Sand," *Athabasca Oil Sands: A Collection of Papers*, M.A. Carrigy (ed.), Research Council of Alberta, Edmonton, Canada (1963) 143-155.

Nelson, T.W. and McNeil, J.S. Jr.: "How to Engineer an In-Situ Combustion Project," *Oil and Gas J.* (June 5, 1961) 58-65.

Nelson, T.W. and McNeil, J.S. Jr.: "Oil Recovery by Thermal Methods, Part I," *Pet. Eng.* (Feb. 1959) B27-B32.

Nelson, T.W. and McNeil, J.S. Jr.: "Past, Present, and Future Development in Oil Recovery by Thermal Methods," *Pet. Eng.*, Part I, (Feb. 1959) B27, Part II (March 1959) B75.

Nelson, W.L.: *Petroleum Refinery Engineering*, fourth edition, McGraw-Hill Book Co. Inc., New York City (1958).

Neuman, C.H.: "A Mathematical Model of the Steam Drive Process Applications," paper SPE 4757 presented at the 1975 SPE California Regional Meeting, Ventura, April 2-4.

Newman, G.H.: "Pore-Volume Compressibility of Consolidated, Friable, and Unconsolidated Reservoir Rocks Under Hydrostatic Loading," *J. Pet. Tech.* (Feb. 1973) 129-134.

Nicholls, J.H. and Luhning, R.W.: "Heavy Oil Sand In-Situ Pilot Plans in Alberta (Past and Present)," *J. Cdn. Pet. Tech.* (July-Sept. 1977) 50-61.

Noran, D.: "Enhanced Recovery Requires Special Equipment," *Oil and Gas J.* (July 12, 1976) 50-56.

Noran, D.: "Growth Marks Enhanced Oil Recovery," *Oil and Gas J.* (March 27, 1978) 113-140.

Noran, D. and Franco, A.: "Production Reports," *Oil and Gas J.* (April 5, 1976) 107-138.

O

Offeringa, J.: "A Mathematical Model of Cyclic Steam Injection," paper PD8(6) presented at the Eighth World Pet. Cong., Moscow, June 13-19, 1971.

Owens, M.E. and Bramley, B.G.: "Performance of Equipment Used in High-Pressure Steam Floods," *J. Pet. Tech.* (Dec. 1966) 1525-1531.

Owens, W.D. and Suter, V.E.: "Steam Stimulation—Newest Form of Secondary Petroleum Recovery," *Oil and Gas J.* (April 26, 1965) 82-87, 90.

Ozen, A.S. and Farouq Ali, S.M.: "An Investigation of the Recovery of the Bradford Crude by Steam Injection," *J. Pet. Tech.* (June 1969) 692-698.

P

Palm, J.W., Kirkpatrick, J.W., and Anderson, W.: "Determination of Steam Quality Using an Orifice Meter," *J. Pet. Tech.* (June 1968) 587-591.

Palmer, G.: "Thermal Conductivity of Liquids," *Ind. and Eng. Chem.* (1948) **40,** 89.

"Panel Says Synfuels Pose No CO_2 Hazard, for Now," *Science* (Aug. 31, 1979) 884.

Parra, D.: "Effects of Gravitational Segregation in Displacement of Petroleum and Water by Steam," paper presented at the Third Soc. Venez. Ing. Pet. Tech. Pet. Conf., Maracaibo, Oct. 14-16, 1971.

Parra, D. and Pujol, L.: "Steam Soak Pilot Tests," "Projectos Experimentales de Injección Alternada de Vapor," *Proc.*, Simposio Sobre Crudos Pesados, U. del Zulia, Inst. de Investigaciones Petroleras, Maracaibo, Venezuela, July 1-3, 1974.

Parrish, D.R. and Craig, F.F. Jr.: "Laboratory Study of a Combination of Forward Combustion and Waterflooding—The COFCAW Process," *J. Pet. Tech.* (June 1969) 753-761; *Trans.*, AIME, **246.**

Parrish, D.R., Pollock, C.B., and Craig, F.F. Jr.: "Evaluation of COFCAW as a Tertiary Recovery Method, Sloss Field, Nebraska," *J. Pet. Tech.* (June 1974) 676-686.

Parrish, D.R., Pollock, C.B., Ness, N.L., and Craig, F.F. Jr.: "A Tertiary COFCAW Pilot Test in the Sloss Field, Nebraska," *J. Pet. Tech.* (June 1974) 667-675; *Trans.*, AIME, **257.**

Parrish, D.R., Rausch, R.W., Beaver, K.W., and Wood, H.W.: "Underground Combustion in the Shannon Pool, Wyoming," *J. Pet. Tech.* (Feb. 1962) 197-205; *Trans.*, AIME, **225.**

Penberthy, W.L. and Bayless, J.H.: "Silicate Foam Wellbore Insulation," *J. Pet. Tech.* (June 1974) 583-588.

Penberthy, W.L. Jr. and Ramey, H.J. Jr.: "Design and Operation of Laboratory Combustion Tubes," *Soc. Pet. Eng. J.* (June 1966) 183-198.

Pérez, R.: "Evaluación Cíclica con Agua Caliente en las Arenas del Grupo I—Camp Morichal," Petróleos de Venezuela—I Simposio de Crudos Extra-Pesados, Maracay, Oct. 13-15, 1976.

Perkins, T.K. and Johnson, D.C.: "A Review of Diffusion and Dispersion in Porous Media," *Soc. Pet. Eng. J.* (March 1963) 70-84; *Trans.*, AIME, **228.**

Perry, G.T. and Warner, W.S.: "Heating Oil Wells by Electricity," U.S. Patent No. 45,584 (July 4, 1865).

Petcovici, V.: "Considerations on the Possibilities of Controlling the Combustion Front in an In-Situ Oil Field Combustion Process," *Rev. Inst. Franc. Pétrol Ann. Combust. Liquides* (Dec. 1970) 1355-1374. (in French).

"Petro-Canada Uses Electricity Before Steam," *Enhanced Recovery Week* (Sept. 14, 1981) 2.

Pierce, B.O. and Foster, R.M.: *A Short Table of Integrals*, fourth edition, Blaisdell Publishing Co., Waltham, MA (1957).

Poettmann, F.H. and Benham, A.L.: "Thermal Recovery Process, An Analysis of Laboratory Combustion Data," *Trans.*, AIME (1958) **213,** 406-410.

Poettmann, F.H., Schilson, R.E. and Surkalo, H.: "Philosophy and Technology of In-Situ Combustion in Light Oil Reservoirs," *Proc.*, Seventh World Pet. Cong., Mexico City (1967) **III,** 487.

Pollock, C.B. and Buxton, T.S.: "Performance of a Forward Steam Project-Nugget Reservoir, Winkleman Dome Field, Wyoming," *J. Pet. Tech.* (Jan. 1969) 35-40.

Pollock, C.B. and Buxton, T.S.: "Winkleman Dome Steam Drive a Success," *Oil and Gas J.* (March 31, 1980) 74-124.

Poston, S.W., Ysrael, S.C., Hossain, A.K.M.S., Montgomery, E.F. III, and Ramey, H.J. Jr.: "The Effect of Temperature on Irreducible Water Saturation and Relative Permeability of Unconsolidated Sands," *Soc. Pet. Eng. J.* (June 1970) 171-180; *Trans.*, AIME, **249.**

Prats, M.: "The Heat Efficiency of Thermal Recovery Processes," *J. Pet. Tech.* (March 1969) 323-332; *Trans.*, AIME, **246.**

Prats, M.: "Peace River Steam Drive Scaled Model Experiments," *The Oil Sands of Canada-Venezuela*, CIM (1977) Special Vol. **17,** 346-363.

Prats, M., Closmann, P.J., Ireson, A.T., and Drinkard, G.: "Soluble-Salt Processes for In-Situ Recovery of Hydrocarbons From Oil Shale," *J. Pet. Tech.* (Sept. 1977) 1078–1088.

Prats, M., Jones, R.F., and Truitt, N.E.: "In-Situ Combustion Away from Thin, Horizontal Gas Channels," *J. Pet. Tech.* (March 1968) 18–32; *Trans.*, AIME, **243.**

Prats, M., Matthews, C.S., Jewett, R.L., and Baker J.D.: "Prediction of Injection Rate and Production History for Multifluid Five-Spot Floods," *J. Pet. Tech.* (May 1959) 98–105; *Trans.*, AIME, **216.**

Price, E.O. and McLaren, G.R.: "Steam Cyclic Operations at Midway-Sunset Sections 15A and 23A," Pet. Ind. Conf. on Thermal Rec., Los Angeles, June 6, 1966.

Proc., CIM Canada/Venezuela Oil Sands Symposium, Edmonton, Alta. (1977) Special Vol. **17.**

Proc., Eleventh Oil Shale Symposium, J.H. Gary (ed.), Colorado School of Mines Press, Golden (1978).

Proc., Second Annual Underground Coal Gasification Symposium, L.Z. Schuck (ed.), Morgantown Energy Research Center, U.S. DOE, Morgantown, WV (1976) MERC/SP-76/3.

Proc., Simposio Sobre Crudos Pesados, U. del Zulia, Inst. de Investigaciones Petroleras, Maracaibo, Venezuela (1974).

Proc., Sixth Annual Underground Coal Gasification Symposium, Laramie Energy Technology Center and Williams Brothers Engineering Co., Shangri-La, OK (1980).

Pujol, L. and Boberg, T.C.: "Scaling Accuracy of Laboratory Steam Flooding Models," paper SPE 4191 presented at the 1972 SPE California Regional Meeting, Bakersfield, Nov. 8–10.

Pursley, S.A.: "Experimental Studies of Thermal Recovery Processes," *Proc.*, Simposio Sobre Crudos Pesados, U. del Zulia, Inst. de Investigaciones Petroleras, Maracaibo, Venezuela, July 1–3, 1974.

R

Raimondi, P., Terwilliger, P.L., and Wilson, L.A. Jr.: "A Field Test of Underground Combustion of Coal," *J. Pet. Tech.* (Jan. 1975) 35–41.

Ramey, H.J. Jr.: "Commentary on the Terms 'Transmissibility' and 'Storage'," *J. Pet. Tech.* (March 1975) 294–295.

Ramey, H.J. Jr.: "Discussion of Reservoir Heating by Hot Fluid Injection," *Trans.*, AIME (1959) **216,** 364–365.

Ramey, H.J. Jr.: "How to Calculate Heat Transmission in Hot Fluid Injection," *Pet. Eng.* (Nov. 1964) 110–120.

Ramey, H.J. Jr.: "Rapid Method of Estimating Reservoir Compressibility," *J. Pet. Tech.* (April 1964) 447–454; *Trans.*, AIME, **231.**

Ramey, H.J. Jr.: "Transient Heat Conduction During Radial Movement of a Cylindrical Heat Source—Applications to the Thermal Recovery Process," *Trans.*, AIME (1959) **216,** 115–122.

Ramey, H.J. Jr.: "Wellbore Heat Transmission," *J. Pet. Tech.* (April 1962) 427–440; *Trans.*, AIME, **225.**

Ramey, H.J. Jr. and Nabor, G.W.: "A Blotter-Type Electrolytic Model Determination of Areal Sweeps in Oil Recovery by In-Situ Combustion," *Trans.*, AIME (1954) **201,** 119–123.

Rattia, A.J. and Farouq Ali, S.M.: "Effect of Formation Compaction on Steam Injection Response," paper SPE 10323 presented at the SPE 56th Annual Technical Conference and Exhibition, San Antonio, Oct. 5–7, 1981.

Reed, M.G.: "Gravel Pack and Formation Sandstone Dissolution During Steam Injection," *J. Pet. Tech.* (June 1980) 941–949.

Reed, R.L., Reed, D.W., and Tracht, J.H.: "Experimental Aspects of Reverse Combustion in Tar Sands," *Trans.*, AIME (1960) **219,** 99–108.

Reyes, R.B. and Siemak, J.B.: "Comparison of Caustic Soda and Soda Ash as Chemical Media for SO_2 Recovery in Oilfield Steam Generators," *J. Pet. Tech.* (May 1980) 751–756.

Reznik, A.A., Dabbous, M.K., Fulton, P.F., and Taber, J.J.: "Air-Water Relative Permeability Studies of Pittsburgh and Pocahontas Coals," *Soc. Pet. Eng. J.* (Dec. 1974) 556–562; *Trans.*, AIME, **257.**

Ricketts, T.E.: "Occidental's Retort 6 Rubblizing and Rock Fragmentation Program," *Proc.*, 13th Oil Shale Symposium, Colorado School of Mines, Golden (1980) 46–61.

Ridley, R.D.: "In-Situ Processing of Oil Shale," *Quarterly*, Colorado School of Mines, Golden (April 1974) 21–24.

Rincón, A.: "Proyectos Pilotos de Recuperación Térmica en la Costa Bolívar, Estado Zulia," *Proc.*, Simposio Sobre Crudos Pesados, U. Del Zulia, Inst. de Investigaciones Petroleras, Maracaibo, Venezuela, July 1–3, 1974.

Rincón, A.C., Diaz-Muñoz, J., and Farouq Ali, S.M.: "Sweep Efficiency in Steamflooding," *J. Cdn. Pet. Tech.* (July–Sept. 1970) 175–184.

Rincón, A.C. and Farouq Ali, S.M.: "Formation Heating by Steam Injection at a Variable Rate," *Prod. Monthly* (May 1968) 28–32.

Rintoul, B.: "Kern River Steam Expansion," *Calif. Oil World* (Sept. 15, 1970) 1, 2, 4, 5.

Rivero, R.T. and Heintz, R.C.: "Field Experiments in Spacing of Steam Soak Wells," *Proc.*, Simposio Sobre Crudos Pesados, U. del Zulia, Inst. de Investigaciones Petroleras, Maracaibo, Venezuela, July 1–3, 1974.

Rivero, R.T. and Heintz, R.C.: "Resteaming Time Determination—Case History of a Steam Soak Well in Midway Sunset," *J. Pet. Tech.* (June 1975) 665–671.

Roberts, J.C. and Williams, J.W.: "Steam Injection, Oil Recovery, Techniques and Water Quality," *J. WPCF* (Aug. 1970) 1437–1445.

Robinson, R.J., Bursell, C.G., and Restine, J.L.: "A Caustic Steamflood—Kern River Field, California," paper SPE 6523 presented at the 1977 SPE California Regional Meeting, Bakersfield, April 13–15.

Rohsenow, W.M. and Choi, H.Y.: *Heat, Mass and Momentum Transfer*, Prentice-Hall Inc., Englewood Cliffs, NJ (1961).

Rohsenow, W.M. and Hartnett, J.P.: *Handbook of Heat Transfer*, McGraw-Hill Publishing Co. Inc. (1973) 3–121.

Rojas G., G., Barrios, T., Scudiero, B., and Ruiz, J.: "Rheological Behavior of Extra-Heavy Crude Oils from the Orinoco Oil Belt," *Proc.*, Oil Sands of Canada-Venezuela, CIM Special Vol. 17, (1977) 284–302.

Romero, E. and Brigham, W.E.: "Optimization of Steam Drive Processes by Geometric Programming," paper presented at the symposium on Heavy Oil Recovery, Maracaibo, July 1–3, 1974.

Romero-Juárez, A.: "A Simplified Method for Calculating Temperature Changes in Deep Oil Well Stimulations," paper SPE 5888, Society of Petroleum Engineers, Richardson, TX (1976).

Rosenbaum, M.J.F. and Matthews, C.S.: "Studies in Pilot Waterflooding" *Trans.*, AIME (1959) **216,** 316–323.

Roupert, R.C., Choate, R., Cohen, S., Lee, A.A., Lent, J., and Spraul, J.R.: "Energy Extraction from Coal In-Situ—A Five-Year Plan," TID27203, prepared by TRW Systems, available from National Technical Information Service (1976).

Rubinshtein, L.L.: "The Total Heat Losses in Injection of a Hot Liquid into a Stratum," *Neft i Gaz* (1959) **2,** No. 9, 41–48.

S

Samaroo, B.H. and Guerrero, E.T.: "The Effect of Temperature on Drainage Capillary Pressure in Rocks Using a Modified Centrifuge," paper SPE 10153 presented at the 1981 SPE Annual Technical Conference and Exhibition, San Antonio, Oct. 5–7.

Sanyal, S.K., Brigham, W.E., Ramey, H.J. Jr., Marsden, S.S. Jr., and Kucuk, F.: "Heavy Oil Research Program at SUPRI," *Proc.*, Fourth Annual DOE Symposium, Tulsa, OK (Aug. 1978) **1B**-Oil, D-5/1 through D-5/28.

Sanyal, S.K., Marsden, S.S. Jr., and Ramey, H.J. Jr.: "Effect of Temperature on Petrophysical Properties of Reservoir Rock," paper SPE 4898 presented at the 1976 SPE Annual Meeting, Houston, Oct. 6–9.

Satman, A., Brigham, W.E., and Ramey, H.J. Jr.: "An Investigation of Steam Plateau Phenomena," paper SPE 7965 presented at the 1979 SPE California Regional Meeting, Ventura, April 18–20.

Satter, A.: "Heat Losses During Flow of Steam Down a Wellbore," *J. Pet. Tech.* (July 1965) 845–851; *Trans.*, AIME, **234.**

Satter, A. and Parrish, D.R.: "A Two-Dimensional Analysis of Reservoir Heating by Steam Injection," *Soc. Pet. Eng. J.* (June 1971) 185–197; *Trans.*, AIME, **251.**

Saucier, R.J.: "Considerations in Gravel Pack Design," *J. Pet. Tech.* (Feb. 1974) 205–212; *Trans.*, AIME, **257.**

Sawyer, D.N., Cobb, W.M., Stalkup, F.I., and Braun, P.H.: "Factorial Design Analysis of Wet-Combustion Drive," *Soc. Pet. Eng. J.* (Feb. 1974) 25–34.

Schafer, J.C.: "Thermal Recovery in the Schoonebeek Oil Field—Fifteen Years of Experience," *Erdoel-Erdgas Zeitschrift* (Oct. 1974) 372–379 (in English).

Schenk, L.: "Analysis of the Early Performance of the M-6 Steam-Drive Project, Venezuela," paper SPE 10710 presented at the 1982 SPE/DOE Enhanced Oil Recovery Symposium, Tulsa, April 5–8.

Schild, A.: "A Theory for the Effect of Heating Oil-Producing Wells," *Trans.*, AIME (1957) **210,** 1–10.

Schoeppel, R.J., Perry, R.H., and Campbell, J.M.: "The Role of Water Vapour in the Oxygen-Carbon Reaction," paper presented at the 55th Natl. AIChE Meeting, Houston, Feb. 7–11, 1965.

Schrider, L.A. and Whieldon, C.E.: "Underground Coal Gasification—A Status Report," *J. Pet. Tech.* (Sept. 1977) 1179–1185.

Schuck, L.A., Pasini, J. III, Martin, J.W., and Bisset, L.A.: "Status of the MERC In-Situ Gasification of Eastern U.S. Coals Project," *Proc.*, Second Annual Underground Coal Gasification Symposium, Morgantown Energy Research Center, U.S. DOE, Morgantown, WV (1976) MERC/SP-76/3.

Scudiero, B. and Rojas G., G.: "Efecto de la Temperatura sobre las Propiedades Reológicas de Varios Crudos de la Faja Petrolífera del Orinoco," *Proc.*, Simposio Sobre Crudos Pesados, U. del Zulia, Inst. de Investigaciones Petroleras, Maracaibo, Venezuela, July 1–3, 1974.

Seba, R.D. and Perry, G.E.: "A Mathematical Model of Repeated Steam Soaks of Thick Gravity Drainage Reservoirs," *J. Pet. Tech.* (Jan. 1969) 87–94; *Trans.*, AIME, **246.**

Sheinman, A.B., Malofeev, G.E., and Sergeev, A.I.: "The Effect of Heat on Underground Formations for the Recovery of Crude Oil—Thermal Recovery Methods of Oil Production," Nedra, Moscow (1969); Marathon Oil Co. Translation (1973) 166.

"Shell In-Situ Pilot Project Opens," *Oilweek* (Dec. 3, 1979) 3.

Shore, R.A.: "The Kern River SCAN Automation System—Sample, Control, and Alarm Network," paper SPE 4173 presented at the 1972 SPE California Regional Meeting, Bakersfield, Nov. 8–10.

Showalter, W.E.: "Combustion Drive Tests," *Soc. Pet. Eng. J.* (March 1963), 53–58; *Trans.*, AIME, **228.**

Showalter, W.E. and MacLean, M.A.: "Fireflood at Brea-Olinda Field, Orange County, California," paper SPE 4763 presented at the 1974 SPE Improved Oil Recovery Symposium, Tulsa, April 22–24.

Shutler, N.D. and Boberg, T.C.: "A One-Dimensional Analytical Technique for Predicting Oil Recovery by Steamflooding," *Soc. Pet. Eng. J.* (Dec. 1972) 489–498.

"The SI Metric System of Units and SPE's Tentative Metric Standard," *J. Pet. Tech.* (Dec. 1977) 1575–1611; also available as pamphlet from SPE, Dallas (1979).

Simm, C.N.: "Improved Firefloods May Cut Steam's Advantages," *World Oil* (March 1972) 59–60, 62.

Simon, R. and Poynter, W.G.: "Downhole Emulsification for Improving Viscous Crude Production," *J. Pet. Tech.* (Dec. 1968) 1349–1353.

Simpson, J.J.: "Financing Enhanced Recovery," *J. Pet. Tech.* (July 1977) 771–775.

Sinnokrot, A.A., Ramey, H.J. Jr., and Marsden, S.S. Jr.: "Effect of Temperature Level Upon Capillary Pressure Curves," *Soc. Pet. Eng. J.* (March 1971) 13–22.

Skinner, W.C.: "A Quarter Century of Production Practices," *J. Pet. Tech.* (Dec. 1973) 1425–1431.

Slobod, R.L.: "Gas Injection Improves Steam-Drive Oil Recovery," *Oil and Gas J.* (March 24, 1969) 138–140.

Slobod, R.L.: "Use of a Permanent Gas Phase to Augment the Benefits of Steam Injection," *Prod. Monthly* (Jan. 1969) 6–9, 15.

Smith, C.R.: "Good Pilot Flood Design Boosts Field Project Profit," *World Oil* (Nov.–Dec. 1973; Jan.–July 1974) Parts 1 through 9.

Smith, C.R.: *Mechanics of Secondary Oil Recovery*, Reinhold Publishing Corp., New York City (1966).

Smith, D.D. and Weinbrandt, R.M.: "Calculation of Unsteady State Heat Loss for Steam Injection Wells Using a TI-59 Programmable Calculator, paper SPE 8914 presented at the 1980 SPE California Regional Meeting, Pasadena, April 9–11.

Smith, H.J.: "Fifteen Practical Ways to Improve Steam Soak Performance," *World Oil* (Dec. 1966) 67–71.

Smith, M.L.: "Waste Water Reclamation for Steam Generator Feed, Kern River Field, California," paper SPE 3689 presented at the 1971 SPE California Regional Meeting, Los Angeles, Nov. 4–5.

Smith, M.W.: "Simultaneous Underground Combustion and Water Injection in the Carlyle Pool, Iola Field, Kansas," *J. Pet. Tech.* (Jan. 1966) 11–18.

Smith, R.V., Bertuzzi, A.F., Templeton, E.E., and Clampitt, R.L.: "Recovery of Oil by Steam Injection in the Smackover Field, Arkansas," *J. Pet. Tech.* (Aug. 1973) 883–889.

Snavely, E.S. Jr. and Bertness, T.A.: "Removal of Sulfur Dioxide Emissions by Scrubbing with Oilfield-Produced Water," *J. Pet. Tech.* (Feb. 1975) 227–232.

Snavely, E.S. Jr. and Blount, F.E.: "Rates of Dissolved Oxygen with Scavengers in Sweet and Sour Brines," *Proc.*, NACE 25th Annual Conf. (1969) 397–404.

Socorro, J.B. and Reid, T.B.: "Comportamiento de Equipos. Proyectos de Estimulación Cíclica con Agua Caliente—Campo Morichal, Estado Monagas," *Proc.*, Third Soc. Venezuela Eng. Pet. Tech. Petroleum Conference, Maracaibo (1972) **1,** 361–368.

Somerton, W.H.: "Some Thermal Characteristics of Porous Rocks," *Trans.*, AIME (1958) **213,** 375-378.

Somerton, W.H. and Boozer, G.D.: "Thermal Characteristics of Porous Rocks at Elevated Temperatures," *J. Pet. Tech.* (June 1960) 77-81.

Somerton, W.H. and Gupta, V.S.: "Role of Fluxing Agents in Thermal Alteration of Sandstones," *J. Pet. Tech.* (May 1965) 585-593.

Somerton, W.H., Keese, J.A. and Chu, S.L.: "Thermal Behavior of Unconsolidated Oil Sands," *Soc. Pet. Eng. J.* (Oct. 1974) 513-521.

Somerton, W.H., Mehta, M.M., and Dean, G.W.: "Thermal Alteration of Sandstone," *J. Pet. Tech.* (May 1965) 589-593.

Somerton, W.H. and Selim, M.A.: "Additional Thermal Data for Porous Rocks—Thermal Expansion and Heat of Reaction," *Soc. Pet. Eng. J.* (Dec. 1961) 249-253.

Sonosky, J.M. and Babcock, C.D.: "Financing Thermal Operations," *J. Pet. Tech.* (Sept. 1965) 999-1006.

"South America—Everybody Is Drilling Almost Everywhere—Venezuela," *World Oil* (Aug. 15, 1980) 103-109.

Spillette, A.G.: "Heat Transfer During Hot Fluid Injection into an Oil Reservoir," *J. Cdn. Pet. Tech.* (Oct.-Dec. 1965) 213-218.

Spillette, A.G. and Nielsen, R.L.: "Two-Dimensional Method for Predicting Hot Waterflood Recovery Behavior," *J. Pet. Tech.* (June 1968) 627-638 and discussion (July 1968) 770; *Trans.*, AIME, **243.**

Standing, M.B.: *Volumetric and Phase Behavior of Oil field Hydrocarbon Systems,* first edition, Reinhold Publishing Corp., New York City (1952).

Standing, M.B.: *Volumetric and Phase Behavior of Oil Field Hydrocarbon Systems,* second edition, SPE, Dallas (1977).

Standing, M.B. and Katz, D.L.: "Density of Natural Gases," *Trans.*, AIME (1942) **146,** 140-149.

Statistical Abstracts of the United States, published annually by the U.S. Dept. of Commerce, Washington, DC.

Steam Tables, Properties of Saturated and Superheated Steam, Combustion Engineering Inc., third edition, 23rd Printing, Windsor, CT (1940).

Steam—Its Generation and Use, 37th edition, The Babcock and Wilcox Co., New York City (1963).

Stegemeier, G.L., Laumbach, D.D., and Volek, C.W.: "Representing Steam Processes With Vacuum Models," *Soc. Pet. Eng. J.* (June 1980) 151-174.

Sterner, T.E. and Campbell, G.G.: "A Case History of Two Steam-Injection Pilot Tests in Pennsylvania," *Prod. Monthly* (Oct. 1968) 10-14.

Stewart, I.McC. and Wall, T.F.: "Reaction Rate Analysis of Borehole 'In-Situ' Gasification Systems," paper presented at 16th Intl. Symposium on Combustion, The Combustion Inst. (1977).

Stewart, R.E.D.: "Utilization of Low BTU Gas From In-Situ Coal Gasification for On-Site Power Generation," *Proc.*, Second Annual Underground Coal Gasification Symposium, Morgantown Energy Research Center, U.S. DOE, Morgantown, WV (1976) MERC/SP-76/3.

Stokes, D.D., Brew, J.R., Whitten, D.G., and Wooden, L.G.: "Steam Drive as a Supplemental Recovery Process in an Intermediate-Viscosity Reservoir, Mount Poso Field, California," *J. Pet. Tech.* (Jan. 1978) 125-131.

Stokes, D.D. and Doscher, T.M.: "Shell Makes a Success of Steam Flood at Yorba Linda," *Oil and Gas J.* (Sept. 2, 1974) 71-76.

Stokes, D.D. and Doscher, T.M.: "Steam Flooding in the Conglomerate Section, Yorba Linda Field, California," *Proc.*, Simposio Sobre Crudos Pesados, U. del Zulia, Inst. de Investigaciones Petroleras, Maracaibo, Venezuela, July 1-3, 1974.

Stovall, S.L.: "Recovery of Oil from Depleted Sands by Means of Dry Steam," *Oil Weekly* (Aug. 13, 1934) 17-24.

Strange, L.K.: "Ignition: Key in Combustion Recovery," *Pet. Eng.* (Dec. 1964) 105.

Strassner, J.E.: "Effects of pH on Interfacial Films and Stability of Crude Oil/Water Emulsions," *J. Pet. Tech.* (March 1968) 303-312; *Trans.*, AIME, **243.**

Suman, G.O. Jr.: *World Oil's Sand Control Handbook,* Gulf Publishing Co., Houston (1975).

Summary of Operations of California Oil Fields, California Div. of Oil and Gas, San Francisco (yearly since 1915).

"Supplement to Letter Symbols and Computer Symbols for Petroleum Reservoir Engineering, Natural Gas Engineering, and Well Logging Quantities," *Trans.*, AIME (1972) **253,** 556-574.

"Supplement to Letter and Computer Symbols for Petroleum Reservoir Engineering, Natural Gas Engineering, and Well Logging Quantities," *J. Pet. Tech.* (Oct. 1975) 1244-1264; *Trans.*, AIME, **259.**

Synthetic Fuels, Cameron Engineers Inc., Denver (June 1976) 2-54 through 2-57.

Synthetic Fuels, Cameron Engineers Inc., Denver (March 1977) 2-11 and 2-12.

Synthetic Fuels, Cameron Engineers Inc., Denver (June 1977) 4-11 and 4-14.

Synthetic Fuels, Cameron Engineers Inc., Denver (Sept. 1977) 2-30 and 2-31.

Synthetic Fuels, Cameron Engineers Inc., Denver (March 1978) 4-2.

Synthetic Fuels Data Handbook, Cameron Engineers Inc., Denver (1975).

T

Tadema, H.J.: "Oil Production by Underground Combustion," *Proc.*, Fifth World Pet. Cong., New York (1959) Sec. II, 279.

Tadema, H.J. and Weijdema, J.: "Spontaneous Ignition in Oil Sands," *Oil and Gas J.* (Dec. 14, 1970) 77-80.

Tarakad, R.R. and Danner, R.P.: "A Comparison of Enthalpy Prediction Methods," *AIChE J.* (March 1976) **22,** No. 2, 409-411.

Tebert, J.A. and Kemp, P.B.: "Cyclic Steam Injection Successful at Wilmington," *Pet. Eng.* (April 1966) 66-81.

Technical Data Book, API, New York City (1964).

Temperature—Its Measurement and Control in Science and Industry, Reinhold Publishing Corp., New York City (1941) 284-314.

Terschak, W.: "Steam Soak Successful in West German Field," *Pet. Eng.* (April 1967) 55-60.

Terwilliger, P.L.: "Fireflooding Shallow Tar Sands—A Case History," paper SPE 5568 presented at the 1975 SPE Annual Technical Conference and Exhibition, Dallas, Sept. 28-Oct. 1.

Terwilliger, P.L., Clay, R.R., Wilson, L.A. Jr., and Gonzalez-Gerth, E.: "Fireflood of the P_{2-3} Sand Reservoir in the Miga Field of Eastern Venezuela," *J. Pet. Tech.* (Jan. 1975) 9-14.

The Minerals Yearbook, published annually by U.S. Dept. of the Interior, Washington, DC.

Thomas, G.W.: "Approximate Methods for Calculating the Temperature Distribution During Hot Fluid Injection," *J. Cdn. Pet. Tech.* (Oct.-Dec. 1967) 123-129.

Thomas, G.W.: "A Study of Forward Combustion in a Radial System Bounded by Permeable Media," *J. Pet. Tech.* (Oct. 1963) 1145-1149; *Trans.*, AIME, **228.**

Thomas, G.W., Buthod, A.P., and Allag, O.: "An Experimental Study of the Kinetics of Dry, Forward Combustion," Fossil Energy Rep. No. BETC-1820-1, U.S. DOE (Feb. 1979).

Thurber, J.L. and Welbourn, M.E.: "How Shell Attempted to Unlock Utah Tar Sands," *Pet. Eng.* (Nov. 1977) 31, 34, 38, and 42.

Tiab, D., Okoye, C.U., and Osman, M.M.: "Caustic Steam Flooding," *J. Pet. Tech.* (Aug. 1982) 1817-1827.

Todd, J.C. and Howell, E.P.: "Numerical Simulation of In-Situ Electrical Heating to Increase Oil Mobility," *Oil Sands of Canada-Venezuela*, CIM (1977) Special Vol. **17**, 19.

Touloukian, Y.S., Kirby, R.K., Taylor, R.E., and Lee, T.Y.R.: "Specific Heat—Nonmetallic Liquids and Gases," *Thermophysical Properties of Matter*, IFI/Plenum, New York City (1970) **6.**

Touloukian, Y.S., Kirby, R.K., Taylor, R.E., and Lee, T.Y.R.: "Specific Heat—Nonmetallic Solids," *Thermophysical Properties of Matter*, IFI/Plenum, New York City (1970) **5.**

Touloukian, Y.S., Kirby, R.K., Taylor, R.E., and Lee, T.Y.R.: "Thermal Conductivity—Nonmetallic Liquids and Gases," *Thermophysical Properties of Matter*, IFI/Plenum, New York City (1970) **3.**

Touloukian, Y.S., Kirby, R.K., Taylor, R.E., and Lee, T.Y.R.: "Viscosity," *Thermophysical Properties of Matter*, IFI/Plenum, New York City (1970)**11.**

Towson, D.E. and Boberg, T.C.: "Gravity Drainage in Thermally Stimulated Wells," *J. Cdn. Pet. Tech.* (Oct.-Dec. 1967) 130-135.

Trantham, J.C. and Marx, J.W.: "Bellamy Field Tests: Oil from Tar by Counterflow Underground Burning," *J. Pet. Tech.* (Jan. 1966) 109-115; *Trans.*, AIME, **237.**

Trimble, A.E.: "Steam Mobility in Porous Media," paper SPE 5018 presented at the SPE 49th Annual Meeting, Houston, Oct. 6-9, 1974.

Trube, A.S.: "Compressibility of Natural Gases," *Trans.*, AIME (1957) **210,** 355-357.

Trube, A.S.: "Compressibility of Undersaturated Hydrocarbon Reservoir Fluids," *Trans.*, AIME (1956) **210,** 341-344.

Trujano, D.N., Garza, B.T., and Lecona, S.C.: "How Ambient Conditions Affect Steam-Line Heat Losses," *Oil and Gas J.* (Jan. 21, 1974) 83-86.

"Twentieth Century Petroleum Statistics," DeGolyer and MacNaughton, Dallas (1976).

U

U.S. Patent No. 3,654,691 (Oct. 20, 1969).

V

Valleroy, V.V., Willman, B.T., Campbell, J.B. and Powers, L.W.: "Deerfield Pilot Test of Recovery by Steam Drive," *J. Pet. Tech.* (July 1967) 956-964.

van der Knaap, W.: "M-6 Steam Drive Project, Preliminary Results of a Large Scale Field Test," paper SPE 9452 presented at the 1980 SPE Annual Technical Conference and Exhibition, Dallas, Sept. 21-24.

van der Knaap, W.: "Nonlinear Behavior of Elastic Porous Media," *Trans.*, AIME (1959) **216,** 179-187.

van Dijk, C.: "Steam-Drive Project in the Schoonebeek Field, The Netherlands," *J. Pet. Tech.* (March 1968) 295-302.

van Ginneken, A.J.J. and Klein, J.P.: "The Desulphurization and Purification of Gases," *Proc.*, Ninth World Pet. Cong., Tokyo (1975) **6,** 107-116.

van Heiningen, J. and Schwarz, N.: "Recovery Increase by 'Thermal Drive'," *Proc.*, Fourth World Pet. Cong., Rome (1955) Sec. II, 299.

van Lookeren, J.: "Calculation Methods for Linear and Radial Steam Flow in Oil Reservoirs," *Soc. Pet. Eng. J.* (June 1983) 427-439.

van Poollen, H.K.: "Transient Tests Find Fire Front in an In-Situ Combustion Project," *Oil and Gas J.* (Feb. 1, 1965) 78.

"Venezuela—Hay Que Echarle Mano al Orinoco," *Petróleo Intl.* (Aug. 1977) 17-19.

Viktorov, P.F. and Teterev, I.G.: "Characteristics of Operation and Possible Oil Recovery from the Sixth Formation of Arlansk Oil Field," *Neft i Gaz* (Jan. 1970) 31-4; Tulsa Abstract No. 131,602.

Villanueva, L.F. and Pittman, G.M.: "Geology of the Kern River Oil Field With a Case History of Its Development and Rejuvenation by Steam," paper presented at the AIME Pacific Southwest Mineral Industry Conference, San Francisco, May 27-29, 1970.

Viloria, G. and Farouq Ali, S.M.: "Rock Thermal Conductivity and Its Variation With Density, Temperature and Fluid Saturation," *Prod. Monthly* (Aug. 1968) 27-30.

Viloria, G. and Farouq Ali, S.M.: "Thermal Properties of Rocks and Fluids," *Prod. Monthly* (July 1968) 20-24.

"Vital Statistics" *Petroleum Independent* (Oct. 1980).

Volek, C.W. and Pryor, J.A.: "Steam Distillation Drive—Brea Field, California," *J. Pet. Tech.* (Aug. 1972) 899-906.

Vonde, T.R.: "Specialized Pumping Techniques Applied to a Very Low-Gravity, Sand-Laden Crude—Cat Canyon Field, California," *J. Pet. Tech.* (Sept. 1982) 1951-1957.

Vossoughi, S., Willhite, G.P., Kritikos, W.P., Guvenir, I.M., and El Shoubary, Y.: "Automation of an In-Situ Combustion Tube and Study of the Effect of Clay on the In-Situ Combustion Process," *Soc. Pet. Eng. J.* (Aug. 1982) 493-502.

W

Walsh, J.W. Jr., Ramey, H.J. Jr., and Brigham, W.E.: "Thermal Injection Well Falloff Testing," paper SPE 10227 presented at the 1981 SPE Annual Technical Conference and Exhibition, San Antonio, Oct. 4-7.

Walter, H.: "Application of Heat for Recovery of Oil; Field Test Results and Possibility of Profitable Operation," *J. Pet. Tech.* (Feb. 1957) 16-22.

Warren, J.E., Reed, R.L., and Price, H.S.: "Theoretical Considerations of Reverse Combustion in Tar Sands," *Trans.*, AIME, **219,** 109-123.

Weijdema, J.: "Studies on the Oxidation Kinetics of Liquid Hydrocarbons in Porous Media With Regard to Subterranean Combustion," *Erdöl & Kohle. Erdgas. Petrochem.* (Sept. 1968) 520-526 (in German).

Weinbrandt, R.M., Ramey, H.J. Jr., and Cassé, F.J.: "The Effect of Temperature on Relative and Absolute Permeability of Sandstones," *Soc. Pet. Eng. J.* (Oct. 1975) 376-384.

Wendt, R.E.: "Review of Stack Gas Scrubber Operation Experience on an Oil Field Steam Generator," paper SPE 7125 presented at the 1978 SPE California Regional Meeting, San Francisco, April 12–14.

White, P.D. and Moss, J.T.: "High Temperature Thermal Techniques for Stimulating Oil Recovery," *J. Pet. Tech.* (Sept. 1965) 1007–1011.

Wieber, P.R. and Sikri, A.P.: "The 1976 ERDA In-Situ Coal Gasification Program," *Proc.*, Second Annual Underground Coal Gasification Symposium, Morgantown Energy Research Center, U.S. DOE, Morgantown, WV (1976) MERC/SP-76/3.

Willhite, G.P.: "Over-all Heat Transfer Coefficients in Steam and Hot Water Injection Wells," *J. Pet. Tech.* (May 1967) 607–615.

Willhite, G.P. and Dietrich, W.K.: "Design Criteria for Completion of Steam Injection Wells," *J. Pet. Tech.* (Jan. 1967) 15–21.

Willhite, G.P., Dranoff, J.S., and Smith, J.M.: "Heat Transfer Perpendicular to Fluid Flow in Porous Rocks," *Soc. Pet. Eng. J.* (Sept. 1963) 185–188; *Trans.*, AIME, **228.**

Willhite, G.P., Kunii, D., and Smith, J.M.: "Heat Transfer in Beds of Fine Particles," *AIChE J.* (Aug. 1962) 340–345.

Willhite, G.P., Wilson, J.M., and Martin, W.L.: "Use of an Insulating Fluid for Casing Protection During Steam Injection," *J. Pet. Tech.* (Nov. 1967) 1453–1456.

Williams, R.L.: "Steamflood Pilot Design for a Massive, Steeply Dipping Reservoir," paper SPE 10321 presented at the 1981 SPE Annual Technical Conference and Exhibition, San Antonio, Oct. 5–7.

Williams, R.L., Brown, S.L., and Ramey, H.J. Jr.: "Economic Appraisal of Thermal Drive Projects—A New Approach," paper SPE 9358 presented at the 1980 SPE Annual Technical Conference and Exhibition, Dallas, Sept. 21–24.

Willman, B.T., Valleroy, V.V., Rumberg, G.W., Cornelius, A.J., and Powers, L.W.: "Laboratory Studies of Oil Recovery by Steam Injection," *J. Pet. Tech.* (July 1961) 681–690; *Trans.*, AIME, **222.**

Wilson, L.A., Reed, R.L., Reed, D.W., Clay, R.R., and Harrison, N.H.: "Some Effects of Pressure on Forward and Reverse Combustion," *Soc. Pet. Eng. J.* (Sept. 1963) 127–137.

Wilson, L.A. and Root, P.J.: "Cost Comparison of Reservoir Heating Using Steam or Air," *J. Pet. Tech.* (Feb. 1966) 233–239; *Trans.*, AIME, **237.**

Wilson, L.A., Wygal, R.J., Reed, D.W., Gergins, R.L., and Henderson, J.H.: "Fluid Dynamics During an Underground Combustion Process," *Trans.*, AIME (1958) **213,** 146–154.

Winestock, A.G.: "Developing A Steam Recovery Technology," *Cdn. Soc. Pet. Geol. Mem.* (Sept. 1974) 190–198.

Wohlbier, R.: "Theoretical Studies of Petroleum Recovery Operations with Forward Combustion, Based on a Computer Model for a Linear System," *Erdoel-Erdgas-Zeitschrift* (1965) No. 11, 435–463.

Wolcott, E.R.: "Method of Increasing the Yield of Oil Wells," U.S. Patent No. 1,457,479 (filed Jan. 12, 1920; issued June 5, 1923).

Wooley, G.R.: "Computing Downhole Temperatures in Circulation, Injection, and Production Wells," *J. Pet. Tech.* (Sept. 1980) 1509–1522.

Wooten, R.W.: "Case History of a Successful Steamflood Project—Loco Field," paper SPE 7548 presented at the 1978 SPE Annual Technical Conference and Exhibition, Houston, Oct. 1–3.

Wu, C.H. and Brown, A.: "A Laboratory Study on Steam Distillation in Porous Media," paper SPE 5569 presented at the 1975 SPE Annual Technical Conference and Exhibition, Dallas, Sept. 28–Oct. 1.

Wu, C.H. and Fulton, P.F.: "Experimental Simulation of the Zones Preceding the Combustion Front of an In-Situ Combustion Process," *Soc. Pet. Eng. J.* (March 1971) 38–46.

Y

Yoelin, S.D.: "The TM Sand Steam Stimulation Project," *J. Pet. Tech.* (Aug. 1971) 987–994; *Trans.*, AIME, **251.**

Young, B.M., McLaughlin, H.C., and Borchardt, J.K.: "Clay Stabilization Agents—Their Effectiveness in High-Temperature Steam," *J. Pet. Tech.* (Dec. 1980) 2121–2131.

Z

Zemanski, M.W.: *Heat and Thermodynamics*, McGraw-Hill Book Co. Inc., New York City (1943).

Zierfuss, H. and van der Vliet, G.: "Laboratory Measurements of Heat Conductivity of Sedimentary Rocks," *Bull*, AAPG (Oct. 1956) **40,** 2475.

Author Index

Y

Z

Subject Index

N

O

P

Q

R

S

T

www.ingramcontent.com/pod-product-compliance
Ingram Content Group UK Ltd.
Pitfield, Milton Keynes, MK11 3LW, UK
UKHW051510290926
15471UKWH00008B/1529

9 781613 995488